MODERN ASPECTS OF ELECTROCHEMISTRY

No. 20

LIST OF CONTRIBUTORS

HÉCTOR D. ABRUÑA
Department of Chemistry
Cornell University
Ithaca, New York 14853

H. P. AGARWAL
Department of Chemistry
M.A. College of Technology
Bhopal, India

ALEKSANDAR DESPÍC
Faculty of Technology
and Metallurgy
University of Belgrade
Belgrade, Yugoslavia

JERRY GOODISMAN
Department of Chemistry
Syracuse University
Syracuse, New York 13244

A. M. KUZNETSOV
A. N. Frumkin Institute for Electrochemistry
Academy of Sciences of the USSR
Moscow V-71, USSR

VITALY P. PARKHUTIK
Department of Microelectronics
Minsk Radioengineering Institute
Minsk, USSR

ISAO TANIGUCHI
Department of Applied Chemistry
Faculty of Engineering
Kumamoto University
Kurokami
Kumamoto 860, Japan

A Continuation Order Plan is available for this series. A continuation order will bring delivery of each new volume immediately upon publication. Volumes are billed only upon actual shipment. For further information please contact the publisher.

MODERN ASPECTS OF ELECTROCHEMISTRY

No. 20

Edited by

J. O'M. BOCKRIS
Department of Chemistry
Texas A&M University
College Station, Texas

RALPH E. WHITE
Department of Chemical Engineering
Texas A&M University
College Station, Texas

and

B. E. CONWAY
Department of Chemistry
University of Ottawa
Ottawa, Ontario, Canada

Springer Science+Business Media, LLC

The Library of Congress cataloged the first volume of this title as follows:

Modern aspects of electrochemistry. no. [1]
Washington, Butterworths, 1954–
v. illus., 23 cm.
No. 1–2 issued as Modern aspects series of chemistry.
Editors: no. 1– J. Bockris (with B. E. Conway, No. 3–)
Imprint varies: no. 1, New York, Academic Press. – No. 2, London, Butterworths.
1. Electrochemistry – Collected works. I. Bockris, John O'M. ed. II. Conway, B. E. ed. (Series: Modern aspects series of chemistry)
QD552.M6 54-12732 rev

ISBN 978-1-4684-8764-0 ISBN 978-1-4684-8762-6 (eBook)
DOI 10.1007/978-1-4684-8762-6

Originally published by Plenum Press, New York in 1989.
Softcover reprint of the hardcover 1st edition 1989

Preface

In Number 20 of *Modern Aspects of Electrochemistry*, we present chapters whose organization is typical for the series: They start with the most fundamental aspects and then work to the more complex.

Thus, Jerry Goodisman gives us an interesting contribution on a subject in which he is one of the pioneers, the electron overlap contribution to the double layer potential difference.

Closely related to this theme, but not always imbued with knowledge of it, is the electron transfer theory, treated in this volume by the experienced author A. M. Kuznetsov of the Frumkin Institute.

H. P. Agarwal is a well-known figure in the field of faradaic rectification, which he originated, and he now tells us about the more recent thinking in the field.

On the other hand, Héctor D. Abruña comes relatively new to us, and his field, that of X-ray interactions with electrodes, is new, too, but probably augers the trend for the future.

The photoelectrochemical reduction of CO_2, described here by Isao Taniguchi from Kumamoto University, is a subject which will have much practical importance as the greenhouse effect continues.

Finally, aluminum in aqueous solutions and the physics of its anodic oxide is a subject which seems ever with us, and is described in its latest guise by Aleksandar Despić and Vitaly P. Parkhutik.

J. O'M. Bockris
Texas A&M University
R. E. White
Texas A&M University
B. E. Conway
University of Ottawa

Contents

Chapter 1

THEORIES FOR THE METAL IN THE METAL-ELECTROLYTE INTERFACE

Jerry Goodisman

Chapter 2

RECENT ADVANCES IN THE THEORY OF CHARGE TRANSFER

A. M. Kuznetsov

Chapter 3

RECENT DEVELOPMENTS IN FARADAIC RECTIFICATION STUDIES

H. P. Agarwal

Chapter 4

X RAYS AS PROBES OF ELECTROCHEMICAL INTERFACES

Héctor D. Abruña

Chapter 5

ELECTROCHEMICAL AND PHOTOELECTROCHEMICAL REDUCTION OF CARBON DIOXIDE

Isao Taniguchi

Chapter 6

ELECTROCHEMISTRY OF ALUMINUM IN AQUEOUS SOLUTIONS AND PHYSICS OF ITS ANODIC OXIDE

Aleksandar Despić and Vitaly Parkhutik

1

Theories for the Metal in the Metal–Electrolyte Interface

Jerry Goodisman

Department of Chemistry, Syracuse University, Syracuse, New York 13244

1. INTRODUCTION

1. The Metal in the Interface

Since the electrochemical interface is usually the interface between a metal and an electrolyte, all properties of the interface may be expected to involve contributions of the metal and of the electrolyte. However, most theories of the electrochemical interface are theories of the electrolyte phase, with no reference to the contributions of the metal. Here, we discuss more recent theoretical work which attempts to redress this inequity. As we shall see, it is not, in general, possible to separate, experimentally, the metal contribution from the electrolyte contribution.

It is instructive to see how this comes about for the potential of a simple electrochemical cell. We consider the potential of zero charge for the cell[1]

$$\mathrm{Hg}|\mathrm{Na}|\mathrm{Na}^+, \mathrm{S}|\mathrm{Hg}'$$

The solution phase S is supposed to contain Na^+. Here, the $Na|Na^+$ electrode, which is reversible, is the reference electrode. This cell is not actually realizable, but its potential of zero charge can be calculated from measurements on other cells.

The cell potential is the difference in electrical potential (inner potential) between the right-hand and left-hand mercury terminals:

$$E = \Delta_{\mathrm{Hg}}^{\mathrm{Hg}'}\phi$$

$$= \Delta_{\mathrm{S}}^{\mathrm{Hg}'}\phi + \Delta_{\mathrm{Na}}^{\mathrm{S}}\phi + \Delta_{\mathrm{Hg}}^{\mathrm{Na}}\phi$$

The third potential difference may be expressed as a difference of chemical potentials by invoking the equality of the electrochemical potentials of the electron in the two metallic phases in contact. The first potential difference may be written as a difference in outer potentials plus a difference in surface potentials of the free surfaces of S and Hg, corresponding to a path from the inside of Hg to the inside of S passing through their free surfaces and through vacuum outside; it may also be written as a difference of surface potentials plus an "ionic" or free-charge contribution, corresponding to a path passing through the interface between Hg and S. At zero electrode charge, the free-charge contribution vanishes, and $\Delta_{\mathrm{S}}^{\mathrm{Hg}}$ is just a difference in surface potentials, so that the potential of zero charge is

$$E_{\mathrm{pzo}} = \chi^{\mathrm{Hg(S)}} - \chi^{\mathrm{S(Hg)}} - \Delta_{\mathrm{S}}^{\mathrm{Na}}\phi + F^{-1}(\mu_e^{\mathrm{Na}} - \mu_e^{\mathrm{Hg}}) \qquad (1)$$

Here, $\chi^{\mathrm{Hg(S)}}$ is the surface potential of mercury in contact with the solution phase S. The difference of metal and solution surface potentials, $\chi^{\mathrm{Hg(S)}} - \chi^{\mathrm{S(Hg)}}$, is the dipolar potential difference, often written[2] as $g_{\mathrm{S}}^{\mathrm{M}}(\mathrm{dip}) = g^{\mathrm{M}}(\mathrm{dip}) - g^{\mathrm{S}}(\mathrm{dip})$, with the metal M here equal to Hg.

The surface potential of mercury in contact with S may be written as the surface potential of the free surface of mercury, χ^{Hg}, plus the change due to the presence of S, $\delta\chi_{\mathrm{S}}^{\mathrm{Hg}}$. Since $F\chi^{\mathrm{Hg}} - \mu_e^{\mathrm{Hg}}$ is the work function of mercury, Φ^{Hg}, we have

$$E_{\mathrm{pzo}} = \Phi^{\mathrm{Hg}}/F - \chi^{\mathrm{S(Hg)}} + \delta\chi_{\mathrm{S}}^{\mathrm{Hg}} - \Delta_{\mathrm{S}}^{\mathrm{Na}}\phi + \mu_e^{\mathrm{Na}}/F \qquad (2)$$

Most of the quantities appearing in this equation are measurable or calculable. Thus, the potential of zero charge for the cell is measured as 2.51 V, and the work function of mercury as 4.51 V, while the chemical potential of the electron in sodium can be reliably calculated as −3.2 eV. Bockris and Khan[1] estimated the free-surface potential χ^S for aqueous solutions as 0.2 V, and the effect of the solution on the mercury surface potential, $\delta\chi_S^{Hg}$, as 0.26 V, to calculate the difference of inner potentials $\Delta_S^{Na}\phi$. Note that $\delta\chi_S^{Hg}$ is larger in magnitude than the surface potential of the electrolyte phase χ^S, but it is generally believed that the latter quantity varies more with surface charge than does the former, so that $\delta\chi_S^M$ is unimportant when capacitances are being considered.

In Eq. (1), $-\Delta_S^{Na}\phi + F^{-1}\mu_e^{Na}$ is a property of the $Na|Na^+$ reference electrode. In order to get information not depending on the reference electrode, one has to study E_{pzc} and Φ^M with changing metal M. For a metal M, the right-hand side of Eq. (2) may be rearranged[3] to $\Phi^M/e + \delta\chi_S^M - \delta\chi_M^S - E_k$, where $\delta\chi_M^S = \chi^{S(Hg)} - \chi^S$ and E_k is the "electrode potential on the vacuum scale" of the reference electrode,[4] which can be determined experimentally.

The quantity $(\delta\chi_S^{Hg} - \delta\chi_{Hg}^S)$ is the difference between the effect of mercury on the surface potential of water and the effect of water on the surface potential of mercury. Experimentally, there is no way to separate[2] these two contributions to g_S^M(dip); theoretically, they are discussed and calculated in radically different ways. This is already clear in Eq. (1), which involves $\chi^{Hg(S)} - \chi^{S(Hg)}$. If one keeps the same solution but substitutes another metal M for mercury, $\chi^{M(S)}$ will replace $\chi^{Hg(S)}$, contributing to the change in the cell potential, but, in addition, $\chi^{S(M)}$ will probably differ from $\chi^{S(Hg)}$, since χ^S is ascribed to the orientation of solvent dipoles at the metal surface, which will be different for a different metal. To separate $\delta\chi_S^M$ and $\delta\chi_M^S$, some assumption or model is required. For example, in interpreting plots of adsorption potential versus chain length for a series of aliphatic alcohols at the water–air and water–mercury interfaces, the difference between the lines may be ascribed[3] to $\delta\chi_M^S$, thus measuring one of the two separately.

When polar or polarizable species are adsorbed on a metal, the work function changes. This is partly due to g^S(dip) but is also due to the change in χ^M and other effects.[2] As for the interfacial potential in the electrochemical cell, the contributions of adsorbate and metal cannot be separated. Usually, the latter gets ignored. It is precisely this term that interests us here.

The basic double-layer model considers the solid as a perfect conductor, so that g^M(dip) is charge independent and the potential drop (metal minus solution) is just $g(\text{ion}) - g^S(\text{dip})$ plus a constant. Although early theories considered penetration of the electric field of the interface into the metal (the work of Rice[5] actually ascribed the entire inner-layer capacitance to the metal, rejecting Stern-type theories of the compact layer), this idea was subsequently dropped. The history of the subject has been well reviewed by Vorotyntsev and Kornyshev.[6] Mott and Watts-Tobin[7] criticized ascribing the inner-layer capacitance entirely to the metal and also argued against using, unchanged, results for the capacitance of the metal–vacuum interface in discussion of the metal–solution interface. Although they did not reject a metal contribution to the capacitance, their work seems to have been interpreted[6–8] to mean that the metal's contribution to the reciprocal of the capacitance is unimportant, so that only solution species should be considered in a model for the interface. Of course, penetration of the electric field into the interface is important for semiconductors, and, in 1979, its possible relevance for semimetals like Bi and Sb was suggested.[9]

If the metal is ideal, the model for the interface is a model for the electrolyte phase. The capacitance of the interface corresponds to two capacitances in series, that of the diffuse layer, for which the Gouy–Chapman theory is considered adequate, and that of the compact layer. The simplest model for the latter is a layer containing N dipoles per unit area, each dipole having an average normal component p_N (p_N depends on the field). Then $g^S(\text{dip}) = Np_N/\varepsilon$, where ε is the dielectric permittivity in the dipole layer. If ε is considered to arise from electrons, the high-frequency dielectric constant of the solvent would be used for it, but other values may be appropriate. Writing the compact-layer capacitance per unit area at the point of zero charge as $C_c = \varepsilon_c/4\pi d$ (ideal parallel-plate capacitor), one can derive[10] values for ε_c from experimental C_c values for various metals. These values increase for different metals along with the (measured) potential drop g across the compact layer. If g measures orientation of solvent dipoles, increased g should be associated with decreased ε_c; the contradiction has been adduced[11] as evidence for a missing metal contribution in the model.

In their classic treatment of the compact-layer capacitances, MacDonald and Barlow[12] affirmed that the thickness of the space charge or penetration region in the metallic electrode is so small for a good conductor that its effect may be neglected. Their theory of the compact layer, together with the Gouy-Chapman theory for the diffuse layer, successfully fitted most capacitance data for Hg/water and Hg/methanol interfaces with nonadsorbing electrolytes (except in the anodic region). Interaction of the solvent dipoles with the mercury was treated by summing multiple ideal image interactions, other forces involved in physisorption being neglected.

Models for the compact layer of the metal–electrolyte interface have become more and more elaborate, providing better and better representations of observed electrocapillary data for different metals, solvents, and temperatures, but almost always leaving the metal itself out of consideration, except for consideration of image interactions of the solvent dipoles. For reviews of these models, see Parsons,[13] Reeves,[14] Fawcett *et al.*,[15] and Guidelli,[16] who gives detailed discussion of the mathematical as well as the physical assumptions used.

Fawcett *et al.*[15] presented a four-state model, involving monomers and clusters, for water in the inner layer of the mercury–aqueous solution interface with no specific adsorption. By treating dipole–dipole interactions on a molecular level, they avoided introduction of an effective dielectric constant. The contribution to the dipole–dipole interaction of dipole images in the metal was included, along with non-nearest-neighbor interactions, by using an effective coordination number. Values for seven parameters, including this effective coordination number, were chosen to give the best agreement with experimental capacitances at various surface charges and temperatures. Recently, Guidelli[17] has developed a statistical mechanical treatment for a monolayer of hydrogen-bonded molecules, interacting with a surface and with each other and orienting in an electric field. Capacitance as a function of charge was calculated and compared to compact-layer capacitances derived from experiment by removing the diffuse-layer contribution. The model accounted qualitatively for some of the observed features, without considering any contribution of the metal itself.

It is perhaps unnecessary to note that the fact that a model, with a suitable choice of parameters, can account for experimental data does not prove that the model is correct or that no physical effects of importance have been left out. Yeager,[18] in his review of nontraditional approaches to the study of the metal–electrolyte interface, stated that compact-layer models that do not adequately take into account the structure of the electrode surface and the interaction of water in the compact layer with water outside will not yield an understanding of the interface. Further, he noted the lack of molecular specificity of traditional electrochemical measurements. Among methods which can distinguish between molecular species in the interface are ellipsometry, Raman spectroscopies, Mössbauer spectroscopy, and electron spin resonance; perhaps these can be of help in disentangling metal and electrolyte contributions to interfacial properties.

The importance of the structure of the metal was also suggested[19] for the capacitances at the point of zero charge of metal–molten salt interfaces. As compared with metal-electrolyte interfaces, the striking difference is that the capacitances of the latter decrease with increasing temperature and those of the former increase sharply with temperature. (It is perhaps just as striking that the sizes of the capacitances for the two systems are quite similar, although the natures of the systems are so different.) According to the Gouy-Chapman theory, the capacitance at the point of zero charge is proportional to $T^{-1/2}$; improved statistical mechanical theories such as the mean spherical approximation[20] give a weaker decrease with T. It is suggested by March and Tosi[20] that the increase in capacitance with temperature is connected with increased penetration of the metal surface by the ions of the electrolyte, this penetration increasing with increase in temperature.

2. Metal–Electrolyte Interaction

One might expect, as a first approximation, that at the same cell potential relative to the potential of zero charge, one would have the same surface charge in the interface, regardless of the metal, but there are many exceptions to this,[21] ascribed to the effect of the metal on the orientation of solvent molecules. Similarly, the increase of the capacitance for the Ga/aqueous solution interface with decreasing negative surface charge is ascribed to specific adsorption of water, with the negative end of the water molecule toward the metal. This chemisorption is likely to be stronger for Ga than for Hg.

The nonelectrostatic forces between metal and solution have sometimes been subsumed[22] into a "natural field," tending to orient the solvent dipoles, whose effect is represented by an energy contribution $-p\Delta E \cos\theta$ where ΔE is the natural field strength, p the orienting dipole moment, and θ the angle between the dipole and the direction normal to the interface; this term, independent of the surface charge, is to be added to the energy of a dipole in an external field (due to surface charge), $-pE_1 \cos\theta$. Experimental evidence[9,16] indeed indicates that, at the point of zero charge, water molecules in the metal-electrolyte interface are preferentially oriented with their negative (oxygen) ends toward the metal. This gives a negative contribution to the metal-solution potential difference $\Delta_S^M\phi$. As the temperature is increased, this contribution should decrease in size, which explains the fact that $\Delta_S^M\phi$ increases with temperature. If this preferential orientaiton is due to nonelectrical forces of adsorption, it is reasonable that a slightly negative surface charge on the metal is necessary to give zero net orientation of dipoles. Below we shall mention another mechanism, due directly to penetration by the metal's conduction electrons of the solvent layer, which produces the same preferential orientation of dipoles at zero surface charge.

In fact, the orientation of water at the potential of zero charge is expected to depend approximately linearly on the electronegativity of the metal.[9] This orientation (see below) may be deduced from analysis of the variation of the potential drop across the interface with surface charge for different metals and electrolytes. Such analysis leads to the establishment of a hydrophilicity scale of the metals ("solvophilicity" for nonaqueous solvents) which expresses the relative strengths of metal-solvent interaction, as well as the relative reactivities of the different metals to oxygen.[23]

In addition to the nonelectrostatic adsorptive force, there is an image force between a dipole and a metal, which will be present whenever charged or dipolar particles in a medium of one dielectric constant are near a region of another dielectric constant. If the metal is treated as an ideal conductor, the image-force contribution to the energy of a dipole in the electrolyte is proportional to p^2/z^3, where z is the distance of the dipole from the plane boundary of the metal (considered ideal, with no surface structure), and to $1 + \cos^2\theta$. This ideal term is, of course, the same for all metals. If the metal structure is considered more realistically, the image-type interactions will differ for different metals.

Specific adsorption of ions (probably anions) of the electrolyte phase on the metal also should depend on the metal. Assuming a Langmuir-type equilibrium, one has[22] for ions of charge q_i and solution concentration c_i

$$Kc_i\Gamma_e = \Gamma_f \exp[(\phi_i + q_i\psi_A)/kT]$$

Here K is a constant, Γ_e and Γ_f are the number of empty and filled sites per unit area on the metal surface, ϕ_i is the adsorption potential, and ψ_A is the electrostatic potential of the empty site; ϕ_i depends on surface charge. The sum $\Gamma_0 = \Gamma_e + \Gamma_f$, total number of sites per unit area, depends on the metal, as does ϕ_i.

The above effects are more familiar than direct contributions of the metal's components to the properties of the interface. In this chapter, we are primarily interested in the latter; these contribute to $\chi^{M(S)}$. The two quantities $\chi^{M(S)}$ and $\chi^{S(M)}$ (or $\delta\chi_S^M$ and $\delta\chi_M^S$) are easily distinguished theoretically, as the contributions to the potential difference of polarizable components of the metal and solution phases, but apparently cannot be measured individually without adducing the results of calculations or theoretical arguments. A model for the interface which ignores one of these contributions to $\Delta_S^M\phi$ may, suitably parameterized, account for experimental data, but this does not prove that the neglected contribution is not important in reality. Of course, the tradition has been to neglect the metal's contribution to properties of the interface. Recently, however, it has been possible to use modern theories of the structure of metals and metal surfaces to calculate, or, at least, estimate reliably, $\chi^{M(S)}$ and $\delta\chi_S^M$ (as well as discuss $\delta\chi_M^S$, which enters some theories of the interface). It is this work, and its implications for our understanding of the electrochemical double layer, that we discuss in this chapter.

We shall discuss first the theoretical separation of metal and solution properties and then turn to the modern theories of metal structure, particularly as they apply to the surface. Then we shall consider the calculation of quantities relevant to the metal in the interface and theories of the metal–solution interface which take the metal, as well as the solution, into account.

II. SEPARATION OF METAL AND ELECTROLYTE PROPERTIES

1. Electrostatics

For an interface between two phases with no common charged or polarizable components, contributions of the two phases to certain properties are easily distinguished theoretically. The charge density and electrical polarization at any point in the interface are each sums of two contributions which can be assigned to the two phases. The overall electroneutrality of a planar interface may be written

$$\int_{-\infty}^{\infty}\left(\sum_i n_i q_i\right) dz = \int_{-\infty}^{\infty}\left(\sum_i^{A} n_i q_i\right) dz + \int_{-\infty}^{\infty}\left(\sum_i^{B} n_i q_i\right) dz = 0$$

where n_i is the density of species i, a function of z, and q_i is the charge of species i. Of the two sums in the second member, one is over charged components of phase A and the other over charged components of phase B. The surface charge density of phase A (in charges per unit area) is

$$q^{\mathrm{A}} = \sum_i^{\mathrm{A}} q_i \int_{-\infty}^{\infty} n_i(z)\, dz \tag{3}$$

and is equal and opposite to the surface charge density of phase B,

$$q^{\mathrm{B}} = \sum_i^{\mathrm{B}} q_i \int_{-\infty}^{\infty} n_i(z)\, dz \tag{4}$$

The charge density at any point is $\sum_i n_i q_i$, the sum being over species of both phases. Note that no "geographical" separation of components of the phases is required, anticipating that, in the real system, it may not be possible to divide the system such that all the components of phase A lie on one side of a geometrical surface and all the components of phase B on the other. It is necessary only to identify each component as belonging to one phase or the other. In a geographical separation, one would write

$$q^{\mathrm{A}} = \int_{-\infty}^{x}\left(\sum_i n_i q_i\right) dz \quad \text{and} \quad q^{\mathrm{B}} = \int_{x}^{\infty}\left(\sum_i n_i q_i\right) dz$$

Because of the overall electroneutrality, q^A will equal $-q^B$ for any choice of X.

All n_i are determined, in principle, by the equations of statistical mechanics, since they are one-particle distributions.[4] As such, each n_i depends on the interactions between particles of species i and all other particles in the system, whether belonging to the metal or to the other phase. If there is a geographical separation of particles of species i from, say, particles of species k (as when i and k belong to different phases), the interaction between particles k and a particle of species i near the surface may be averaged over positions of particles k, i.e., no correlation is assumed between the particles of the two species, so that the particles k become a source of external field for particles i. For a particle i far from the surface, the interaction is probably unimportant (unless it is a long-range electrostatic interaction).

For electrochemistry, of course, the most important properties of the n_i involve the electric field and potential in the system. In general, to find the electromagnetic field in a medium, one has to solve the basic equation

$$\nabla \cdot \mathbf{E} = 4\pi(\rho + \rho_{\text{ext}}) \tag{5}$$

for the statistically averaged electric field $\mathbf{E}$. Here, $\rho_{\text{ext}}(\mathbf{r})$ is the charge density of external field sources and $\rho(\mathbf{r})$ is the averaged density of induced charge due to polarization of the medium.[24] The charge density ρ is $\sum n_i q_i$, the sum extending over species of particles belonging to both phases; there is no separation in this general formalism. One requires, in addition to Eq. (5), a phenomenological relationship between ρ and the electrical potential ϕ, where $\mathbf{E} = -\nabla\phi$. Alternatively, one can deal with the polarization field $\mathbf{P}(\mathbf{r})$, where $\rho = -\nabla \cdot \mathbf{P}$, and assume a dependence of $\mathbf{P}$ on $\mathbf{E}$: $\mathbf{P}$, like ρ, involves contributions of particles from both phases. The relation between ρ and ϕ is often taken as linear and local, so that ρ at any point is proportional to the electrostatic potential at that point (although the constant of proportionality may vary from point to point). However, it may be necessary to go to a nonlinear relationship, to take into consideration spatial correlations between charge density fluctuations. Then $\rho(\mathbf{r})$ is given, in the linear approximation, by $\int d\mathbf{r}'\, \alpha(\mathbf{r}, \mathbf{r}')\phi(\mathbf{r}')$. If $\phi(\mathbf{r})$ is more slowly varying than the polarizability function $\alpha(\mathbf{r}, \mathbf{r}')$, one recovers the local limit.

The Poisson equation in electrostatic units for planar symmetry may be written

$$dD/dz = 4\pi\rho(z) = 4\pi \sum_i n_i(z)q_i$$

where $\mathbf{D} = \mathbf{E} + 4\pi\mathbf{P}$, ρ is the free-charge density, and the polarization $\mathbf{P} = \sum_i n_i(z)\mu_i(z)$, with μ_i the average z-component of the dipole moment of a molecule of species i, which may depend on z. Since the z-component of the electric field E is $-dV/dz$,

$$-d^2V/dz^2 = 4\pi \sum_i n_i(z)q_i - 4\pi\, dP/dz$$

Integrating twice to get the potential drop across the interface, we have

$$\begin{aligned} V(\infty) - V(-\infty) &= -4\pi \int_{-\infty}^{\infty} dz \int_{-\infty}^{z} dz' \left(\sum_i n_i q_i - dP/dz' \right) \\ &= 4\pi \int_{-\infty}^{\infty} dz \left(z \sum_i q_i n_i + \sum_i \mu_i n_i \right) \qquad (6) \end{aligned}$$

We have used the electroneutrality of bulk phases $[\sum q_i n_i(\pm\infty) = 0]$ and the fact that there is no electrical polarization in bulk phases.

The two terms in the potential drop are the dipole moments of free charges and the permanent dipole moments. Each may be divided into contributions of components of the two phases. Equation (6) may be rewritten in terms of the surface charge densities of Eqs. (3) and (4):

$$V(\infty) - V(-\infty) = 4\pi(q^A z^A + q^B z^B) + 4\pi \int_{-\infty}^{\infty} dz \left(\sum_i^{A} \mu_i n_i + \sum_i^{B} \mu_i n_i \right)$$

Here, the centroids of the charge distributions of phases A and B have been introduced:

$$z^A = \left[\int_{-\infty}^{\infty} dz\, z \sum_i^{A} q_i n_i \right] \bigg/ \left[\int_{-\infty}^{\infty} dz \sum_i^{A} q_i n_i \right]$$

The term $4\pi(q^A z^A + q^B z^B)$ is a free-charge term, resembling that of an ideal capacitor of surface charge density $q^A = -q^B$ and interplanar spacing $z^A - z^B$.

In another notation, and supposing the metal appears for $z \to -\infty$ and the electrolyte for $z \to \infty$, the potential difference across the interface may be written as

$$\Delta V = V(-\infty) - V(\infty) = \Delta_S^M \phi$$
$$= g_S^M(\text{ion}) - g^S(\text{dip}) + g^M(\text{dip}) \tag{7}$$

where $g_S^M(\text{ion})$ is the free-charge contribution, vanishing for $q^M = q^S = 0$, and the other terms are dipolar contributions due to components of solution and metal. In the free-charge term, some fixed position z^0 in, say, phase M may be substituted for z^M. Then the difference

$$4\pi q^M(z^M - z^0) = 4\pi \int_{-\infty}^{\infty} dz\,(z - z^0) \sum_i^M n_i(z)$$

is assigned to the dipolar term for metal M.

It is usually assumed that the components of a metal are ions (with tightly bound charge) and electrons, so that there are no polarizable species in the metal phase. The contribution of the metal to the potential drop across the interface is then

$$\Delta V^M = 4\pi \int_{-\infty}^{\infty} dz(z - z^0) \sum_i^M q_i n_i(z)$$

At the point of zero charge, the metal is neutral:

$$\int_{-\infty}^{\infty} dz \sum_i^M q_i n_i(z) = 0$$

but this does not make ΔV^M zero if the positive and negative species (ions and electrons) are distributed differently in space. One knows that the electrons spill over from a metal surface, corresponding to a surface dipole or double layer with its negative and toward the outside of the metal. This is responsible for the surface potential of a metal, which is several volts in size, and makes an important contribution to the work function. Although the presence of an electrolyte phase on the outside of the metal will certainly have an effect on the distribution of the metal's electrons, one should not expect the large surface dipole and large potential difference to disappear when the metal is in an interface.

It may be noted that for a polycrystalline surface of a solid metal electrode, there is a complication in defining the potential of zero charge because one does not have zero charge density on all crystal faces simultaneously. The work function is different for different crystal faces but, if the grains are in contact, the electrochemical potential of the electron has a single value. Thus, the Volta or outer potential is different for different faces, so different surface charge densities are present. An average work function can be defined simply, but the definition of an average E_{pzc} is not as easy.[2] Similarly, the surface structure of a real solid metal makes the electronic structure more complicated than for a liquid metal or for a single-crystal face. Certainly, the use of density profiles depending only on distance perpendicular to the interface is an oversimplification.

The reciprocal of the capacitance, C^{-1}, is the change of potential drop with the surface charge q^{M}. According to Eq. (7), C^{-1} is a sum of three contributions, each an inverse capacitance, corresponding to three capacitances in series. Like the charge density and ΔV, C^{-1} may be divided into metal and solution contributions. However, if the electronic distribution of the metal does not change relative to the metal's ionic distribution as the interface is charged, ΔV^{M} will be independent of q^{M}, and the metal will not contribute to the capacitance of the interface. This is the case for an ideal metal, in which, furthermore, there is no penetration of electric field so that $\Delta V^{\mathrm{M}} = 0$. Of course, the metal may contribute indirectly, since the interaction between the metal and the solution components may affect the ability of the latter to reorient as the electric field is changed.

Differentiating Eq. (7), the inverse capacitance is

$$\begin{aligned} C^{-1} &= (\partial \Delta V / \partial q^{\mathrm{M}}) \\ &= [C(\text{ion})]^{-1} - [C^{\mathrm{S}}(\text{dip})]^{-1} + [C^{\mathrm{M}}(\text{dip})]^{-1} \end{aligned} \tag{8}$$

In the notation of Eqs. (1) and (2), $\Delta V = \chi^{\mathrm{M(S)}} - \chi^{\mathrm{S(M)}}$, so that

$$\begin{aligned} E &= \Delta V + \Phi^{\mathrm{Hg}}/F - \chi^{\mathrm{Hg}} - E_k \\ &= \Phi^{\mathrm{Hg}}/F + \delta\chi_{\mathrm{S}}^{\mathrm{Hg}} - g^{\mathrm{S}}(\text{dip}) + g_{\mathrm{S}}^{\mathrm{M}}(\text{ion}) - E_k \end{aligned}$$

with E_k the contribution of the reversible reference electrode, $\Delta_{\mathrm{S}}^{\mathrm{Na}}\phi - F^{-1}\mu_e^{\mathrm{Na}}$. Differentiating,

$$dE/dq^{\mathrm{M}} = d\delta\chi_{\mathrm{S}}^{\mathrm{Hg}}/dq^{\mathrm{M}} - dg^{\mathrm{S}}(\text{dip})/dq^{\mathrm{M}} + dg_{\mathrm{S}}^{\mathrm{M}}(\text{ion})/dq^{\mathrm{M}}$$

Then $1/C = 1/C^{M} - 1/C^{S} + 1/C^{ion}$, the terms arising from the metal electronic polarizability, the reorientation of solvent dipoles, and the electric field. C^{S} is sometimes written as C^{dip}.

The electric field or ionic term corresponds to an ideal parallel-plate capacitor, with potential drop $g_S^M(\text{ion}) = q^M d/4\pi\varepsilon$. It includes a contribution from the polarizability of the electrolyte, since the dielectric constant is included in the expression. The distance d between the layers of charge is often taken to be from the outer Helmholtz plane (distance of closest approach of ions in solution to the metal in the absence of specific adsorption) to the position of the image charge in the metal; a model for the metal is required to define this position properly. The capacitance per unit area of the ideal capacitor is a constant, $\varepsilon/4\pi d$, often written as K_{ion}. The contribution to $1/C$ is $1/K_{ion}$; this term is much less important in the sum (larger capacitance) than the other two contributions.[2]

2. Experimental Results

According to Eq. (2), the work function for a metal M is related to the potential of zero charge for the M/aqueous solution interface by

$$E_{pzc} = \Phi^M/F - \chi^{S(M)} + \delta\chi_S^M + E_k$$

where the constant E_k involves quantities relevant to the reference electrode only. The relation between the potential of zero charge and the work function has been discussed in detail by Trasitti,[10,23,25] for nonaqueous as well as aqueous solvents. If experimental values for Φ^M are plotted versus E_{pzc} for the same reference electrode, the points for different metals fall mostly along two lines.[10,25] The slope of the line for the *sp* metals differs noticeably from unity, indicating that $-\chi^{S(M)} + \delta\chi_S^M$ is not independent of M. The point for Ga is far off this line; it is in fact close to the line for the *d* metals, which has a slope close to unity. This means that $-\chi^{S(M)} + \delta\chi_S^M$ is about the same for all *d* metals. At a pzc corresponding to Hg, the line for the *d* metals is higher than the Hg work function by about 0.25 V. Thus, for a *d* metal and an *sp* metal having the same work function, E_{pzc} is significantly higher for the latter, i.e., the quantity $-\chi^{S(M)} + \delta\chi_S^{(M)}$ is more negative for the *d* metal.

An interpretation of these facts is that the orientation of water dipoles is essentially complete on all *d* metals, with the negative ends of the dipoles pointed toward the metal, giving $\chi^{S(M)}$ the same positive value for all M (and that $\delta\chi_S^M$ is the same as well). The slope of the line for the *sp* metals implies that $-\chi_S^{(M)} + \delta\chi_S^M$ becomes more negative in the order {Hg Sn Bi Pb In Cd}, which could reflect increasing ordering or strength of adsorption of water on the metal. Thus, one is disregarding the difference between $\delta\chi_S^M$ from one metal to another.[3] The direct effect of the metal's components, through $\delta\chi_S^M$, cannot be separated experimentally from the indirect one, adsorption or orientation forces on the solvent dipoles by the metal; only by theoretical reasoning or calculations can one judge which is dominant.[3,26] Thus, one can measure $\delta\chi_{H_2O}^{Hg} - \delta\chi_{Hg}^{H_2O}$ as -0.25 V and estimate from adsorption data and a reasonable assumption[26] that $\delta\chi_{Hg}^{H_2O} \simeq -0.06$ V so that $\delta\chi_{H_2O}^{Hg} \simeq -0.31$ V. [Gratifyingly (see below), various theories for the metal in the interface have found values close to this one for this quantity.] From a number of experiments, it is suggested that χ^{H_2O} is between 0.08 and 0.13 V, which means that $g^{H_2O}(\text{dip})$ is between 0.02 and 0.07 V on a mercury electrode at the point of zero charge (oxygen end of the water molecules toward the metal).

For transition or *d* metals, the value of the quantity $X = \delta\chi_S^M - g^S(\text{dip})$ or $\delta\chi_S^M - \chi^{S(M)}$ at the point of zero charge is large and negative and about the same (-0.67 ± 0.07 V) for all the metals for which data are available.[26] At the same time, the heat of formation of the metal oxide, a measure of the strength of the physisorption of water on the metal in the interface, varies from 300 to 900 kJ/mol. For *sp* metals, the heat of formation of the oxide is 300 kJ/mol or less, while the value of X for Hg is estimated as +0.06 V. The interpretation of the results for the *d* metals is that the interaction between metal and solvent is strong enough in all cases to completely orient the water molecules, giving $g^S(\text{dip})$ its maximum value, while $\delta\chi_S^M$ does not change much. It is less likely, though possible, that $\delta\chi_S^M$ and $g^S(\text{dip})$ both vary from metal to metal, but in a compensating way, so as to keep their difference constant. A strong metal dependence of $\delta\chi_S^M$ would require a reconsideration of the hydrophilicity scale for metals.[3]

For the *sd* metals (Au, Ag, and Cu) the situation is more complicated. Higher X is associated with lower heat of oxide formation. A number of explanations for this can be given,[26] including values of $\delta\chi_S^M$ exceeding $g^S(\text{dip})$ on structured surfaces, and protruding *d* orbitals at the positions of strong interaction with water. The data on the effect of crystal face on *X*, which would be useful to resolve some ambiguities, have been questioned.[26]

The experimental data bearing on the question of the effect of different metals and different crystal orientations on the properties of the metal–electrolyte interface have been discussed by Hamelin *et al.*[27] The results of capacitance measurements for seven *sp* metals (Ag, Au, Cu, Zn, Pb, Sn, and Bi) in aqueous electrolytes are reviewed. The potential of zero charge is derived from the maximum of the capacitance. Subtracting the diffuse-layer capacitance, one derives the inner-layer capacitance, which, when plotted against surface charge, shows a maximum close to $q^M = 0$. This maximum, which is almost independent of crystal orientation, is explained in terms of the reorientation of water molecules adjacent to the metal surface. Interaction of different faces of metal with water, ions, and organic molecules inside the outer Helmholtz plane are discussed, as well as adsorption.

Since the surface dipole potential of a metal χ^M is due to spillover of electrons from the metal toward the outside (electrolyte phase in the interface), the positive end of the dipole always points toward the interior of the metal. The electron density must also show a lateral smoothing as compared to the atomically distributed positive charge density. This decreases the surface dipole potential because the ion cores protrude into the electronic tail. The rougher the crystal face, the smaller is the magnitude of χ^M. Because of this, the electronic work function of a metal, which is $e\chi^M - \mu_e^M$ (μ_e^M = chemical potential of electrons in the metal), differs for different crystal faces. The rougher faces have lower values of χ^M and hence lower work functions, whereas the faces with higher number densities of ions (which have smoother positive charge densities in the directions parallel to the interface) have higher work functions.[28]

For the metal in the interface, the surface potential of the electrolyte phase is nearly the same for all crystal orientations.[29] Therefore, referring to Eq. (2), the potential of zero charge varies with the surface potential or the work funtion and is larger for the most densely packed faces. Correspondingly, atomic irregularities lower the potential of zero charge. Trasatti gives a table (Table 2 in Ref. 26) which shows the experimental results. There is also a parallelism between the potential of zero charge and the calculated surface energy, which is larger for more densely packed faces. Because the difference of pzc's should be the same as the difference of electronic work functions in going from one crystal orientation to another, the equilibrium or Nernst potential of an electrochemical reaction, when there is no adsorption, is independent of crystal orientation.[28]

It may be noted that the statement made above—that the surface potential in the electrolyte phase does not depend on the orientation of the crystal face—is necessarily an assumption, as is the neglect of $\delta\chi_S^M$. It is another example of separation of metal and electrolyte contributions to a property of the interface, which can only be done theoretically. In fact, a recent article[29] has discussed the influence of the atomic structure of the metal surface for solid metals on the water dipoles of the compact layer. Different crystal faces can allow different degrees of interpenetration of species of the electrolyte and the metal surface layer. Nonuniformities in the directions parallel to the surface may be reflected in the results of capacitance measurements, as well as optical measurements.

The capacitances of electrode-solution interfaces vary with the metal, for *sp* metals, in the same order as their attraction for oxygen (the "hydrophilicity," measured, for example, by the heat of formation of the metal oxide). This is true for nonaqueous solvents as well as for water, although a change in the solvent changes the size of the capacitance. There are few experimental results for capacitances of interfaces involving *d* metals. One interpretation[26] is that, for those metals for which the water molecules are strongly oriented by their interaction with the surface, the effect of the electric field is to give a large modification of their orientation. One may also reason that if the molecules interact strongly with the surface, a large electric field (large change in the surface charge) is required to change their orientation, which gives the solvent contribution to the potential drop across the interface.

Calculations, which we shall discuss later, show[30-32] that the direct contribution of the metal to the interfacial capacitance (from the conduction electrons) increases with the electron density, in the same way as the experimental capacitance. Indeed, the measured capacitances for different metal–aqueous solution interfaces can be "calculated" by combining calculated capacitances for the metal with the same value for the solution contribution in all cases. The model used in these calculations seems more reasonable for liquid metals than for polycrystalline solid metals. Furthermore, these calculations (see below) treat the electrolyte phase very crudely and probably overestimate the importance of the metal in the interface.[26] This is shown by consideration of the changes in capacitances when the solvent is changed, some of which are easily explained in terms of the properties of the solvent molecules.

Kalyuzhnaya *et al.*[33] argued that the reason for the higher capacitances of In and In–Ga alloy as compared to Ga and Hg in the same solution was due to the different electron density distributions in the surface layer, rather than to adsorption of ions. In acetonitrile, for which the separation of the plates of the ideal ionic capacitor is larger than for water, the difference in capacitances is smaller. As we will see later, the penetration depth of the electric field into the metal, and hence the effective dimension of the capacitor corresponding to g(ion), should depend on the metal's electron density. However, from analysis of capacitances of various interfaces with different metals, it can be inferred[9] that the capacitance associated with g_S^M(ion) is independent of metal, so that g_S^M(ion) itself is independent of metal. Capacitance curves for different metals in the same solution often coincide at large negative charge densities,[34] so that a large effect of metal on the capacitance is unlikely.

From capacitance–potential curves for various aqueous electrolytes on the (110) crystal face of silver, Valette[35] derived the adsorption for F^-, ClO_4^-, PF_6^-, and BF_4^-. The order of adsorption strengths was inverse to that found for mercury, and a possible explanation is in terms of the alteration of the potential of zero charge on the surface defects where the adsorption takes place. Furthermore, the inner-layer capacitance at negative potentials, assumed due to oriented water molecules, was found to have a value almost twice as high as what is found for other metals; this, too, was explained by a model of atomic-scale surface topography. If this is important, one cannot necessarily use capacitances at the point of zero charge to classify the strengths of metal–solvent interactions. Further, a correct model of the interface must necessarily describe the components of both the metal and the electrolyte phase.

Information about the metal in the electrochemical interface can also be obtained[36] from measurements on metal surfaces. Thus, information on the potential drop across the metal-vacuum interface is obtained from measurements of the work function (see next section), which are normally done in vacuum. To simulate the solution side of the interface, one can adsorb H_2O and ions, and still make measurements in high vacuum. There are problems with measuring surface charge, however, and with controlling the potential drop. The latter must be done by adsorbing electropositive or electronegative species. For example, adsorption of alkalis produces a positively charged adsorbate and a negative charge on the metal, thus behaving like a cathodic potential and decreasing ΔV. This leads to a decrease in the work function of several volts, when several monolayers are adsorbed.

The change in work function as water was adsorbed on a particular crystal face of Ag or Cu was measured[36]; the work function decreases with increasing water adsorption. From the initial slope of the curves of work function versus amount adsorbed, one can deduce a dipole moment of 0.9 D for H_2O on Cu(110) and 0.6 D for H_2O on Ag(110), the negative (oxygen) end toward the metal. The work function became independent of water adsorption after about two monolayers. Adding other high-vacuum techniques to work function measurements, one can find out a lot about the way water is adsorbed on these surfaces. However, one still cannot experimentally distinguish the dipole orientation contribution to the change in surface potential from the effects of charge transfer and intermolecular interactions. If the adsorbed water has an effect on the conduction-electron tail, the resulting contribution to the change in surface potential can likewise not be detected and separated out.

From measurements of the rate of change in work function with adsorption of Cs on Ag, one can infer a surface dipole of 8.3 D. Assuming the Cs is completely ionized, this gives the distance between the positive charge of Cs^+ and its image charge in the metal as $d = 1.73$ Å. From this, one can calculate the differential capacitance ($C_{\text{diff}} = dq^{\text{M}}/d\Delta V$) of the metal surface according to the ideal-capacitor formula, as $1/4\pi d = 4.6 \times 10^6\ \text{cm/cm}^2$ or $5.1\ \mu\text{F/cm}^2$. Similarly, from the decrease in work function with Br adsorption one can infer a dipole moment of 0.4 D. If Br is assumed to ionize completely, this means a distance d of 0.08 Å and a differential capacitance of $110\ \mu\text{F/cm}^2$.

III. METAL STRUCTURE

1. Electrons in Metals

We begin with a presentation of the ideas of the electronic structure of metals. A liquid or solid metal of course consists of positively charged nuclei and electrons. However, since most of the electrons are tightly bound to individual nuclei, one can treat a system of positive ions or ion cores (nuclei plus "core" electrons) and free electrons, bound to the metal as a whole. In a simple metal, the electrons of the latter type, which are treated explicitly, are the conduction electrons, whose parentage is the valence electrons of the metal atoms; all others are considered as part of the cores. In some metals, such as the transition elements, the distinction between core and conduction electrons is not as sharp.

If there are N metal atoms, each of which can easily lose z electrons to form an ion M^{z+}, one would like to find the energy and wave function of a system of zN electrons interacting with N ions and with each other. The total energy is the electronic energy plus the interionic repulsion; it depends on the positions of the ions, taken as fixed. From knowledge of the total energy as a function of ionic configuration, one could calculate the equilibrium arrangement of the ions (crystal structure for a solid metal) and the force constants associated with displacement of ions from their equilibrium positions (crystal vibrations). The electron density which enters the equations of electrostatics is obtained by averaging over ionic configurations of low energy.

Since the equilibrium ionic configuration is determined by the energy of the conduction electrons, it is not surprising that ions at a metal surface, which are in a different electronic environment from ions in the bulk and hence experience different forces, are arranged somewhat differently from ions in the bulk metal. For some metals, including gold and platinum, the two-dimensional arrangement of the outermost layer of ions differs from that of layers beneath it, as revealed[37] by low-energy electron diffraction (LEED) from the surface. For others, LEED shows that the spacing between the outermost layer and the layer beneath it is as much as 10% smaller than interlayer spacings in bulk. The falloff of the electronic charge density from the uniform bulk value, discussed below, produces forces on a surface ion which cause its displacement to a position for which the electronic forces are compensated by forces due to other ions. This, in turn, leads to distortion in the structure of several layers, not only the outermost, so that subsequent interlayer spacings also differ from their bulk values. In binary alloys, the same mechanism produces segregation (compositions differing from bulk composition) in surface and near-surface layers and different displacements of ions of the two components.

One can expect that the electron density corresponding to the electronic state of lowest energy is roughly constant in the interior of the metal and decreases to zero outside the metal. This means that the potential seen by an electron, due to the ion cores and the other electrons, is roughly constant inside the metal, with a value significantly lower than the potential outside. The simplest model for electrons in a metal, the Sommerfeld[38] model, takes this potential as $-V_0$ inside and 0 outside. One is then led to consider the one-dimensional Schrödinger equation

$$(-\hbar^2/2m)(d^2\psi/dx^2) + V\psi = \varepsilon\psi \qquad \text{with } V = \begin{cases} -V_0 & \text{for } x < 0 \\ 0 & \text{for } x \geq 0 \end{cases} \tag{9}$$

There are acceptable solutions to this equation for all $\varepsilon \geq -V_0$. By combining the electron densities for all solutions with ε between $-V_0$ and some upper limit ε_1, one gets the electron density $\rho(x)$, which approaches a constant, the bulk electron density ρ_b, for $x \to -\infty$, and which approaches zero for $x \to \infty$ (if V_0 is large enough so that $\varepsilon_1 < 0$). The difference in energy between the highest filled electronic state and the lowest, $\varepsilon_1 + V_0$, is called the Fermi energy (although other definitions of the term are sometimes used).

The solution to the Schrödinger equation (9) is a plane wave, $e^{\pm ikx}$, for $x < 0$ and a decreasing exponential, e^{-Kx}, for $x \geq 0$.

Substituting into the Schrödinger equation, we find $(\hbar^2k^2/2m) - V_0 = \varepsilon$ and $-\hbar^2K^2/2m = \varepsilon$. A suitable combination of e^{ikx} and e^{-ikx} must be taken so that the value and slope of the wave function are continuous at $x = 0$ with e^{-Kx}. The electron density is obtained by adding $|\psi|^2$ for all states occupied by electrons. Taking each state as either occupied (if $\varepsilon < \varepsilon_1$) or empty (if $\varepsilon \geq \varepsilon_1$) corresponds to the Fermi distribution (Fermi–Dirac statistics) for absolute temperature $T = 0$. If we suppose that $V_0 \to \infty$, ε is large and negative and K large and positive, so that $e^{-Kx} = 0$ and the continuity conditions on the wave function are replaced by $\psi = 0$ at $x = 0$.

For the infinite-barrier problem, then, $\psi = 0$ for $x \geq 0$ and $\psi = A \sin kx$ for $x < 0$, where A is a normalization constant. The normalization may be determined by considering the metal as a one-dimensional box of length L and later letting $L \to \infty$. Then the acceptable wave functions are

$$\psi = (2/L)^{1/2} \sin n\pi x/L$$

($\psi = 0$ for $x = 0$ and for $x = L$), with the familiar particle-in-a-box energies

$$\varepsilon + V_0 = n^2\hbar^2\pi^2/2mL^2 = n^2h^2/8mL^2$$

Because of electron spin, each state can hold two electrons. Adding together contributions for all n from 1 to $n_1 = (8mL^2/h^2)^{1/2}(\varepsilon_1 + V_0)^{1/2}$, we should obtain the total number of electrons:

$$Nz = \sum_1^{n_1} 2 = 2(8mL^2/h^2)^{1/2}(\varepsilon_1 + V_0)^{1/2}$$

The Fermi energy $\varepsilon_F = \varepsilon_1 + V_0$ is then

$$\varepsilon_F = (Nz/L)^2(h^2/32m)$$

which is proportional to the square of the one-dimensional electron density Nz/L.

In three dimensions, the normalized eigenfunctions may be taken as

$$\psi = (2/L)^{3/2}(\sin n_x\pi x/L)(\sin n_y\pi y/L)(\sin n_z\pi z/L)$$

(cubical box of volume L^3) and the energies are $-V_0 + n^2h^2/8mL^2$ where $n^2 = n_x^2 + n_y^2 + n_z^2$. All states with $n^2h^2/8mL^2 \leq \varepsilon_1 + V_0$ are occupied with two electrons. The sum over states may be approximated by an integral, since the energies are very closely spaced:

$$Nz = \int dn_x \int dn_y \int dn_z(2) = 2(4\pi/8) \int_0^{n_1} n^2\, dn \qquad (10)$$

We have gone from Cartesian to spherical coordinates in the last member, as a convenience in imposing the limit on the integrations, and carried out the integration over angles, giving the factor of 4π; the division by 8 is because only positive values of n_x, n_y, and n_z are to be considered. Now we have

$$Nz = \pi(n_1)^3/3 = (\pi/3)[8mL^2(\varepsilon_1 + V_0)/h^2]^{3/2}$$

The average electron density is now $\rho_b = Nz/L^3$, so that $3\rho_b/\pi = (8m\varepsilon_F/h^2)^{3/2}$ or

$$\varepsilon_F = (h^2/8m)(3\rho_b/\pi)^{2/3} = (\hbar^2/2m)(3\pi^2\rho_b)^{2/3} \qquad (11)$$

Usually,[39] periodic boundary conditions on the electronic wave functions are imposed instead of requiring that they vanish on the boundary; the result of Eq. (11) is unchanged.

In addition to calculating the total number of electrons and average electron density ρ_b for a large system, we may calculate the variation of electron density with distance near the metal surface (electron density profile) using this model by summing over the electron densities of the occupied wave functions. We suppose that the compensating positive charge density due to the ion cores is given by

$$\rho_+(z) = \begin{cases} \rho_b & z < 0 \\ 0 & z \geq 0 \end{cases}$$

The position of the infinite barrier at which the electronic eigenfunctions vanish is given by $z = z_W$. The value of z_W will be determined to make the system electrically neutral overall. The wave functions for electrons at an infinite barrier are products of imaginary exponentials (free-particle wave functions) in the x- and y-directions, and sine functions vanishing at z_W in the z-direction. The electron density $n_-(z)$ is obtained by adding together the squares of the electronic wave functions, starting with zero energy and including enough to give the density of Eq. (11) for $z \to -\infty$ (bulk). Then, for $z < z_W$,

$$n_-(z) = \rho_b \left\{ 1 + \frac{3\cos[2k_F(z - z_W)]}{4k_F^2(z - z_W)^2} - \frac{3\sin[2k_F(z - z_W)]}{8k_F^3(z - z_W)^3} \right\} \quad (12)$$

where $\hbar^2 k_F^2/2m = \varepsilon_F$ so $k_F^3 = 3\pi^2 \rho_b$. For electrical neutrality, since $\rho_+(z) = \rho_b$ for $z < 0$, we have

$$\rho_b \int_{-L}^{z_W} dz \left\{ 1 + \frac{3\cos[2k_F(z - z_W)]}{4k_F^2(z - z_W)^2} - \frac{3\sin[2k_F(z - z_W)]}{8k_F^3(z - z_W)^3} \right\}$$

$$= \int_{-L}^{0} dz \, \{\rho_b\}$$

where $L \to \infty$, or

$$\int_{z_W}^{0} dz = (3/2k_F) \int_{-\infty}^{0} dx \, (x^{-2} \cos x - x^{-3} \sin x)$$

$$= -(3/4k_F) \int_{-\infty}^{0} dx \, x^{-1} \sin x$$

where $x = 2k_F(z - z_W)$. This gives

$$z_W = 3\pi/8k_F \quad (13)$$

to guarantee electroneutrality.

Note that the interactions between electrons are not taken into account in this calculation, which corresponds to a single-particle model, each electron interacting with the average field. One could go on to calculate the response to an external potential $U(\mathbf{r})$ in the same approximation. If U were spatially homogeneous, each electronic energy would be changed by the same constant. Thus, the change in the electron density due to U would be the same as the change due to a shift in the Fermi level, and the linear response to a spatially homogeneous potential is obtained[40] by differentiating the electron density with respect to the Fermi level, obtaining $\delta\rho = (Uk_F/\pi^2 a_0)(1 - \sin x/x)$.

By integrating the Poisson equation

$$d^2V/dz^2 = -4\pi(\rho_+ - n_-)$$

from $-\infty$ to $+\infty$ we find

$$V(\infty) - V(-\infty) = 4\pi \int_{-\infty}^{\infty} dz \, z(\rho_+ - n_-)$$

We can now calculate a surface potential as follows:

$$V(\infty) - V(-\infty) = 4\pi \int_{-L}^{0} dz\, z\rho_b - 4\pi \int_{-L}^{z_W} dz\, zn_-(z)$$

$$= 4\pi\rho_b\left\{-\int_0^{z_W} z\,dz - (3/4k_F^2)\int_{-\infty}^{0} dx\,(x + 2k_F z_W) \times [x^{-2}\cos(x) - x^{-3}\sin(x)]\right\}$$

$$= 4\pi\rho_b\left[\frac{-z_W^2}{2} - \frac{3}{4k_F^2} + \frac{3z_W k_F \pi}{8k_F^2}\right]$$

which, on substituting for ρ_b and z_W, is

$$V(\infty) - V(-\infty) = (k_F/\pi)[(3\pi^2/32) - 1] \tag{14}$$

The surface potential of the metal,

$$\chi = V(-\infty) - V(\infty)$$

is thus positive. For example, if $\rho_b = 0.01\ e/(a_0)^3$ and $k_F = 0.67(a_0)^{-1}$, χ is 0.016 esu = 4.8 V according to this model.

The surface potential comes about because of the spillover toward the outside of the metal of the conduction electron wave functions, which are delocalized over the metal. As we will see below, this result follows from more sophisticated models for the metal surface as well. The core electrons, which are localized, have not been considered explicitly. In fact, the electronic energy levels in a crystal form bands. That is, for certain ranges of energy there are continua of energy eigenvalues (bands), while for others, the energy gaps, there are no allowed electronic energy eigenvalues. A crystal acts as a metal if there is a partly filled band, so that low-energy excitations of electrons in the band are possible. The continuum of free-particle energy levels we have been discussing is supposed to represent the highest-energy band, with all bands of lower energy filled.

2. Band Structure

To have free-electron energy levels, the potential in which the electrons move must be constant ($V = -V_0$); in a crystalline metal there are ion cores arranged in a regular array or lattice, which makes the potential in which the electrons move periodic. The effect of such a potential may be understood as diffraction of the electron waves by the lattice. In a one-dimensional crystal the free-electron wave functions are e^{ikx} with energy $\hbar^2k^2/2m$, where $k = 2n\pi/L$ with L the size of the crystal and n an integer. (This corresponds to periodic boundary conditions; box normalization, as above, has $\sin kx$ with $kL = n\pi$ and n non-negative.) The energy is quadratic in k. The Bragg equation for reflection by a lattice of a wave perpendicular to the layers is $2d = m\lambda$ where d is the lattice spacing, m is an integer, and the electron wavelength is given by the de Broglie relationship

$$\lambda = h/p = h/k\hbar = 2\pi/k$$

Thus, the free-electron wave functions will be seriously disturbed when $d = m\pi/k$. The corresponding values of k,

$$k = m\pi/d \tag{15}$$

give the positions of the breaks in the energy versus k plot, i.e., the gaps in the band structure.

For these values of k the functions $e^{\pm ikx}$ for $k = m\pi/d$ are replaced by the linear combinations, $\sin kx$ and $\cos kx$. It is easy to see that the potential of the ion cores makes the energies of these two waves quite different, since one corresponds to zero probability density at the positions of the ion cores and the other to a maximum probability density at these positions. The former will have the higher energy since the positive ion cores represent an attractive potential for the electrons. The difference in energies of the two functions is the band gap. For a perturbation theory treatment of the effect of the ion cores, see Appendix A of Kittel[39] or Ziman.[41] The result is that the energy as a function of k follows the parabola $E = \hbar^2k^2/2m$ except near $k = m\pi/d$, where there are distortions such that the energy for k just above $m\pi/d$ differs by a finite amount (the band gap) from the energy for k just below $m\pi/d$.

The origin of allowed bands and band gaps may also be understood in terms of the electronic energy levels of neutral atoms which are allowed to come together so that their charge distributions overlap (tight-binding approximation). The overlap of valence-shell atomic orbitals leads to binding and antibinding molecular orbitals with energies above and below the energies of the atomic orbitals. For N atoms, each contributing one valence orbital, there will be N molecular orbitals of different energies. The spread in energy, since it comes from the overlap, increases as the atoms get closer together. Such a band of energy levels will result from each atomic orbital energy level in the separated atoms. Thus, for a collection of lithium atoms, there will be a $1s$ band, a $2s$ band, $2p$ bands, etc. The $1s$ band will be lower in energy than the $2s$ band and the $2s$ band lower in energy than the $2p$ band, but if the interatomic distances decrease enough, nothing will prevent the bands from overlapping: the band gaps necessarily decrease as the widths of the bands increase. For a given interatomic distance, the width of the $1s$ band will be less than that of the $2s$ band because $1s$ orbitals on different atoms, being more tightly bound, overlap less than $2s$ orbitals. It is clear that a regular array of atoms, such as is found in a crystalline solid, is not necessary for the formation of bands in the tight-binding picture; one expects band structure for liquid metals as well.

If, in each atom, the atomic orbital giving rise to a band is filled with two electrons, there will be $2N$ electrons in the N molecular orbitals of the band formed from N atoms. Then the band will be filled, provided there is no overlap with other bands (e.g., the $1s$ band for Li). These electrons will not respond to an applied electric field since it will require some minimum energy (the band gap) for them to change their state, so they do not contribute to conductivity. If there is only one electron per atom in an atom orbital, as for the $2s$ orbital of Li, the band will be half-filled (N electrons in the $2N$ states of the Li $2s$ band). An electric field will induce a current in these bands, since electrons at the Fermi level can gain momentum (change k) by going from one state of the band to another with little cost in energy. A solid will be an insulator if all bands are either completely filled or completely empty, and if the band gap between the highest filled and lowest empty bands is much greater than kT. Thus solid hydrogen is an insulator. The alkaline earth metals are not, because of the overlap in energy of the valence s band and the valence p bands.

The band structure (widths of bands, energy gaps) will obviously depend on the arrangement of the atoms as well as on their nature. For the one-dimensional crystal lattice, the positions of the gaps depend on the interionic spacing d according to $k = m\pi/d$ in the free-electron picture, but the size of the gap, as calculated by perturbation theory, depends on the ion-core potential. The ranges of k corresponding to allowed energies are called Brillouin zones. Thus, the first Brillouin zone in the one-dimensional crystal extends from $k = -\pi/d$ to $k = \pi/d$. The second Brillouin zone includes two ranges of k corresponding to the same energies: $-2\pi/d \leq k \leq -\pi/d$ and $\pi/d \leq k \leq 2\pi/d$. In three dimensions, a plane wave with energy $E = \hbar^2k^2/2m$ has wave function $\exp(i\mathbf{k} \cdot \mathbf{r})$. The values of the vector $\mathbf{k}$ corresponding to the boundaries of Brillouin zones are found from

$$2\mathbf{k} \cdot \mathbf{g} + 2\pi g^2 = 0 \qquad (16)$$

where $\mathbf{g}$ is a reciprocal lattice vector ($g = 1/d$ for a one-dimensional lattice). The orientations of the planes defined by Eq. (16), which bound the Brillouin zones, and hence the shape of the zones, depend on the crystal structure through the reciprocal lattice vectors.

For instance, the first Brillouin zone of the simple cubic lattice is a cube of edge length $2\pi/d$, where d is the interatomic distance. The volume of the zone in $\mathbf{k}$-space is $(2\pi/d)^3$. According to Eq. (10), the number of states, including electron spin, with $k \leq k_1$ ($k_1 = n_1\pi/L$) is $(\pi/3)(k_1L/\pi)^3$. Dividing by the volume in $\mathbf{k}$-space, $4\pi k_1^3/3$, and by the volume of the crystal, L^3, we find that the number of states per unit volume in $\mathbf{k}$-space per unit volume of crystal is $2/(2\pi)^3$. Thus, the number of states in the first zone of the crystal per unit volume of crystal is $2/d^3$, which corresponds to two states for each atom. The first zone for the body-centered cubic structure is a rhombic dodecahedron, and the first zone for the face-centered cubic structure is a truncated octahedron.[39] There are two states per atom in the first zone in each case. To the extent that the energy levels are like those for free particles, the constant-energy surfaces in $\mathbf{k}$-space are spheres. This obviously does not hold near zone boundaries. The Fermi surface is the boundary, in $\mathbf{k}$-space, between filled and unfilled levels. Being a constant-energy surface, the Fermi surface will be spherical as long as it is not too near the zone boundaries.

For a periodic lattice, it can be shown (Bloch theorem) that the solutions to the one-electron Schrödinger equation are of the form

$$\psi = u_k(\mathbf{r})\, e^{ik\cdot r} \tag{17}$$

where $u_k(\mathbf{r})$ is periodic with the same period as the potential of the lattice, so that a translation of $\mathbf{r}$ by one of the lattice vectors $\mathbf{v}$ just multiplies ψ by $e^{ik\cdot v}$. The function u_k depends on $\mathbf{k}$. If $\mathbf{k}$ is outside the first Brillouin zone, one may write the wave function in Eq. (17) as

$$\begin{aligned}\psi &= u_k(\mathbf{r}) \exp(2\pi i\mathbf{g}\cdot\mathbf{r}) \exp[i(\mathbf{k} - 2\pi\mathbf{g})\cdot\mathbf{r}] \\ &= u'_{k'}(\mathbf{r})\, e^{ik'\cdot r}\end{aligned}$$

Here $u' = u \exp(2\pi i\mathbf{g}\cdot\mathbf{r})$ is, like u, periodic with the period of the lattice, and $\mathbf{k}' = \mathbf{k} - 2\pi\mathbf{g}$ is a reduced wave vector. Repeating this as necessary, one may reduce $\mathbf{k}'$ to a vector in the first Brillouin zone. In this "reduced zone scheme," each wave function is written as a periodic function multiplied by $e^{ik\cdot r}$ with $\mathbf{k}$ a vector in the first zone; the periodic function has to be indexed, say $u_{jk}(\mathbf{r})$, to distinguish different families of wave functions as well as the $\mathbf{k}$ value. The index j could correspond to the atomic orbital if a tight-binding scheme is used to describe the crystal wave functions.

The conduction band normally corresponds to the valence electrons and is the highest-energy band containing electrons. As mentioned above, for alkali metals this band is an s-band and is half-filled, since each atom contributes one valence orbital and one electron. The crystal structure is body-centered cubic; with one electron per atom, the Fermi surface should be quite spherical. The divalent metals, including the alkaline earths, exhibit hexagonal closest-packing, face-centered and body-centered cubic, and other structures. With two electrons in the outermost s orbital of each atom, one would have a filled band of highest energy in the solid, and insulator behavior, were it not for the mixing of the s- and p-bands. For beryllium, there is less mixing than for the other elements, so that the conductivity is small.

The overlap between s- and p-bands also occurs for the alkali metals and for the monovalent noble metals copper, silver, and gold, which have face-centered cubic structures. The noble metals differ from the alkalis because of the filled d-shell just below the s-shell in energy; the d-band and the s-band overlap in the solid.

Furthermore, the electrons of the *d*-shells, through the exclusion principle, produce added repulsions between atoms, giving these metals a much lower cohesive energy and a much lower compressibility than the alkalis. The shape of the Fermi surface is complicated, and it is thought to touch the zone boundary. The transition elements have partly filled *d*-shells, and there is significant overlap between the *d*-band and the *s*-band just above it in energy. The state density (number of one-electron states per unit energy range) may be quite a complicated function of energy.

3. Ion–Electron Interactions

The very use of the energy-band description implies that each electron is in a one-electron state, whose wave function is the solution of a one-electron Schrödinger equation like Eq. (9), except with a more complicated potential V. V must represent the effect of the positively charged nuclei, arranged at the points of a lattice or according to some distribution, and of the electrons other than the one considered in the Schrödinger equation. The potential V is then a self-consistent potential of Hartree–Fock theory; it depends on all the electronic wave functions other than the one being determined in the one-electron Schrödinger equation and is determined along with the wave functions as part of an iterative procedure. The electronic contribution to V includes the coulomb potential of the electronic cloud and an operator representing exchange.

For the conduction electrons, it is reasonable to consider that the inner-shell electrons are all localized on individual nuclei, in wave functions very much like those they occupy in the free atoms. The potential V should then include the potential due to the positively charged ions, each consisting of a nucleus plus filled inner shells of electrons, and the self-consistent potential (coulomb plus exchange) of the conduction electrons. However, the potential of an ion core must include the effect of exchange or antisymmetry with the inner-shell or core electrons, which means that the conduction-band wave functions must be orthogonal to the core-electron wave functions. This is the basis of the *orthogonalized-plane-wave method*, which has been successfully used to calculate band structures for many metals.[41]

Formally, each orthogonalized-plane-wave basis function may be written as $(1 - P)\psi_k$, where ψ_k is a plane wave and P is the projection operator such that $P\psi_k$ gives the core-state component of ψ_k:

$$P\psi_k = \sum \psi_{tj} \int \psi_{tj}^* \psi_k \, \mathbf{dr}$$

with ψ_{tj} the wave function of the jth core state on ion t. The expansion in orthogonalized plane waves converges rapidly, so that the wave function $\sum c_k(1 - P)\psi_k$ has contributions only from small $\mathbf{k}$ and $\phi = \sum c_k\psi_k$ is a smooth function. Note that ϕ, called the pseudo wave function, differs from the true wave function $(1 - P)\phi$ only in the core regions. Inserting $(1 - P)\phi$ into the Schrödinger equation, one can derive [42] an equation for ϕ:

$$-(\hbar^2/2m)\nabla^2\phi + W\phi = E\phi \tag{18}$$

where W is called a *pseudopotential.* W includes the true potential V plus terms involving the projection operator P, making W a nonlocal operator. Since the effect of P is to introduce oscillations into a smooth wave function, it raises the kinetic energy. This cancels off part of the strongly attractive potential V, making W a much less attractive potential. W may be small enough for its effect to be considered by perturbation theory, which is not the case for V itself.

Although the pseudopotential is, from its definition, a nonlocal operator, it is often represented approximately as a multiplicative potential. Parameters in some chosen functional form for this potential are chosen so that calculations of some physical properties, using this potential, give results agreeing with experiment. It is often the case that many properties can be calculated correctly with the same potential.[43] One of the simplest forms for an atomic model effective potential is that of Ashcroft[44]: $r^{-1}\theta(r - R_c)$, where the parameter is the "core radius" R_c and θ is a step-function.

One now has a picture of conduction electrons in the potential of the ions, which is really a collection of pseudopotentials. The energy of the electronic system obviously depends on the positions of the ions. From the electronic energy as a function of ionic positions, say $U_{el}(\mathbf{R})$, one could determine the equilibrium ionic configuration (interionic spacing in a crystal or ion density profile in a liquid). From $U_{el}(\mathbf{R})$ one could also calculate the force exerted by the electrons on each ion. It is clear that the effective interionic force in a metal differs from the force between ions in a vacuum, because a displacement of one ion produces a change in the electronic wave functions and hence a changed potential felt by the other ions. In fact, the energy of the ion–electron system can approximately (but only approximately) be written[45] as a sum of pair-interaction terms plus a term independent of ion positions.

One can consider an assembly of ions, each the source of a potential and a pseudopotential which act on the electrons, giving rise to the induced electronic charge density $\rho^{\text{ind}}(\mathbf{R})$. The interaction of $\rho^{\text{ind}}(\mathbf{R})$ with the ions gives the *band-structure energy*, which depends on the arrangement of the ions. This, combined with the Madelung energy of ions in a uniform, negatively charged background, gives the part of the energy of the system which depends on the ionic arrangement. The result can be written, for a homogeneous metal, as a volume-dependent term plus a structural term, $\frac{1}{2}\sum \Phi(\mathbf{R}_i - \mathbf{R}_j)$; the effective interionic interaction potential Φ also depends on volume, since it includes the effects of the interacting electron gas. In an inhomogeneous system, such as one involving a surface, Φ could depend on position as well as on the interionic vector $\mathbf{R}_i - \mathbf{R}_j$, but for homogeneous metals, the effective interionic interaction Φ depends on interionic distance R only. It is strongly repulsive at small R, but includes a small attraction as well, so that there is a region for which the force $-d\Phi/dR$ is negative and a minimum, for which Φ is less than its value at $R = \infty$. There are oscillations in Φ at large R, whose origin is the dielectric function of the electrons (see below), which expresses the screening of the coulomb potential by the electron gas.

For simple monovalent metals, the pseudopotential interaction between ion cores and electrons is weak, leading to a uniform density for the conduction electrons in the interior, as would obtain if there were no point ions, but rather a uniform positive background. The arrangement of ions is determined by the ion–electron and interionic forces, but the former have no effect if the electrons are uniformly distributed. As the interionic forces are mainly coulombic, it is not surprising that the alkali metals crystallize in a body-centered cubic lattice, which is the lattice with the smallest Madelung energy for a given density.[46] Diffraction measurements of the melts of Na and K indicate little change in local coordination from the solid. Thus, the interionic forces (coulomb and hard-sphere) dominate in determining structure in the liquid. For polyvalent simple metals, it is necessary to take into account the concentration of conduction electron density around the multiply charged ions to understand[46] the stable ionic arrangements in the crystal and in the melt.

4. Screening

The effect of screening by the electrons in a metal is to convert a bare coulombic interaction q^2/r, where r is the distance between two charges q, to an interaction $(q^2/r)\, e^{-r/l}$, l being the screening length. The Fourier transform of q^2/r is $4\pi e^2/k^2$, and the Fourier transform of $(q^2/r)e^{-r/l}$ is $4\pi q^2/(k^2 + l^{-2})$. The ratio of Fourier components, $1 + (kl)^{-2}$, is the k-dependent dielectric function $\varepsilon(k)$, which describes the screening of a sinusoidally varying field of wavelength $1/k$. The value of l for a system of ions screening each other, as in the Debye–Hückel model for an ionic liquid,[47] is $(kT/4\pi nq^2)^{1/2}$ if n is the number density. The screening by a degenerate Fermi gas of electrons is characterized by a different screening length, which may be estimated from the Debye–Hückel result by noting that $(kT/m)^{1/2}$ is the average velocity of a particle in a classical fluid while (some fraction of) $\hbar k_F/m$ is the corresponding velocity in the quantum fluid; k_F is the Fermi momentum, related [see Eqs. (12) and (13)] to the density by $3\pi^2 n = k_F^3$. Thus, for $(kT)^{1/2}$ in the Debye–Hückel screening length, one should substitute $m^{-1/2}\hbar k_F$, and $q = e$, giving a quantum screening length $m^{-1/2}\hbar k_F(4\pi ne^2)^{-1/2}$. In terms of the Bohr radius $a_0 = \hbar^2/me^2$, the screening length is

$$\left(\frac{a_0}{4\pi n}\right)^{1/2} k_F = \left(\frac{a_0 3\pi^2}{4\pi k_F^3}\right)^{1/2} k_F = \left(\frac{3\pi a_0}{4k_F}\right)^{1/2}$$

The screening length is of the order of an angstrom in good metals.

It is, of course, usual in discussing the electrochemical interface to use a dielectric *constant*, which is the ratio of the electric displacement to the electric field. By Fourier transforming the dielectric function $\varepsilon(k)$, one would obtain an effective dielectric constant, which would, however, depend on position. In fact,[48] the screening in the interface is nonlocal, and the effective dielectric constant is actually an integral of a dielectric response function. It can be shown[47] that it is not always possible to replace the exact nonlocal theory by a local theory, even with a position-dependent dielectric constant.

An expression for $\varepsilon(k)$ in the case of a Fermi gas of free electrons can be obtained by considering the effect of an introduced point charge potential, small enough so the arguments of perturbation theory are valid. In the absence of this potential, the electronic wave functions are plane waves $V^{-1/2}\exp(i\mathbf{k}\cdot\mathbf{r})$, where V is the volume of the system, and the electron density is uniform. The point charge potential is screened by the electrons, so that the potential felt by an electron, Φ, is due to the point charge and to the other electrons, whose wave functions are distorted from plane waves. The electron density and the potential are related by the Poisson equation,

$$\nabla^2\Phi = -4\pi e(n_b - n)$$

where n_b is the density of the positive background charge. Another equation relating Φ and n is required.

In the Thomas-Fermi model,[49] the kinetic energy density of the electron gas is written as

$$\tfrac{3}{10}(3\pi^2)^{2/3}n^{5/3}e^2a_0$$

and the potential energy density as $-en\Phi$, where n may vary with position. The energy of the most energetic electron must be the same everywhere, so

$$\tfrac{1}{2}(3\pi^2)^{2/3}n^{2/3}e^2a_0 - e\Phi = \text{constant}$$

Far from the introduced point charge, $n = n_b$ and $\Phi = 0$. Assuming Φ is small, the deviation of n from n_b is small, and

$$\begin{aligned} e\Phi &= \tfrac{1}{2}(3\pi^2)^{2/3}(n^{2/3} - n_b^{2/3})e^2a_0 \\ &\simeq \tfrac{1}{2}(3\pi^2)^{2/3}(\tfrac{2}{3}n_b^{-1/3})(n - n_b)e^2a_0 \end{aligned}$$

Combining this with the Poisson equation, we get

$$\nabla^2\Phi = 12\pi e^2(3\pi^2)^{-2/3}n_b^{1/3}\Phi/(e^2a_0) \tag{19}$$

The solution corresponding to a point charge q at the origin is

$$\Phi = (q/r)\, e^{-r/l},$$

with

$$l^2 = (12\pi)^{-1}(3\pi^2)^{2/3} n_b^{-1/3} a_0 \tag{20}$$

Since $n_b^{-1/3} = (3\pi^2)^{1/3}/k_F$, we find $l = (\pi a_0/4k_F)^{1/2}$.

In the Hartree-Fock or self-consistent field picture, Φ also enters the Schrödinger equation which determines the electronic wave functions. One thus has to solve the Schrödinger equation

$$\nabla^2\psi_k + (k^2 - 2m\Phi/\hbar^2)\psi_k = 0$$

for the electronic wave functions. This equation is equivalent to the integral equation

$$\psi_k = V^{-1/2}\exp(i\mathbf{k}\cdot\mathbf{r}) + (2m/\hbar^2)\int G(\mathbf{r}-\mathbf{r}')\Phi(\mathbf{r}')\psi_k(\mathbf{r}')\,d\mathbf{r}' \tag{21}$$

where $G(\mathbf{r}-\mathbf{r}')$, a Green's function, is $-\exp(ik|\mathbf{r}-\mathbf{r}'|)/(4\pi|\mathbf{r}-\mathbf{r}'|)$, because

$$\nabla^2|\mathbf{r}-\mathbf{r}'|^{-1} = -4\pi\delta(\mathbf{r}-\mathbf{r}')$$

In the integral of Eq. (21), one can substitute the unperturbed wave function $V^{-1/2}\exp(i\mathbf{k}\cdot\mathbf{r}')$ for $\psi_k(\mathbf{r}')$ since Φ is a small perturbation. With the resulting expression for ψ_k, one calculates the electron density as a sum of densities for the occupied orbitals (with $|\mathbf{k}| \le k_F$). Replacing the sum by an integral and subtracting the unperturbed electron density, one finds[50] for the change of electron density the Lindhard result

$$\delta n(\mathbf{r}) = -(mk_F^2/2\pi^3\hbar^3)\int j_1(2k_F|\mathbf{r}-\mathbf{r}'|)|\mathbf{r}-\mathbf{r}'|^{-2}\Phi(\mathbf{r}')\,d\mathbf{r}' \tag{22}$$

where $j_1(x) = (\sin x - x\cos x)/x^2$. Since Φ arises from the introduced point charge and from δn,

$$\nabla^2\Phi = -4\pi e^2\delta(\mathbf{r}) + 4\pi e^2\delta n$$

Solving the resulting equation by Fourier transformation, one obtains for the Fourier components

$$\Phi(k) = 4\pi e^2/k^2\varepsilon(k)$$

with

$$\varepsilon(k) = 1 + \frac{2mk_F e^2}{\pi\hbar^2 k^2}\left(1 + \frac{k^2 - 4k_F^2}{4k_F k}\ln\left|\frac{k - 2k_F}{k + 2k_F}\right|\right) \tag{23}$$

A noteworthy feature of $\varepsilon(k)$ is the singularity at $k = 2k_F$, the diameter of the Fermi sphere.

Writing this as $1 + (kl)^{-2}$, one can extract the screening length l. For long wavelengths (small k), the second term in parentheses vanishes and the screening length obeys

$$l^2 = \pi\hbar^2/2mk_F e^2 = \pi a_0/2k_F$$

which is twice the estimate of Eq. (20). Alternatively, one notes that the small-k limit corresponds to a slowly varying potential. As shown above, the introduction of a small point charge Ze into an electron plasma results in a potential $V(\mathbf{r})$ which obeys

$$\nabla^2 V = 4\pi Ze^2\delta(\mathbf{r}) + \frac{2mk_F^2 e^2}{\pi^2\hbar^3}\int \frac{j_1(2k_F|\mathbf{r} - \mathbf{r}'|)}{|\mathbf{r} - \mathbf{r}'|^2} V(\mathbf{r}')\,\mathbf{dr}'$$

If one assumes V is slowly varying in space, compared to other factors in the integral, one can put $V(\mathbf{r})$ for $V(\mathbf{r}')$ and take it out of the integral as a constant; what remains is

$$\int j_1(2k_F s)s^{-2}\,\mathbf{ds} = 4\pi(2k_F)^{-1}\int_0^\infty (\sin u - u\cos u)u^{-4}u^2\,du$$
$$= 2\pi/k_F$$

Then

$$\nabla^2 V = 4\pi Ze^2\delta(\mathbf{r}) + 4k_F V/\pi a_0$$

which has the solution $V(r) = -(Ze^2/r)e^{-qr}$ with $q^2 = 4k_F/\pi a_0$; this is clearly a screened coulomb potential.

The change in the electronic charge density according to Eq. (22) is

$$(mk_F^2/2\pi^3\hbar^2)\int j_1(2k_F|\mathbf{r} - \mathbf{r}'|)|\mathbf{r} - \mathbf{r}'|^{-2}V(\mathbf{r}')\,\mathbf{dr}'$$

If V is localized, say, near the origin, then for locations far from the origin, this behaves like $j_1(2k_Fr)/r^2$, which means as $\cos(2k_Fr)/r^3$. These damped oscillations of frequency $2k_F$ are the Friedel oscillations, which always arise when an electron gas is perturbed; the frequency of oscillation comes from the "kink" in the dielectric function at $2k_F$. We see the Friedel oscillations (in planar rather than in spherical geometry) for the electron gas at a hard wall [Eq. (12) *et seq.*] and for the electron density at the surface of a metal.

More generally, one considers an external test-charge density $n_0(\mathbf{r}, t)$ acting on a plasma (here, the electron density) in the presence of a uniform charge density of opposite sign which is fixed. The polarization of the plasma is described by a change in the electron density $\delta n(\mathbf{r}, t)$ which is assumed to be related linearly to the potential V_0 of the density n_0. The most general linear relation is

$$\delta n(\mathbf{r}, t) = \int\int d\mathbf{r}'\, dt'\, K(|\mathbf{r}-\mathbf{r}'|, t-t')\, V_0(\mathbf{r}', t') \tag{24}$$

The response function K can depend only on $|\mathbf{r}-\mathbf{r}'|$ for a uniform isotropic system. On Fourier transformation,

$$\delta n(\mathbf{k}, \omega) = \chi(\mathbf{k}, \omega)\, V_0(\mathbf{k}, \omega) \tag{25}$$

where $\delta n(\mathbf{k}, \omega)$ and $V_0(\mathbf{k}, \omega)$ are the Fourier transforms of $\delta n(\mathbf{r}, t)$ and $V_0(\mathbf{r}, t)$. Since V_0 is the electrostatic potential of the charge distribution n_0, $V_0(\mathbf{k}, \omega) = (4\pi e^2/k^2) n_0(\mathbf{k}, \omega)$ where $n_0(\mathbf{k}, \omega)$ is the Fourier transform of $n_0(\mathbf{r}, t)$. The Poisson equation for the electrical displacement is

$$\nabla \cdot \mathbf{D}(\mathbf{r}, t) = -4\pi e n_0(\mathbf{r}, t)$$

In the presence of the plasma, the electrical displacement $\mathbf{D}$ is the same, but the electric field now obeys

$$\nabla \cdot \mathbf{E}(\mathbf{r}, t) = -4\pi e[n_0(\mathbf{r}, t) + \delta n(\mathbf{r}, t)]$$

On Fourier transforming these equations, one finds

$$\frac{\mathbf{k} \cdot \mathbf{E}(\mathbf{k}, \omega)}{\mathbf{k} \cdot \mathbf{D}(\mathbf{k}, \omega)} = 1 + \frac{\delta n(\mathbf{k}, \omega)}{n_0(\mathbf{k}, \omega)} = 1 + \frac{4\pi e^2}{k^2}\chi(\mathbf{k}, \omega)$$

the ratio of $\mathbf{k} \cdot \mathbf{D}$ to $\mathbf{k} \cdot \mathbf{E}$ is just the wavelength- and frequency-dependent dielectric constant, $\varepsilon(\mathbf{k}, \omega)$. The calculation of χ and ε at various levels of approximation, for classical and quantum plasmas, is discussed by March and Tosi.[50]

Treating the free electrons in a metal as a collection of zero-frequency oscillators gives rise[51] to a complex frequency-dependent dielectric constant of $1 - \omega_p^2/(\omega^2 - i\omega/\tau)$, with $\omega_p = (4\pi ne^2/m)^{1/2}$ the plasma frequency and τ a collision time. For metals like Ag and Au, and with frequencies ω corresponding to visible or ultraviolet light, this simplifies to give a real part

$$\varepsilon_f' = 1 - \omega_p^2/\omega^2$$

and an imaginary part $\varepsilon_f'' = 0$. The complex refractive index, which is the square root of the dielectric constant, determines the reflectivity of a solid. This is the basis for UV-visible reflectance spectroscopy. The electroreflectance is the change in reflectivity of the metal–electrolyte interface with electrode potential.

The electroreflectance of a Cu single-crystal surface in 1 N H_2SO_4 is discussed by Kolb.[51] The gross features are explained in terms of a potential-induced change in the metal surface's optical constants. For more anodic potentials, the free-electron concentration near the metal surface is lowered compared to the bulk, changing the plasma frequency to $[4\pi(n + \Delta n)^2 e^2/m]^{1/2}$ where $\Delta n = C_{DL}\Delta V/ed$, with C_{DL} the double-layer capacitance of the interface, ΔV the potential drop across the interface, and d the penetration depth of a static electric field into the metal. To explain the electroreflectance in more detail, one must consider bound electrons as well. Surface states must also be taken into account: their energies depend strongly on the applied field. It is assumed that, because they are localized at the surface, they sample a fraction of the potential drop across the interface. The surface states of adsorbates at the electrode–solution interface can be studied by *in situ* time-modulated reflection spectroscopy, which also shows the effects of modification on the conduction-electron tail.[52]

The interaction between two ions in a metal is screened by the gas of conduction electrons. Although corrections for exchange and correlation are required, the features of the screened interaction are what one would expect from the preceding calculation of the screened potential of a single charge.[53] It is coulombic at small interionic separations R, almost completely screened away at large R, and oscillatory at intermediate R. From such an effective interionic interaction potential, one can calculate the stability of various liquid and crystal structures, interionic distances, force constants, etc. Much of the theory requires, as we have seen above, that the potential of an ion be a weak perturbation on the electron gas. The coulombic potential of an ion core would not seem to be weak, but the ion core is not really a point charge, and thus is a source of a pseudopotential as discussed above. The total effective potential is indeed weak.

IV. METAL SURFACES

1. Density Functionals

We now consider the electron density at the surface of a metal. An often-used approach to surface properties of many-particle systems is the density-functional approach, which supposes that the free energy of such a system can be written as the integral of a free-energy density. In the simplest such theories, the free-energy density is a function of the local particle density. By requiring that the chemical potential (electrochemical potential, for a system of charged particles) be independent of z, one derives[49] an equation for $n(z)$. This approach has been applied to a one-component plasma, with a background step-function density of compensating charge, and gives results which agree well with the results of computer simulation for this problem.

For liquid metals, one has to set up density functionals for the electrons and for the particles making up the positive background (ion cores). Since the electrons are to be treated quantum mechanically, their density functional will not be the same as that used for the ions. The simplest quantum statistical theories of electrons, such as the Thomas–Fermi and Thomas–Fermi–Dirac theories, write the electronic energy as the integral of an energy density $\varepsilon(n)$, a function of the local density n. Then, the actual density is found by minimizing $\varepsilon(n) + vn$, where v is the potential energy. Such calculations produce surface energies and other properties which are not in good agreement with experiment. Better theories[49] include a term depending on the gradient of the particle density. Thus, if $n(z)$ is the density as a function of position for a planar interface, the free energy per unit surface area would be $f_1[n(z)] + f_2[n(z), n'(z)]$. On general grounds, f_2 cannot depend on the sign of $n'(z)$.

For a density which is not too rapidly varying, the electronic energy can be written as the integral over all space of a sum of three terms. In addition to the potential energy term, vn, there is[49] a term, $\varepsilon(n)$, which represents the density of kinetic energy (including exchange and correlation) of a homogeneous electron fluid of density n, and the inhomogeneity term, proportional to $|\nabla n|^2/n$. For a system containing a planar surface, n is a function of z. The density profile $n(z)$ is determined[19] by minimizing the integral over all space of the energy density, $\varepsilon(n) + c|\nabla n|^2/n + vn$, with respect to $n(z)$. For simple liquid metals, an electron density profile can be calculated by minimizing this energy functional with respect to $n(\mathbf{r})$.

Smith[54] used this approach to calculate electron density profiles, and from them, surface potentials and work functions, for 26 metals. The positive charge density was modeled as a jellium (step-function) and the quantity $E_v[n] - \mu N$ was made stationary to variations in parameters in the density profile function n. Here, E_v is the surface energy, μ is a Lagrange multiplier, and $N = \int n(z)\,dz$ is the number of electrons per unit surface area. The functional for E_v included the interelectronic repulsion, the interaction of $n(z)$ with the potential of the positive background, and terms representing kinetic, exchange, and correlation energy. For these last three contributions, Smith used

$$\tfrac{3}{10}(3\pi^2)^{2/3}\int n^{5/3}\,dz - \tfrac{3}{4}(3/\pi)^{1/3}\int n^{4/3}\,\mathrm{d}z$$
$$- 0.056\int n^{4/3}(0.079 + n^{1/3})^{-1}\,dz$$

which are the corresponding energy densities of a uniform electron gas of density n. He also included an inhomogeneity term,

$\int (\nabla n)^2(72n)^{-1}\,dz$. The density profile was taken as

$$n = \begin{cases} n_+ - \frac{1}{2}n_+e^{bz}, & z < 0 \\ \frac{1}{2}n_+e^{-bz}, & z \geq 0 \end{cases} \tag{26}$$

with n_+ the positive charge density and b a parameter determined by variation. Surface barriers in agreement with experiment (error of about 10%, with barriers from 4 to 30 eV) were obtained.

Schmickler and Henderson[55] have improved on Smith's results by using the trial density profile

$$n(x) = \begin{cases} n_+[1 - Ae^{ax}\cos(cx + d)], & x < 0 \\ n_+(Be^{-bx}), & x \geq 0 \end{cases}$$

There are four conditions that mut be satisfied: continuity of $n(x)$ and $n'(x)$, charge balance between $n(x)$ and the background, and the Budd–Vannimenus half-moment condition (see below) relating the potential difference between $-\infty$ and 0 to the derivative of the energy per electron in bulk. The two remaining free parameters were determined by minimizing the same density functional used by Smith. The results for work functions are closer to those obtained by Lang and Kohn[56] (see below) than Smith's (taking advantage of the Budd–Vannimenus condition, only the potential drop between the jellium edge and vacuum at infinity needs to be calculated). The effect of an external field, due to a low charge density on a sheet at infinite distance from the jellium, was also calculated. The resulting profile of induced surface charge density, as well as the position of the effective image plane (center of mass of induced surface charge), agreed well with that of Lang and Kohn.[57]

One knows, however, that the simple density-functional theories cannot produce an oscillatory density profile. The energy obtained by Schmickler and Henderson[55] is, of course, lower than that of Smith[54] because of the extra parameters, but the oscillations in the profile found are smaller than the true Friedel oscillations. Further, the density-functional theories often give seriously inexact results. The problem is in the incorrect treatment of the electronic kinetic energy, which is, of course, a major contributor to the total electronic energy. The electronic kinetic energy is not a simple functional of the electron density like $\varepsilon(n) + c|\nabla n|^2/n$, but a quantum mechanical expectation value over a wave function of the kinetic energy operator $-(\hbar^2/2m)\nabla^2$. Lang and Kohn[56] developed a theory in which the electronic kinetic energy is treated correctly, as a sum of expectation values over occupied one-electron states (orbitals). These states are determined self-consistently, as in the Hartree–Fock theory, but, as we will see, the theory of Lang and Kohn includes interelectronic correlation, absent from Hartree-Fock.

Suppose one seeks the electron density in the presence of an external potential $V(\mathbf{r})$. The density is given by

$$n(\mathbf{r}) = N \int \Psi^*\Psi \, d\mathbf{r}' \tag{27}$$

where N is the number of electrons, the integration is over all electronic coordinates but one, and Ψ is the (antisymmetric) eigenfunction of the N-electron Hamiltonian. This Hamiltonian includes operators for electronic kinetic energy, interaction of the electrons with V, and interelectronic repulsion. In the theory of Lang and Kohn,[56,58] one seeks a potential V', different from V, such that the density of *noninteracting* electrons in the presence of V' is equal to n, the density of *interacting* electrons in the presence of V. In the absence of interelectronic interaction, the Hamiltonian is a sum of one-electron Hamiltonians, $H = \sum_i h(i)$, where

$$h(i) = -(\hbar^2/2m)\nabla_i^2 + V'(\mathbf{r}_i)$$

The eigenfunctions of H are determinants formed from one-electron functions which are eigenfunctions of h, say,

$$h(i)\psi_j(i) = \varepsilon_j\psi_j(i) \tag{28}$$

Then, the electron density is just

$$n(\mathbf{r}) = \sum_j |\psi_j(\mathbf{r})|^2$$

the sum being over the eigenfunctions corresponding to occupied one-electron states. The problem is the determination of V'.

It can be shown rigorously[59] that for a system of interacting electrons in a compensating charge background, the energy can be written as an electrostatic part plus a functional of the electron density $G[n]$. The energy of the system is

$$E = \int V(\mathbf{r})n(\mathbf{r})$$
$$+ (e^2/2) \int n(\mathbf{r})n(\mathbf{r}')\int |\mathbf{r} - \mathbf{r}'|^{-1}\, \mathbf{dr}\, \mathbf{dr}' + G[n]$$

so that the density could be determined by minimizing E with respect to n. However, the form of $G[n]$ is not known, and is likely to be nonlocal and extremely complex. Theories which use simple approximate forms for $G[n]$ do not give very good densities. The exchange-correlation energy functional is defined as

$$\varepsilon_{xc}[n] = G[n] - \sum_j \int \psi_j^*(-\hbar^2/2m)\nabla^2\psi_j\, d\tau \tag{29}$$

the last term being the kinetic energy of the system of noninteracting electrons which gives the same density as the system of interest. Although the exact form of $\varepsilon_{xc}[n]$ is not known either, it is a relatively small part of the total energy, and sufficiently accurate approximations to it can be constructed. The self-consistent theory then takes the effective potential $V'(\mathbf{r}_i)$ as $\phi(\mathbf{r}) + \delta\varepsilon_{xc}/\delta n$, where ϕ is the electrostatic potential (due to positive charges and electrons); if there is an external potential, it is also included in V'. This corresponds to choosing the $\{\psi_j\}$ to minimize E, with G being ε_{xc} + the correct kinetic energy in terms of the $\{\psi_j\}$.

2. Self-Consistent Theories

The scheme for obtaining the electron density is thus as follows. (i) Given an electron density $n(\mathbf{r})$, one can construct the electrostatic potential and $\delta\varepsilon_{xc}/\delta n$. (ii) This gives the potential V' and the Hamiltonian h. (iii) The eigenfunctions of h are found and combined into a new electron density. (iv) One seeks self-consistency; i.e., input and output electron densities should be identical. In principle, though not in practice, this involves iteration of steps (i)–(iii). With the self-consistent functions, $G[n]$ is written as

$$\varepsilon_{xc}[n] + \sum_j \int \psi_j^*(-\hbar^2/2m)\nabla^2\psi_j\, d\mathbf{r} = \varepsilon_{xc}[n] + \sum_j \int \psi_j^*(h - V')\psi_j\, \mathbf{dr}$$

so that, using Eq. (29), the energy of the system is

$$E = \int V(\mathbf{r})n(\mathbf{r})\,\mathbf{dr} - \int V'(\mathbf{r})n(\mathbf{r}) + (e^2/2)\int n(\mathbf{r})n(\mathbf{r}')|\mathbf{r}-\mathbf{r}'|^{-1}\,\mathbf{dr}\,\mathbf{dr}' + \varepsilon_{xc}[n] + \sum_j \varepsilon_j$$

Thus, for a planar interface one determines the one-electron wave functions according to

$$[(-\hbar^2/2m)(d^2/dz^2) + V_{\text{eff}}(z)]\psi_k(z) = \varepsilon_k\psi_k(z) \qquad (30)$$

where V_{eff} includes the external potential, the coulombic (Hartree) potential of the electrons, and the exchange-correlation potential, $\delta\varepsilon_{xc}/\delta n$. V_{eff} depends on n through this last term. The one-electron wave functions are actually $\psi_k(z)\,e^{i\mathbf{k}'\cdot\mathbf{s}}$ where $\mathbf{k}'$ is $\{k_x, k_y\}$ and $\mathbf{s} = \{x, y\}$, so the one-electron energy is actually $\hbar^2k'^2 + \varepsilon_k$. In combining the squared wave functions to form $n(z)$, one integrates over all values of k_x and k_y such that $\varepsilon_k + \hbar^2k'^2/2m \leq \hbar^2k_F^2/2m$, where $\hbar k_F$ is the Fermi momentum.

For a surface problem, the external potential is usually that of a semi-infinite slab of positive charge, called *jellium*. In calculations for the solid (crystalline) metal, the stepfunction charge distribution must be replaced by a lattice of ions, each carrying a pseudopotential for the electrons which has a contribution from inner-shell exchange repulsion as well as from the coulombic attraction. The effect of this replacement on the electronic energy is calculated by perturbation theory. For liquid metals, one expects that there will be an ion density profile different from a step function.[60] It is possible to perform the calculation of electronic wave functions using, in the Schrödinger equation, the potential of a non-step distribution of ion cores, each carrying inner-shell repulsion as well as coulombic attraction. If more gradually varying (broader) ionic profiles are assumed, the electronic profile also broadens out, remaining broader than the ionic one. The surface dipole potential, which depends on the difference between the ionic and electronic profiles, decreases in size.

Calculations using oscillatory ion profiles have also been performed[61] and give better surface profiles and work functions for liquid metals. Such oscillatory profiles are expected on the following argument: The electronic profile provides the potential in which the ions are supposed to move (Born-Oppenheimer or adiabatic approximation); the potential due to an electron density which decreases to zero outside the metals acts like a wall, keeping the ions inside the metal, and this should produce oscillations. Other arguments have been given, and oscillatory ion profiles also result from simulations.[62] It is always true that the electrons "spill over" toward the outside of the metal, i.e., that the electronic density profile is broader than the positive-charge (ionic) background profile. This, of course, is the origin of the surface dipole potential, which is an important part of the work function.

It should be noted that the ionic profile referred to is an average profile, and electronic and ionic profiles should actually be calculated together, in a self-consistent manner, as follows: With the ions fixed at some configuration, the electron density is determined, perhaps by solution of the Schrödinger equation by the Lang–Kohn theory.[58] This yields, in addition to an electron density profile, the energy of the ion–electron system. The energy as a function of ionic configuration, say, $E(\mathbf{R})$, becomes the potential energy which determines the arrangement of the ions. In principle, each ionic configuration is weighted by a Boltzmann factor, $\exp[-E(\mathbf{R})/kT]$. Then integration or averaging, with the Boltzmann factors as weights, gives the average distribution, and averaging the electron densities with the same weights gives the average electronic profile. The preceding models for liquid metals short-circuit this by assuming that the average electronic profile, calculated in the above way, can be replaced by the electronic profile calculated for the average ionic configuration.

Since the positive background is in the form of ions, it is not really a continuous density profile depending only on the coordinate perpendicular to the interface. The electron density, though always "smoother" than the positive charge density, is likewise more complicated than a one-dimensional profile. However, it is rare to see calculations for the electron density which take into account inhomogeneities in the parallel directions, except when adsorbed atoms are being considered.

The atomic distribution of positive charge must be taken into account when considering work functions for different crystal faces of solid metals. The chemical potential μ_e^M is a bulk property and its value is independent of the crystal face through which the leaving electrons are supposed to pass, so that the difference in work functions observed for different crystal faces is due to the changed surface potential. As a first approximation, one may take it that the smoothing of the electron density in directions parallel to the surface means that the electron density profile is almost the same for all crystal faces, so that the different surface potentials come about because of the way in which the electron density cloud is penetrated by the ions. Any penetration of the negative charge density by positive charges will lower χ^M, so that crystal faces for which the positive ions are distributed most uniformly, i.e., the faces of higher ion density, will have higher χ^M and higher work functions. Similarly, defects and other inhomogeneities in the surface will reduce χ^M and Φ. Because of the connection between work function and potential of zero charge when the metal is in an electrochemical cell, a similar explanation can be given for the differences in E_{pzc} for different crystal faces of solid metals.

3. Screening

We now consider calculations of screening of external charges by the electrons at the surface of a metal. It may be noted at the outset that one cannot simply transfer results for the metal–vacuum surface into the electrochemical interface, although some general ideas will carry over. Thus, one can expect that the screening length is greater at the metal surface than in bulk because of the decreased electron density. This was used by Kuklin[63] in a theory for the contribution of the metal to the capacitance of the electrochemical interface, which has been criticized[8] on a number of points, including mathematical errors; it leads to serious discrepancies with experimental results.

Newns[40] calculated the response to an external electric field of an idealized metal, plane waves at an infinite barrier, with density profile [see Eq. (12)]

$$\rho(x) = \rho_0(1 + 3\cos x'/x'^2 - 3\sin x'/x'^3), \qquad x' = 2k_F x$$

The response was calculated in the Hartree approximation, and only a linear response was considered. Suppose an external potential $U(\mathbf{r})$ leads to a change in the electron density $\delta\rho$. This then produces an additional electrical potential which has an effect on the electron density. In the Hartree model, one requires a function $\delta\rho$ which is self-consistent, so that the total electrical potential, due to U and to $\delta\rho$, produces the change $\delta\rho$. In the linear approximation, one calculates only the change in ρ which is proportional to U. This approach is like that of Eqs. (19)-(23) for bulk metal.

Calculations were performed for an external field due to a point charge as well as for a uniform external field. In discussing the results, a screening length was defined as $d = V(0)/V'(0)$, where the potential outside the metal is $V(0) + V'(0)x$ and the potential inside the metal, which is screened by the electrons, approaches zero for $x \to -\infty$. Since $-V'(0)$ is the external electric field, proportional to the surface charge density of a capacitor, and $V(0)$ is the potential drop across the metal from 0 to $-\infty$, the screening length is the inverse of a capacitance. For a perfect conductor, $d = 0$, since there is no penetration of electric field into the metal and no potential drop in the metal. In the Thomas–Fermi theory, $d = \lambda^{-1}$, where the Thomas–Fermi screening length λ is defined by the Thomas–Fermi equation [Eq. (19)], $\nabla^2 V = \lambda^2 V$. Schiff[64] also performed an approximate calculation of the shielding of the field of an external charge distribution by the conduction electrons at a metal surface and estimated the penetration into the metal of the electric field; the calculation was done by a variational method, and the Thomas–Fermi approximations were used to describe the electrons.

As expected, because the Thomas–Fermi theory is valid for high densities, $d = \lambda^{-1}$ is valid only for $r_S \ll 1$, where $4\pi r_S^3/3$ is the inverse of the electron density. For higher r_S (lower k_F, lower bulk density) Newns' calculations[40] show that d is well approximated (to a few percent out to $r_S = 6$) by $\lambda^{-1} + \pi/4k_F$. Experimental measurement of d for the metal-vacuum interface may be made via the change of the work function due to a layer of adsorbed ions. If these ions have charge q and are at a distance a from the metal surface, the change in work function is $\Delta\Phi = 4\pi n_S q(a + d)$, where n_S is the surface concentration of ions. This assumes that there is negligible penetration of ions into the tail of the electron density, and neglects nonlinear effects. Penetration would reduce d. Experiments for Cs^+ adsorbed on tungsten give results varying widely with crystal face and with the experimenter,[40] and their interpretation is complicated by the band structure of tungsten, but it is concluded that d is about 1.35–1.85 Å. This is greatly in excess of any reasonable Thomas–Fermi screening length but is compatible with Newns' results[40] if r_S is 3–4. The large increase in screening length relative to Thomas–Fermi theory is related to the low electron density near the surface of the metal.

Beck and Celli[65] also calculated the linear response of a metal to an external charge distribution by a method equivalent to that of Newns,[40] with the electrons confined to a half-space by an infinite barrier. It was concluded that, for most electrical properties, the Friedel oscillations, associated with the discontinuity in the linear-response function at $2k_F$, are not important. Rudnick[66] also calculated the screening of a static charge distribution in the self-consistent Hartree approximation, but with a more transparent formalism than that of Newns. He emphasized the Friedel oscillations in the screening charge density and how they were affected by the electron gas boundary. For a very dense electron gas, it was argued that, since the infinite-barrier model does not give a good description of the electron gas near the surface, it should not be a valid model for the response of the electrons in a metal to an external charge distribution.

Ying *et al.*[67] applied the density-functional formalism to the linear-response problem. One starts with an inhomogeneous electron gas at $T = 0$ in an external potential $V_0^{\text{ext}}(\mathbf{r})$ and with electron density $n_0(\mathbf{r})$. Assuming a perturbing potential $V_1^{\text{ext}}(\mathbf{r}) = \int \rho_1^{\text{ext}}(\mathbf{r}')|\mathbf{r} - \mathbf{r}'|^{-1}\,d\mathbf{r}^1$, one seeks the change $n_1(\mathbf{r})$ in the electron density. Here, the Lang–Kohn formalism[58] is used for both the perturbed and the unperturbed system. The equation for the total electrostatic potential $V(\mathbf{r})$,

$$-V(\mathbf{r}) + \delta G/\delta n = \mu$$

is linearized with respect to V_1^{ext} and n_1, which yields equations for n_1 and V_1. These equations were solved by Ying *et al.*[67] using the Thomas–Fermi functional for G,

$$G[n] = \tfrac{3}{10}(3\pi^2)^{2/3} \int n^{5/3}(\mathbf{r})\,d\mathbf{r}$$

In this case, the unperturbed solution outside the metal is $V_0 = 400/(b + x)^2$, with $b = 5.0813$.

Heinrichs[68] used the Thomas-Fermi model, with several phenomenological approximations added, to study the response of the metal surfaces to external charge distributions. This includes the image-potential interaction with a static point charge. An adsorbed ion and its induced screening charge constitute a dipole, which is involved in the change in the work function of the metal as well as in other phenomena, which are reviewed by Heinrichs.[68] The Thomas-Fermi response is discussed in detail, as well as the solutions to the relevant equations. Results are compared to those from self-consistent calculations (see below).

The Thomas-Fermi-Dirac model, which adds an exchange energy density to the kinetic energy density of the Thomas-Fermi model,[49] has also been used to describe the electrons of a metal surface.[65] One can solve the Thomas-Fermi-Dirac equation numerically to obtain the electron density profile for any surface charge density, thus obtaining the capacitance of the metal surface as well as the surface potential. Here, one is not limited to the linear-response regime. Making the metal one plate of a capacitor representing the interface, one finds that the inverse capacitance of the metal increases as q^{M} becomes negative.

Lang and Kohn[57] used their self-consistent density-functional theory to calculate the screening charges induced in a metal surface by a uniform electric field or an external point charge. In the latter case, one is calculating the image potential. The positive charges are described by the uniform-background model (step function or jellium). The theory was worked out for the linear-response regime; i.e., one seeks the induced charge profile which is proportional to the external charge distribution. It may be noted that one can simply add an external field to the potential of the Schrödinger equation and solve the self-consistent equations; it is not necessary to use a perturbation-type expansion and limit oneself to the linear response.

Noting the symmetry of the background in the y- and z-directions, the external charge distribution was written[57] as a sum of terms of the form

$$\rho^{e}(\mathbf{r}) = \delta(x - x_1)\, e^{i\mathbf{p}\cdot\mathbf{s}}$$

where $\mathbf{s} = (y, z)$ and $\mathbf{p} = (p_y, p_z)$; $\mathbf{p}$ specifies whether one has a sheet of charge, a point charge, or some other distribution in the plane $x = x_1$. Since x_1 is supposed to be well outside values of x

for which the electron density is appreciable, the external potential is

$$\psi^e = (2\pi/p)\exp(-p|x - x_1|)$$

on solving the Poisson equation. The total perturbing potential is

$$\psi(\mathbf{r}) = \psi^e(\mathbf{r}) + \int \delta n(\mathbf{r}')|\mathbf{r} - \mathbf{r}'|^{-1}\,d\mathbf{r}'$$

where δn is the induced charge density.

Of great interest is the center of gravity of the induced charge,

$$z_0 = \int_{-\infty}^{\infty} z\delta n(z)\,dz \bigg/ \int_{-\infty}^{\infty} \delta n(z)\,dz$$

which is the effective position of the added or subtracted charge in the metal. (For larger charge densities, the center of gravity will change with charge density. Like the change in the shape of δn, this will contribute to the dipolar capacitance of the metal surface.) The charge induced in the metal by an external point charge has its centroid at z_0 if the point charge is far enough away. If the end of the metal-ion (step-function) distribution is at $z = 0$, and the external charge (which would be in the electrolyte phase if the metal were in an interface) is at $z = Z$, the "ideal capacitor" has a width, not of Z, but of $Z - z_0$ [see Eqs. (6), (7) *et seq.*]. It turns out that $z_0 = 1.9a_0$ for $r_S = 1.5a_0$ (bulk density of $0.0236a_0^{-3}$) and $z_0 = 1.2a_0$ for $r_S = 6a_0$ (density of $0.000368a_0^{-1}$). As the point charge approaches the metal, z_0 decreases slightly until for some position d from the jellium edge, $z_0(d)$ becomes equal to d and remains equal to d inside the metal.

Some exact results for the charge distribution induced at a jellium surface by a static applied field were derived by Budd and Vannimenus.[70] These authors calculated the linear response of a jellium slab of finite thickness to a charged plane parallel to the surface, either for isolated jellium (for which total charge is conserved) or grounded jellium (for which the charge of the jellium is equal and opposite to that of the perturbing charged plane). They showed that the half-moment of the charge distribution induced in grounded jellium is related to the unperturbed electric field E_0 by

$$E_0(z) = 4\pi\rho_b e[z\theta(-z) - \int_{-\infty}^{0} dz'\,z'n(z'|z)] \tag{31}$$

Here, $n(z'|z)$ is the charge density induced at z' by a charged plane, of unit surface charge density, located at z, and ρ_b is the bulk jellium density. The Hellmann–Feynman theorem was used. The full moment of $n(z'|z)$ obeys

$$\int_{-\infty}^{\infty} dz'\, z'n(z|z') = \int_{0}^{z} dz'\, z'n(z'|\infty) + z\int_{z}^{\infty} dz'\, n(z'|\infty)$$

where $n(z'|\infty)$ is the response of the jellium (for $z < 0$) to a charged plane at infinity.

In general (beyond linear response), the half-moment of $n(z|\infty)$ is given by

$$\tfrac{1}{2}\sigma^2 = \frac{1}{2}\left[\int_{-\infty}^{\infty} dz\, n(z|\infty)\right]^2 = \rho_b \int_{-\infty}^{0} dz\, zn(z|\infty)$$

where σ is the surface charge density. The half-moment is thus inversely proportional to the jellium density and is quadratic in the applied surface charge density. This is shown by considering the force on the plane infinitely far away, $(\sigma e)(-2\pi\sigma e)$, where $-2\pi\sigma e$ is the field at the plane. This must be equal and opposite to the force on the jellium background, $-\rho_b e \int_{-L}^{0} dz'\, E'(z'|\infty)$, where $E'(z'|\infty)$ is the electric field at z' induced by the charged plane. Since $E' \to 0$ as $z' \to -\infty$, one can let L approach ∞. Integrating by parts, one finds

$$-2\pi\sigma^2 e^2 = -\rho_b e\{[z'E']_{-\infty}^{0} - \int_{-\infty}^{0} z'\, dE'/dz'\, dz'\}$$

$$= \rho_b e \int_{-\infty}^{0} z'\{-4\pi e n(z'|\infty)\}$$

Therefore, $\tfrac{1}{2}\sigma^2 = \rho_b \int_{-\infty}^{0} dz\, zn(z|\infty)$.

Perhaps of greater interest to us are results derived by the same authors[71] that relate surface and bulk electronic properties of jellium. Considering two jellium slabs, one extending from $-L$ to $-D$ and the other from D to L, they calculated the force per unit area exerted by one on the other. According to the Hellmann–Feynman theorem, this is just the sum of the electric fields acting on each slice of positive background, multiplied by the positive chage density of the slice. Then,

$$F/A = \int_D^L \rho_+(z)(-\partial V/\partial z)\, dz = -\rho_b[V(L; D) - V(D; D)]$$

where $V(x; D)$ is the electrostatic potential at x when the slabs are separated by $2D$. When the separation distance is zero, the force per unit area can be shown to be simply the pressure inside the jellium. If L is large, the latter can be calculated in uniform bulk jellium, and $V(L; 0) - V(0; 0)$ is just the difference in electrostatic potential between a point in bulk jellium ($z = 0$) and a point at the jellium surface ($z = L$).

Now suppose the energy density of the electrons in bulk jellium is given by $nf(n)$, so that the density functional $f(n)$ is the energy per electron of a uniform electron gas of density n in the positive background. (The integral of $nf(n)$, plus inhomogeneity terms, is the quantity one minimizes to obtain the density profile.) The pressure in bulk jellium is then $n(\partial f/\partial n)$, so that

$$[n(\partial f/\partial n)]_{n=\rho_b} = \phi(-\infty) - \phi(0) \tag{32}$$

We have returned to a former notation, denoting the electrostatic potential by ϕ, so that $\phi(0)$ is the potential at the jellium surface and $\phi(-\infty)$ is the potential in bulk jellium. The left member of Eq. (32) is, of course, a bulk property and the right member a surface property. Another proof of Eq. (32) was given by Vannimenus and Budd.[72] They also derived a related important theorem,

$$\mathrm{d}E_s/d\rho_b = \int_{-\infty}^{0} [\phi(z) - \phi(-\infty)]\, dz$$

Here E_s is the surface energy and the left side refers to the change of E_s with change in bulk electron density.

Equation (32) is extremely important to the calculation of work functions. The work required to remove an electron from the interior of the metal to vacuum outside is

$$\Phi = -\mu_e + e\chi^{\mathrm{M}}$$

where μ_e, the chemical potential of the electron, is a bulk property and the surface potential, χ^{M}, is a surface property. In the density-functional model,

$$\mu_e = d[nf(n)]/dn$$

and

$$\chi^{M} = \phi(-\infty) - \phi(\infty)$$

Equation (32) shows that there is some cancellation between the bulk and surface contributions; combining Eq. (32) with the above equations, we have

$$\Phi = -f(n) + e[\phi(0) - \phi(+\infty)]$$

This helps explain[58,73] why simple variational calculations can give good work functions: only the electronic tail, spilling over from the positive background, is involved in the surface contribution.

Equation (32) also holds[72] for the charged interface, in the form

$$[n(\partial f/\partial n)]_{n=\rho_b} = \phi(-\infty) - \phi(0) - q^2/2\rho_b \tag{33}$$

where $q = q^{M} = -q^{S}$ is the surface charge density. The last term is the contribution to the force on the jellium of a sheet of charge density q^{S} far outside the metal electrons. The dipolar contribution of the metal to the capacity is [see Eq. (8)] given by $1/C_{M} = d[\phi(\infty) - \phi(-\infty)]/dq^{M}$. Using Eq. (33) for $\phi(-\infty) - \phi(0)$ and the Poisson equation in the form

$$\begin{aligned}\phi(0) - \phi(\infty) &= 4\pi \int_{\infty}^{0} dz' \int_{\infty}^{z'} dz'' [n(z'') - \rho_+(z'')] \\ &= -4\pi \int_{\infty}^{0} dz'\, z'[n(z') - \rho_+(z')]\end{aligned}$$

we have

$$1/4\pi C_{M} = -q^{M}/\rho_b - d\left[\int_0^{\infty} zn(z)\, dz\right] \Big/ dq^{M}$$

Thus, only the tail of the electron density (outside the jellium) contributes. The last term above may further be related[11] to the position of the image plane z_{im} so, at the point of zero charge,

$C_M = -\frac{1}{4}\pi z_{im}$. It should be noted that the Budd–Vannimenus theorems are obeyed by the electron density $n(z)$ which is the exact solution to the jellium problem, and not necessarily by any approximate (e.g., variational) solution. There are extensions of the Budd–Vannimenus theorems to a bimetallic surface.[74]

V. METAL ELECTRONS IN THE INTERFACE

1. Metal Nonideality

As we have mentioned, traditional theories of the electrochemical interface almost always assumed that the metal is ideal, with a sharp boundary, although it was sometimes recognized that water molecules and other species in the interface feel the electron density tail. This should affect the distance of approach of the solvent molecules and the behavior of the solvent dipoles in the surface. For example, the penetration of the solvent layer by the tail of the electron density means that the field at the position of the first layer of solvent dipoles is not zero when the surface charge density is zero. The fact that the metal is not ideal also means that the image plane for charges of the electrolyte phase is about 1.0 Å in front of the metal surface (position of the outermost ions, or of the jellium representing them). Although this has little effect[75] on the plate spacing of an ordinary (macroscopic) capacitor, it is important in changing the spacing of the ideal capacitor representing the inner layer of the metal-electrolyte interface.[55] This decreased spacing raises the capacitance of the interface above what it would be for an idealized metal. Note that, if there is negligible electron exchange between the ions of the electrolyte and the metal, the image interaction is the only interaction (electron exchange implies a chemical interaction).

Any theory which includes an infinite barrier to metal electrons at the interface will make the reciprocal capacitance too large because it makes the effective interplanar spacing of the inner-layer capacitor too large.[76] This is why Rice's early[5] results (see below) were so incorrect. The fact that the electron tail penetrates a region of higher dielectric constant further reduces the calculated inverse capacity.[30,77]

The interface is, from a general point of view, an inhomogeneous dielectric medium. The effects of a dielectric permittivity, which need not be local and which varies in space, on the distribution of charged particles (ions of the electrolyte), were analyzed and discussed briefly by Vorotyntsev.[78] Simple models for the system include, in addition to the image-force interaction, a potential representing interaction of ions with the metal electrons.

The image interaction for a nonideal metal was discussed in more detail by Kornyshev and Vorotyntsev.[24] Let $q\phi_0(x', x)$ be the electrostatic potential produced at a point $\{x', 0, 0\}$ in a homogeneous medium by a charge q on an ion at $\{x, 0, 0\}$, and $q\phi(x', x)$ the potential produced at $\{x', 0, 0\}$ by this charge if a metal is present for $x < 0$. The energy of the ion in either case is calculated by integrating $[c\phi(x', x)]\, dc$ from $c = 0$ to $c = q$ (Guntelberg charging process), so that the difference in energy, due to the metal, is

$$\Delta E = \tfrac{1}{2}q^2 \lim_{x' \to x}[\phi(x', x) - \phi_0(x', x)]$$

and the image potential is obtained by dividing ΔE by q.

The classical result for the image potential is $-q/4x$, independent of the metal, but various theories of the metal which assume an infinite potential barrier for the metal electrons give potentials which are reduced in size near the metal boundary, so that the interaction energy is actually finite[24] at $x = 0$. An interpolation formula which reproduces this behavior is

$$W = -q[4(x + \alpha/\kappa_{TF})]^{-1}$$

where α is a fitting coefficient and $1/\kappa_{TF}$ is the Thomas–Fermi screening length. Classically, the image charge for a charge at $x = x_0$ is at $x = -x_0$, while the above formula puts the image charge at $-(x_0 + \alpha/\kappa_{TF})$. This is interpreted[24] to mean that the distribution of induced charge in the metal is that of a disk of thickness $\kappa_{TF}{}^{-1}$ and radius $x + \kappa_{TF}^{-1}$.

The conduction-electron tail is expected to move into the metal as the surface charge q^M becomes more positive, and away from the metal as q^M becomes more negative. It has been suggested[18] that, for large positive q^M, the tail could contract enough to deshield the inner-shell d orbitals at the surface, which would have a strong effect on the interaction of solvent molecules with the metal. Furthermore, the variation of the position of the effective image plane, which determines the size of the inner-layer capacitor, will change with surface charge.[55] This variation is not symmetrical: it is easier to extend the tail outward than to push electrons into the metal. Thus, the position of the image plane changes more rapidly with q^M as q^M becomes more negative.

The movement and distortion of the conduction-electron density profile give the contribution of the metal to the interfacial capacitance. According to the discussion after Eq. 7, the free-charge or ionic contribution to the potential drop across the compact layer ΔV is $4\pi q^M(z^0 - z^S)$, and the contribution of the metal, assuming no dipolar species, is $4\pi q^M(z^M - z^0)$, where z^M is the centroid of the metal's charge distribution. Here, z^0 is fixed but z^M changes with q^M. Then, $C(\text{ion})^{-1}$ in Eq. (8) is simply $4\pi(z^0 - z^S)$, while the metal contribution is

$$C^M(\text{dip})^{-1} = 4\pi(z^0 - z^S) + 4\pi q^M \frac{d}{dq^M} \frac{\int dz\, z \sum_i^M n_i(z)}{\int dz \sum_i^M n_i(z)} \tag{34}$$

This contribution involves the positive-ion and electron density profiles of the metal, and the former is often assumed not to change with charging of the interface. In 1983 and 1984, several workers[30-32,79] showed how certain features of the interfacial capacity curves should depend on the metal.

In addition to the effect of the nonideality of the metal on the electrolyte phase, one must consider the influence of the electrolyte phase on the metal. This requires a model for the interaction between conduction electrons and electrolyte species. Indeed, this interaction is what determines the position of electrolyte species relative to the metal in the interface. Some of the work described below is concerned with investigating models for the electrolyte-electron interaction. Although we shall not discuss it, the penetration of water molecules between the *atoms* of the metal surface may be related[3] to the different values of the free-charge or ionic contribution to the inner-layer capacitance found for different crystal faces of solid metals. Rough calculations have been done to show the effect of different atomic packings on the effective thickness d of the inner-layer capacitor which enters $K_{\text{ion}} = \varepsilon/d$.

2. Density-Functional Theories

Attempts to include the electrons of the metal in a description of the electrochemical interface go back to Rice[5] in 1928. Rice attempted to explain the measured electrocapillary curves for the mercury-aqueous solution interface by combining the Gouy-Chapman theory for the ions on the solution side of the interface with a quantum mechanical theory (the Thomas-Fermi model) for the conduction electrons of the metal. The Thomas-Fermi energy functional is $\frac{3}{10}(3\pi^2)^{2/3}n^{5/3} + en\phi$. The energy of the most energetic electron in a metal must be the same everywhere, which means that the electron density n must vary with the potential so that $\frac{1}{2}(3\pi^2)^{2/3}n^{2/3}e^2a_0 - e\phi$ is constant, or

$$d\phi = (h^2/8me)(3/\pi)^{2/3}\, d(n^{2/3}) \equiv A\, d(n^{2/3}) \qquad (35)$$

(A more general relation between potential and electronic pressure for a density-functional treatment of a metal-metal interface has been given.[74]) For two metals, 1 and 2, in contact, equilibrium with respect to electron transfer requires that the electrochemical potential of the electron be the same in each. Ignoring the contribution of chemical or short-range forces, this means that $-e\phi + (h^2/8m) \times (3n/\pi)^{2/3}$ should be the same for both metals. In the Sommerfeld model for a metal[38] (uniformly distributed electrons confined to the interior of the metal by a step-function potential), there is no surface potential, so the difference of *outer* potentials, which is the contact potential, is given by

$$\phi_2 - \phi_1 = (h^2/8me)[(3n_2/\pi)^{2/3} - (3n_1/\pi)^{2/3}]$$

This gives contact potentials of the right magnitude but the wrong sign, because of the neglect of chemical forces and surface potentials.

Now, consider a planar metal surface, n depending on the single coordinate x. In the interior of the metal, $x \to -\infty$, $\phi \to 0$, and $n \to n^\infty$, which is equal to the positive-charge (ion) density. At the surface, $\phi = \phi_0$ and $n = n_0$. Poisson's equation is

$$d^2\phi/dx^2 = 4\pi e(n - n^\infty)/\varepsilon$$

where ε is the dielectric constant. Using Eq. (35) to eliminate ϕ, we find

$$A\, d^2(n^{2/3})/dx^2 = 4\pi e(n - n^{\infty})/\varepsilon$$

Rice[5] used this Thomas–Fermi theory for the metal in the interface. This equation is multiplied by $2\, d(n)^{2/3}/dx$, and then integrated from $x = 0$ to $x = -\infty$. Since $dn(x)/dx \to 0$ when $x \to -\infty$ one gets

$$-A\{[d(n^{2/3})/dx]_0\}^2 = (16\pi e/5\varepsilon)[(n^{\infty})^{5/3} - n_0^{5/3}] \\ - (8\pi e n^{\infty}/\varepsilon)[(n^{\infty})^{2/3} - n_0^{2/3}] \quad (36)$$

Now, $n_0^{2/3} - (n^{\infty})^{2/3} = \phi_0/A$ and $[d(n^{2/3})/dx]_0 = A^{-1}(d\phi/dx)_0$. The electric field at the metal surface, $-(d\phi/dx)_0$, is equal to $4\pi q^{\mathrm{M}}/\varepsilon$, with q^{M} the surface charge density of the metal.

For small potentials ϕ_0, $\delta n = n_0 - n^{\infty}$ is small. Then, $n_0^{5/3} - (n^{\infty})^{5/3}$ is approximated as

$$(n^{\infty})^{5/3}[(1 + \delta n/n^{\infty})^{5/3} - 1] \simeq (n^{\infty})^{5/3}[\tfrac{5}{3}(\delta n/n^{\infty}) + \tfrac{5}{9}(\delta n/n^{\infty})^2]$$

and

$$\phi_0/A = n_0^{2/3} - (n^{\infty})^{2/3} = (n^{\infty})^{2/3}[(1 + \delta n/n^{\infty})^{2/3} - 1] \\ \simeq (n^{\infty})^{2/3}[\tfrac{2}{3}(\delta n/n^{\infty}) - \tfrac{1}{9}(\delta n/n^{\infty})^2]$$

For small $\delta n/n^{\infty}$, the solution is

$$\delta n/n^{\infty} = \tfrac{3}{2}A^{-1}\phi_0(n^{\infty})^{-2/3} + \tfrac{3}{8}A^{-2}\phi_0^2(n^{\infty})^{-4/3}$$

so that $n_0^{5/3} - (n^{\infty})^{5/3}$ becomes $\tfrac{5}{2}A^{-1}\phi_0 n^{\infty} + \tfrac{15}{8}A^{-2}\phi_0^2(n^{\infty})^{1/3}$ and Eq. (36) reads

$$-A^{-1}(4\pi q^{\mathrm{M}}/\varepsilon)^2 = -(8\pi e/\varepsilon)[A^{-1}\phi_0 n^{\infty} + \tfrac{3}{4}A^{-2}\phi_0^2(n^{\infty})^{1/3}] \\ + (8\pi e n^{\infty}/\varepsilon)(\phi_0/A) \quad (37)$$

Rice showed[5] that these approximations are good as long as ϕ_0 is less than about 1 V. According to Eq. (37), $4\pi q^{\mathrm{M}}/\varepsilon = (6\pi e/\varepsilon A)^{1/2}\phi_0(n^{\infty})^{1/6}$. This range of potentials thus corresponds to a constant capacity for the metal surface. Since the capacity per unit area of a plane capacitor with interplanar spacing d is ε/d, the effective capacitor width for the metal electrons in this model is $(8\pi\varepsilon A/3e)^{1/2}(n^{\infty})^{-1/6}$. This is just the Thomas–Fermi screening length.

The metal surface capacitance is simply

$$C_M = \varepsilon_M \kappa_{TF}/4\pi$$

where $\kappa_{TF} = (6\pi n_0 e^2/\varepsilon\varepsilon_F)^{1/2}$ is the inverse Thomas–Fermi screening length, ε_F is the Fermi energy, and ε is the background dielectric constant.[8] Using $\varepsilon = 8.9$, which would now be considered as unrealistically large for a metal, Rice was able to explain experimental electrocapillary data by coupling the diffuse-layer theory for the solution with this model for the metal. He rejected the compact or Stern layer as part of the explanation: modern theory, of course, generally rejects the contribution of the metal and explains electrocapillary curves and capacitances by combining diffuse-layer theory with a theory for the ions and solvent of the compact layer.

The background dielectric constant ε for the metal arises from the polarizability of the ion cores and the contribution of interband transitions.[11] For mercury and other simple metals, with a large band gap and relatively unpolarizable ion cores, one expects a background dielectric constant close to unity. With $\varepsilon = 1$ and $n^{\infty} = 8.17 \times 10^{22}\ \text{cm}^{-3}$ (mercury), the capacitance per unit area is

$$q/\phi_0 = em^{1/2}h^{-1}(3n^{\infty}/\pi)^{1/6} = 2.10 \times 10^6\ \text{cm}^{-1} = 2.33\ \mu\text{F/cm}^2.$$

Since this capacitance is supposed to be in series with that of the solution and since capacitances of mercury–solution interfaces are much larger than $2\ \mu\text{F/cm}^2$, this number is too low. The Thomas–Fermi theory as well as the neglect of interactions between metal electrons and the electrolyte are at fault. To reduce the metal's contribution to the inverse capacitance, a model must include[6] penetration of the electron tail of the metal into the solvent region, where the dielectric constant is higher, as the models discussed below do.

Another model for the metal in the interface, also employing the Thomas–Fermi approximation for the electrons, was presented by Kuklin.[63] It posited, in addition, a sharp boundary between metal and solution, and made other assumptions which have been criticized.[6] In spite of its errors, it was one of the first attempts since Rice[5] at a model for the interface which treated the metal as well as the electrolyte. Related work by Salem[80] also contains errors, as discussed by Vorotyntsev and Kornyshev[6].

A better density-functional model for the electrons includes, in addition to the Thomas–Fermi kinetic energy density, exchange and correlation energies, as well as an inhomogeneity term. Such a model was used by Smith[54,81] for the bare metal surface [see above, Eq. (26)]. Capacitances for the Thomas–Fermi–Dirac model, which includes only kinetic and exchange energy densities, have recently been calculated[69] for the metal surface, representing the electrolyte as a charged plane 5 Å from the edge of the metal's positive charge density, represented as a step function. Although no interaction between metal electrons and solvent species was included, the results suggest that the metal can make a significant contribution to the capacitance of the interface.

With the addition of a pseudopotential interaction between electrons and metal ions, the density-functional approach has been used[82] to calculate the effect of the solvent of the electrolyte phase on the potential difference across the surface of a liquid metal. The solvent is modeled as a repulsive barrier or as a region of dielectric constant greater than unity or both. Assuming no specific adsorption, the metal is supposed to be in contact with a monolayer of water, modeled as a region of 3-Å thickness (diameter of a water molecule) in which the dielectric constant is 6 (high-frequency value, appropriate for nonorientable dipoles). Beyond this monolayer, the dielectric constant is assumed to take on the bulk liquid value of 78, although the calculations showed that the dielectric constant outside of the monolayer had only a small effect on the electronic profile.

The ionic profile of the metal was modeled as a step function, since it was anticipated that it would be much narrower than the electronic profile, and the distance d_1 from this step to the beginning of the water monolayer, which reflects the interaction of metal ions and solvent molecules, was taken as the crystallographic radius of the metal ions, R_c. Inside the metal, and out to d_1, the relative dielectric constant was taken as unity. (It may be noted that these calculations, and subsequent ones[83] which couple this model for the metal with a model for the interface, take the position of the outer layer of metal ion cores to be on the jellium edge, which is at variance with the usual interpretation in terms of Wigner–Seitz cells[58] wherein the metal centers are half a cell diameter in.) In some calculations, d_1 was reduced to a fraction of the crystallographic radius, to model interpenetration of the metal ions and solvent molecules.

The electron density profile was assumed to have the exponential form

$$n(z) = \begin{cases} \rho_b(1 - A\,e^{az}), & z < z_0 \\ \rho_b B\,e^{-bz}, & z \geq 0 \end{cases} \tag{38}$$

The requirements that n and its slope be continuous at z_0 allow the expression of A and B in terms of a, b, and z_0; a third condition, that the total charge density of the interface be equal to zero, gives

$$z_0 = 1/a - 1/b + q^S/\rho_b$$

The remaining parameters, a and b, are determined variationally for any value of q^S by minimizing the electronic surface energy, which is the difference between the total electronic energy of the system and the electronic energy of a homogeneous electron gas of density ρ_b extending from $z = -\infty$ to $z = 0$, the position of the positive-charge background profile. The total electronic energy itself is infinite for this system.

In the electronic surface energy, the kinetic, exchange, correlation, and inhomogeneity energies of the electrons were taken as functionals of the local electronic density (and, in the case of the inhomogeneity energy, of its gradient). Using the Thomas–Fermi form for the density of electronic kinetic energy, $\kappa_k n^{5/3}$ with $\kappa_k = 3(3\pi^2)^{2/3}/10$, the contribution of electronic kinetic energy to the surface energy is

$$\kappa_k \int_{-\infty}^{\infty} [n_-(z)^{5/3} - \rho_b \theta(-z)]\, dz$$

where $\theta(z)$ is a step function. The electronic exchange energy density is $-\kappa_a n^{4/3}$ with $\kappa_a = \frac{3}{4}(3/\pi)^{1/3}$, the electronic correlation energy density is $-\frac{1}{2}[0.115n - 0.0311n \ln(3n/4\pi)]^{1/3}$, and the inhomogeneity energy density is $(72n)^{-1}(dn/dz)^2$. The electrostatic energy includes the interaction of the electrons with each other and with the charges of the ion cores (in a step-function distribution). A pseudopotential ion–electron interaction was included by writing the total ion–electron interaction potential in the Heine–Animalu form[84]

$$W(r) = \begin{cases} A_0, & r < R_m \\ -Z/r, & r > R_m \end{cases}$$

with given values for the parameters A_0 and R_m, and averaging $W(r)$ over the ion distribution. Finally, a direct repulsive interaction between the conduction electrons and the cores of the water molecules was included in the electronic energy, as

$$\lambda\delta(z - \tfrac{1}{2}d_1 - \tfrac{1}{2}d_2)$$

Values for the parameter λ were chosen without much justification.

With this model, electron density profiles were calculated[82] for the bare metal surface and for the metal in the presence of the solvent. From these profiles, surface potentials were computed by the integrated form of the Poisson equation. For the seven metals investigated by this model, the effect of the dielectric film is to lower χ^M by some tenths of a volt, as shown in Table 1. This is a direct effect of the penetration of the tail of the electron density into the region of higher dielectric constant; the change in the electron density profile is very small. The effect of the repulsive barrier is also to lower χ^M by some tenths of a volt, but this lowering is due to the deformation of the electronic profile by the repulsion: the profile is pushed inward toward the metal, which reduces the difference between electronic and ionic densities and lowers the size of the surface potential. The lowering, which was found to be

Table 1
Change in Metal Surface Potential Due to Dielectric Film and Barrier

Metal	$\delta\chi_S^M$ (film) (V)	$\delta\chi_S^M$ (barrier) (V)
Hg	−0.24	−0.21
Cd	−0.45	−0.53
In	−0.63	−0.68
Zn	−0.67	−0.68
Pb	−0.71	−0.71
Ga	−1.01	−0.83
Al	−1.47	−0.92

proportional to the value chosen for λ, was greatest for the metals of highest electron density and smallest R_c. Although high electron density and low R_c normally go together, it was established by calculations[82] that the value of R_c, which determines the distance of the water monolayer from the metal, was the more important factor.

It is interesting that the lowering found for Hg and Cd was close to the values suggested for these metals from experimental results.[85] Trasatti[3,26] derived a value of −0.31 V for $\delta\chi_S^{Hg}$ near the point of zero charge from adsorption measurements of aliphatic alcohols on the mercury–water and mercury–air interfaces. (He found no reliable information on adsorption on other metals and noted that his assumption that the adsorbing polymers of different chain lengths had a constant orientation on the surface might be unreliable for more hydrophilic metals; as we have emphasized, it is necessary to make *some* assumption to separate metal and solvent contributions.) This is quite close to the value of −0.24 V obtained from the dielectric film model.

However, consideration, for a series of metals, of the difference of $X = \delta\chi_S^M - g^S(\text{dip})$ from its value for Hg, which can be derived from measurements of potentials of zero charge and work functions, suggests[6] that these calculations[82] overestimate the importance of the metal electrons to the properties of the interface. If X is plotted versus metal electron density, the data follow no trend, as they do if X is plotted versus some quantity measuring the affinity of the metal for oxygen; this shows the importance of adsorption and $g^S(\text{dip})$ in X. Trasatti[3] plotted the calculated $\delta\chi_S^M$ versus $X(\text{M}) - X(\text{Hg})$. The plot suggests that these calculations overestimate the size of $\delta\chi_S^M$ for higher-density metals; for calculated $\delta\chi_S^M$ to be consistent with experiment, $g^S(\text{dip})$ would have to be *less* for other metals than for mercury.

The successes of hydrophilicity scales in correlating much data mean that one should not underestimate the importance of $g^S(\text{dip})$. A plot of ΔH_f^0 (heat of formation of metal oxide, a measure of hydrophilicity) versus $X(\text{M}) - X(\text{Hg})$ shows two lines. Preferential orientation increases with oxygen affinity. Correlations between ΔH_f^0 and $X(\text{M}) - X(\text{Hg})$ exist also for solvents other than water, with the rate of increase of $X(\text{M}) - X(\text{Hg})$ with ΔH_f^0 being stronger in the sequence acetonitrile $<$ H_2O $<$ DMSO, with increasing orientational polarizability and dielectric constant; the polarizability of the metal should not be very different for the three solvents. Note that, in the dielectric film model, the solvent dielectric constant is taken as the electronic dielectric constant, which should not vary much from solvent to solvent.

It may be noted that the dielectric film model seems to give the correct $\delta\chi_S^M$ for Hg, and too large a size for other metals: this means the electrons stick out too far for the higher-density metals. As will be seen below, an important term, the repulsion due to inner-shell electrons of solvent molecules, is left out in this model. For transition metals, $X = \delta\chi_S^M - g^S(\text{dip})$ is constant while ΔH_f^0 changes, suggesting that $\delta\chi_S^M$ is of minor importance and $g^S(\text{dip})_0$ is metal independent. (Note that, for transition metals, ionic radii are fairly constant, as are electron densities.) This is ascribed[3] to the different nature of the bond to water, which is here strong enough to hold the water in its orientation (ΔH_f^0 is also much higher). For *sp* metals, the surface bond is of an ionic type, whose strength is a strong function of surface electronegativity. For *d* metals, chemisorption prevents much reorientation of molecules by an electric field.

The model for the metal electrons was used[30] in a consideration of the capacitance, i.e., the effect of charging of the system on the electronic distribution. The solution charge density q^S was supposed to be located in the plane which bounds the dielectric layer of water: in concentrated solutions, Gouy-Chapman theory shows that the width of the ionic charge distribution becomes small. The dielectric constant was supposed to be unity from $z = 0$ to $z = d_1$ (related to the crystallographic radius of the metal ions), equal to $\varepsilon_1 = 6$ for z between d_1 and d_2 ($d_2 - d_1$ = width of a water monolayer), and equal to $\varepsilon_2 = 78$ for z beyond z_2. For each choice of q^S, the parameters a and b of Eq. (38) were found by minimizing the surface energy with, of course, the proper value of $q^M = -q^S$ imposed. (For $a = b$, the explicit form of the profile function for $q^M \neq 0$ was given by Partenskii and Smorodinskii.[86]) The electron-water interaction is a δ-function pseudopotential, with a value of the multiplying parameter chosen to be "reasonable."

The potential drop across the double layer, ΔV, was then calculated in terms of a, b, q^S, ρ_b, d_1, and d_2. All the terms in ΔV which depended on a and b were assigned to χ_S^M or $g_M(\text{dip})$ [see

Eq. (7)]; the remainder, which depends explicitly on q^S, was taken as the ionic or free-charge term:

$$g_S^M(\text{ion}) = -4\pi q^S[(-q^S/2\rho_b) + d_1 + (d_2 - d_1)/\varepsilon_1] \qquad (39)$$

The quantity $g^M(\text{dip})$ vanishes when a and b become infinite, corresponding to step-function profiles for electrons as well as for ions. It is assumed that d_1 and d_2 are unchanged on charging (no electrostriction). The reciprocal of the capacitance involves only two terms, since solvent is not considered explicitly:

$$C^{-1} = (\partial\Delta V/\partial q^M) = C(\text{ion})^{-1} + C(\text{dip})^{-1}$$

where

$$C(\text{ion})^{-1} = (4\pi/\varepsilon_1)(d_2 - d_1) + 4\pi d_1 + 4\pi q^M/\rho_b$$

($q^M = -q^S$). It may be noted that $C(\text{dip})$ turns out to be negative, because an increase in q^M (charge density on the metal) and in ΔV is associated with a decrease in the surface dipole of the metal surface χ_S^M, which makes a negative contribution to ΔV.

With $\varepsilon_1 = \varepsilon_2 = 1$, capacitances were very small compared to experimental ones, so that $C^S(\text{dip})$ would have to be much smaller than generally accepted values to get agreement with experiment. With a dielectric present but no δ-function barrier, more reasonable results were obtained. The addition of the barrier changed $\delta\chi^M$, but had little effect on capacitances. Of course, these calculations are of interest only in comparing the contributions of two metals, or in investigating the importance of modifications of the model, since there is no solvent.

The main result of this work was stated to be that larger capacitances are found for Ga than for Hg because Ga has a smaller crystallographic radius, allowing closer approach of the Helmholtz planes in the electrolyte to the metal. Thus, capacitances for mercury were about half those for gallium, the difference being due to the larger ρ_b and smaller d_1 for the latter metal. This means that, even neglecting the contribution of the dipoles of the electrolyte phase, the capacitance of the gallium–solvent interface would be greater than that of the mercury–solvent interface, as is observed experimentally. It is interesting that, combined with a solvent-dipole contribution of 25 $\mu F/cm^2$, the capacitances calculated in this model for the metal give capacitances for the interface about the same as those measured for the Hg–solvent and Ga–solvent interfaces at the point of zero charge.

However, one cannot reproduce the capacitance as a function of charge by combining this model with a constant solution capacitance. The explanation for the different capacitances found for different metals in the interface is usually considered to lie in the different interactions of solvent dipoles with different metals. Frumkin *et al.*[34,87] have ascribed the observed larger capacitances for the gallium interface to the larger interaction of the dipoles of the solvent with the gallium surface. The present work does not invalidate this but mainly points out that a direct effect of the metal ought also to be considered.

The effect of decreasing d_1 from $d_1 = R_c$ was investigated: the capacitance of the metal surface is much increased when d_1 decreases, as this gives rise to a substantial repulsion due to the electron–molecule interaction. Apparently, there is then a large barrier to deformation of the electronic profile by an electrostatic field, so that ΔV is less changed by a change in q^S, and the capacitance is increased. The variation of metal capacitance with surface charge was also shown to be substantial. However, the large difference between mercury and gallium capacitances remained over the entire range of q^M investigated. The fact that, experimentally, capacitances of the metal-solution interface are found to become independent of the nature of the metal for $q^M < -15\ \mu C/cm^2$ was noted. The explanation for this, it was stated, must come from a model including both solvent and metal.

It is also interesting that if some of the simple models for the bare metal surface are used to calculate the metal's contribution to the capacitance, a fit to experimental results would require unreasonable values for the solution contribution. Thus, the simple Thomas–Fermi result[88] of $C(\text{dip}) = 4\pi/\lambda_{TF}$ (λ_{TF} = Thomas–Fermi screening length) is greater than $C(\text{experiment})^{-1}$, and the same is true for the improved Thomas–Fermi results of Newns[40] and the model of free electrons at an infinitely repulsive wall [see Eq. (12)]. These models are thus considered to be less realistic than the model of this work.[30]

It was also noted that for any model including the extension of the conduction-electron tail into the interface, the electric field at the dipoles of the first solvent layer will not be zero for zero surface charge, because some of the metal's electron density will be found on the other side of these dipoles from the metal. To have zero electric field acting on the solvent dipoles, one requires a slightly negative value of q^M. It is, in fact, found experimentally[89] that the maximum entropy of formation of the compact layer occurs for q^M equal to about $-5\ \mu C/cm^2$. Assuming that the mercury makes a negligible contribution to the charge dependence of the entropy of formation, the entropy must be associated with disorder of the water in the inner layer. A simple picture of water molecules orienting in a field would associate maximum disorder with zero field. These calculations emphasize that, at $q^M = 0$, one does *not* have zero field at the dipoles, precisely because of the penetration of the solvent layer by the electron tail. In fact, the value of q^M required to give zero field at the position of the dipoles turns out to be negative, several $\mu C/cm^2$, in agreement with the position of maximum entropy. Of course, it is possible for models for the interface which ignore the metal to obtain a similar result: the four-state model of Fawcett *et al.*[15] found the maximum entropy of the solvent monolayer (not the configurational entropy) to occur at about the same charge density.

Trasatti[3] suggested that the contribution of the metal to the capacitance was also exaggerated by these calculations. He wrote the reciprocal of the capacitance of the interface as the sum of three terms:

$$1/C = 1/C^M - 1/C_{dip} + 1/K_{ion}$$

with the free-charge contribution, K_{ion}, usually considered to be independent of q^M, equal to ε/d_2 (ε = dielectric constant of the inner layer). Interfaces involving Hg and Ga(l) exhibit the same capacitance for large negative q^M, although the electronic polarizabilities for the metal should be quite different. At large negative q^M, if the contribution of the metal is ignored, the experimental inner-layer capacitance is just K_{ion}, since the orientation of dipoles can make little contribution. For Hg, the experimental inner-layer capacitance for $q^M < -10\ \mu C/cm^2$ is $16.8\ \mu F/cm^2$. If d_2 is taken as 0.31 nm (diameter of a water molecule), ε comes out to be 5.9, a reasonable value for the electronic or high-frequency dielectric constant. Thus, it seems that the metal contribution is unimportant.

(Defenders of the importance of the metal may respond by noting that for q^M large and negative, one is trying to push the electronic tail far out from the metal, which becomes increasingly difficult because of the repulsive barrier due to the inner shells of the solvent molecules. Eventually, making q^M more negative will produce little change in the electronic tail and $\partial\chi^M/\partial q^M$, which is $1/C^M$, will be small for all metals.)

It must be noted that there are a number of more or less arbitrary assumptions made in this work[30,31] which need justification, as well as parameters whose values should be calculated rather than assumed. For instance, the importance of the distance d_1, taken as equal to R_c, has been mentioned. In principle, the value of this distance is a consequence of the forces between components of the metal and molecules of solvent, and would be calculated in a consistent model of the complete interface. This was pointed out by Yeager,[18] who noted that the electron density tail of the metal determines the distance of closest approach of solvent in the interface, as well as the behavior of the solvent dipoles on the surface. Since changing q^M will move the electron density tail in and out, d_1 should depend on the state of charge of the interface. In fact, it turns out[31] that if d_1 varies linearly with surface charge according to

$$d_1 = R_c - (R_c/80)q^M$$

the dependence of the metal's capacitance on q^M is removed.

3. Metal–Solvent Distance

That d_1 should depend on the state of charge of the electrode may be argued[36] as follows: Charging of the electrode shifts the center of gravity of the conduction-electron tail but not the center of gravity of the image charge corresponding to a charged adsorbed species. Thus, adsorption of an electropositive species causes the electron tail to move outward, and adsorption of an electronegative species causes it to move inward, toward the metal. The "effective medium theory" of adsorption states that the position of an adsorbate atom relative to the metal surface is determined by the optimum electronic charge density for the bonding interaction, so that an adsorbed species will move with the electron tail.

The length d_1 is just the distance between the metal surface (positive background density) and the layer of adsorbed water molecules. The change of this distance with surface charge of the electrode was adduced by Kornyshev and Vorotyntsev[90] as the cause of the "capacitance hump." Explanation of this maximum in the plot of capacitance as a function of surface charge, occurring for positive values of q^M for many metal–liquid electrolyte interfaces, has long been a preoccupation of theoretical electrochemists[91]; naturally, past explanations have been in terms of the displacements and orientations of solvent species.

Recently, another explanation for the hump, involving the distortion of the metal's conduction-electron tail, has been proposed[92] by Schmickler and Henderson. (Although previous workers[30,31] had found a metal contribution to the capacitance which increased monotonically with q^M, it was suggested that this was because the assumption for the electron density profile was too simple.) Schmickler and Henderson[92] noted that preferential adsorption of water with the oxygen end toward the metal would shift the maximum in the capacitance toward negative charges, since a negative value of q^M would be required to give zero net orientation, for which the capacitance is largest. Differentiating $V(-\infty) - V(\infty)$ [see Eq. (6)] with respect to q^M, one obtains $1/4\pi C$ as a sum of three terms, of which the metal contribution is

$$1/4\pi C_M = -d\left[\int_{-\infty}^{\infty} z\rho_M(z)\,dz\right]\Big/dq^M$$

where ρ_M is the charge density of the metal. One writes ρ_M as $\rho_M^0 + \delta\rho_M$, with ρ_M^0, the density profile for $q^M = 0$, being independent of q^M by definition. Then, invoking the Budd–Vannimenus theorem [Eq. (31) *et seq.*], one has

$$1/4\pi C_M = -(q^M/e\rho_b) - d\left[\int_0^{\infty} z\delta\rho_M(z)\,dz\right]\Big/dq^M$$

The first term tends to make the capacitance greater for increasing q^M. For the interface in vacuum, this term is outweighed by the other, so that the calculated capacitance decreases with q^M, reflecting the fact that it is easier to spill more electrons out into the vacuum than to push them back into the metal againtst the repulsive forces. In the metal-solution interface, however, it is surmised[92] that the short-range repulsion of the solvent molecules makes it harder for the electrons to expand outward, restoring the original situation of C_M increasing with q^M. This will lead to a capacitance maximum for positive q^M.

The dependence of d_1 on q^M is central in a model, proposed by Price and Halley,[93] for the metal surface in the double layer which is related to that discussed above. The positively charged ion background profile $\rho_+(z)$ is assumed uniform, with a value equal to the bulk density ρ_b, from $z = -\infty$ to $z = 0$, with the electronic density profile $n(z)$ more diffuse. In contrast to the previous model[30] which emphasizes penetration by the conduction electrons of the region of solvent, this model[93] supposes that the density profile $n(z)$ is zero for $z \geq d_1$, where $z > d_1$ defines the region of the electrolyte. Then the potential at d_1 is given by

$$\phi(d_1) = \phi_0 - 4\pi \int_{-\infty}^{d_1} dz \int_{-\infty}^{z} dz'[\rho_+(z) - n(z)]$$

where ϕ_0 is the potential inside the metal, for $z \to -\infty$. Integrating by parts and using the fact that the field at d_1 is given by

$$E(d_1) = 4\pi \int_{-\infty}^{d_1} dz[\rho_+(z) - n(z)]$$

and that the field vanishes at $z \to -\infty$, one has

$$\phi(d_1) = \phi_0 - d_1 E(d_1) + 4\pi \int_{-\infty}^{d_1} z[\rho_+(z) - n(z)]\, dz \qquad (40)$$

The last (dipole) term is supposed to be approximated by the truncated Taylor series

$$4\pi \int_{-\infty}^{d_1} z[\rho_+(z) - n(z)]dz = -\Delta\phi_0 + \lambda[E(d_1)]^2$$

and d_1 is written as $d_1(0) - \lambda' E(d_1)$, emphasizing the dependence of d_1 on the field $E(d_1)$.

Using $q^M = E(d_1)/4\pi$, one thus has

$$[C(\text{dip})]^{-1} = -[\partial\phi(d_1)/\partial q^M] = 4\pi d_1 - (\kappa/e\rho_b)32\pi^2 q^M$$

where κ is a dimensionless parameter, since $\phi(d_1) - \phi_0$ is the potential drop across the metal. This predicts[93] that $[C(\text{dip})]^{-1}$, which is identified as the compact-layer capacitance by these authors, should be linear in q^M for small q^M, the slope being inversely proportional to ρ_b and the intercept being related to the parameter d_1, which is the separation between the edge of the metal (as defined by the ion cores) and the solution. (Alternatively, one can say that one is studying the capacitance for q^M small enough to neglect higher terms in a power series for C^{-1} in q^M.) It was also argued that $d_1(0)$ should increase with ρ_b. Two formulas for κ were presented on the basis of "slide" and "swing" models: $\kappa = 1/8\pi$ and $\kappa = 1/3\pi$. The "slide" model assumed that the profile is displaced without changing its shape and the "swing" model assumed that the electron density shifts out in such a way that the most distant part of the charge moves the most. Experimental results for five metals, plotted as $1/C_c$ versus ρ_b^{-1}, show that neither is well obeyed. It was suggested[93] that band structure effects and surface states were playing a role. It was also noted that this model, which neglects the compact layer of the solvent phase, could not explain the temperature dependence of the capacitance and thus should be useful only above 50°C.

Guidelli,[16] reviewing work on the capacitance of the metal-electrolyte interface, writes the equation for calculating capacitances from models for the inner layer as:

$$1/C_1 = d(\Delta_S^M \phi)/dq^M = (K_{\text{ion}})^{-1} + dg^{H_2O}(\text{dip})/dq^M$$

with K_{ion} independent of q^M. This assumes that the surface dipole of the metal is independent of q^M, in contradiction to the results just discussed,[31,93] which give an important contribution of the electron tail to the capacitance. However, Guidelli points out that the model of Badiali *et al.*[31] predicts a capacitance which decreases as q^M becomes more positive, whereas the experimental curve for Hg/H_2O shows the opposite behavior, while the slide and swing models[93] do show a capacitance which increases with more positive q^M but incorrectly predict that C_1 is temperature independent. The moral is perhaps that a good model for one part of the interface can only show whether or not there is a large contribution of that part to properties of the interface but cannot hope to reproduce these properties unless the other part of the interface is also well described.

The interaction between the metal surface and a collection of water molecules needs to be considered more carefully.[94] The interaction of conduction electrons of the metal with molecules of the electrolyte includes an electrostatic interaction with a fixed charge distribution, which is attractive since the electrostatic potential is largest near the nuclei, an exchange interaction, which is attractive if it is approximated by an exchange-correlation potential[95] and is apparently more important[94] then the electrostatic potential, and a repulsive interaction representing the effect of orthogonality of the metal electron wave functions to the closed-shell cores, which may be represented by a pseudopotential. There is a also a polarization interaction, which has been represented by the use of a dielectric constant different from unity. The exchange-correlation potential is a local-density approximation, derived from the uniform electron gas result, i.e., $d(n\varepsilon_{xc})/dn$, where $\varepsilon_{xc}(n)$ is the exchange-correlation energy per electron in a uniform electron gas of density n. It has been shown[31] that the electrostatic potential (or the charge distribution of the water molecules which gives rise to it) does not need to be known very accurately in the variational calculations based on density-functional theory: the presence of a layer of dipoles or a double layer of charges at $\frac{1}{2}(d_1 + d_2)$ (representing the solvent molecules) hardly affects the capacitance calculated for the metal. The representation of the other potentials must be considered carefully, as the effects on electron density profiles can be large. Because of their repulsive interaction with the solvent molecules, the electrons of the metal extend out less from the ions when the metal is in the electrochemical interface than for the free metal surface.

4. Models for Metal and Electrolyte

Several workers have now presented models of the interface as a whole, treating metal and electrolyte phases simultaneously. In principle, one should treat the entire interface, including species of both metal and electrolyte phases. Treatments attempting to do this [see below, around Eq. (46)] have proved too difficult. Fortunately, it seems[96] that the details of the metal's electronic profile do not much affect the distribution of solvent species and vice versa. This allows separate solution of the problems for the two phases, obtaining the distribution of particles of each one in the field due to the other.

The metal-solvent interactions were put into the model of Price and Halley[93] in a later paper by Halley and co-workers,[97] which also remedied some of the deficiencies of the original model, such as the inability to calculate the slope of a plot of $(C_c)^{-1}$ versus q^M and the dependence of the compact-layer capacitance on crystal face. One can show in general [see Eq. (40)] that

$$\phi_0 - \phi(x_2) = 4\pi x_2 q^M - 4\pi q^M \int_{-\infty}^{x_2} x(\rho_+ - n)\, dx$$

where the position where the metal electron density vanishes and the solvent-molecule density begins is here denoted by x_2. Therefore, the inverse of the compact-layer capacitance C_c, which in this model arises from the metal alone, is given by

$$(C_c)^{-1} = d[\phi_0 - \phi(x_2)]/dq^M = 4\pi[x_2 - \bar{x} + q^M\, d(x_2 - \bar{x})/dq^M]$$

where $\bar{x}$ is the first moment of the charge induced on the metal:

$$\bar{x} = \int_{-\infty}^{\infty} x'\, \delta\rho(x')\, dx' \bigg/ \int_{-\infty}^{\infty} \delta\rho(x')\, dx' \qquad (41)$$

Here $\delta\rho$ is the difference between the charge density at surface charge q^M and the charge density at $q^M = 0$. Both x_2 and $\bar{x}$ depend on q^M.

In Ref. 97, the value of x_2 is found by minimizing the surface energy of the interface, written as a sum of five contributions:

(i) The surface energy of the electron layer in the presence of the positive background (a semi-infinite jellium slab), calculated self-consistently according to the Lang–Kohn theory with the local-density approximation.

(ii) The long-range interaction between solvent and metal, including the closed-shell pseudopotential repulsion, acting on the electrons, and an image-force attraction of the solvent dipoles. Without the image term, no solvent–metal binding was found.

(iii) The electrostatic interaction of the ions of the electrolyte, supposed to be located at $x_2 + d$, with the metal, given by $-\frac{1}{2}q^M\phi(x_2 + d)$.

(iv) The contribution of the metal-ion pseudopotentials, acting on the metal electrons.

(v) The interaction between the solvent molecules and the metal-ion cores. Electron–solvent repulsions come from averaging a pseudopotential repulsion over the distribution of solvent molecules. Since the latter is a step function, the resulting potential acting on the electrons is also taken as a step function.

Solvent and other contributions to the surface energy that are independent of x_2 need not be considered, since the surface energy is used only to find x_2 by minimization. It is assumed that no change in the electronic structure of the metal-ion cores or in their distribution occurs during charging.

The electron density profile is calculated by solving

$$(-\tfrac{1}{2}\nabla^2 + v)\phi_n = \varepsilon_n\phi_n \tag{42}$$

where v includes a Hartree (interelectronic repulsion) term, a local-density approximation to the exchange-correlation potential, an attraction due to the jellium background, and a short-range repulsion from the solvent molecules. Thus, the presence of the solvent is felt only through the repulsion term, which is written $V_0\theta(x - x_2)$, where θ is a step function and V_0 is a parameter. Since this term is considered large enough to exclude electron density from the region $x > x_2$, one can neglect screening of the Hartree term by the solvent, which was represented by the dielectric constant 6 in the model of Badiali *et al.*[82] The procedure is to solve the Lang-Kohn equations [see Eq. (30)] self-consistently for fixed values of x_2 and q^M. For each q^M, the proper value of x_2 is found by minimizing numerically the surface energy. Then the inverse capacitance is found as $4\pi[x_2 - \bar{x} + q^M\, d(x_2 - \bar{x})/dq^M]$. To compare with experiment, the Gouy-Chapman part of the capacitance is subtracted from measured capacitances, since it is assumed that the compact-layer capacitance arises wholly from the metal.

For the solid metals Cu, Ag, and Au, x_2 has been calculated[97] for surface charges between -0.003 a.u. and 0.003 a.u. (1 a.u. = 5700 $\mu C/cm^2$). The slopes of $1/C_c$ versus q^M at $q^M = 0$ are of the right size for Cu and several times too high for Ag and Au. The inverse capacitances have values of about 10 a.u. and decrease with q^M near $q^M = 0$; they continue to decrease for positive q^M but level off at about 20 a.u. for $q^M \simeq -0.002$ a.u. For positive q^M, near 0.002 a.u., the inverse capacitances show maxima and minima as a function of q^{M}, which are stated by the authors to be real. The experimentally observed dependence of capacitance on crystal face is reproduced.

Values for several parameters appearing in the ion pseudopotential and the solvent-dipole image forces, which appear in the surface energy, need to be chosen. In some cases, such as the ion–electron pseudopotential, their values could be obtained from other workers. For other parameters, such as the constant in the image-force solvent–metal interaction, values had to be chosen to fit experiment and were not always the expected ones: the constant for the solvent–metal interaction corresponded to a solvent dipole moment three times the actual one for water, suggesting that other metal-solvent interactions were being "covered" by the functional form used. It was also noted[97] that a theory including the polarization of the solvent molecules in the compact layer must be included along with a treatment of the metal to explain the temperature dependence of the compact-layer capacitance. However, it was emphasized that the contribution of the metal to $1/C_c$ is of the same size as the experimental inverse capacitance. Furthermore, the metal contribution gives, at least qualitatively, the correct charge dependence of capacitance as well as the correct trends with crystal face.

In later work by some of these authors,[98,99] the model for the repulsive effect of solvent on metal electrons, previously taken as a square barrier starting at $z = x_2$ and with the height a parameter to be chosen, is improved. This allows the use of the correct value for the solvent dipole in the image-force attraction of solvent to metal, rather than treating it as a fitting parameter. Two terms used in the previous calculation are dropped as unimportant: the pseudopotential correction to the jellium potential and the solvent-metal ion-core interaction. The interaction between a metal electron and a solvent molecule is described by an effective potential depending on the distance from the position of the center of mass of a solvent molecule and on the orientation of the solvent molecule. It includes the long-range dipole potential and the electronic pseudopotential calculated by replacing the actual charge distribution of the water molecule by that of the isoelectronic neon atom,[94] solving the Lang–Kohn equations to obtain wave functions for all electrons of a neon atom, and deriving pseudo wave functions [Eq.

(18) above] which resemble the true wave functions far from the atom. From the equations satisfied by the pseudo wave functions, a pseudopotential, depending on the local electron density, is derived, with some approximations. A single parameter entering the pseudopotential is chosen so that a calculation using the pseudopotential yields the binding energy of Ne on jellium in agreement with the result of an all-electron calculation. This potential is then averaged over the distribution of solvent molecule positions and orientations, obtained from statistical mechanical calculations[100] for a liquid of atoms, each of which is a hard sphere containing a point dipole, at a hard wall. The solvent distribution is no longer a step function. The position of the hard wall with which the hard dipolar spheres are supposed to interact, x_w, replaces x_2 as a parameter which must be found for each q^M by minimization of the surface energy.

For calculation of capacitances, another parameter is required—the point at which an ideal Gouy-Chapman system, which would have the same potential drop as the actual system, would begin. It is assumed that the difference between this point and x_w is independent of surface charge; for $q^M = 0$, it can be obtained from the hard-dipole distribution. For Cd, this model gave inverse capacitances which, for the range of q^M investigated, differed from the experimental capacitances by only half as much as did those obtained with the previous model.[96] The variation of C_c with q^M is correctly reproduced, although the errors in the values of C_c are still substantial. It was expected that the improvement in the calculation of long-range interactions and inclusion of saturation effects in the dipolar response of the solvent would help improve agreement with experiment.

Kornyshev *et al.*[76] proposed several models of the interface, including both orienting solvent dipoles and polarizable metal electrons, to calculate the position of the capacitance hump. Although it had been shown[32,79,101] that this was one of the features of the interfacial capacity curves that should depend on the nature of the metal, available calculations did not give the proper position of the hump. The solvent molecules in the surface layer were modeled as charged layers, associated with the protons and the oxygen atoms of molecules oriented either toward or away from the surface. These layers also carried Harrison-type pseudopotentials[84] associated with the oxygen cores and the electrons around the protons, which interact with the metal conduction electrons. Parameters related to the metal-electron profile and to the solvent-molecule layers were chosen by minimization of the surface energy. Capacitance–charge curves were calculated and discussed. However, due to the crude way in which the solvent was represented, the shapes of the capacitance–charge curves were quite incorrect.

Improving the model, Schmickler[32] considered a jellium for $x \le 0$, a layer of point dipoles at x_1, and a layer of counterion charges at x_2 (outer Helmholtz plane). The inner layer of solvent molecules was represented by a hexagonal lattice of point dipoles, each of magnitude p_0, and with lattice constant a. Schmickler used 6.12×10^{-30} Cm for p_0 and 3×10^{-10} m for a, giving a surface density of $10^{19}\,\text{m}^{-2}$. Each dipole had three possible orientations and so could be represented as a spin-1 system. The interaction of each dipole with its neighbors was written as a spin coupling, S_1S_2J, and the interaction of a dipole with non-nearest neighbors was represented as a mean-field interaction. In the mean-field theory, which was shown to give results identical to a Monte Carlo method for fields below 10^{10} V/cm, the field of the dipoles is just times $(p_0/4\pi\varepsilon_0)$ $\langle s\rangle \sum_r' r_i^{-3}$, where $\langle s\rangle$ is the average of the cosine of the angle between a dipole axis and the field direction, and r_i is the distance of dipole i from that being considered (the sum is over all *other* dipoles). The quantity $\langle s\rangle$ is given by the Langevin expression, $\coth x - x^{-1}$, where $x = (p_0/kT)E_{\text{ext}}$ and the field E_{ext} is that of the dipoles plus that due to free charges. The mean-field approximation seems to work quite well for fully orientable dipoles at small fields, although it is known to give poor results for two- and three-state dipole models.

The electrons of the metal were treated as a plasma of density $n(x)$, described by the one-parameter profile $n_+(1 - \frac{1}{2}e^{bx})$ for $x < 0$ and $\frac{1}{2}n_+ e^{-bx}$ for $x \ge 0$. The parameter b was calculated by minimizing the electronic surface energy for a particular value of surface charge or dipole field. The interaction energy of the electrons with the point dipoles of the solvent turned out to be quite small.[32] The surface energy was written as a density functional, including gradient terms, and with an external field, but not including a metal-ion pseudopotential. The external field was that of the surface charge density plus that generated by the solvent dipoles. The latter was found by calculating the potential difference across the point-dipole layer. The potential drop across the solvent layer, $-Np_0\langle s\rangle/\varepsilon_0$, was spread out over the region $0 \leq x \leq a$ to take into account the finite size of solvent molecules. Self-consistent solutions for the electron distribution and the dipole field were found. The parameters in the model included the value of the dipole on a solvent molecule, the lattice parameter for the solvent molecules, and the locations of the inner and outer Helmholtz planes (these positions were also assumed to be independent of surface charge density). From the electronic charge distribution and calculated $\langle s\rangle$, the potential drop across the interface, ΔV^{M}, was calculated[32] as a function of surface charge, and in particular for zero surface charge.

This model was first used[102] to investigate the dependence of the potential of zero charge on the metal. Only the bulk-metal electron density changes from metal to metal. For a metal electrode connected to a standard electrode to form an electrochemical cell, the potential of zero charge is given by Eq. (2) or, for a metal M,

$$E_{\mathrm{pzc}} = -\chi^{\mathrm{M}} + \chi^{\mathrm{M(S)}} - \chi^{\mathrm{S(M)}} + F^{-1}(\Phi^{\mathrm{M}} - \Phi^{\mathrm{ref}}) \qquad (43)$$

where Φ^{ref}, the work function of the reversible electrode (for emission to solution), is given by $F\Delta_S^{\mathrm{ref}}\phi - \mu_e^{\mathrm{ref}}$ and is independent of M. The potential drop across the interface, $g_S^{\mathrm{M}}(\mathrm{dip}) = \chi^{\mathrm{M(S)}} - \chi^{\mathrm{S(M)}}$, was calculated.[102] The quantities χ^{M} and Φ^{M} were calculated for the jellium using the density-functional model in the absence of external fields, as had been done by Smith,[54] who first suggested it. The value of Φ^{ref} was taken from experiments: the reference electrode was taken to be H^+/H_2, so Φ^{ref} was taken as -4.5 eV, and χ^{S} was taken as -0.13 V.

Calculated work functions, potentials of zero charge, and absolute values of surface potentials all increase in size with bulk electron density (the only parameter changing from metal to metal). Calculated values for the potential of zero charge were too low by about a volt, but when experimental values for Φ^{M} were substituted for calculated values, agreement of E_{pzc} with experiment was good, with errors of only several tenths of a volt. Furthermore, the changes from metal to metal were generally reproduced, as shown in Table 2. In general, the potential of zero charge becomes more negative as the electron density of the metal increases. Note that there is no

Table 2
Potentials of Zero Charge for $M|sol, H^+|H_2|M'$

Metal	E^{exp}(V)	E^{calc}(V)
Hg	−0.19	−0.26
Ga	−0.69	−0.79
Cd	−0.75	−0.67
In	−0.65	−0.80
Pb	−0.56	−0.70
Sn	−0.38	−0.52
Tl	−0.71	−1.06
Ag	−0.60	−0.56
Au	0.19	0.28

repulsive interaction between metal electrons and the solvent in this model, so that the effect of the solvent dipoles is to pull the electrons out of the metal and increase the metal's surface potential relative to χ^M, its value for the free metal surface.

As noted by Badiali *et al.*,[30,82] there is a positive electric field at the position of the dipoles when the interface is at the point of zero charge because of the spillover of the electron density past the dipoles. This produces a small negative dipole field, corresponding to preferential orientation of the oxygen end of the water molecules toward the metal. Because of this dipole field, $\phi_{met} - \phi_{sol}$ at the potential of zero charge is smaller than the surface potential χ^M for the bare metal surface, even though the attractive effect of the solvent dipoles on the metal electrons makes the *metal* contribution to $\phi_{met} - \phi_{sol}$ larger in the interface than in the free surface. The solvent dipole terms are important: neglecting them would make $E_{pzo} = \Phi/e - 4.63$ V [see Eq. (43)], which, for all the metals examined but one, is several tenths of a volt higher than the experimental value.

Capacitance as a function of charge was calculated.[79] The capacitance curves showed a single hump, near $q^M = 0$, and leveled off for q^M about 10 $\mu C/cm^2$ on either side of the potential of zero charge, due to the dielectric saturation of the dipole system. The limiting values of the capacitance increased with increasing electron density of the metal. The nonideality of the metal was shown to increase the capacitance in two ways: because of the penetration of the electric field into the metal (increase in the effective value of the width of the capacitor d), and because of the distortion of the electronic cloud of the metal by the dipole field (change of parameters in the electronic profile).

A more realistic model of the interface, which combines the jellium model for the metal with a modern picture of the electrolyte solution, was used by Schmickler and Henderson[103] to calculate the capacitance of the interface. The ions of the electrolyte are charged hard spheres and the solvent molecules hard spheres with point dipoles at their centers. An approximate solution for the distribution functions of hard-sphere dipoles and ions bounded by a charged hard plane, valid for small surface charge densities, had previously been derived.[102] The electrons of the metal are described by the usual density functional, with three terms representing their interaction with the solvent. Two of these are the interaction with the charge density and polarization (dipole density) of the electrolyte; as had been noted by others,[31,96] the details of the interaction of the metal electrons with the solution species are unimportant, only the total field, or charge density, being important. The third term is a repulsive potential energy barrier, represented as a step function at $z = 0$ of height V_b. The value of V_b, which should be related to the effective barrier for tunneling of metal electrons through a water layer, was taken as 3 eV for Hg.

In addition to V_b, the parameters for which values need to be chosen are the sizes of the hard spheres; other parameters are well defined. Since the distance of closest approach of spheres to the metal is just the jellium edge, the distance of closest approach d_1, which was a crucial parameter in other work,[31,83] does not enter. (It is noted that, usually, the jellium edge is not considered to represent the position of the last layer of metal ions. Rather the ions are at the center of the Wigner–Seitz cell, or half a lattice spacing behind the jellium edge,[11,58] which ensures electroneutrality of the Wigner–Seitz cell. Vorotyntsev and Kornyshev[6] have also argued that the edge of the jellium is one ionic radius beyond the last plane of ions, giving the example of a lattice consisting of two monolayers: if this is to be represented by a jellium slab, its thickness should be about two ionic diameters, whereas the choice of Badiali *et al.*[31,83] would make it half this.)

Calculations of the capacitance of the mercury/aqueous electrolyte interface near the point of zero charge were performed[103] with all hard-sphere diameters taken as 3 Å. The results, for various electrolyte concentrations, agreed well with measured capacitances as shown in Table 3. They are a great improvement over what one gets[104] when the metal is represented as ideal, i.e., a perfectly conducting hard wall. The temperature dependence of the compact-layer capacitance was also reproduced by these calculations.

For metals other than mercury, it was assumed that V_b was 3.0 eV plus the difference in work functions $\Phi^M - \Phi^{Hg}$. The changes in V_b had little effect on calculated capacitances compared to the effect of the changed bulk electron density n_b. The capacitance increases with n_b, eventually diverging and becoming negative for $n_b > 0.025$ a.u. This unrealistic behavior may be related to the jellium model or to the form of the profile function used.[96] The increase of capacitance with n_b is not borne out by experiment, the deviations between theory and experiment being worst for Sn, Pb, Bi, and Sb. This is ascribed[103] to the fact that not all the valence electrons are free for these metals, so that the correct value to use for n_b should be less than what is calculated by multiplying the number of atoms per unit volume by the number of valence electrons per atom.

For solvents other than water, the model predicts, even in the absence of calculations, that interfacial capacitances in any solvent should increase in the order Hg < In < Ga because of the increasing electron densities.[103] This is, in fact, the case for DMSO and acetonitrile as well as for water. From the model used for the

Table 3
Capacitances for Mercury–Solution Interface

Concentration (M)	Capacitance (μF/cm^2)		
	Calc.	Expt.	Ideal
10^{-3}	6.0	6	5.1
10^{-2}	13.6	13	9.8
10^{-2}	22.6	21	13.8
1	28.7	26	15.9

distributions of hard spheres, one can also expect that, for a particular metal, the capacitances should be smaller for DMSO or acetonitrile than for water, because the smaller size of the molecules of the latter allows a closer metal–solvent approach. A value for the barrier height V_b of 1.8 eV was used for Hg/DMSO and a value of 2.25 eV for Hg/acetonitrile, and V_b for other metals was obtained from that for Hg using the calculated jellium work functions. Calculated inner-layer capacitances for these other solvents agreed reasonably well with experiment. In these calculations, the interaction of the polarization with the electric field of the electrons spilling past the first layer of solvent molecules was neglected, as it played only a minor role in the water calculations.

The model was also extended[11] to single-crystal surfaces of silver. Although the calculated inner-layer capacitances varied in the right way from one face to another, the values were much too low. The problem was suspected to be due to the importance of the *d* electrons. What is still needed in this model is a better treatment of the solvent phase, valid at higher charge density, and a better way of deriving the repulsive potential of the solvent on the electrons, perhaps by a direct pseudopotential calculation, as done by Price and Halley.[98,99]

Another model which combined a model for the solvent with a jellium-type model for the metal electrons was given by Badiali *et al.*[83] The metal electrons were supposed to be in the potential of a jellium background, plus a repulsive pseudopotential averaged over the jellium profile. The solvent was modeled as a collection of equal-sized hard spheres, charged and dipolar. In this model, the distance of closest approach of ions and molecules to the metal surface at $z = 0$ is fixed in terms of the molecular and ionic radii. The effect of the metal on the solution is thus that of an infinitely smooth, infinitely high barrier, as well as charged surface. The solution species are also under the influence of the electronic tail of the metal, represented by an exponential profile.

Previously derived results for the charge-density profile and polarization profile in this model solution, valid only for small fields, were used. Although these did not consider the penetration of the electrons into the solution, the change in the field is small. A Harrison-type pseudopotential[84] was used to represent the effect of core electrons of the solution species on the metal electrons.

This was averaged over the total distribution of ionic and dipolar spheres in the solution phase. Parameters in the calculations were chosen to simulate the Hg/DMSO and Ga/DMSO interfaces, since the mean-spherical approximation, used for the charge and dipole distributions in the solution, is not suited to describe hydrogen-bonded solvents. Some parameters still had to be chosen arbitrarily. It was found that the calculated capacitance depended crucially on d, the metal–solution distance. However, the capacitance was always greater for Ga than for Hg, partly because of the different electron densities on the two metals and partly because d depends on the crystallographic radius. The importance of d is specific to these models, because the solution is supposed (perhaps incorrectly; see above) to begin at some distance away from the jellium edge.

It was later shown[94] that the use of a density-functional theory led to errors in the calculated surface potentials, although Smith[81] had shown that variational calculations within the density-functional formalism could give profiles close to those obtained from more exact calculations. Furthermore, the variational method, which assumed a two-parameter functional form for $n_-(z)$ and allowed variation only of these two parameters, could also be a cause of error. For the bare surface of Ga, the value of χ^M obtained from the Lang–Kohn self consistent theory is 9.48 V, more than a volt below the value obtained from the two-parameter variational calculation.[94] Furthermore, if one uses a non-step density profile for the ions, results are also changed significantly.[61,105] In fact, it seems necessary to use oscillatory ion-density profiles[62] to obtain work functions for the bare metal surface in agreement with experiment. Calculations for the capacitance of the metal surface in the interface which use such profiles have not been performed, partly because one would have to make an assumption about how the profiles change with charging.

5. General Dielectric Formalism

We have already mentioned a more general approach[24] to calculating the properties of the interface, which is elegant but apparently difficult to realize. The Poisson equation for the statistically averaged electric field **E**,

$$\nabla \cdot \mathbf{E} = 4\pi(\rho + \rho_{\text{ext}}) \tag{44}$$

where ρ_{ext} is the charge density of the external field sources and $\rho(\mathbf{r})$ the averaged density of induced charge due to polarization of the medium, describes the entire interface and adjoining phases. One requires, in addition, a phenomenological relationship between ρ and the electrical potential ϕ, where $\mathbf{E} = -\nabla \cdot \phi$. Alternatively, one can deal with the polarization field $\mathbf{P}(\mathbf{r})$, where $\rho = -\nabla \cdot \mathbf{P}$, and assume a dependence of $\mathbf{P}$ on $\mathbf{E}$. The relation between ρ and ϕ is usually taken as linear and local, so that ρ at any point is proportional to the electrostatic potential at that point. Kornyshev and Vorotyntsev[24] noted that, although nonlinear behavior of the solvent phase of the interface had generally been considered, no work had been published in the electrochemical literature on nonlinear response of the metal because the metal was considered as merely a region of constant potential, as in the original Gouy-Chapman theory.[106] Although models of the interface generally assume a local dielectric constant, it may not always be possible[48] to replace the exact nonlocal theory by a local one, even if the dielectric constant is allowed to be position dependent. In a nonlocal theory,

$$\rho(\mathbf{r}) = \int \mathbf{dr}'\alpha(\mathbf{r}, \mathbf{r}')\phi(\mathbf{r}')$$

The mathematical methods for treating problems in nonlocal electrostatics were discussed by Kornyshev *et al.*[107]

For a metal in contact with a dielectric, the fields arise from species of both phases. Kornyshev and Vorotyntsev[24] speculated on the effect of the dielectric on the electron density, pointing out that there should be a long-range interaction with the polarization of the dielectric, as well as a quantum interaction associated with overlap between the metal electron's tail and the occupied electronic orbitals of the dielectric. The former tends to spread out the electronic tail as compared to the tail in vacuum, since it screens the interactions between the electronic charge and the ion cores of the metal, while the latter pushes the electronic tail back toward the metal. Other authors have incorporated these two effects in their models, as discussed above. In general, it is impossible to predict *a priori* which effect will dominate, but Kornyshev and Vorotyntsev suggested[24] that the spreading will win out if the electronic affinity for solvent molecules is greater than the work function. In the linear approximation, both effects may be described in terms of the (nonlocal) dielectric function of the system, $\varepsilon(z, z', \mathbf{r} - \mathbf{r}')$, where the dielectric function[108] relates the electric displacement to the electric field according to

$$D_i(\mathbf{r}) = \sum_j \int d\mathbf{r}' \, \varepsilon_{ij}(\mathbf{r}, \mathbf{r}') E_j(\mathbf{r}')$$

Thus, the electric displacement at $\mathbf{r}$ is linearly related to the electric field at positions other than $\mathbf{r}$; a local dielectric constant makes $\mathbf{D}(\mathbf{r})$ proportional to $\mathbf{E}(\mathbf{r})$. If one has planar symmetry, so that only the x-components of $\mathbf{D}$ and $\mathbf{E}$ are nonzero and these depend on x only,

$$D(x) = \int \varepsilon_x(x, x') E(x') \, dx'$$

after integrating over z' and y' where ε_x is an integrated dielectric function.

For small deviations from electroneutrality, the charge density at x is proportional to $-\phi(x)/kT$, where ϕ is the difference of the electrostatic potential from its (constant) value when there is no charge density (the density of a species of charge z is proportional to $1 - z\phi kT$ on linearizing the Boltzmann exponential). Then the Poisson equation [Eq. (44)] becomes the linearized Poisson–Boltzmann equation:

$$(d/dx)\left\{\int \varepsilon_x(x, x')[-d\phi(x')/dx'] \, dx'\right\} = \kappa_0^2 \phi(x) \qquad (45)$$

With the proper definitions of ε_x and κ_0, this equation is applicable to the metal as well as to the electrolyte in the electrochemical interface.[24] Kornyshev *et al.*[109] used this approach to calculate the capacitance of the metal-electrolyte interface. In applying Eq. (45) to the electrolyte phase, ε_x is the dielectric function of the solvent, x' extends from 0 to ∞, and x extends from L, the distance of closest approach of an ion to the metal (whose surface is at $x = 0$), to ∞, so that κ_0^2 is replaced by $\kappa_0^2 \theta(x - L)$. Here κ_0 is the inverse Debye length for an electrolyte with dielectric constant of unity, since the dielectric constant is being taken into account on the left side of Eq. (45). For the metal phase ($x < 0$) one takes ε_x as the dielectric function of the metal and limits the integration over x'

to $-\infty < x \leq 0$. One thus has a sharp-boundary model, with no metal electron density entering the region of solvent.

One has to solve for $\phi(x)$ with $\phi(-\infty) = V$, $\phi(\infty) = 0$, and ϕ and D continuous at $z = 0$. Since the effect of the metal electrons is incorporated into the dielectric function, there are no free charges to consider in the metal, so that D is constant inside the metal, and the equation becomes

$$\int_{-\infty}^{0} \varepsilon_{\mathrm{M}}(x, x')(d\phi/dx')\, dx' = \text{constant } (x \leq 0)$$

with ε_{M} the dielectric function of the metal. The equations are solved[109] for small κ_0, after insertion of a model which gives $\varepsilon_{\mathrm{M}}(x, x')$ in terms of the bulk-metal dielectric function, $\varepsilon_{\mathrm{bM}}(x, x')$, with Thomas–Fermi screening for the metal. The electron density profile for $q^{\mathrm{M}} = 0$ is approximated as $n_0\,(1 + Ae^{ax/3})^{-3}$ with a a variable parameter. In the electrolyte, the basic equation to be solved is

$$\frac{d}{dx}\int dx'\varepsilon(x, x')\frac{d\phi(x')}{dx'} = \varepsilon\kappa^2\phi(x)\theta(x - L)$$

where ε is the macroscopic dielectric constant, κ^{-1} is the Debye length, and L is the distance of closest approach of electrolyte ions to the metal. This is solved with a simplified form for $\varepsilon(x, x')$. The meaning of the various parameters entering the model and their effect on the results of calculations are discussed.[109]

The capacitance of the system can be written as that of four capacitances in series, i.e.,

$$C^{-1} = C_s^{-1} + C_S^{-1} + C_{\mathrm{M}}^{-1} + C_{\mathrm{GC}}^{-1} \tag{46}$$

where C_s arises from the dielectric function of the solution phase, C_{M} is the contribution of the metal phase, C_{GC} is the Gouy–Chapman capacitance, and C_S is the capacitance of the solvent layer between $z = 0$ and the distance of closest approach for ions (Stern layer). The inverse capacitance of the metal then turns out to be $4\pi L_{\mathrm{M}}$, where

$$L_{\mathrm{M}} = (2/\pi)\int_0^{\infty} dk\,[k^2\varepsilon_{\mathrm{M}}(k)]$$

with $\varepsilon_M(k)$ the Fourier transform of the bulk-metal dielectric function $\varepsilon_{bM}(x - x')$. This characteristic length may be calculated by any of the models for screening in bulk metal, and turns out to be about half an angstrom. In fact, Kornyshev *et al.*[109] found that C_M was close to the Thomas–Fermi value for mercury when any of a number of approximations to the dielectric function of mercury was used. The resulting inverse capacitance is much larger than experimental capacitances, so that no positive values for the other terms in Eq. (46) can give agreement with experiment.

It was concluded, on the basis of these calculations, that this and related models are incapable of giving capacitances for the mercury–aqueous solution interface in agreement with experiment. The explanation was considered most likely to be the assumption of a sharp boundary between metal and electrolyte phases, which does not permit the electron density tails to penetrate the first layer of solvent, where they would feel a dielectric constant considerably larger than unity.[24] Model calculations, using a single simplified dielectric function $\varepsilon(x, x')$ such that $\varepsilon(x, x')$ reduces to the dielectric functions of solvent or metal far from the interface, showed that the penetration of the electron tails into the solvent region indeed reduces the contribution of the electrons to the capacitance. This is obvious from the dependence of the capacitance of a plane capacitor on the interplate spacing. The capacitance is then given by

$$C^{-1} = C_{GC}^{-1} + 4\pi \int_{-\infty}^{\infty} dx' \left[\int^{\infty} dx\, \varepsilon^{-1}(x, x') - (1/\varepsilon)\theta(x' - L) \right] \tag{47}$$

Here θ is a step function, L is the distance of closest approach of ions to $x = 0$, ε is the static dielectric constant [the $k = 0$ Fourier component of $\varepsilon(x, x')$], and the inverse dielectric function is defined by

$$\int_{-\infty}^{\infty} dz\, \varepsilon(x'', x)\varepsilon^{-1}(x, x') = \delta(x'' - x)$$

Only by smearing of the conduction-electron tail from the metal into the electrolyte phase can the integral in Eq. (47) be made small enough to obtain capacitances for the interface in agreement with experiment.

The calculations were subsequently extended to "moderate" surface charges and electrolyte concentrations.[8] The compact-layer capacitance, in this approach, clearly depends on the nature of the solvent, the nature of the metal electrode, and the interaction between solvent and metal. The work[8,109] describing the electrode-solvent system with the use of nonlocal dielectric functions $\varepsilon(x, x')$ is reviewed and discussed by Vorotyntsev, Kornyshev, and co-workers.[6,77] With several assumptions for $\varepsilon(x, x')$, related to the Thomas-Fermi model, an explicit expression[6] for the compact-layer capacitance could be derived:

$$4\pi C_c = k_0[1 + \varepsilon_0^{1/2} \tanh(k_0\delta\varepsilon_0^{-1/2})]/[1 + \varepsilon_0^{-1/2} \tanh(k_0\delta\varepsilon_0^{-1/2})]$$

Here, k_0^{-1} is the Thomas-Fermi screening length in bulk metal, and ε_0 is the effective dielectric constant of the region, of thickness δ, in which metal electrons and the polar liquid interpenetrate. Since $\varepsilon_0 > 1$, C_c is increased relative to the Thomas-Fermi estimate, $k_0/4\pi$. This increase is thus due to the polarizability of the interlayer penetrated by the electron tail of the metal, as was already included in the first model of Badiali *et al.*[82] Various limiting cases of this equation are discussed.[76] For increasingly negative q^M, δ increases and the solvent parameter ε_0 becomes more important than the metal parameter k_0; this is proposed as a reason why the compact-layer capacitance is observed to become independent of the metal for large negative q^M. A related model for the interface[110] is discussed by Vorotyntsev *et al.* and compared to theirs. Perspectives for development of self-consistent theories for metal and electrolyte of the compact layer are also discussed.

VI. CONCLUSIONS

One can say that the metal in the interface has come a long way in the last ten years, from a featureless conducting medium to a component with its own complicated structure and role. Since separation of properties of the interface into metal and electrolyte contributions cannot be done unambiguously by experiment alone, theoretical calculations on the metal surface have been necessary to establish the importance of the contribution of the metal. It is clear that the metal itself, aside from its influence on the electrolyte, makes a contribution to the capacitance of the interface and to the change of the potential drop across the interface as one substitutes one metal for another. Accurate models of the metal in contact with electrolyte are thus necessary for a correct understanding of the electrochemical interface.

Modern theories of electronic structure at a metal surface, which have proved their accuracy for bare metal surfaces, have now been applied to the calculation of electron density profiles in the presence of adsorbed species or other external sources of potential. The spillover of the negative (electronic) charge density from the positive (ionic) background and the overlap of the former with the electrolyte are the crucial effects. Self-consistent calculations, in which the electronic kinetic energy is correctly taken into account, may have to replace the simpler density-functional treatments which have been used most often. The situation for liquid metals, for which the density profile for the positive (ionic) charge density is required, is not as satisfactory as for solid metals, for which the crystal structure is known.

For the metal in the electrochemical interface, one requires a model for the interaction between metal and electrolyte species. Most important in such a model are the terms which are responsible for establishing the metal-electrolyte distance, so that this distance can be calculated as a function of surface charge density. The most important such term is the repulsive pseudopotential interaction of metal electrons with the cores of solvent species, which affects the distribution of these electrons and how this distribution reacts to charging, as well as the metal-electrolyte distance. Although most calculations have used parameterized simple functional forms for this term, it can now be calculated correctly *ab initio*.

What one requires is a self-consistent picture of the interface, including both metal and electrolyte, so that, for a given surface charge, one has distributions of all species of metal and electrolyte phases. Unified theories have proved too difficult but, happily, it seems that some decoupling of the two phases is possible, because the details of the metal-electrolyte interaction are not so important. Thus, one can calculate the structure of each part of the interface in the field of the other, so that the distributions of metal species are appropriate to the field of the electrolyte species, and vice versa.

Quantum mechanical calculations are appropriate for the electrons in a metal, and, for the electrolyte, modern statistical mechanical theories may be used instead of the traditional Gouy-Chapman plus orienting dipoles description. The potential and electric field at any point in the interface can then be calculated, and all measurable electrical properties can be evaluated for comparison with experiment.

The calculations involved in a fully consistent model of the interface are difficult, but, as may be seen in the work discussed above, quite practicable. It may well be that when a complete model, calculating all terms properly and giving properties in agreement with experiment, is available, it will appear that some of the simplifications and assumptions used in the less complete models of the past, such as separation into diffuse and compact layers, will still be valid. However, it will no longer be possible to neglect the structure of the metal and reduce it to an uninteresting ideal conductor.

REFERENCES

[1] J. O'M. Bockris and S. U. M. Khan, *Quantum Electrochemistry*, Plenum Press, New York, 1979, p. 17.

[2] S. Trasatti, in *Trends in Interfacial Electrochemistry*, Ed. by A. Fernando Silva, Reidel, Dordrecht, 1986, Chap. I.

[3] S. Trasatti, *J. Electroanal. Chem.* **150** (1983) 1.

[4] J. Goodisman, *Electrochemistry: Theoretical Foundations*, Wiley, New York, 1987, Chap. 3.

[5] O. K. Rice, *Phys. Rev.* **31** (1928) 1051.

[6] M. A. Vorotyntsev and A. A. Kornyshev, *Elektrokhimiya* **20** (1984) 3. [Engl. transl.: *Sov. Electrochem.* **20** (1984) 1.]

[7] N. F. Mott and R. J. Watts-Tobin, *Electrochim. Acta* **4** (1961) 79.

[8] A. A. Kornyshev and M. A. Vorotyntsev, *Can. J. Chem.* **59** (1981) 2031.

[9] S. Trasatti, *Modern Aspects of Electrochemistry*, No. 13, Ed. by B. E. Conway and J. O'M. Bockris, Plenum Press, New York, 1979, p. 81.

[10] S. Trasatti, *Advances in Electrochemistry and Electrochemical Engineering*, Vol. 10, Ed. by H. Gerischer and C. W. Tobias, Wiley-Interscience, New York, 1976.

[11] W. Schmickler, in *Trends in Interfacial Electrochemistry*, Ed. by A. Fernando Silva, Reidel, Dordrecht, 1986.

[12] J. R. MacDonald and C. A. Barlow, Jr., *J. Chem. Phys.* **36** (1962) 3062.

[13] R. Parsons, in *Modern Aspects of Electrochemistry*, No. 1, Ed. by J. O'M. Bockris and B. E. Conway, Plenum Press, New York, 1954, p. 103; R. Parsons, *J. Electroanal. Chem.* **59** (1975) 229.

[14] R. Reeves, in *Comprehensive Treatise of Electrochemistry*, Vol. 1, Ed. by J. O'M. Bockris, B. E. Conway, and E. Yeager, Plenum Press, New York, 1980, p. 83.

[15] W. R. Fawcett, S. Levine, R. M. deNobriga, and A. D. McDonald, *J. Electroanal. Chem.* **111** (1980) 163.
[16] R. Guidelli, in *Trends in Interfacial Electrochemistry*, Ed. by A. Fernando Silva, Reidel, Dordrecht, 1986, p. 387.
[17] R. Guidelli, *J. Electroanal. Chem.* **123** (1981) 59; **197** (1986) 77.
[18] E. Yeager, *Surf. Sci.* **101** (1980) 1.
[19] N. H. March and M. P. Tosi, *Coulomb Liquids*, Academic, London, 1984, Chap. 8.
[20] N. H. March and M. P. Tosi, Ref. 19, Section 8.3.
[21] A. N. Frumkin, B. Damaskin, N. Grigoryev, and I. Bagotskaya, *Electrochim. Acta* **19** (1974) 69.
[22] M. J. Sparnaay, *The Electrical Double Layer*, Pergamon Press, Oxford, 1972, Sections 3.2 and 3.3.
[23] S. Trasatti, *Colloids and Surfaces* **1** (1980) 173.
[24] A. A. Kornyshev and M. A. Vorotyntsev, *Surf. Sci.* **101** (1980) 23.
[25] S. Trasatti, in *Comprehensive Treatise of Electrochemistry*, Vol. I, Ed. by J. O'M. Bockris, B. E. Conway, and E. Yeager, Plenum Press, New York, 1980.
[26] S. Trasatti, in *Trends in Interfacial Electrochemistry*, Ed. by A. Fernando Silva, Reidel, Dordrecht, 1986, Chap. II.
[27] A. Hamelin, T. Vitanov, E. Sevastyanov, and A. Popov, *J. Electroanal. Chem.* **145** (1983) 225.
[28] A. Hamelin, in *Trends in Interfacial Electrochemistry*, Ed. by A. Fernando Silva, Reidel, Dordrecht, 1986.
[29] S. H. Liu, *J. Electroanal. Chem.* **150** (1983) 305.
[30] J.-P. Badiali, M.-L. Rosinberg, and J. Goodisman, *J. Electroanal. Chem.* **143** (1983) 73.
[31] J.-P. Badiali, M.-L. Rosinberg, and J. Goodisman, *J. Electronal. Chem.* **150** (1983) 25.
[32] W. Schmickler, *J. Electroanal. Chem.* **150** (1983) 19.
[33] A. M. Kalyuzhnaya, N. B. Grigoryev, and I. A. Bagotskaya, *Elektrokhimya* **10** (1974) 1717. [Engl. transl: *Sov. Electrochem.* **10** (1974) 1628.]
[34] A. N. Frumkin, N. B. Polinovskaya, N. Grigoryev, and I. A. Bagotskaya, *Electrochim. Acta* **10** (1965) 793.
[35] G. Valette, *J. Electroanal. Chem.* **122** (1981) 285.
[36] E. M. Stuve, K. Bange, and J. K. Sass, in *Trends in Interfacial Electrochemistry*, Ed. by A. Fernando Silva, Reidel, Dordrecht, p. 255.
[37] J. R. Noonan and H. L. Davis, *Science* **234** (1986) 310.
[38] A. Sommerfeld, *Naturwissenschaften* **15** (1927) 826.
[39] C. Kittel, *Elementary Solid State Physics*, Wiley, New York, 1962, Chap. 5.
[40] D. M. Newns, *Phys. Rev. B* **1** (1970) 3304.
[41] J. M. Ziman, *Principles of the Theory of Solids*, Cambridge University Press, Cambridge, 1964, Chap. 3.2.
[42] W. A. Harrison, *Solid State Theory*, Dover Publications, New York, 1979, Chap. II.
[43] M. L. Cohen and V. Heine, *Solid State Phys.* **24** (1970).
[44] N. W. Ashcroft, *Phys. Lett.* **23** (1966) 48.
[45] N. W. Ashcroft, in *Interaction Potentials and Simulation of Lattice Defects*, Ed. by P. C. Gehlen, J. R. Beeler, and R. I. Jaffee, Plenum Press, New York, 1972.
[46] N. H. March and M. P. Tosi, Ref. 19, Sections 2.1 and 3.1.
[47] J. Goodisman, Ref. 4, Chap. 4. B.
[48] M. A. Vorotyntsev and A. A. Kornyshev, *Electrokhimya* **15** (1979) 660. [Engl. transl.: *Sov. Electrochem.* **15** (1979) 560.]

[49] J. Goodisman, *Diatomic Interaction Potential Theory*, Vol. I, Academic, New York, 1973, Chap. III.F.
[50] N. H. March and M. P. Tosi, Ref. 19, Chap. 4.
[51] D. M. Kolb, in *Trends in Interfacial Electrochemistry*, Ed. by A. Fernando Silva, Reidel, Dordrecht, 1986.
[52] J. D. McIntyre, *Surf. Sci.* **37** (1973) 658.
[53] Ref. 19, Section 6.1.
[54] J. R. Smith, *Phys. Rev.* **181** (1969) 522.
[55] W. Schmickler and D. Henderson, *Phys. Rev. B* **30** (1984) 3081.
[56] N. D. Lang and W. Kohn, *Phys. Rev. B* **1** (1970) 4555; N. D. Lang, *Solid State Commun.* **7** (1969) 1047.
[57] N. D. Lang and W. Kohn, *Phys. Rev. B* **7** (1973) 3541.
[58] N. D. Lang, in *Theory of the Inhomogeneous Electron Gas*, Ed. by S. Lundqvist and N. H. March, Plenum Press, New York, 1981.
[59] P. Hohenberg and W. Kohn, *Phys. Rev. B* **136** (1964) 864.
[60] J. Goodisman and M.-L. Rosinberg, *J. Phys. C* **16** (1983) 1143.
[61] J. Goodisman, *Phys. Rev. B* **32** (1985) 4835.
[62] M. P. D'Evelyn and S. A. Rice, *J. Chem. Phys.* **78** (1983) 5081.
[63] R. N. Kuklin, *Elektrokhimiya* **14** (1978) 381.
[64] L. I. Schiff, *Phys. Rev. B* **1** (1970) 4649.
[65] D. E. Beck and V. Celli, *Phys. Rev. B* **2** (1970) 2955.
[66] J. Rudnick, *Phys. Rev. B* **5** (1972) 2863.
[67] S. C. Ying, J. R. Smith, and W. Kohn, *J. Vac. Sci. Technol.* **9** (1972) 575.
[68] J. Heinrichs, *Phys. Rev. B* **8** (1973) 1346.
[69] J. Goodisman, *J. Chem. Phys.* **86** (1987) 882.
[70] H. F. Budd and J. Vannimenus, *Phys. Rev. B* **12** (1975) 509.
[71] H. F. Budd and J. Vannimenus, *Phys. Rev. Lett.* **31** (1973) 1218, 1430.
[72] J. Vannimenus and H. F. Budd, *Solid State Commun.* **15** (1974) 1739.
[73] G. D. Mahan and W. L. Schaich, *Phys. Rev. B* **12** (1975) 5585.
[74] J. Heinrichs and N. Kumar, *Phys. Rev. B* **12** (1975) 802.
[75] A. K. Theophilou and A. Modinos, *Phys. Rev. B* **6** (1972) 801.
[76] A. A. Kornyshev, M. B. Partenskii, and W. Schmickler, *Z. Naturforsch. A* **39** (1984) 1122.
[77] M. A. Vorotyntsev, V. Yu. Isotov, A. A. Kornyshev, and W. Schmickler, *Elektrokhimiya* **19** (1983) 295. [Engl. transl.: *Sov. Electrochem.* **19** (1983) 260.]
[78] M. A. Vorotyntsev, *Elektrokhimiya* **14** (1978) 911, 913. [Engl. transl.: *Sov. Electrochem.* **14** (1978) 781, 783.]
[79] W. Schmickler, *J. Electroanal. Chem.* **149** (1983) 15.
[80] R. R. Salem, *Zh. Fiz. Khim.* **54** (1980) 212.
[81] J. R. Smith, in *Interactions on Metal Surfaces*, Ed. by R. Gomer, Springer, Berlin, 1975, Chap. 1.
[82] J.-P. Badiali, M.-L. Rosinberg, and J. Goodisman, *J. Electroanal. Chem.* **130** (1981) 31.
[83] J.-P. Badiali, M.-L. Rosinberg, F. Vericat, and L. Blum, *J. Electroanal. Chem.* **158** (1983) 253.
[84] W. Harrison, *Pseudopotentials in the Theory of Metals*, Benjamin, New York, 1966, p. 57.
[85] J. O'M. Bockris and M. A. Habib, *J. Electronal. Chem.* **68** (1976) 367; S. Trasatti, *J. Electroanal. Chem.* **33** (1971) 351.
[86] M. B. Partenskii and Ya. G. Smorodinskii, *Fiz. Tverd. Tela* **16** (1974) 644. [Engl. transl.: *Sov. Phys.—Solid State* **16** (1974) 423.]

[87] A. N. Frumkin, B. Damaskin, N. Grigoryev, and I. Bagotskaya, *Electrochim. Acta* **19** (1974) 75; B. B. Damaskin, *J. Electroanal. Chem.* **75** (1977) 359.
[88] R. N. Kuklin, *Elektrokhimiya* **13** (1977) 1182, 1796.
[89] J. A. Harrison, J. E. B. Randles, and D. J. Schiffrin, *J. Electroanal. Chem.* **48** (1973) 359.
[90] A. A. Kornyshev and M. A. Vorotyntsev, *J. Electroanal. Chem.* **167** (1984) 1.
[91] M. A. Habib and J. O'M. Bockris, in *Comprehensive Treatise of Electrochemistry*, Vol. I, Ed. by J. O'M. Bockris, B. E. Conway, and E. Yeager, Plenum Press, New York, 1980, Chap. 4.
[92] W. Schmickler and D. Henderson, *J. Electroanal. Chem.* **176** (1984) 383.
[93] D. Price and J. W. Halley, *J. Electroanal. Chem.* **150** (1983) 347.
[94] J. Goodisman, *Theor. Chim. Acta* **68** (1985) 197.
[95] E. P. Wigner, *Phys. Rev.* **46** (1934) 1002.
[96] D. Henderson, in *Trends in Interfacial Electrochemistry*, Ed. by A. Fernando Silva, Reidel, Dordrecht, 1986.
[97] J. W. Halley, B. Johnson, D. Price, and M. Schwalm, *Phys. Rev. B* **31** (1985) 7695.
[98] J. W. Halley and D. Price, preprint, 1986.
[99] D. Price, Thesis, University of Minnesota, August 1986.
[100] S. L. Carnie and D. Y. Chan, *J. Chem. Phys.* **73** (1980) 2949.
[101] W. Schmickler, *J. Electroanal. Chem.* **157** (1984) 1.
[102] W. Schmickler, *Chem. Phys. Lett.* **99** (1983) 135.
[103] W. Schmickler and D. Henderson, *J. Chem. Phys.* **80** (1984) 3381.
[104] L. Blum and D. Henderson, *J. Chem. Phys.* **74** (1981) 1902.
[105] J. Goodisman, *J. Chem. Phys.* **82** (1985) 560.
[106] G. Gouy, *J. Phys. (Paris)* **9** (1910) 457; D. L. Chapman, *Phil. Mag.* **25** (1913) 1475.
[107] A. A. Kornyshev, A. I. Rubinshtein, and M. A. Vorotyntsev, *J. Phys. C* **11** (1978) 3307.
[108] K. L. Kliewer and R. Fuchs, *Adv. Chem. Phys.* **27** (1974) 356.
[109] A. A. Kornyshev, W. Schmickler, and M. A. Vorotyntsev, *Phys. Rev. B* **25** (1982) 5244.
[110] V. A. Kiryanov, Electrokhimiya **17** (1981) 286.

2

Recent Advances in the Theory of Charge Transfer

A. M. Kuznetsov

A. N. Frumkin Institute for Electrochemistry, Academy of Sciences of the USSR, Moscow V-71, USSR

I. INTRODUCTION

Development of the quantum mechanical theory of charge transfer processes in polar media began more than 20 years ago. The theory led to a rather profound understanding of the physical mechanisms of elementary chemical processes in solutions. At present, it is a good tool for semiquantitative and, in some cases, quantitative description of chemical reactions in solids and solutions. Interest in these problems remains strong, and many new results have been obtained in recent years which have led to the development of new areas in the theory. The aim of this paper is to describe the most important results of the fundamental character of the results obtained during approximately the past nine years. For earlier work, we refer the reader to several review articles.[1-4]

II. INTERACTION OF REACTANTS WITH THE MEDIUM

At present, it is understood that the change in the configuration of the medium molecules due to thermal or quantum fluctuations plays a major role in charge transfer processes. This may be clearly seen by considering the electron transfer reaction between the simple ions A^{z_1} and B^{z_2} located at some fixed distance R from each other. According to the Franck–Condon principle, electron transfer is possible if the electron energies in the donor ε_A and in the acceptor ε_B are approximately equal to each other. In the reaction under consideration, the matching of the electron energy levels can occur only due to the interaction of the electron with the medium molecules. Thus, the activation of the reactants in this case is directly effected by the fluctuations of the medium molecules. If the reactants have complex intramolecular structure, the intramolecular vibrations may also have some role in the activation process. However, the fluctuations in the medium play an essential role in this case as well since the change in the energy of the intramolecular vibrations is due to their interaction with the vibrations of the medium molecules.

To show more clearly the difference between this new approach and that used earlier, we will briefly summarize the model which was widely used for the calculation of the probability of the elementary act of charge transfer processes in polar media.

1. Adiabatic and Diabatic Approaches: A Reference Model

There are two basic approaches to the calculation of the probability of the electron transition. One of them, called the adiabatic approach, consists in that the electron states ϕ_α of the total Hamiltonian of the system, H, are chosen as the zeroth-order electron states in the Born–Oppenheimer approximation. These states and the corresponding energies ε_α depend on the nuclear coordinates Q_k. In the simplest case, the potential energy surface $U_\alpha(Q_k)$ corresponding to the ground electron state ϕ_α has two minima at the points Q_{k0i} and Q_{k0f}. The regions near Q_{k0i} and Q_{k0f} correspond to the reactants and reaction products, respectively. They are separated by a potential barrier with the top at the point Q_k^* defined as the saddle point on the surface $U_\alpha(Q_k)$ (Fig. 1).

The potential energy surface (PES) $U_\beta(Q_k)$ for the excited electron state ϕ_β has its minimum near the point Q_k^* (Fig. 1). In the classical limit, the electron transition may be treated as a continuous motion of the system on the lower PES, U_α, from the

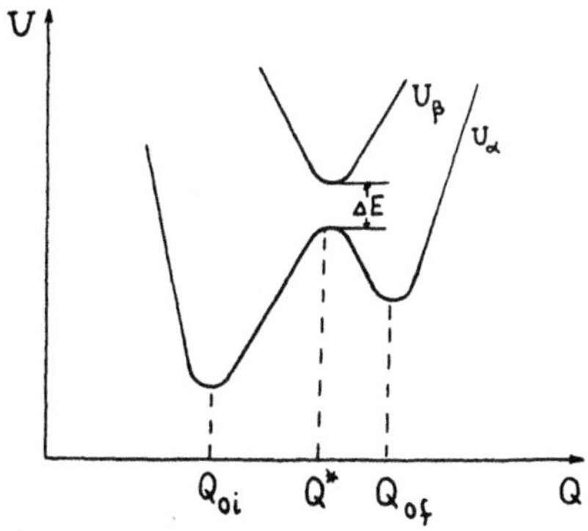

Figure 1. Adiabatic potential energy surfaces.

point Q_{k0i} to Q_{k0f}. During this motion, the adiabatic redistribution of the electron density from the donor site to the acceptor site takes place. However, in the region near the top of the potential barrier, the transition of the system to the upper PES U_β is possible due to the effects of nonadiabaticity. In this case, the electron remains in the donor ion.

If the probability for the system to jump to the upper PES is small, the reaction is an adiabatic one. The advantage of the adiabatic approach consists in the fact that its application does not lead to difficulties of fundamental character, e.g., to those related to the detailed balance principle. The activation factor is determined here by the energy (or, to be more precise, by the free energy) corresponding to the top of the potential barrier, and the transmission coefficient, κ, characterizing the probability of the rearrangement of the electron state is determined by the minimum separation ΔE of the lower and upper PES. The quantity ΔE is the same for the forward and reverse transitions.

However, if the PES are multidimensional, as is the case for reactions in the condensed phase, the adiabatic approach is inconvenient for practical calculations, especially for nonadiabatic reactions.

Another approach widely used for nonadiabatic reactions is the diabatic one. The channel Hamiltonians H_i^e and H_f^e determining the zeroth-order Born–Oppenheimer electron states of the donor ϕ_A and acceptor ϕ_B and the perturbations $\hat{V}_i$ and $\hat{V}_f$ leading to the forward and reverse electron transitions, respectively, are separated from the total Hamiltonian in this approach:

$$H = H_N + H_i^e + \hat{V}_i^{0d} = H_N + H_f^e + \hat{V}_f^{0d} \tag{1}$$

where H_N is the Hamiltonian of the nuclei.

The diabatic potential energy surfaces U_i and U_f corresponding to the zeroth-order electron states ϕ_i and ϕ_f cross along a hypersurface S_* (Fig. 2). In the diabatic approach, the electron transition is usually treated as follows. Moving along the nuclear degrees of freedom Q_k on the initial PES, U_i, at constant initial distribution of the charge, the system reaches the point of minimum energy Q_k^* on the intersection surface S_*. In this configuration, referred to as the transitional configuration, the change of the electron state from ϕ_i to ϕ_f is possible. This corresponds to electron transfer from the donor to the acceptor and to transition of the system from the potential energy surface U_i to the potential energy surface U_f followed by the relaxation of the nuclear configuration to the equilibrium final position.

The activation energy for the nonadiabatic reaction, $E_a^{\text{n.ad}}$, is determined by the point of minimum energy on the intersection surface of PES U_i and U_f, and the transmission coefficient κ is determined by the electron resonance integral

$$V_i = \int \phi_A \hat{V}_i^{0d} \phi_B \, d^3x \tag{2}$$

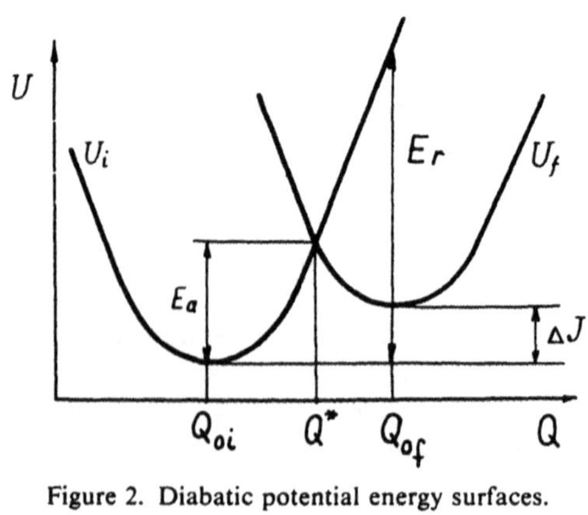

Figure 2. Diabatic potential energy surfaces.

where $\hat{V}_i^{0d}$ is the off-diagonal part of the operator for the interaction of the electron with the acceptor: $\hat{V}_i^{0d} = V_{eB}^{0d} = V_{eB} - V_{eB}^{d}$.

The electron resonance integral for the reverse transition has the form

$$V_f = \int \phi_A \hat{V}_f^{0d} \phi_B \, d^3x \tag{3}$$

where $\hat{V}_f^{0d}$ is the off-diagonal part of the operator for the interaction of the electron with the donor: $\hat{V}_f^{0d} = V_{eA}^{0d} = V_{eA} - V_{eA}^{d}$.

For the adiabatic reactions, the activation energy $E_a^{\text{n.ad.}}$ must be reduced by the quantity δE_a, determined by the resonance splitting of the potential energy surfaces, ΔE, and by the slopes of U_i and U_f at the point Q_k^*, and $\kappa = 1$.

In calculating the transition probability for the nonadiabatic reactions, it is sufficient to use the lowest order of quantum mechanical perturbation theory in the operator $\hat{V}_i^{0d}$. For the adiabatic reactions, we must perform the summation of the whole series of the perturbation theory.[5] (It is insufficient to retain only the first term of the series that appeared in the quantum mechanical perturbation theory.) Correct calculations in both adiabatic and diabatic approaches lead to the same results, which is evidence of the equivalence of the two approaches.

The diabatic approach will be mainly used below, since it is more convenient for nonadiabatic reactions. However, in Section VI adiabatic reactions will be also considered using the adiabatic approach.

A harmonic approximation has usually been used for the description of the nuclear potential energy,

$$U_N = \tfrac{1}{2} \sum_k \hbar\omega_k Q_k^2 \tag{4}$$

where the Q_k are dimensionless normal coordinates of the harmonic oscillators characterized by the frequencies ω_k.

For reactants having complex intramolecular structure, some coordinates Q_k describe the intramolecular degrees of freedom. For solutions in which the motion of the molecules is not described by small vibrations, the coordinates Q_k describe the effective oscillators corresponding to collective excitations in the medium. Summation rules have been derived which enable us to relate the characteristics of the effective oscillators with the dielectric properties of the medium.[5]

The interaction of the electrons and other charged particles with the vibrations of the nuclei in the reference model was usually considered to be linear in the coordinates Q_k:

$$V_{zQ} = -\sum_k \gamma_k Q_k \tag{5}$$

It is convenient to split this interaction into two parts: the interaction with the intramolecular vibrations and with the nearest medium molecules, V_{zv}, and the interaction with the rest of the solvent, V_{zP}, which may be described in terms of the inertial polarization per unit volume $\mathbf{P}(\mathbf{r})$:

$$V_{zQ} = V_{zv} + V_{zP} \tag{6}$$

The interaction V_{zP} may be written in the form

$$V_{zP} = -\int \mathbf{P}(\mathbf{r})\mathbf{E}_z^v(\mathbf{r})\, d^3r \tag{7}$$

where $\mathbf{E}_z^v(\mathbf{r})$ is the electric field in vacuum due to the charge ez.

Since the interaction of the electron with the medium polarization is strong, in the reference model it was usually included in the zeroth-order Hamiltonians determining the Born–Oppenheimer electron states:

$$\begin{aligned}(T_e + V_{eA} + V_{ev}^{A} + V_{eP} + V_{eB}^{d} + V_{ev}^{Bd})\phi_A &= \varepsilon_A\phi_A\\ (T_e + V_{eB} + V_{ev}^{B} + V_{eP} + V_{eA}^{d} + V_{ev}^{Ad})\phi_B &= \varepsilon_B\phi_B\end{aligned} \tag{8}$$

where T_e is the kinetic energy of the electron and the superscript d denotes the diagonal part of the operator, which does not lead to the electron transition but leads to a distortion of the electron state.

Note that in the reference model all the interactions of the electron with the medium polarization V_{eP} are included in Eqs. (8) determining the electron states. The dependence of ϕ_A and ϕ_B on the polarization and intramolecular vibrations was entirely neglected in most calculations of the transition probability [the approximation of constant electron density (ACED)]. This approximation, together with Eqs. (4)–(7), resulted in the parabolic shape of the diabatic PES U_i and U_f. The latter differed only by the shift of equilibrium coordinates Q_{k0i} and Q_{k0f} and by the energies of the effective oscillators at the minima, J_i and J_f.

The result for the transition probability in the classical limit had the form

$$W = (\omega_{\text{eff}}/2\pi)\kappa \exp(-H^{\ddagger}/kT) \tag{9}$$

where

$$H^{\ddagger} = (E_r + \Delta J)^2/4E_r \tag{10}$$

Here $\Delta J = J_f - J_i$ and E_r is the total reorganization energy of the vibrational subsystem:

$$E_r = \tfrac{1}{2}\sum_k \hbar\omega_k(Q_{k0i} - Q_{k0f})^2 = \sum_k E_{rk} \tag{11}$$

Equation (9) was obtained using the assumption that the vibrational subsystem is in the state of thermal equilibrium corresponding to the initial electron state. The expression for the effective frequency ω_{eff} has the form[5]

$$\omega_{\text{eff}}^2 = \sum_k \omega_k^2 E_{rk}/E_r \tag{12}$$

2. A New Approach to the Interaction of the Electron with the Polarization of the Medium in Nonadiabatic Reactions

As noted above, in the reference model the dependence of the electron wave functions ϕ_A and ϕ_B on the nuclear coordinates was entirely neglected and the wave functions ϕ_A and ϕ_B for the isolated ions or the wave functions calculated for corresponding equilibrium nuclear configurations Q_{k0i} and Q_{k0f} according to Eqs. (8) were usually used in the calculations.

Recent analysis has shown that this approximation is, in general, insufficient.[6] This is due to the long-range character of the interaction of the electron with the medium polarization. The zeroth-order states determined from Eqs. (8) taking into account the total interaction of the electron with the total inertial polarization of the medium V_{eP} may not describe the states of the electron localized in the donor or in the acceptor sites. Since the polarization varies due to thermal fluctuations, at certain configurations of the polarization field the potential energy of the electron may involve several potential wells localized in different points. Therefore, the solutions of Eqs. (8), taking into account all the interactions of the electron with the inertial polarization, may give states which, in general, do not correspond to the electron localization in the donor or in the acceptor.

It was suggested that the zeroth-order electron states be calculated using equations similar to Eqs. (8) at initial equilibrium values of the polarization $\mathbf{P}_{0i}$.[7] However, it may be seen that if the acceptor is an anion, even in the initial equilibrium configuration the equilibrium polarization of the medium near the acceptor may create a potential well for the electron.

To solve this problem, it was suggested[6] that the zeroth-order electron states be found from the equations

$$\begin{aligned}(T_e + V_{eA} + V_{ev}^{A} + V_{eP}^{A} + V_{eB}^{d} + V_{ev}^{Bd} + V_{eP}^{Bd})\phi_A = \varepsilon_A\phi_A \\ (T_e + V_{eB} + V_{ev}^{B} + V_{eP}^{B} + V_{eA}^{d} + V_{ev}^{Ad} + V_{eP}^{Ad})\phi_B = \varepsilon_B\phi_B\end{aligned} \tag{13}$$

Unlike Eqs. (8), the first of Eqs. (13) involves only part of the interaction of the electron with the medium polarization V_{eP}^{A}, which, together with V_{eA}, creates the potential well for the electron near the donor A. As for the interaction with the polarization V_{eP}^{B}, which, together with V_{eB}, creates the potential well for the electron near the acceptor, the first of Eqs. (13) involves only the diagonal part of this interaction, V_{eP}^{Bd}, leading to a distortion of the state ϕ_A without a change in the electron localization. The state ϕ_B is determined in a similar way.

The perturbation operators must be modified in accordance with the new definition of the zeroth-order electron states:

$$\begin{aligned}\hat{V}_i^{0d} &= (V_{eB})_{0d} + (V_{eP}^{B})_{0d} + (V_{ev}^{B})_{0d} \\ &= V_{eB} - V_{eB}^{d} + V_{eP}^{B} - V_{eP}^{Bd} + V_{ev}^{B} - V_{ev}^{Bd} \\ \hat{V}_f^{0d} &= (V_{eA})_{0d} + (V_{eP}^{A})_{0d} + (V_{ev}^{A})_{0d} \\ &= V_{eA} - V_{eA}^{d} + V_{eP}^{A} - V_{eP}^{Ad} + V_{ev}^{A} - V_{ev}^{Ad}\end{aligned} \tag{14}$$

The method of separating the interaction V_{eP} into V_{eP}^{A} and V_{eP}^{B} is discussed in the next section.

This new approach enables us to consider all the physical effects due to the interaction of the electron with the medium polarization and local vibrations and to take them into account in the calculation of the transition probability. These physical effects are as follows:

1. *Effect of diagonal dynamic disorder* (*DDD*). Fluctuations of the polarization and the local vibrations produce the variation of the positions of the electron energy levels $\varepsilon_A(Q)$ and $\varepsilon_B(Q)$ to meet the requirements of the Franck-Condon principle.

2. *Effect of off-diagonal dynamic disorder* (*off-DDD*). The interaction of the electron with the fluctuations of the polarization and local vibrations near the other center leads to new terms $V_{eP}^{B} - V_{eP}^{Bd}$, $V_{ev}^{B} - V_{ev}^{Bd}$ and $V_{eP}^{A} - V_{eP}^{Ad}$, $V_{ev}^{A} - V_{ev}^{Ad}$ in the perturbation operators $\hat{V}_i^{0d}$ and $\hat{V}_f^{0d}$ [see Eqs. (14)]. A part of these interactions corresponding to the equilibrium values of the polarization $\mathbf{P}_{0i}$ and $\mathbf{P}_{0f}$ results in the renormalization of the electron interactions with ions A and B, due to their partial screening by the dielectric medium. However, at arbitrary values of the polarization $\mathbf{P}$, there is another part of these interactions which is due to the fluctuating electric fields. This part of the interaction depends on the nuclear coordinates and may exceed the renormalized interactions of the electron with the donor and the acceptor. The interaction of the electron with these fluctuations plays an important role in processes involving solvated, trapped, and weakly bound electrons.

3. *Effect of diagonal-off-diagonal dynamic disorder* (*D-off-DDD*). The polarization fluctuations and the local vibrations give rise to variation of the electron densities in the donor and the acceptor, i.e., they lead to a modulation of the electron wave functions ϕ_A and ϕ_B. This leads to a modulation of the overlapping of the electron clouds of the donor and the acceptor and hence to a different transmission coefficient from that calculated in the approximation of constant electron density (ACED). This modulation may change the path of transition on the potential energy surfaces.

4. *Additional effect of diagonal dynamic disorder*. The variations of the electron densities near the centers A and B due to polarization fluctuations and local vibrations lead to changes in the interaction of the electron with the medium and, hence, to changes in the shape of the potential energy surfaces U_i and U_f as compared

to that in the reference model. As a result, the activation energy will be different from that calculated in the ACED.

III. NONADIABATIC ELECTRON TRANSFER REACTIONS

The above effects manifest themselves both in electron transfer reactions and in reactions involving the transfer of heavy particles. However, before discussing these effects in electron transfer reactions, we will consider some problems arising in the reference model.

1. Energy of Activation or Free Energy of Activation?

The expression in Eq. (10) for the exponent in Eq. (9) is quite similar to that for the activation free energy in electron transfer reactions derived by Marcus using the methods of nonequilibrium classical thermodynamics[8]:

$$F_a = (E_r + \Delta F)^2/4E_r \tag{15}$$

where ΔF is the free energy of the transition.

Instead of the quantity given by Eq. (15), the quantity given by Eq. (10) was treated as the activation energy of the process in the earlier papers on the quantum mechanical theory of electron transfer reactions. This difference between the results of the quantum mechanical theory of radiationless transitions and those obtained by the methods of nonequilibrium thermodynamics has also been noted in Ref. 9. The results of the quantum mechanical theory were obtained in the harmonic oscillator model, and Eqs. (9) and (10) are valid only if the vibrations of the oscillators are classical and their frequencies are unchanged in the course of the electron transition (i.e., $\omega_k^i = \omega_k^f$). It might seem that, in this case, the energy of the transition and the free energy of the transition are equal to each other. However, we have to remember that for the solvent, the oscillators are the effective ones and the parameters of the system Hamiltonian related to the dielectric properties of the medium depend on the temperature. Therefore, the problem of the relationship between the results obtained by the two methods mentioned above deserves to be discussed.

It was shown in Ref. 5 that the quantity ΔJ involved in Eq. (10) includes, in addition to the difference of the electron energies, quantities of the type

$$E_f^{aq} - E_i^{aq} = -(I/8\pi)(I/2\pi)^3 \int d^3k\,[|\mathbf{D}_f(\mathbf{k})|^2 - |\mathbf{D}_i(\mathbf{k})|^2](2/\pi)$$

$$\times \int_0^\infty d\omega\, Im\,\varepsilon(k,\omega)/\omega|\varepsilon(k,\omega)|^2$$

where $\mathbf{D}_i(\mathbf{k})$ and $\mathbf{D}_f(\mathbf{k})$ are the Fourier components of the electrostatic inductions due to the initial and final charge distributions in the medium.

One may easily see that the quantities E^{aq} represent the generalization of the expressions for the electrostatic contribution to the solvation free energy for the case of spatial dispersion of the dielectric function $\varepsilon(k,\omega)$. Thus, it has been shown in Ref. 5 that the quantity ΔJ in Eqs. (9) and (10) for the transition probability represents the free energy of the transition. A similar result was obtained later in Ref. 10.

The problem of the physical meaning of the quantity $H^\ddagger$ and of the reorganization energy of the medium E_s has been analyzed in Ref. 11. Following Ref. 11, we write the expression for the transition probability per unit time in the form[3]

$$W = (1/i\hbar kT)\exp(F_{i0}/kT)\int_{c-i\infty}^{c+i\infty} d\theta\,\mathrm{Tr}[V_i\rho_i(1-\theta)V_i^+\rho_f(\theta)] \quad (16)$$

where $\rho_i(1-\theta) = \exp[-\beta(1-\theta)H_i]$ and $\rho_f(\theta) = \exp[-\beta\theta H_f]$, with $\beta = 1/kT$, are the statistical operators (the density matrices) of the initial and final states, and F_{i0} is the free energy of the initial state.

In the Condon approximation, Eq. (16) is transformed to the form[11]

$$W = (1/i\hbar kT)|V_i|^2 \int_{c-i\infty}^{c+i\infty} d\theta\, f_m(\theta) \quad (17)$$

$$f_m(\theta) = \exp[-\beta F(\theta)]\cdot\exp(-\beta\theta E_s - \beta\theta\Delta F) \quad (18)$$

where ΔF is the free energy of the transition, $\beta = 1/kT$,

$$E_s = -\tfrac{1}{2}\int d^3r\, d^3r V' \sum_{\alpha,\beta} \Delta E^v_\alpha(\mathbf{r}) \times \Delta E^v_\beta(\mathbf{r}') D^R_{\alpha\beta}(\mathbf{r}, \mathbf{r}'; \omega = 0) \qquad \alpha, \beta = x, y, z \tag{19}$$

$$\begin{aligned}\exp[-\beta F(\theta)] &= \left\langle T_\tau \exp\left[\int_0^{\beta\theta} d\tau \int d^3r\, \delta\mathbf{P}_i(\mathbf{r}, \tau)\Delta\mathbf{E}^v(\mathbf{r})\right]\right\rangle_i \\ &= \exp(\beta F_{i0})\, \mathrm{Tr}\left\{\exp(-\beta H^a_i) T_\tau \exp\left[\int_0^{\beta\theta} d\tau \times \int d^3r\, \delta\mathbf{P}_i(\mathbf{r}, \tau)\Delta\mathbf{E}^v(\mathbf{r})\right]\right\}\end{aligned} \tag{20}$$

where $\Delta\mathbf{E}^v(\mathbf{r}) = \mathbf{E}^v_f(\mathbf{r}) - \mathbf{E}^v_i(\mathbf{r})$ is the difference of the electric fields in vacuum due to the reaction products and the reactants, T_τ is the operator of time ordering, $\delta\mathbf{P}_i(\mathbf{r}, \tau) = \mathbf{P}(\mathbf{r}, \tau) - \mathbf{P}_{0i}(\mathbf{r}, \tau)$ is the operator in the Heisenberg representation describing the deviation of the inertial polarization from its initial equilibrium value $\mathbf{P}_{0i}(\mathbf{r})$, and $D^R_{\alpha\beta}(\mathbf{r}, \mathbf{r}'; \omega)$ is the Fourier component of the retarded Green's function:

$$D^R_{\alpha\beta}(\mathbf{r}, \mathbf{r}'; t) = -i\theta(t)\langle[P_\alpha(\mathbf{r}, t), P_\beta(\mathbf{r}', 0)]\rangle \tag{21}$$

In the limit where the vibrational spectrum of the inertial polarization lies in the classical region ($\omega < kT/\hbar$), the calculation of the expression in Eq. (20) and of the integral over θ in Eq. (17) gives[11]

$$W = (|V_i|^2/\hbar(kTE_s/\pi)^{1/2}) \exp(-\beta|F(\theta^*)|) \tag{22}$$

where

$$F(\theta) = -\theta^2 \cdot E_s \tag{23}$$

$$\theta^* = \tfrac{1}{2} + \Delta F/2E_s \tag{24}$$

From a comparison of Eqs. (9) and (22) we see that $H^\ddagger = |F(\theta^*)|$. To elucidate the physical meaning of the exponent in Eq. (22), we consider first the case when $\theta^* = 1$ (barrierless reaction). In this case Eq. (20) determines the change of the free energy of the system $F(1)$ when it is polarized by the electric field $\Delta\mathbf{E}^v = \mathbf{E}^v_f - \mathbf{E}^v_i$ (only the free energy related to the inertial polarization is considered). It may be easily seen that the absolute value of $F(1)$ is equal to the energy of the reorganization of the medium E_s (>0).

This equality determines the physical meaning of E_s as the change in the free energy of the system due to a change in the inertial polarization by the quantity $\Delta\mathbf{P}(\mathbf{r}) = \mathbf{P}_{0f}(\mathbf{r}) - \mathbf{P}_{0i}(\mathbf{r})$:

$$E_s = \frac{1}{2}\int d^3r\, \Delta\mathbf{E}^v(\mathbf{r})\Delta\mathbf{P}(\mathbf{r}) \tag{25}$$

For $\theta^* \neq 1$ we note that in the classical limit Eq. (20) may be written in the form

$$\exp[-\beta F(\theta)] = \left\langle T_t \exp\left\{\int_0^\beta dt \int d^3r\, \delta\mathbf{P}_i(\mathbf{r}, t)[\theta\Delta\mathbf{E}^v(\mathbf{r})]\right\}\right\rangle_i \tag{26}$$

Thus, in this case $F(\theta^*)$ represents the change in the free energy of the medium due to the application of the electric field $\theta^*\Delta\mathbf{E}^v(\mathbf{r})$. The absolute value of $F(\theta^*)$ is equal to the energy of the reorganization of the medium when the inertial polarization is changed by the value $\Delta\mathbf{P}_{\theta^*}(\mathbf{r}) = \theta^*\Delta\mathbf{P}(\mathbf{r})$:

$$E_s(\theta^*) = \frac{1}{2}\int d^3r[\theta^*\Delta\mathbf{E}^v(\mathbf{r})]\Delta\mathbf{P}_{\theta^*}(\mathbf{r}) \tag{27}$$

Using Eq. (27), we can write the transition probability in the form

$$W = (|V_i|^2/\hbar(kTE_s/\pi)^{1/2})\exp[-E_s(\theta^*)/kT] \tag{28}$$

Equation (28) gives the physical meaning of the activation factor $H^\ddagger$ as the free energy required for the system to reach the transitional configuration.

In the case when the vibrational spectrum of the system spreads out in the quantum region and the vibrational frequencies of the reaction complex are unchanged in the course of the transition, the following approximate formula can be obtained instead of Eqs. (9) and (10)[3,12]:

$$\begin{aligned} W = (kT/\hbar)\kappa \exp\Big\{&-(\tilde{E}_r + \Delta F)^2/4\tilde{E}_r kT - \sum_l (E_{rl}/\hbar\omega_l) \\ &\times [\cosh(\beta\hbar\omega_l/2) - 1 - \tfrac{1}{2}(\beta\hbar\omega_l/2)^2]/\sinh(\beta\hbar\omega_l/2) \\ &- \sigma^c + \ln(\hbar\omega_{\text{eff}}/kT)\Big\} \end{aligned} \tag{29}$$

where

$$\tilde{E}_r = E_r^{\text{cl}} + \sum_l E_{rl}(\hbar\omega_l/2kT)/\sinh(\hbar\omega_l/2kT) \tag{30}$$

$$\sigma^c = \sum_m E_{rm}/\hbar\omega_m \tag{31}$$

The symmetry factor has the form

$$\alpha = \theta^* = \tfrac{1}{2} + \Delta F \Big/ \left[2E_r^{\text{cl}} + \sum_l E_{rl}\beta\hbar\omega_l/\sinh(\beta\hbar\omega_l/2)\right] \tag{32}$$

The summation over l in Eqs. (30) and (32) means summation over all nonclassical discrete local intramolecular vibrations and integration (using an appropriate weighting function) over the frequencies describing the spectrum of the polarization fluctuations beyond the classical region which satisfy the condition

$$\beta\hbar\omega_l(1-2\alpha)/2 \ll 1 \tag{33}$$

The quantity E_r^{cl} is the energy of the reorganization of all the classical degrees of freedom of the local vibrations and of the classical part of the medium polarization, and σ^c is the tunneling factor for quantum degrees of freedom ($\beta\hbar\omega_m \gg 1$) which do not satisfy the condition given by Eq. (33).

An expression of the type in Eq. (29) has been rederived recently in Ref. 13 for outer-sphere electron transfer reactions with unchanged intramolecular structure of the complexes where essentially the following expression for the effective outer-sphere reorganization energy $\tilde{E}_{rs}$ was used:

$$\tilde{E}_{rs} = (1/8\pi^2)(\hbar/kT)\int_0^\infty d\omega[Im\,\varepsilon(\omega)/|\varepsilon(\omega)|^2] \times \frac{1}{\sinh(\hbar\omega/2kT)} \cdot \int (\mathbf{D}_i - \mathbf{D}_f)^2\,d^3r \tag{34}$$

This formula is easily obtained from Eq. (30) if we use the summation rules relating the parameters of the effective oscillators with the dielectric properties of the medium.[5]

The physical reason for the appearance of the free energies in the formulas for the transition probability consists in the fact that the reactive vibrational modes interact with the nonreactive modes.

Averaging over the states of the nonreactive modes leads to the appearance of free energies. This problem was recently analyzed independently in Refs. 14 and 15.

Following Ref. 14, in the classical limit in the Condon approximation we transform Eq. (16) to the form

$$W = \frac{2\pi}{\hbar}|V_i|^2 \exp(F_{i0}/kT) \int dq_1\, dq_2 \cdots dQ_1\, dQ_2 \cdots$$
$$\times \exp\{-[\tilde{U}(q_1, q_2, \ldots, Q_1, Q_2) + \varepsilon_i(q_1, q_2, \ldots)]/kT\}$$
$$\times \delta[\varepsilon_i(q_1, q_2, \ldots) - \varepsilon_f(q_1, q_2, \ldots)] \quad (35)$$

where $\tilde{U}$ is the potential energy of the nuclear subsystems, q_1, $q_2, \ldots$ are the coordinates of the reactive modes directly interacting with the electron, $Q_1, Q_2, \ldots$ are the coordinates of the nonreactive modes, and ε_i and ε_f are the electron energies in the initial and final states, respectively.

The integration over the coordinates of the nonreactive modes $Q_1, Q_2, \ldots$ gives

$$W = \frac{2\pi}{\hbar}|V_i|^2 \exp(F_{i0}/kT) \int dq_1\, dq_2 \cdots$$
$$\times \exp\{-[F(q_1, q_2, \ldots) + \varepsilon_i(q_1, q_2, \ldots)]/kT\}$$
$$\times \delta[\varepsilon_i(q_1, q_2, \ldots) - \varepsilon_f(q_1, q_2, \ldots)] \quad (36)$$

where $F(q_1, q_2, \ldots)$ is the configurational part of the *free energy* of the system as a function of the coordinates of the reactive modes.

Introducing the *diabatic free energy surfaces* of the initial and final states,

$$U_i(q_1, q_2, \ldots) = F(q_1, q_2, \ldots) + \varepsilon_i(q_1, q_2, \ldots)$$
$$U_f(q_1, q_2, \ldots) = F(q_1, q_2, \ldots) + \varepsilon_f(q_1, q_2, \ldots) \quad (37)$$

the expression for W may be transformed to the form

$$W = \frac{2\pi}{\hbar}|V_i|^2 \exp(F_{i0}/kT) \int dq_1\, dq_2 \cdots$$
$$\times \exp[-U_i(q_1, q_2, \ldots)/kT]\delta[U_i(q_1, q_2, \ldots) - U_f(q_1, q_2, \ldots)] \quad (38)$$

Thus, to calculate the transition probability for the nonadiabatic reaction it is sufficient to know the diabatic *free energy surfaces* of the system, representing the configurational parts of the free energy of the system as a function of the coordinates of the reactive modes at fixed states of the quantum particles.

2. Effects of Diagonal and Off-Diagonal Dynamic Disorder in Reactions Involving Transfer of Weakly Bound Electrons (A Configurational Model)

The effects of the modulation of electron density by local vibrations and polarization fluctuations are most pronounced for reactions involving transfer of weakly bound electrons. These effects were investigated in Ref. 16 for the transfer of weakly bound electrons from a donor A^{z_1} to an acceptor B^{z_2} in a polar medium.

The total wave function of the system was presented in the form

$$\psi = \phi_A(x; q) \sum_n C_n^i(t)\chi_n^i(q) + \phi_B(x; q) \sum_m C_m^f(t)\chi_m^f(q) \quad (39)$$

where χ^i and χ^f are the wave functions of the vibrational subsystem.

The Schrödinger equation for the vector $C\chi$ having the components

$$C\chi = \begin{pmatrix} C^i\chi^i \\ C^f\chi^f \end{pmatrix};$$
$$C^i\chi^i = \begin{pmatrix} \vdots \\ C_n^i\chi_n^i \\ \vdots \end{pmatrix}; \qquad C^f\chi^f = \begin{pmatrix} \vdots \\ C_m^f\chi_m^f \\ \vdots \end{pmatrix} \quad (40)$$

was written in matrix form as follows:

$$i\hbar\, \partial C\chi/\partial t = HC\chi \quad (41)$$

where

$$H = (1 - S^2)^{-1}\begin{pmatrix} H_{AA} - SH_{BA} & V_f \\ V_i & H_{BB} - SH_{AB} \end{pmatrix} \quad (42)$$

$$S = \langle \phi_A | \phi_B \rangle; \qquad V_i = H_{BA} - SH_{AA}; \qquad V_f = H_{AB} - SH_{BB} \quad (43)$$

The elements of the matrix H involve the electron matrix elements of the Hamiltonian of the system. The general expression for the transition probability in Eq. (16) is transformed to the form

$$W \simeq \frac{\beta}{i\hbar}\Delta^2 \int dq |\hat{V}_i(x^*, q)|^2 n_A(x^*, q) n_B(x^*, q) \int d\theta \int ds$$
$$\times \exp(\beta F_{i0}) \exp(-\beta\theta\Delta F)$$
$$\times \rho_i\left(q - \frac{s}{2}; q + \frac{s}{2}; 1 - \theta\right)\rho_f\left(q + \frac{s}{2}; q - \frac{s}{2}; \theta\right) \quad (44)$$

where the approximate formula for the electron matrix element

$$V_i = \int \phi_A^*(x; q)\hat{V}_i(x, q)\phi_B(x; q)\, d^3x$$
$$\simeq \Delta\hat{V}_i(x^*, q)\phi_A(x^*; q)\phi_B(x^*; q) \quad (45)$$

was used and the electron densities

$$n_A(x; q) = |\phi_A(x; q)|^2, \qquad n_B(x; q) = |\phi_B(x; q)|^2 \quad (46)$$

are introduced.

The usual Condon approximation (CA) is obtained from Eq. (44) if we assume that the dependence of the electron factors on q is weaker than that of the other factors and use the approximation $n_A = |\phi_A(x; q_{0i})|^2$ and $n_B = |\phi_B(x; q_{0f})|^2$ in calculating the density matrices of heavy particles ρ_i^0 and ρ_f^0:

$$W_{CA} = \Delta^2|\hat{V}_i(x^*, q^*)|^2 \cdot n_A(x^*; q^*)n_B(x^*; q^*)$$
$$\times \frac{\beta}{i\hbar}\int d\theta\, e^{\beta F_{i0} - \beta\theta\Delta F}\,\mathrm{Tr}[\rho_i^0(1-\theta)\rho_f^0(\theta)] \quad (47)$$

where q^* is the point for which the integrand in Eq. (47) is maximum.

Equation (47) shows that in the Condon approximation the probabilities of forward and reverse transitions satisfy the detailed balance principle since the point q^* corresponds to the intersection of the potential energy surfaces (and free energy surfaces) where $H_{AA} = H_{BB}$. Therefore, at the point q^* we have

$$V_i(q^*) = V_f(q^*) \quad (48)$$

This result was obtained in Ref. 17 and was rederived in a number of subsequent papers.[6,16,18,19]

If the dependence of n_A and n_B on q is taken into account in the calculation of the statistical operators for heavy particles, we obtain the improved Condon approximation (ICA) which differs from Eq. (17) only by the change of ρ_i^0 and ρ_f^0 to ρ_i and ρ_f, respectively. In the classical limit for ρ_i and ρ_f, the expression for the transition probability takes the form

$$W^{cl} \simeq \frac{2\pi}{\hbar}\Delta^2 e^{\beta F_{i0}} \int dq |\hat{V}_i(x^*, q)|^2 n_A(x^*, q) n_B(x^*, q) \times e^{-\beta U_i(q)} \delta[U_i(q) - U_f(q)] \tag{49}$$

Using Eq. (49) one may consider all the cases of the Condon approximation (CA, ICA), and in some models go beyond the Condon approximation (BCA).

(*i*) *Condon Approximation*

It has been shown[16] that in the Condon approximation the value of the polarization in the transitional configuration $\mathbf{P}^*$ is equal to

$$\begin{aligned} \mathbf{P}^*(\mathbf{r}) &= (1-\theta^*)\mathbf{P}_{0i} + \theta^*\mathbf{P}_{0f} = \mathbf{P}_A^* + \mathbf{P}_B^* \\ \mathbf{P}_A^*(\mathbf{r}) &= (c/4\pi)[\mathbf{D}_A(\mathbf{r}) + (1-\theta^*)\mathbf{D}_e^{(A)}(\mathbf{r}; \mathbf{P}_{0i})] \\ \mathbf{P}_B^*(\mathbf{r}) &= (c/4\pi)[\mathbf{D}_B(\mathbf{r}) + \theta^* \cdot \mathbf{D}_e^{(B)}(\mathbf{r}; \mathbf{P}_{0f})] \end{aligned} \tag{50}$$

where θ^* is the symmetry factor determined as the saddle point in the integral over θ in Eq. (44).

The interactions V_{eP}^A and V_{eP}^B involved in Eqs. (13) determining the zeroth-order electron states in the transitional configuration represent the interactions of the electron with the polarizations $\mathbf{P}_A^*$ and $\mathbf{P}_B^*$, respectively. At long transfer distances the perturbation leading to the electron transfer has the form

$$\hat{V}_i(x^*, \mathbf{P}^*) = -\frac{z_2 e^2}{\varepsilon_s|\mathbf{x}^* - \mathbf{R}|} - \frac{\theta^* c e^2}{|\mathbf{x}^* - \mathbf{R}|} - \langle\phi_A| - \frac{z_2 e^2}{\varepsilon_s|\mathbf{x} - \mathbf{R}|} - \frac{\theta^* c e^2}{|\mathbf{x} - \mathbf{R}|}|\phi_A\rangle \tag{51}$$

The interaction leading to the reverse transition has a similar form. The first term on the right-hand side of Eq. (51) describes

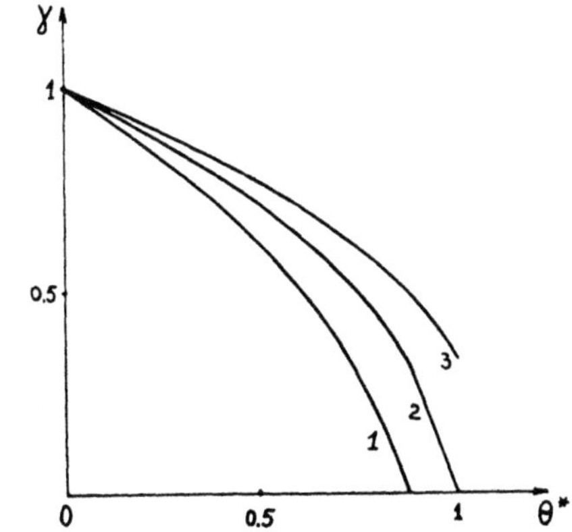

Figure 3. The dependence of the inverse radius of the localization of the electron density ($\gamma = \lambda/\lambda_0 = (r/r_0)^{-1}$) near the initial trap (donor of the electron) on the symmetry factor θ^*: (1) $P/S = -0.5$; (2) $P/S = 0$; (3) $P/S = 0.5$. $P = ze^2/\varepsilon_s$; $S = 5ce^2/16$.

the interaction of the electron with the ion in the static dielectric. The second term describes the interaction with the fluctuation of the polarization near the acceptor site corresponding to the transitional configuration.

Equations (50) and (51) show that for $0 < \theta^* < 1$ the potential well for the electron near the donor site is more shallow than that in the initial equilibrium configuration. This leads to the fact that the radius of the electron density distribution in the transitional configuration is greater than in the initial equilibrium one (Fig. 3). A similar situation exists for the electron density distribution near the acceptor site. This leads to an increased transmission coefficient as compared to that calculated in the approximation of constant electron density (ACED).

(ii) *Improved Condon Approximation*

To take into account the additional effect of diagonal dynamic disorder in the improved Condon approximation it was suggested in Ref. 16 that fluctuations of the polarization of the type

$$\mathbf{P}(\mathbf{r}) = (c/4\pi)[\mathbf{D}_A + \mathbf{D}_B + \zeta\mathbf{D}_e^{(A)}(\mathbf{r}; \mathbf{P}_{0i}) + \eta\mathbf{D}_e^{(B)}(\mathbf{r}; \mathbf{P}_{0f})] \quad (52)$$

where ζ and η are the independent variables, be considered.

The free energy surfaces of the initial and final states, $U_i(\zeta, \eta)$ and $U_f(\zeta, \eta)$ were calculated taking into account the dependence of the electron energies ε_A and ε_B on ζ and η. The latter were calculated using a direct variational method similar to that used in the polaron theory.[20]

The transitional configuration (ζ^*, η^*) was found as the solution of two coupled equations for ζ and η:

$$\begin{aligned} U_i(\zeta, \eta) &= U_f(\zeta, \eta) \\ dU_i/d\zeta &= 0 \end{aligned} \tag{53}$$

In the limit when n_A and n_B are independent of the medium polarization, ζ^* and η^* are related to each other by the equation $\zeta^* = 1 - \eta^* = 1 - \theta^*$ [see Eqs. (50)]. However, in general, they are independent quantities. Table 1 shows the results for the symmetric transition.[16] In this case we have $\zeta^* = \eta^*$.

Table 1 shows that the smaller the binding energy of the electron, determined by the quantity P/S, the stronger are the effects of the modulation of the electron density by the polarization fluctuations on the value of ζ^*, on the radius of the electron density distribution, and on the activation free energy. However, since at high binding energy values the electron is more localized, even relatively small changes in the radius of the electron density produce a greater effect on the transition probability than at small values of the binding energy. The effect of off-diagonal dynamic disorder is characterized by the quantity $n_A n_B / n_A^0 n_B^0$ and an additional effect of diagonal dynamic disorder is characterized by the quantity

Table 1
Kinetic Parameters[a] for the Symmetric Transition in the Improved Condon Approximation

				$2\lambda_p R = 4$		$2\lambda_p R = 8$	
p/S	γ	ζ^*	F_a/F_a^0	$n_A n_B/n_A^0 n_B^0$	W/W_0	$n_A n_B/n_A^0 n_B^0$	W/W_0
0	0.613	0.376	0.811	4.70	56.99	22.11	268.0
0.5	0.725	0.418	0.864	5.21	76.92	27.11	400.5
1	0.786	0.439	0.893	5.54	93.39	30.69	517.4
2	0.850	0.459	0.925	6.05	117.29	36.60	713.4
−0.1	0.547	0.359	0.795	5.11	58.34	26.09	298.0

[a] The meaning of the parameters is as follows: $W/W_0 = (n_A n_B / n_A^0 n_B^0) \exp[-(F_a - F_a^0)/kT]$, $\lambda_p = 5mce^2/16\hbar^2$, $\gamma = \lambda/\lambda_0$, $P = ze^2/\varepsilon_s$, $S = 5ce^2/16$, $c = 0.5$.

F_a/F_a^0. Table 1 shows that these quantities vary in the opposite direction with the variation of the binding energy.

(*iii*) *Beyond the Condon Approximation*

Equations (49) and (52) enable us to go beyond the Condon approximation. The deviations from the Condon approximation are due to the fact that for long-distance electron transfer, the overlapping of the electron wave functions is exponentially small (as small as $\exp(-aR)$ where R is the transfer distance) and even small changes in the behavior of the decreasing tail of the electron wave function produce large changes in the values of the electron resonance integral (i.e., of the quantity $n_A n_B$). This leads to a strong dependence of $n_A n_B$ on the polarization which must be taken into account in the calculation of the transitional configuration (ζ^*, η^*):

$$kTd\ln(n_A n_B)/d\zeta - dU_i/d\zeta = 0; \qquad U_i(\xi, \eta) = U_f(\zeta, \eta) \quad (54)$$

The results for the symmetric system are given in Table 2. A comparison of Tables 1 and 2 shows that the dependence of $n_A n_B$ on ζ and η influences the position of the transitional configuration and this effect increases with increase in the transfer distance. The physical reason for the change of the path of the transition in this case is that the system prefers to shift from the saddle point to the

Table 2
Kinetic Parameters for the Symmetric Transition Taking Account of the Modulation of the Zeroth-Order Electron Densities (Beyond the Condon Approximation)

P/S	$2\lambda_p R$	γ	ζ^*	F_a/F_a^0	$\exp\left(-\frac{\Delta F_a}{kT}\right)$	$n_A n_B/n_A^0 n_B^0$	W/W_0
0	1	0.599	0.360	0.812	9.560	1.493	14.28
	4	0.541	0.301	0.825	8.20	6.27	51.46
1	1	0.783	0.433	0.893	12.90	1.54	19.91
	4	0.774	0.414	0.895	12.31	6.10	75.10
	8	0.761	0.388	0.902	10.50	45.78	480.8
2	1	0.849	0.455	0.925	14.67	1.57	23.08
	4	0.846	0.445	0.926	14.36	6.35	91.10
	8	0.840	0.427	0.929	12.90	46.52	600.5

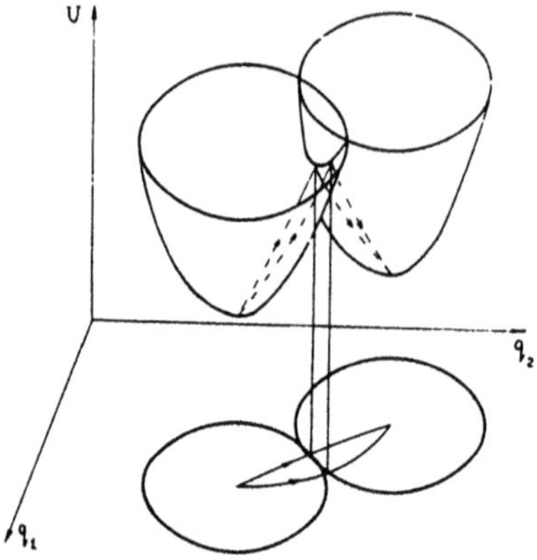

Figure 4. Path of the transition on the potential energy surfaces in the Condon approximation and beyond the Condon approximation.

point of higher free energy to increase the electron resonance integral (Fig. 4).

The configurational model was used for the calculation of the elementary act in the reactions of solvated electrons[21] and in the electrochemical generation of solvated electrons.[22] The results for the activation free energy of the process of electrochemical generation of solvated electrons as a function of the reaction free energy

Table 3
Dependence of the Kinetic Parameters for the Reaction of Electrochemical Generation of Solvated Electrons on the Free Energy of the Transition

	$P/S = 0$		$P/S = 1$	
$\Delta F/E_r$	γ	F_a/F_a^0	γ	F_a/F_a^0
−0.5	0	0	0.638	0.487
−0.4	0.368	0.295	0.678	0.629
−0.3	0.483	0.507	0.714	0.731
−0.2	0.563	0.653	0.746	0.804
−0.1	0.627	0.756	0.775	0.858
0	0.680	0.828	0.802	0.897
+0.1	0.726	0.880	0.827	0.927
+0.2	0.767	0.917	0.851	0.948
+0.3	0.804	0.944	0.873	0.965
+0.4	0.838	0.964	0.894	0.976
+0.5	0.870	0.977	0.914	0.985

are given in Table 3. Table 3 shows that the allowance for the distortion of the electron wave functions by the polarization fluctuations in ICA leads to essential differences between the activation free energy F_a and the quantity F_a^0 calculated in the approximation of constant electron density. This difference increases on going from endothermic ($\Delta F > 0$) to exothermic ($\Delta F < 0$) processes, which considerably affects the shape of the current versus potential dependence.

3. Feynman Path Integral Approach

Polarization fluctuations of a certain type were considered in the configuration model presented above. In principle, fluctuations of a more complicated form may be considered in the same way. A more general approach was suggested in Refs. 23 and 24, where Eq. (16) for the transition probability has been written in a mixed representation using the Feynman path integrals for the nuclear subsystem and the functional integrals over the electron wave functions of the initial and final states $\psi_i(x, t)$ and $\psi_f(x, t)$ for the electron:

$$
\begin{aligned}
W = {} & \frac{\beta}{i\hbar} \exp(\beta F_{i0}) \int d\theta \int dx\, dx'\, dq\, dq' \\
& \times \int D\psi_i D\psi_f Dq(t) D\bar{q}(t) A_i A_f \hat{V}_i(x, q^*) \\
& \times V_f(x', q^*) \psi_i^*(x, 0) \psi_i(x', 0) \psi_f(x, 0) \psi_f^*(x', 0) \\
& \times \exp\left(\frac{i}{\hbar} \{ S_i[\psi_i, q(t); q, q', T_i] + S_f[\psi_f, \bar{q}(t); q, q', T_f] \} \right)
\end{aligned} \tag{55}
$$

where

$$
\begin{aligned}
& T_i = T/(1-\theta); \qquad T_f = T/\theta \\
& A_\alpha^{-1} = \int D\psi_\alpha \exp\left(\frac{i}{h} S_\alpha \right), \qquad \alpha = i, f \\
& S_\alpha = \langle \psi_\alpha | \hat{S}_\alpha | \psi_\alpha \rangle \\
& \hat{S}_\alpha = \hat{S}_\alpha^e + \hat{S}_\alpha^{eq} + \hat{S}_\alpha^q
\end{aligned} \tag{56}
$$

where $\hat{S}_\alpha$ is the total action of the system in the state α ($\alpha = i, f$), $\hat{S}^e_\alpha$ and $\hat{S}^q_\alpha$ are the actions of the electron and nuclear subsystems, respectively, and $\hat{S}^{eq}_\alpha$ is the contribution to the action from the interaction of the electrons with the nuclei.

The calculation of the integrals in Eq. (55) in the classical limit in the improved Condon approximation (for the nuclear subsystem) using the saddle point method leads to two coupled equations for the electron wave functions of the donor and the acceptor in the transitional configuration:

$$\left\{H_0^i + 2\frac{ce}{8\pi}\int d^3x \frac{(\mathbf{r}-\mathbf{x})}{|\mathbf{r}-\mathbf{x}|^3}\left[\left[\int_0^{\tau_i} d\tau'\, \mathbf{D}_{eA}(\mathbf{x}, \tau')\right.\right.\right.$$
$$\left.\left.\left. + \int_0^{\tau_f} d\tau'\, \mathbf{D}^d_{eB}(\mathbf{x}, \tau') + \mathbf{D}_A(\mathbf{x}) + \mathbf{D}^d_B(\mathbf{x})\right]\right]\right\}\phi_A(\mathbf{r}, \tau)$$
$$= E_A(\tau)\phi_A(\mathbf{r}, \tau) \tag{57}$$

$$\left\{H_0^f + 2\frac{ce}{8\pi}\int d^3x \frac{(\mathbf{r}-\mathbf{x})}{|\mathbf{r}-\mathbf{x}|^3}\left[\left[\int_0^{\tau_f} d\tau'\mathbf{D}_{eB}(\mathbf{x}, \tau')\right.\right.\right.$$
$$\left.\left.\left. + \int_0^{\tau_i} d\tau'\mathbf{D}^d_{eA}(\mathbf{x}, \tau') + \mathbf{D}^d_A(\mathbf{x}) + \mathbf{D}_B(\mathbf{x})\right]\right]\right\}\phi_B(\mathbf{r}, \tau)$$
$$= E_B(\tau)\phi_B(\mathbf{r}, \tau) \tag{58}$$

where H_0^i and H_0^f are the electron Hamiltonians of the donor and the acceptor when the interaction of the electron with the medium is omitted, $\tau_i = 1 - \theta^*$, $\tau_f = \theta^*$, $\mathbf{D}_A$ and $\mathbf{D}_B$ are the electrostatic inductions due to the ions A^{z_1} and B^{z_2}, and the superscript d denotes the diagonal parts of the operators.

A direct variational method was used in Refs. 23 and 24 to go beyond the Condon approximation. Functions of the type

$$\phi_\rho = (\lambda_\rho^3/\pi)^{1/2}\exp(-\lambda_\rho|\mathbf{r}-\mathbf{R}|); \qquad \rho = i, f \tag{59}$$

were used as the probe electron wave functions. Here, the λ_ρ are variable parameters.

The calculation for the symmetric system gave[23,24]

$$\gamma = \lambda/\lambda_0 = (1 + 2P/S_0 - 1/\bar{\mu})/2(1 + P/S_0) \tag{60}$$

with $S_0 = 5ce^2/16$, $\lambda_0 = \lambda_p(1 + P/S_0)$, $\lambda_p = 5mce^2/16\hbar^2$, and $P = ze^2/\varepsilon_s$, and

$$F_a/F_a^0 = (3 + 4P/S_0 + 1/\bar{\mu}^2)/4(1 + P/S_0) \tag{61}$$

where $F_a^0 = (5/32)ce^2\lambda_0$ and $\bar{\mu} = \beta F_a^0/2R\lambda_0$.

Equation (60) shows that the quantity λ characterizing the decrease of the electron wave function in Eq. (59) decreases with the increase of the transfer distance R. The activation free energy increases with R, and the state of the electron in the transitional configuration becomes less localized. Equations (60) and (61) take into account both the dependence of the electron matrix element on the medium polarization and the distortion of the shape of the diabatic free energy surfaces U_i and U_f. For strongly bound electrons ($P/S_0 \gg 1$), the latter effect is small and the result is similar to that obtained in Ref. 25. The comparison of the results obtained by this method in ICA with the results of the configurational model showed that they do not differ greatly from each other. This means that the fluctuation of the polarization corresponding to the transitional configuration is close to that described by Eq. (52). Effects due to deviation from the Born-Oppenheimer approximation in the sub-barrier region are also possible in nonadiabatic long-distance electron transfer processes. However, such effects are more important in processes involving transfer of heavy particles, which are considered in Section V.

4. Lability Principle in Chemical Kinetics

Based on the results obtained in the investigation of the effects of modulation of the electron density by the nuclear vibrations, a lability principle in chemical kinetics and catalysis (electrocatalysis) has been formulated in Ref. 26. This principle is formulated as follows: the greater the lability of the electron, transferable atoms or atomic groups with respect to the action of external fields, local vibrations, or fluctuations of the medium polarization, the higher, as a rule, is the transition probability, all other conditions being unchanged. Note that the concept "lability" is more general than the concept "polarizability" which is often used in the discussion of the reactivity of atoms and molecules.[27,28] The change of the electron density and the shift of the atoms in the reactants may occur, in general, without any significant change in their dipole moments. For example, in the models of electron transfer considered above, almost spherically symmetric vibrations of the electron density in the donor and in the acceptor occurred in the activation process. The lability of the particles manifests itself in various physical phenomena (polarizability, the dependence of the dipole moment of the optical transition on the nuclear vibrations, anharmonicity of the intramolecular vibrations, the formation of charge transfer complexes, etc.).

The lability of the particles has two aspects. One of them is related to the interaction of the particles with the other degrees of freedom of the reacting system. The more labile the particle, the greater is the change of its state due to a change in the configuration of the particles and molecules interacting with it. As a rule, this lability essentially facilitates the transition of the system to the transitional configuration corresponding to the optimum configuration of particles participating in the elementary act of the process.

The other aspect of the lability is the fact that the more labile particles are usually less localized. The examples considered in Sections III(2) and III(3) show that weakly bound electrons whose wave function decreases more slowly with increasing distance from the center of the electron localization are more labile. In the case of heavy particles, they are more labile the smaller their force constants and hence the smaller their vibrational frequencies. Therefore, the amplitude of vibrations of more labile particles is usually greater than that of less labile particles. The smaller degree of localization of more labile particles also facilitates the transition even in the absence of their interaction with other degrees of freedom due to both the increase of the overlapping of the wave functions and the decrease of the energy (or free energy) required to reach the transitional configuration in the classical case.

The lability of the transferable electrons or atoms influences all the factors in the expression for the reaction rate constant,

$$k \simeq \exp[-u_i(R^*)/kT] \cdot \kappa \cdot \exp(-F_a/kT) \tag{62}$$

where κ is the transmission coefficient, F_a is the activation free energy for the elementary act of the reaction, and the first factor describes the probability for the reactants to be separated by the distance R^*.

For nonadiabatic reactions, the higher lability of the transferable electrons or atoms leads to the following effects:

1. Smaller values of the activation free energy, F_a, due to the distortion of the shape of the free energy surfaces.

2. Larger values of the transmission coefficient, κ, due to improved overlapping of the wave functions of quantum particles (electrons, protons, etc.).

3. A smaller repulsion between the reactants due to their mutual "polarization."

The lability principle is valid also for adiabatic reactions. For adiabatic reactions, the higher lability of the transferable particles leads to the following effects:

1. Smaller values of the activation free energy due to (i) the distortion of the shape of the free energy surfaces and (ii) the increase of the resonance splitting, ΔE, of the potential free energies for the classical subsystem due to the increased overlapping of the wave functions of the quantum particles.

2. Smaller values of the energy of the repulsion between the reactants.

Note that the lability principle is formulated first of all for transferable electrons and atoms. An increase in their lability leads as a rule to an increase in the overlapping of the wave functions. For atoms the latter means a decrease in the Franck–Condon barrier.

As for the other atomic and molecular species (both reactants and solvent) which play the role of the "effective medium" for the transition, the influence of their lability on the reaction rate is not always unambiguous. The higher lability of the "medium" particles usually leads to the increase of the Franck–Condon barrier and thus the increase of the reorganization energy. However, the repulsion of the reactants decreases at the same time. One of the manifestations of the lability of the "medium" is the dielectric polarizability of the solvent, characterized by the dielectric constant, ε_s. In the simplest case the dependence of the reorganization energy on ε_s is determined by the factor $E_s \simeq e^2(1/\varepsilon_0 - 1/\varepsilon_s)$. Thus, at large values of ε_s the reorganization energy depends rather weakly on ε_s. At the same time the repulsion energy $u_i(R^*) \simeq z_1 z_2/\varepsilon_s$ decreases with increasing ε_s and this effect may be stronger than the effect due to a weak increase of E_s (i.e., due to the increase of the Franck-Condon barrier). For processes in the abnormal region, both effects act in the same direction and the increase of the lability of the "medium" particles leads to the increase of the reaction rate constant.

For reactions in which the approach of the reactants to each other is not very important (e.g., for the transfer between two centers located at a fixed distance in a rigid structure) an increase in the lability of the medium particles leads to a decrease in the rate of the transition.

In Sections III(1) and III(2) the lability principle has been illustrated for processes involving the transfer of weakly bound electrons, including the reactions of solvated and trapped electrons and F-centers and processes of electrochemical generation of solvated electrons. In Sections IV and V, it will be illustrated also by atom transfer reactions and, in particular, by reactions involving adsorbed atoms.

5. Effect of Modulation of the Electron Density on the Inner-Sphere Activation

Since new experimental data concerning the change in structure of complex ions in the course of the electron transition have recently appeared,[29] interest in the estimation of the inner-sphere contribution to the transition probability has increased again. In Refs. 29–31 this problem was considered assuming classical behavior of the ligands in the course of the electron transfer. It was concluded[29] that the outer-sphere contribution to the activation free energy is dominant for fast reactions in the systems $Fe(phen)_3^{2+/3+}$, $Ru(bpy)_3^{2+/3+}$, and $Co(bpy)^{+/2+}$. For slow reactions (e.g., $Cr(H_2O)_6^{2+/3+}$, $Co(en)_3^{2+/3+}$) the inner-sphere contribution is dominant. The conclusion about the dominant contribution of the inner-sphere reorganization has also been made in Refs. 30 and 31. However, we should keep in mind that in electron transfer reactions between complex ions, quantum effects may take place.[32-34] Estimations show that although the quantum effects are not very large, they can amount to an order of magnitude.[32,33]

The effects of modulation of the electron density by the intramolecular vibrations on the process of inner-sphere activation and on the transmission coefficient of the nonadiabatic reaction may be illustrated using a simple model. Let us consider electron transfer from the complex ion AL to an acceptor B located at some fixed distance R in the model of a linear complex where it is assumed that the ligand L is located between atoms A and B (Fig. 5). If the energy of the unoccupied acceptor orbital of the ligand L (ε_L) lies considerably higher than the energy of the orbital occupied by the electron in the atom A (ε_A), the electron will be mainly localized on this atom. However, if the energies ε_A and ε_L depend on the interatomic distance Q in the complex AL, the intramolecular vibrations can produce redistribution of the electron density.

If the size of the complex is rather small and the intramolecular vibrations along the coordinate Q may be described in the harmonic approximation, the free energy surfaces of the initial and final states may be written in the form

$$\begin{aligned} U_i &= \tfrac{1}{2}\sum_k \hbar\omega_k(q_k - q_{k0i})^2 + \tfrac{1}{2}\hbar\Omega(Q - Q_0)^2 + \varepsilon_i(Q) + J_i^0 \\ U_f &= \tfrac{1}{2}\sum_k \hbar\omega_k(q_k - q_{k0f})^2 + \tfrac{1}{2}\hbar\Omega(Q - Q_0)^2 + J_f \end{aligned} \tag{63}$$

where Q_0 is the equilibrium length of the chemical bond A—L of the complex AL in the oxidized form, and, $\varepsilon_i(Q)$ is the energy of the electron in the complex AL, which, with certain approximations, may be written in the form

$$\begin{aligned} \varepsilon_i = \tfrac{1}{2}(&\varepsilon_A^0 + \varepsilon_L^0 - \gamma_+(Q - Q_0) \\ &- \{[\varepsilon_A^0 - \varepsilon_L^0 + \gamma_-(Q - Q_0)]^2 + 4V^2\}^{1/2}) \end{aligned} \tag{64}$$

Here, the dependence of ε_A and ε_L on Q is assumed to be linear,

$$\begin{aligned} \varepsilon_A &= \varepsilon_A^0 - \gamma_A(Q - Q_0) \\ \varepsilon_L &= \varepsilon_L^0 - \gamma_L(Q - Q_0) \end{aligned} \tag{65a}$$

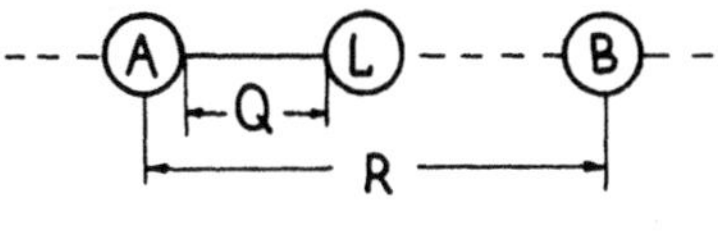

Figure 5. Linear reaction complex for electron transfer reactions in the system AL/B.

and

$$\gamma_+ = \gamma_A + \gamma_L; \qquad \gamma_- = \gamma_L - \gamma_A \tag{65b}$$

The equilibrium value of the coordinate Q for the complex in the reduced form Q_r is determined by the equation

$$dU_i/dQ = 0 \tag{66}$$

In principle, the free energy surface U_i may have two minima. We restrict ourselves to the case when there is only one minimum (Fig. 6), and $Q_r < Q_0$, i.e., the length of the chemical bond A—L in the reduced form is shorter than that in the oxidized form. Then, if $\gamma_L > \gamma_A > 0$, the free energy surface U_i has the form shown in Fig. 6. Figure 6 and Eq. (65a) show that an increase in the length of the chemical bond A—L leads to a change in the localization of the electron energy levels ε_A and ε_L. For $\gamma_L > \gamma_A$ the quantity $\varepsilon_L - \varepsilon_A$ decreases with an increase in $Q - Q_0$ and, therefore, the electron density on the ligand increases and that on the atom A decreases. This redistribution of the electron density leads to two effects:

1. The height of the potential barrier separating the initial and final states of the nuclear subsystem decreases and, hence, the Franck-Condon factor increases (Fig. 6). In the classical limit, this results in a decrease of the activation free energy.

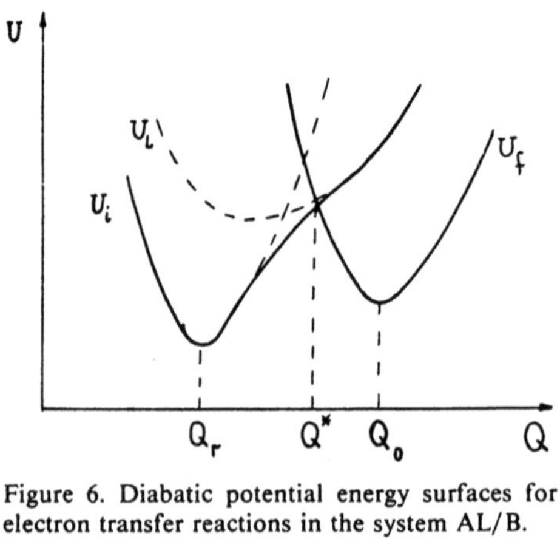

Figure 6. Diabatic potential energy surfaces for electron transfer reactions in the system AL/B.

2. The electron resonance integral V_i increases. In this model V_i has the form

$$V_i = \int \phi_i \hat{V}_i \phi_{\mathrm{B}} \, d^3x$$
$$= C_{\mathrm{A}}(Q) \int \phi_{\mathrm{A}}(x) \hat{V}_i \phi_{\mathrm{B}}(x) \, d^3x$$
$$+ C_{\mathrm{L}}(Q) \int \phi_{\mathrm{L}}(x; Q) \hat{V}_i \phi_{\mathrm{B}}(x) \, d^3x \qquad (67)$$

where $\phi_{\mathrm{A}}(x)$ and $\phi_{\mathrm{L}}(x)$ are the atomic orbitals for the electrons, and $C_{\mathrm{A}}(Q)$ and $C_{\mathrm{L}}(Q)$ are the coefficients characterizing the contribution of the atomic orbitals ϕ_{A} and ϕ_{L} to the molecular orbital $\phi_i = \phi_{\mathrm{AL}}$.

Equation (67) shows that in addition to a direct overlapping of the electron wave functions of the donor A and the acceptor B, the electron resonance integral involves a term related to the overlapping of the wave functions of the acceptor B and the ligand L. In the initial equilibrium configuration the contribution of the atomic orbital of the ligand to the wave function ϕ_{AL} is small ($C_{\mathrm{L}} \ll C_{\mathrm{A}}$); however, the overlap integral $\langle \phi_{\mathrm{A}} | \phi_{\mathrm{B}} \rangle$ is exponentially small as compared to the overlap integral $\langle \phi_{\mathrm{L}} | \phi_{\mathrm{B}} \rangle$ so that in the initial equilibrium configuration the contribution of the atomic orbital of the ligand to the resonance integral can be considerable.

When the system approaches the transitional configuration two effects take place: (1) the coefficient C_{L} increases and at $Q = Q^*$ it may be of the order of, or even greater than, the coefficient C_{A} ($C_{\mathrm{L}} \gtrsim C_{\mathrm{A}}$), and (2) the distance between the ligand L and atom B decreases with increasing Q at a fixed distance R between atoms A and B, leading to an exponential increase of the overlap integral $\langle \phi_{\mathrm{L}} | \phi_{\mathrm{B}} \rangle$.

To make the estimations we use hydrogen-like 1*s* functions for ϕ_{A}, ϕ_{L}, and ϕ_{B} and assume that the orbital exponents are approximately equal to each other, i.e., $\alpha_{\mathrm{A}} \simeq \alpha_{\mathrm{L}} \simeq \alpha_{\mathrm{B}} = \alpha$, then we obtain

$$V_i \simeq B_{\mathrm{A}} C_{\mathrm{A}}(Q^*) \exp(-\alpha R) + B_{\mathrm{L}} C_{\mathrm{L}}(Q^*) \exp[-\alpha(R - Q^*)] \quad (68)$$

where B_{A} and B_{L} are constants and the coefficients $C_{\mathrm{A}}(Q)$ and

$C_L(Q)$ are determined by the equations

$$C_A(Q) = [\tfrac{1}{2}(1 - [\varepsilon_A(Q) - \varepsilon_L(Q)]\{[\varepsilon_A(Q) - \varepsilon_L(Q)]^2 + 4V^2\}^{-1/2})]^{1/2}$$
$$C_L(Q) = [\tfrac{1}{2}(1 + [\varepsilon_A(Q) - \varepsilon_L(Q)]\{[\varepsilon_A(Q) - \varepsilon_L(Q)]^2 + 4V^2\}^{-1/2})]^{1/2} \quad (69)$$

In the classical limit, the transitional configuration q_k^*, Q^* in the Condon approximation is determined by the equations

$$q_k^* = (1 - \theta^*)q_{k0i} + \theta^* q_{k0f}$$

$$\hbar\Omega(Q - Q_0) = (1 - \theta^*)\frac{1}{2}\left[\gamma_+ + \frac{\varepsilon_A^0 - \varepsilon_L^0 + \gamma_-(Q - Q_0)}{\{[\varepsilon_A^0 - \varepsilon_L^0 + \gamma_-(Q - Q_0)]^2 + 4V^2\}^{1/2}}\right] \quad (70)$$

$$\varepsilon_i(Q) - \sum_k \hbar\omega_k(q_{k0i} - q_{k0f})[(1 - \theta^*)q_{k0i} + \theta^* q_{k0f}] = J_f - J_i^0 + \tfrac{1}{2}\sum_k \hbar\omega_k(q_{k0f}^2 - q_{k0i}^2)$$

where θ^* is the symmetry factor.

Equations (68)-(70) and Fig. 6 show that, according to the lability principle, the greater the change of the electron wave function of the donor complex AL due to intramolecular vibrations, the higher is the value of the transmission coefficient κ (which depends on the electron resonance integral V_i). The higher also is the value of the activation factor since, in the classical limit, the value of the activation free energy F_a is smaller than the quantity F_a^0 calculated neglecting the effect of the modulation of the electron wave function of the donor ϕ_A (see Fig. 6).

A simple model was considered above. A more refined theory taking into account the modulation of the electron wave function of the complex AL by fluctuations of the medium polarization is given in Ref. 35.

IV. ELEMENTARY ACT OF THE PROCESS OF PROTON TRANSFER

The development of the theory of the processes of proton transfer has taken more than 50 years and the description of earlier approaches may be found in review articles cited previously.[1-5] Some points of earlier models continue to be of interest. However, methods have been developed in recent years which enable us to take into account a number of new physical effects playing certain roles in these processes.

1. Physical Mechanism of the Elementary Act and a Basic Model

To formulate the basic model, we consider the transfer of a proton from a donor AH^{z_1+1} to an acceptor B^{z_2} in the bulk of the solution. For reactions in the condensed phase, at any fixed distance R between the reactants, the transition probability per unit time $W(R)$ may be introduced. Therefore, we will consider first the transition of the proton at a fixed distance R and then we will discuss the dependence of the transition probability on the distance between the reactants.

Unlike the simplest outer-sphere electron transfer reactions where the electrons are the only quantum subsystem and only two types of transitions are possible (adiabatic and nonadiabatic ones), the situation for proton transfer reactions is more complicated. Three types of transitions may be considered here[5]:

1. Entirely nonadiabatic transitions, in which the electrons cannot adiabatically follow the change in the positions of the proton and the medium molecules.

2. Partially adiabatic transitions, in which the electrons follow adiabatically the motion of the nuclei but the state of the proton cannot adiabatically follow the change in the state of the medium polarization.

3. Entirely adiabatic transitions, in which both the electrons and the proton adiabatically follow the change in the configuration of the medium molecules.

The transition probability for entirely nonadiabatic transitions may be calculated in the framework of the diabatic approach considering the diabatic potential or free energy surfaces[36] (Fig. 7). The resonance splitting of these surfaces, $2V_e$, is small and Eq. (16) is applicable where ρ_i and ρ_f now represent the density matrices describing the proton vibrations and the fluctuations of the medium polarization in the initial (the proton is in the donor molecule) and final (the proton is in the acceptor molecule) states.

For partially adiabatic reactions, the resonance splitting of the diabatic potential energy surfaces, $2V_e$, is large (Fig. 7) and only lower potential energy surfaces corresponding to the ground state electrons need be considered. Since the proton is a quantum particle and the probability of its tunneling from one potential well to the other is small, the zeroth-order states describing the localization of the proton in each potential well, $\phi^i_m(r)$ and $\phi^f_n(r)$, may be defined. This means that instead of the lower adiabatic potential energy surface, we consider two new diabatic surfaces, $U^p_i(r)$ and $U^p_f(r)$, and some new perturbation, $\hat{V}$, leading to the proton transfer. Since the probability of a proton transition between weakly excited vibrational states is small, Eq. (16) may be also used here. However, the meaning of the quantities involved in Eq. (16) should be changed in accordance with the new definition of the zeroth-order states and of the perturbation operator. This method of calculation for partially adiabatic reactions was developed in Ref. 5. A similar approach was subsequently used in Ref. 37.

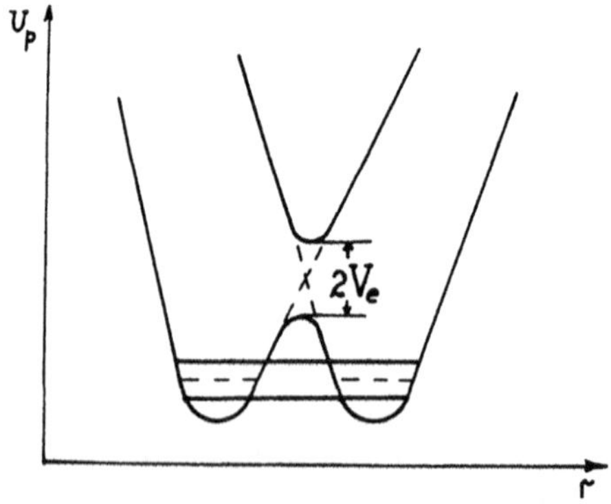

Figure 7. Potential energy surfaces for the proton at the transitional configuration for the medium molecules.

Note that since the profile of the lower adiabatic potential energy surface for the proton depends on the coordinates of the medium molecules, the zeroth-order states and the diabatic potential energy surfaces depend also on the coordinates of the medium molecules. The double adiabatic approximation is essentially used here: the electrons adiabatically follow the motion of all nuclei, while the proton zeroth-order states adiabatically follow the change of the positions of the medium molecules.

The physical mechanism of entirely nonadiabatic and partially adiabatic transitions is as follows. Due to the fluctuation of the medium polarization, the matching of the zeroth-order energies of the quantum subsystem (electrons and proton) of the initial and final states occurs. In this transitional configuration, $\{q_k^*\}$, the sub-barrier transition of the proton from the initial potential well to the final one takes place followed by the relaxation of the polarization to the final equilibrium configuration.

The activation free energy of the transition between two fixed vibrational states of the proton in each potential well is determined by the free energy of the fluctuation of the polarization to the transitional configuration corresponding to matching of proton energies. Possible activation of the chemical bond A—H by means of the excitation of the proton to various vibrational levels *is also taken into account in the theory* since calculations using equations of the type of Eq. (16) take into account transitions between *any* vibrational energy levels of the proton. It is assumed that the molecule AH is in *thermal equilibrium* with the medium so that the distribution over vibrational states of the proton is described by the Gibbs formula.

From the discussion of this basic model in the literature,[38,39] we note two points:

1. Due to strong interaction of the reactants with the medium, the influence of the latter may not be reduced only to the widening of the vibrational levels of the proton in the molecules AH and BH. The theory takes into account the Franck-Condon factor determined by the reorganization of the medium during the course of the reaction.

2. The calculations show that for high, narrow barriers the transitions between unexcited vibrational states of the proton give the main contribution to the transition probability. This result is

not explained simply by low occupation of the excited vibrational states. The reason is that the probabilities of the transitions involving the excited states, taking due account of their occupancies, make a small contribution to the total transition probability as compared to the transition probability between the unexcited states.

The height of the potential barrier decreases with the decrease of the transfer distance. Therefore, the contribution of the transitions between excited vibrational states increases and so does the transition probability. However, short-range repulsion between the reactants increases with a decrease of R, and the reaction occurs at an optimum distance R^* which is determined by the competition of these two factors. In principle, we may imagine the situation when the optimum distance R^* corresponds to the absence of a potential barrier for the proton. However, we should keep in mind that the transitions between certain excited states may become entirely adiabatic at short distances.[40,41] In this case, the further increase of the transition probability with the decrease of R becomes quite weak, and it cannot compensate for the increased repulsion between the reactants, so that even for the adiabatic transition, the optimum distance R may correspond to sub-barrier proton transfer.

2. Distance-Dependent Tunneling in the Born–Oppenheimer Approximation

First calculations of the optimum distance between the reactants, R^*, taking into account the dependence of the probability of proton transfer between the unexcited vibrational energy levels on the transfer distance have been performed in Ref. 42 assuming classical character of the reactant motion. Effects of this type were considered also in Ref. 43 in another model. It was shown that R^* depends on the temperature and this dependence leads to a distortion of the Arrhenius temperature dependence of the transition probability.

A general method for the calculation of the transition probability in the harmonic approximation developed in Ref. 44 enabled us to take into account, in a rigorous way, both the dependence of the tunneling of the quantum particles on the coordinates of other degrees of freedom of the system and the effects of the inertia and nonadiabaticity of the tunneling particle, taking into account the mixing of the normal coordinates of the system in the initial and final states. The calculation has been generalized to the case of a nonparabolic shape of the potential energy describing the relative motion of the centers of mass of the reactants, taking into account the changes during the course of the transition in the coordinates, describing the motion of the center of mass and the intramolecular vibrations.[5] For the limiting case of quantum particles (protons), the equations for the transitional configuration have been obtained for both classical and quantum motion of the reactants as a whole. It was also taken into account that the change in the probability of the tunneling of the quantum particle is due not only to the change in the distance between the centers of mass of the reacting molecules but also to a possible dependence of the equilibrium length of the chemical bond on the distance between the reactants.[5] In Refs. 45 and 46 the dependence of the probability of the tunneling on the transfer distance was taken into account in the calculation of the hydrogen isotope effect.

The dependence of the proton resonance integral J for the unexcited vibrational states on the vibrations of the crystal lattice was taken into account recently in Ref. 47 for proton transfer reactions in solids. The dependence of J on the nuclear coordinates was chosen phenomenologically as an exponential Gaussian function.

Below we will use Eq. (16), which, in certain models in the Born-Oppenheimer approximation, enables us to take into account both the dependence of the proton tunneling between fixed vibrational states on the coordinates of other nuclei and the contribution to the transition probability arising from the excited vibrational states of the proton. Taking into account that the proton is the easiest nucleus and that proton transfer reactions occur often between heavy donor and acceptor molecules we will not consider here the effects of the inertia, nonadiabaticity, and mixing of the normal coordinates. These effects will be considered in Section V in the discussion of the processes of the transfer of heavier atoms.

First, we shall consider the model where the intermolecular vibrations A—B and intramolecular vibrations of the proton in the molecules AH^{z_1+1} and BH^{z_2+1} may be described in the harmonic approximation.[48] In this case, using the Born-Oppenheimer approximation to separate the motion of the proton from the motion of the other atoms for the symmetric transition, Eq. (16) may be transformed to the form[48]

$$W = (\beta V^2/i\hbar) \int d\theta \int dQ\, dQ' \int dr\, dr'\, g(\theta) \exp[-H(\theta)] \quad (71)$$

where $g(\theta)$ is a nonexponential function of θ, and

$$\begin{aligned} H(\theta) = \tfrac{1}{4}\sum_k [&(Q_k + Q'_k - 2Q^i_{k0})^2 \tanh[\beta\hbar\omega_k(1-\theta)/2] \\ &+ (Q_k + Q'_k - 2Q^f_{k0})^2 \tanh[\beta\hbar\omega_k\theta/2] \\ &+ (Q_k - Q'_k)^2 \operatorname{cotanh}[\beta\hbar\omega_k(1-\theta)/2] \\ &+ (Q_k - Q'_k)^2 \operatorname{cotanh}[\beta\hbar\omega_k\theta/2]] \\ &+ \tfrac{1}{4}\{[r - r_{0i}(Q) + r' - r_{0i}(Q')]^2 \tanh[\beta\hbar\Omega_p(1-\theta)/2] \\ &+ [r - r_{0f}(Q) + r' - r_{0f}(Q')]^2 \tanh[\beta\hbar\Omega_p\theta/2] \\ &+ [r - r_{0i}(Q) - r' + r_{0i}(Q')]^2 \operatorname{cotanh}[\beta\hbar\Omega_p(1-\theta)/2] \\ &+ [r - r_{0f}(Q) - r' + r_{0f}(Q')]^2 \operatorname{cotanh}[\beta\hbar\Omega_p\theta/2]\} \end{aligned} \quad (72)$$

Here $Q_k = \{q_k, R\}$ is the set of dimensionless normal coordinates describing the medium polarization (q_k) and intermolecular vibrations A—B (R), and r is the proton coordinate. The vibrational frequencies are assumed to be unchanged during the course of the transition. The vibrational frequency of the proton and its initial and final equilibrium positions may depend on the coordinates Q. The dependence of Ω_p on Q can, in general, lead to a deviation from the harmonic approximation. However, for large tunneling barriers, a small change in the frequency Ω_p strongly affects the probability of tunneling of the particle. The effect of this factor on the other part of the vibrational subsystem is weaker.

The region near $Q \simeq Q'$ and $r \simeq r'$ gives the major contribution to the integrals. For the symmetric transition, the saddle point is equal to $\theta^* = \frac{1}{2}$. It is convenient to choose the origin of the coordinate system in such a way that the equality $r_{0i} = -r_{0f} = r_0$ is fulfilled. After integration over r and θ, the transition probability may be written as

$$W \simeq \hat{V}^2 \int \prod_k dq_k \, dR \, B(q_k, R) \times \exp\Big\{-\sum_k [(Q_k - Q_{k0}^i)^2 + (Q_k - Q_{k0}^f)^2] \times \tanh(\beta\hbar\omega_k/4) - 2r_0^2(Q_k)\tanh[\beta\hbar\Omega_p(Q_k)/4]\Big\} \quad (73)$$

where $B(q_k, R)$ is a slowly varying function.

Equation (73) involves various limiting cases. In some situations the integrand in Eq. (73) has a sharp maximum at a point Q_k^*. Then W may be approximately written in the form

$$W \simeq \text{const.} \cdot \hat{V}^2 \exp\Big\{-\sum_k [(q_k^* - q_{k0i})^2 + (q_k^* - q_{k0f})^2] \tanh(\beta\hbar\omega_k/4) - [(R^* - R_{0i})^2 + (R^* - R_{0f})^2] \tanh(\beta\hbar\Omega_k/4) - 2r_0^2(q_k^*, R^*)\tanh[\beta\hbar\Omega_p(q_k^*, R^*)/4]\Big\} \quad (74)$$

where $Q_n^* = \{q_k^*, R^*\}$ is determined by the set of equations

$$[(Q_n - Q_{n0i}) + (Q_n - Q_{n0f})]\tanh(\beta\hbar\omega_n/4) + \frac{\partial}{\partial Q_n}\{2r_0^2(Q_n)\tanh[\beta\hbar\Omega_p(Q_n)/4]\} = 0 \quad (75)$$

where $\omega_n = \{\omega_k, \Omega_R\}$.

If the behavior of the medium atoms and the relative motion of the reactants are classical ($\beta\hbar\omega_n \ll kT$), we have

$$W \simeq \hat{V}^2 \int \prod_k dq_k \, dR \, B(q_k, R) \times \exp\{-[U_i(q_k, R) + U_f(q_k, R)]/2kT - 2r_0^2(q_k, R)\tanh[\beta\hbar\Omega_p(q_k, R)/4]\} \times \delta(U_i - U_f) \quad (76)$$

where $U_i(q_k, R)$ and $U_f(q_k, R)$ are the free energy surfaces describing the motion of the medium atoms and the reactants in the initial and final states, respectively.

In this form, Eq. (76) is also valid in the case when Ω_p depends strongly on q_k and R and when the free energy surfaces $U_i(q_k, R)$ and $U_f(q_k, R)$ are nonparabolic. In particular, if r_0 and Ω_p are independent of q_k and the repulsion potential for the reactants is the same in the initial and final states, i.e., $U_{Ri}(R) = U_{Rf}(R) = U(R)$, we obtain from Eq. (76)

$$W \simeq \hat{V}^2 \exp(-E_s/4kT) \int dR\, \tilde{B}(R) \times \exp\{-\beta U(R) - 2r_0^2(R) \tanh[\beta\hbar\Omega_p(R)/4]\} \quad (77)$$

Equation (77) shows that if $\beta\hbar\Omega_p(R^*)/4 \gg 1$ at an optimum distance R^* between the reactants, proton transfer occurs by means of tunneling between the unexcited states. However, the distance of the proton jump, $2r_0(R^*)$, is not equal to the distance between the points of minima of the potential wells of the proton in the equilibrium nuclear configuration. This case is a generalization of the results obtained in an earlier model by Dogonadze, Kuznetsov, and Levich[36] (DKL model).

If at the optimum distance R^*, the frequency Ω_p is considerably smaller than at large distances ($\beta\hbar\Omega_p(R^*)/4 \simeq 1$), then along with the shift of the proton equilibrium position when the system goes to the transitional configuration, the contribution of transitions between the excited vibrational states of the proton increases and the proton transition may occur from the levels located near the top of the potential barrier. This case corresponds to the Kreevoy type of transition.[49] In the limit $\beta\hbar\Omega_p(R^*)/4 \ll 1$ we have the case of the overbarrier proton transition. However, the formulas of the nonadiabatic theory become inapplicable in this case and the reaction is an adiabatic one.

3. Hydrogen Ion Discharge at Metal Electrodes

The basic model presented above is applicable to hydrogen ion discharge reactions at metals. A characteristic feature of these processes is that one of the reactants is the metal electrode and the reaction rate depends exponentially on the overpotential with a constant value of the symmetry factor $\alpha = \frac{1}{2}$ in a wide region of potential variation (the Tafel law). The results of the quantum mechanical model suggested by Dogonadze, Kuznetsov, and Levich for this reaction[36] remain of interest. According to this model, the discharge of the ion H_3O^+ located at some fixed distance from the electrode occurs in the following way.[3,36] In the initial state, the proton vibrates in the H_3O^+ ion in various vibrational states according to the thermal distribution. In the initial equilibrium configuration of the medium molecules, the vibrational energy levels of the proton are not equal to those for the proton in the adsorbed state at the electrode. A classical fluctuation of the molecular surroundings leads to matching of a given pair of proton energy levels. In this configuration, a quantum (sub-barrier) proton transition from the vibrational level in the H_3O^+ ion to a corresponding vibrational level of the adsorbed state occurs. A change in the state of the electrons in the metal takes place when the proton goes under the barrier in the region of values of its coordinate near the point of intersection of the potential energy curves for the proton in the initial (proton in the H_3O^+ ion) and final (adsorbed hydrogen atom) states. This change in the electron state results in a redistribution of the electron density and the formation of a chemical bond between the proton and the metal. We may conventionally say that electron transfer from the metal to the Me—H chemical bond occurs.

Thus, for a transition between any two vibrational levels of the proton, the fluctuation of the molecular surrounding provides the activation. For each such transition, the motion along the proton coordinate is of quantum (sub-barrier) character. Possible intramolecular activation of the H—O chemical bond is taken into account in the theory by means of the summation of the probabilities of transitions between all the excited vibrational states of the proton with a weighting function corresponding to the thermal distribution.[3,36] Incorporation in the theory of the contribution of the excited states enabled us in particular to improve the agreement between the theory and experiment with respect to the independence of the symmetry factor of the potential in a wide region of $\delta\varphi$.[50] A similar approach has been used recently in Refs. 51 and 52, where the dependence of the probability of the electron tunneling to the H_3O^+ ion on the electrode potential was also taken into account.

With respect to the electrochemical reactions of hydrogen ions where the transfer of two particles (electron and proton) occurs, it is of interest to discuss the problem of whether the electron transfer and the process of breaking or formation of the chemical bond are simultaneous or sequential.[53] Since the characteristic times of the motions of light and heavy particles are rather different, we may determine along the coordinates of which particles the motion of the system occurs at a given part of the complex trajectory in the hyperspace leading from the initial equilibrium configuration to the final one. For the transition of the proton between unexcited vibrational states, this trajectory, in a crude approximation, is as follows: (1) classical motion along the coordinates of the medium molecules q_k from q_{k0i} to q_k^*, (2) sub-barrier motion of the proton at a fixed value of q_k from r_{0i} to r^*, (3) sub-barrier motion of the electron from x_{0i} to x_{0f} at fixed values of q_k and r, (4) sub-barrier motion of the proton from r^* to r_{0f}, and (5) classical motion along the medium coordinates q_k from q_k^* to q_{k0f}.

Thus, the change of the electron state ends earlier than the change of the proton state. However, since the change of the localization in space of the electron itself says nothing about the rearrangement of the chemical bonds, this does not mean that the electron transfer step occurs first and then the breaking or formation of a chemical bond takes place. The electron transfer and the rearrangement of the chemical bond may be a unified step. The character of the process depends on the potential energy surface of the system after the change of the electron state.

If prior to the electron transition the potential energy surface along the proton coordinate r had a minimum corresponding to a stable chemical bond, various situations are possible after the change of the electron state due to the electron transition:

1. The new potential energy surface has no minima. This means that the electron transfer leads to cleavage of the chemical bond. The possibility of the formation of a new chemical bond depends, in this case, on the location of the other PES (see below).

2. The new potential energy surface has a minimum corresponding to localization of the proton near another atom or molecule (or near the surface of a solid). This means that the

electron transfer results in the breaking of one chemical bond and the formation of another one, both processes proceeding in a single step.

3. The new surface has a slightly shifted minimum. In this case, the result depends on the location of the other PES. If the new PES intersects another PES corresponding to the localization of the proton near another molecule (or a solid), a dynamic or a fluctuation-relaxation transition to this PES is possible, leading to the formation of a new chemical bond. However, the rate of the whole process depends on the characteristics of the chemical bond formed only if the latter transition is the rate-determining one. If it is fast, the rate of formation of the new chemical bond will be independent of its characteristics and will be determined only by the characteristics of the original molecule.

Further development of the basic model and the detailed analysis of the dependence of the symmetry factor on the potential and the temperature[54] have shown that there are additional factors which can affect the elementary act of this reaction. These investigations led to the formulation of the charge variation model (CVM)[55] which will be discussed in the next section.

4. Charge Variation Model

In the basic model presented above, it was assumed that the hydrogen atom in the adsorbed state is neutral and weakly influences the state of the medium molecules. In this model the free energy surfaces of the solvent, determining the activation free energy of the transition between two fixed vibrational states of the proton, *m* and *n* were of parabolic shape,

$$\begin{aligned} U_{mi} &= (2\pi/c)\int \mathbf{P}^2(\mathbf{r})\,d^3r - \int \mathbf{P}(\mathbf{r})\mathbf{E}^{v}_{H_3O^+}(\mathbf{r})\,d^3r \\ U_{nf} &= (2\pi/c)\int \mathbf{P}^2(\mathbf{r})\,d^3r + \text{const.} \end{aligned} \tag{78}$$

where $\mathbf{P}(\mathbf{r})$ is the inertial polarization of the medium and $\mathbf{E}^{v}_{H_3O^+}$ is the electric field due to the H_3O^+ ion. This resulted in the linear dependence of the symmetry factor for this local transition, α_{mn}, on the free energy of the transition, ΔF_{nm},

$$\alpha_{mn} = \tfrac{1}{2}[1 + \Delta F_{nm}/E_s] \tag{79}$$

and only by allowing for transitions between various excited states of the proton was this dependence made weaker.

A model taking into account the variation of the charge of the adsorbed hydrogen in the activation–deactivation process has been suggested in Ref. 55. It may be called the charge variation model (CVM). It is known from the theory of chemisorption that the charge of an adsorbed atom depends on the position of the electron energy level in the adsorbed atom.[56] Due to the interaction of the electron with the medium, its energy level in the adatom varies. Therefore, with the fluctuations of the molecular surroundings, the charge of the adsorbed atom varies and hence so does its interaction with the medium. The latter becomes nonlinear in the coordinates of the medium species. This leads to a distortion of the shape of the free energy surfaces of the solvent at fixed vibrational states of the proton, m and n, as compared to that for the basic model and leads to a change of the dependence of the activation free energy of the local transition on ΔF_{nm}.

If we describe the state of the medium molecules by one effective configurational coordinate q, the free energy surfaces of the solvent U_{im} and U_{fn} have the form[55]

$$\begin{aligned} U_{im}(q) &= \tfrac{1}{2}\hbar\omega(q - q_{0i})^2 + J_{im} \\ U_{fn}(q) &= \tfrac{1}{2}\hbar\omega q^2 + \varepsilon_f(q) + V_{\mathrm{H}s}(q) + E_n^f + J_0^f \end{aligned} \tag{80}$$

where $V_{\mathrm{H}s}(q)$ is the energy of the interaction of the proton with the solvent, and E_n^f is the energy of the nth vibrational state of the adsorbed atom H.

The electron energy $\varepsilon_f(q)$ in the Anderson model[56] has the form[55]

$$\begin{aligned} \varepsilon_f(q) \simeq (2\Delta/\pi)\{&[(\varepsilon_a - \varepsilon_F)/\Delta]\tan^{-1}[\Delta/(\varepsilon_a - \varepsilon_F)] \\ &- 1 + \pi[(\varepsilon_a - \varepsilon_F)/\Delta]\theta(\varepsilon_F - \varepsilon_a)\} - \hat{U}n^2/4 \end{aligned} \tag{81}$$

where $\varepsilon_a(q)$ and Δ are, respectively, the energy and the width of the electron level in the adatom, ε_F is the Fermi energy in the metal, and $\hat{U}$ is the repulsion energy of the electrons in the adatom.

The number of electrons in the adatom, $n(q)$, is determined by the equation

$$n = (2/\pi)\{\tan^{-1}[(\varepsilon_F - \varepsilon_a^0 - \hat{U}n/2)/\Delta] + \pi/2\} \tag{82}$$

where ε_a^0 is the zeroth-order electron energy level in the adatom, and ε_a is related to ε_a^0 by the relationship

$$\varepsilon_a = \varepsilon_a^0 + \tfrac{1}{2}\hat{U}n(q) \tag{83}$$

From Eqs. (80) and (81) we obtain for the activation free energy F_a of the local transition $m \to n$[55]

$$F_a/E_r = \{\gamma[q_0 + \bar{U}x + b\tan(\pi x/2)] - 1\}^2 \tag{84}$$

where the excess number of electrons in the adatom, $x = n - 1$ is related to the local free energy of the transition, ΔF_{nm}, by the equation

$$2\gamma[\bar{U}x + q_0 + b(1 - x)\tan(\pi x/2) - \bar{U}x^2/2] = 1 - \Delta F_{nm}/E_r \tag{85}$$

and the following dimensionless quantities are introduced:

$$\gamma = (\hbar\omega/2E_r)^{1/2}; \qquad \bar{U} = \hat{U}/2(2E_r\hbar\omega)^{1/2}; \qquad b = \Delta/(2E_r\hbar\omega)^{1/2}$$

$$q_0 = [\varepsilon_0 - \varepsilon_F + \hat{U}/2]/(2E_r\hbar\omega)^{1/2}; \qquad E_r = (eg)^2/2\hbar\omega \tag{86}$$

where ε_0 is the electron energy in the isolated atom, and g is the coupling constant for the interaction of the electron and the proton with the medium.

The local symmetry factor α_{mn} is given by the equation

$$\begin{aligned}\alpha_{mn} = dF_a/d\Delta F_{nm} &= \{1 - \gamma[b\tan(\pi x/2) + q_0 + \bar{U}x]\}\\ &\times (1 - x - 2\tan(\pi x/2)/\pi[1 + \tan^2(\pi x/2)]\\ &\times \{1 + 2\bar{U}[1 + \tan^2(\pi x/2)]/\pi b\})^{-1}\end{aligned} \tag{87}$$

Calculations with the use of Eqs. (84)–(87) show that α_{mn} can remain equal to $\frac{1}{2}$ over a wide range of the ΔF_{nm} values for certain values of the physical parameters (Fig. 8). The results depend rather weakly on the value of $b/\bar{U} = 2\Delta/\hat{U}$ for b/J values between 0.001 and 0.1 and are quite sensitive to the values of q_0 and $\bar{U}$. If we

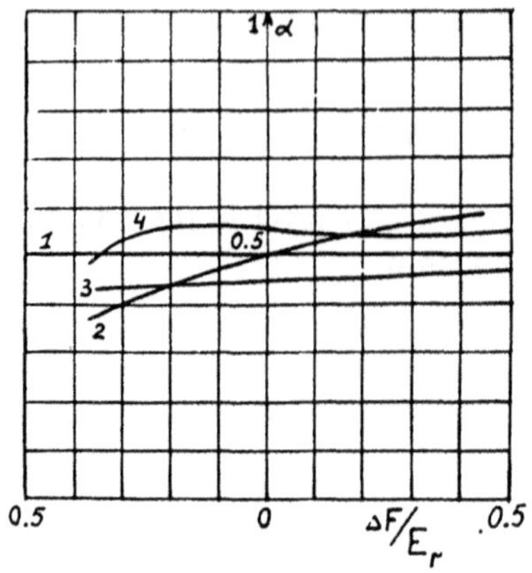

Figure 8. Dependence of the symmetry factor α on the free energy of the transition for the reaction of hydrogen ion discharge on a metal electrode.

neglect the terms in b, Eqs. (84), (85), and (87) may be simplified as follows:

$$F_a/E_r = [\gamma\bar{U}x - \tfrac{1}{2}]^2$$

$$\alpha_{mn} = (1 - 2\gamma\bar{U}x)/2(1-x) = (1 - \hat{U}x/2E_r)/2(1-x)$$

$$2\gamma(\bar{U}x - \bar{U}x^2/2) = -\Delta F_{nm}/E_r \tag{88}$$

where the value $1/(2\gamma)$ is chosen for q_0.

For small x-values, we obtain from Eq. (88)

$$\alpha_{mn} \simeq \tfrac{1}{2}[1 + (1 - 2E_r/\hat{U})\Delta F_{nm}/E_r] \tag{89}$$

Equation (89) shows that the allowance for the variation of the charge of the adsorbed atom in the activation-deactivation process in the Anderson model leads to the appearance of a new parameter $2E_r/\hat{U}$ in the theory. If $\hat{U} \simeq 2E_r$, the dependence of α_{mn} on ΔF_{nm} becomes very weak as compared to that for the basic model [see Eq. (79)]. In the first papers on chemisorption theory, a $\hat{U}$ value of ~13 eV was usually accepted for the process of hydrogen adsorption on tungsten. However, a more refined theory gave values of ≤6 eV.[57] For the adsorption of hydrogen from solution we may expect even smaller values for this quantity due to screening by the dielectric medium.

It should be noted that a weaker dependence of the symmetry factor on the free energy of the transition was obtained without considering contributions from the excited vibrational states of the proton. The latter make this dependence even weaker. The physical meaning of the result obtained is that the charge of the adsorbed atom varies in the activation-deactivation process, leading to the distortion of the shape of the free energy surfaces (FES) of the solvent. When the coordinate q varies from q_{0f} to q^*, the dependence of U_{fn} on q deviates from the parabolic shape assumed in the basic model. This deviation influences both the value of the activation free energy, which is smaller here than in the basic model, and its dependence on the free energy of the transition.

Note that the results obtained are in accordance with the lability principle. The smaller $\hat{U}$ is, the more labile are the electrons in the adatom and the stronger is the distortion of the shape of the free energy surfaces, leading to a decrease of the activation free energy and to an increase of the transition probability.

The modulation of the charge of the adsorbed atom by the vibrations of heavy particles leads to a number of additional effects. In particular, it changes the electron and vibrational wave functions and the electrostatic energy of the adatom. These effects may also influence the transition probability and its dependence on the electrode potential.

The CVM provides, to some extent, a theoretical explanation for the constancy of the symmetry factor over a wide range of potential variation. However, the problem of the sharp change of the symmetry factor α between normal and barrierless regions[58] still remains unsolved in the framework of the existing models. It was suggested in Ref. 54 that such a sharp change of α may be due to a local structural transition near the reaction zone due to a phase transition in the near-electrode layer. Transitions of this type usually occur in a narrow range of values of external parameters (pressure, temperature, electric field, etc.). Together with the CVM, these transitions may, in principle, explain the experimentally observed dependence of α on the overpotential.

The effects of charge variation may also play a certain role in other processes involving adsorbed atoms, in particular, in electron transfer processes.[59] The physical nature of these effects is to some extent similar to that of the effects of polarization of the electron plasma of the metal by vibrations in a polar medium considered in Ref. 60.

V. PROCESSES INVOLVING TRANSFER OF ATOMS AND ATOMIC GROUPS

Reactions involving transfer of atoms and atomic groups represent a more complicated theoretical problem since they are often partially or entirely adiabatic and, in addition, a number of effects which are not very important in electron transfer reactions must be considered. These effects are:

1. fluctuational preparation of the potential barrier for tunneling
2. inertia
3. nonadiabaticity of the motion
4. deviations from the Condon approximation
5. interrelation of the motions of the atoms and mixing of the normal coordinates
6. anharmonicity of the vibrations.

Furthermore, there are some effects related to the interaction of the reactants with the medium. We shall first consider the effects of the fluctuational preparation of the potential barrier in nonadiabatic reactions.

1. Fluctuational Preparation of the Barrier and Role Played by the Excited Vibrational States in the Born–Oppenheimer Approximation

The effects of transfer of atoms by tunneling may play an essential role in a number of phenomena involving the transfer of atoms and atomic groups in the condensed phase. One may expect that these effects may exist not only in the proton transfer reactions considered above but also in such processes as the diffusion of hydrogen atoms and other light ions (e.g., Li^+) in liquids, tunnel inversion and isomerization in some molecules, quantum diffusion of defects and light atoms in the electrode at cathodic incorporation of the ions, ion transfer across the liquid/solid interface, and low-temperature chemical reactions.

A model for the diffusion of light ions in structured liquids has been suggested recently by Schmidt.[61] The elementary act of diffusion is considered in this model to be the transfer of the ion from one cage formed by solvent molecules to a neighboring one.

A formalism similar to that used for partially adiabatic proton transfer reactions was applied in the calculation of the transition probability. This model of the diffusion jump is similar to the model of the diffusion of light defects in solids which was first considered in Ref. 62.

The probability of the tunneling of a heavy particle from one cage of the condensed medium to another depends on the shape of the potential barrier formed by other atoms, which, to certain degrees, can block the transfer of a given particle. This effect is similar to the modulation of the electron resonance integral by fluctuations of the medium molecules considered above (see also Ref. 63). The tunneling of the particle between unexcited vibrational states has been considered in Ref. 62, and the dependence of the resonance integral J_{00} on the coordinates of the symmetric vibrational modes of the medium, q_s, was taken into account. This was done essentially by averaging the transition probability with the quantum distribution function for the coordinates of the symmetric modes.[62] It will be shown below that this procedure is valid only in the high-temperature limit [see also Eqs. (76) and (77)]. This problem was considered also in Ref. 64, where only the coherent transitions were taken into account. Therefore, the result obtained in Ref. 64 in the high-temperature limit differs from that obtained in Ref. 62. A more general method has been developed in Ref. 65. It takes into account the deviations from the Born–Oppenheimer approximation which will be discussed later.

Below we will restrict ourselves to the Born–Oppenheimer approximation and, unlike Refs. 62, 64, and 65, we will take into account the contribution from the excited vibrational states of the tunneling particle and consider the role played by the transverse quantum vibrations of the tunneling particle itself in the preparation of the potential barrier.[48]

(i) *Role of Quantum Fluctuations of the Tunneling Particle*

First, consider the symmetric transition of a particle between unexcited vibrational states assuming classical behavior of the medium atoms which form the microstructure near the tunneling particle and determine its potential energy. The states of the system corresponding to the localization of the particle in the initial and final potential wells are characterized by different shifts of the equilibrium positions of the medium atoms. The shifts of the equilibrium positions of the medium atoms lead to shifts of the energy levels of the tunneling particle ("polaron" effect). However, in addition to the polaron effect, the tunneling at equilibrium configurations of the atoms may be inhibited due to the blocking effect of the medium atoms. In this case, the symmetric vibrational modes of the medium atoms play an essential role by leading to the shift of the blocking atoms and hence to a decrease of the barrier for tunneling. However, if a face-to-face blocking (or a blocking close to this type) takes place such that the blocking atom is located on a line connecting the equilibrium positions of tunneling particles or if the blocking atoms are not located in symmetric positions with respect to this line, the barrier may be considerably decreased due to the quantum fluctuations of the tunneling particle in a direction perpendicular to the tunneling direction. We will consider this effect for the case of face-to-face blocking.[48]

In this case, we may consider that the resonance integral V depends on $\rho = |\mathbf{q}_s - \mathbf{q}_p|$, where $\mathbf{q}_s$ is the shift of the blocking atom in the direction perpendicular to the tunneling axes, q_a, and $\mathbf{q}_p$ is the vector determining the position of the tunneling particle in the symmetry plane which is perpendicular to the axes q_a. Assuming that the motion of the particle along the coordinate q_a can be adiabatically separated from that along the coordinate q_p, in the harmonic approximation for the medium atoms we obtain for the transition probability

$$W = W_{00}\langle V^2\rangle / V_{00}^2 \tag{90}$$

where V_{00} is the resonance integral calculated at the equilibrium position of the blocking atom,

$$W_{00} = V_{00}^2(\pi/\hbar^2 kTE_r)^{1/2} \exp(-E_r/4kT) \tag{91}$$

where E_r is the reorganization energy for the antisymmetric modes q_a, and

$$\langle V^2\rangle = \int d^2q_s f(\mathbf{q}_s)\left|\int d^2q_p\, V(|\mathbf{q}_s - \mathbf{q}_p|)|\psi_0(q_p)|^2\right|^2 \tag{92}$$

where $\psi_0(q_p)$ is the wave function of the ground state for the transverse vibrations of the tunneling particle, and $f(q_s)$ is the distribution function for the symmetric modes in the classical limit.

In the harmonic approximation for the transverse vibrations of the tunneling particle, we obtain from Eq. (92)

$$\langle V^2\rangle = \frac{4}{\Delta_p^2\Delta^2}\int d\rho\, d\rho'\, \rho\rho'\, V(\rho)V(\rho')I_0(2\rho\rho'\Delta_s^2/\Delta_p^2\Delta^2)$$
$$\times \exp[-(\Delta_p^2+\Delta_s^2)(\rho^2+\rho'^2)/\Delta_p^2\Delta^2] \tag{93}$$

where I_0 is the Bessel function of imaginary argument, and

$$\Delta^2 = \Delta_p^2 + 2\Delta_s^2; \qquad \Delta_p^{-2} = m_p\Omega_p/\hbar;$$
$$\Delta_s^{-2} = \left(\frac{M\omega}{\hbar}\right)\tanh\frac{\hbar\omega}{2kT} = \frac{M\omega^2}{2kT} \qquad (\hbar\omega \ll kT) \tag{94}$$

Various limiting cases can be obtained from Eq. (93).[48] If the dependence of V on ρ is an exponential one:

$$V(\rho) = V_0\exp[-B(\rho)] \tag{95}$$

where $B(\rho) \gg 1$ and $B(\rho)$ decreases with an increase of ρ, then at $2\Delta_s^2 \gg \Delta_p^2$ we obtain[48]

$$\langle V^2\rangle \sim V^2(\rho^*)\exp[-M\omega^2\rho^{*2}/2kT] \tag{96}$$

where ρ^* is determined by the equation

$$\partial B/\partial\rho + \rho/\Delta_s^2 = 0 \tag{97}$$

Thus, in this limit the activation energy E_a in the transition probability involves, in addition to the quantity $E_r/4$, the term $M\omega^2\rho^{*2}/2$ related to the shift of the blocking atom with respect to the axis q_a. This result coincides with that obtained by Flynn and Stoneham[62] in the classical limit.

In the opposite limit ($\Delta_p/\sqrt{2} \gg \Delta_s$) we have

$$\langle V^2\rangle \sim V^2(\rho^*)\exp(-2m_p\Omega_p\rho^{*2}/\hbar) \tag{98}$$

In this case the preparation of the barrier is performed mainly by the quantum fluctuations of the tunneling particle in the transverse direction. Note that the width of the distribution here is $1/\sqrt{2}$ of that in the distribution function for the coordinates q_p. This is due to the fact that in this case the fluctuations of the particle are of quantum character and a coherent averaging of the resonance integral over the wave function of the ground state occurs. The equation for the optimum configuration q^* in this case has the form

$$\partial B/\partial \rho + 2\rho^*/\Delta_p = 0 \tag{99}$$

(ii) *Role Played by the Excited States of the Tunneling Particle and Quantum Effects for the Vibrations of the Medium Atoms*

At not very low temperatures, the excited vibrational states of the tunneling particle make some contribution to the transition probability. Furthermore, at high enough frequencies of the vibrations of the medium atoms, quantum effects may be important for the medium. In the harmonic approximation for the tunneling particle we obtain, in a similar way as in the case of proton transfer, the expression for the transition probability in the Condon approximation[48]

$$\begin{aligned} W \simeq{}& V_i^2 \exp\left[-2\sum_k (q_{k0}^a)^2 \tanh(\beta\hbar\omega_k^a/4)\right] \\ &\times \int \prod_k dq_k^s\, dq_p B(q) \exp(-\{[r - r_0(q,0)]^2 + [r + r_0(q,0)]^2\} \\ &\times \tanh[\beta\hbar\Omega(q,0)/4])\phi_s(q_k^s) \exp\left[-2\sum_p q_p^2 \tanh(\beta\hbar\Omega_p^{\perp}/4)\right] \end{aligned} \tag{100}$$

where $q = \{q_k^s, q_p\}$, and r is the tunneling coordinate. The origin of the coordinate system is chosen in such a way that $q_{k0}^{ai} = -q_{k0}^{af} = q_{k0}^a$, $q_{k0}^{si} = q_{k0}^{sf} = 0$, and $r_{0i} = -r_{0f} = r_0(q, q_k^a)$.

The value $g_k^a = 0$ in $r_0(q, q_k^a)$ and $\Omega(q, q_k^a)$ corresponds to the transitional configuration for the antisymmetric modes. The function $\phi_s(q_k^s)$ for the symmetric modes

$$\phi_s(q_k^s) = Z_s^{-1} \exp\left[-2\sum_k (q_k^s)^2 \tanh(\beta\hbar\omega_k^s/4)\right] \tag{101}$$

does not coincide, in general, with the distribution function over the coordinates q_s^k,

$$f_s(q_k^s) = Z_{s0}^{-1} \exp\left[-\sum_k (q_k^s)^2 \tanh(\beta\hbar\omega_k^s/2)\right] \tag{102}$$

The functions $\phi_s(q_k^s)$ and $f_s(q_k^s)$ coincide with each other only in the classical limit ($\hbar\omega_k^s \ll kT$).

After the integration over r, we obtain from Eq. (100)[48]

$$W \simeq \hat{V}_i^2 \exp\left[-2\sum_k (q_{k0}^a)^2 \tanh(\beta\hbar\omega_k^a/4)\right]$$

$$\times \int \prod_k dq_k^s\, dq_p\, B(q) \exp[-2r_0^2(q,0)\tanh(\beta\hbar\Omega(q,0)/4)$$

$$\left. - 2\sum_k (q_k^s)^2 \tanh(\beta\hbar\omega_k^s/4) - 2\sum_p q_p^2 \tanh(\beta\hbar\Omega_p^{\perp}/4)\right] \tag{103}$$

Thus, the allowance for the dependence of the resonance integral on q_k^s may not be reduced in general to averaging the transition probability over the distribution function in Eq. (102). The function $\phi_s(q_k^s)$ plays the role of the distribution function for the coordinates q_k^s in the case of the symmetric transition. In the classical limit, the results of Flynn and Stoneham[62] can be obtained from Eq. (103), and in the low-temperature limit, the result of Kagan and Klinger[64] can be obtained.

2. Role of Inertia Effects in the Sub-Barrier Transfer of Heavy Particles

In the calculation of the transfer by tunneling of light atomic species, it is usually assumed that the potential energy of the particle $u(r)$ depends on the distance $r - R$ between the particle and the atom to which it is bound. The dynamic interaction between the motion of the atom creating the potential $u(r - R)$ and the motion of the tunneling particle has usually been neglected. This approximation is good if the mass of the particle is considerably smaller than the mass of the atom, as is the case for proton transfer in a potential field of heavy atoms. However, if the masses of the species are comparable, effects of nonadiabaticity and inertia can take place.[65,66] In this section we shall consider the inertia effect in a model similar to that used in Ref. 66. However, we will take into account some additional effects[67] which were not included in the earlier theory.

Let us consider the transition of a particle of mass m (for brevity, we shall call it a "proton") in a two-minimum potential $u(r - R)$ formed by an atom of mass M_0 in a polar medium. The total Hamiltonian of the system taking account of the interaction, V_{pL}, of the tunneling particle with the vibrations of the medium atoms (phonons) has the form

$$H = P^2/2M_0 + M_0\omega_0^2R^2/2 + p^2/2m + u(r - R) + V_{pL} + H_L \quad (104)$$

where H_L is the Hamiltonian of the medium atoms, P and p are the momentum operators of the atom and the proton, respectively, and ω_0 is the frequency of vibrations of the atom ($\omega_0^2 = \bar{k}/M_0$, where $\bar{k}$ is the force constant).

The Hamiltonian in Eq. (104) may describe both the process of tunnel inversion or isomerization of a molecule and the inertia effects arising from the symmetric vibrations of the reaction complex AH···B in the cage of the solvent or solid matrix (Fig. 9). In the latter case, the coordinate and the frequency of the symmetric vibration correspond to R and ω_0.

Introducing the coordinates of the center of mass and of the relative motion of the proton and the atom

$$\begin{aligned} &x = r - R, \qquad X = (M_0R + rm)/M, \qquad M = M_0 + m \\ &\hat{p} = (M_0p - mP)/M, \qquad \hat{P} = P + p, \qquad \mu = mM_0/M \end{aligned} \quad (105)$$

we may transform the Hamiltonian in Eq. (104) to the form[66]

$$\begin{aligned} H = \hat{P}^2/2M + M\omega^2X^2/2 - \mu\omega_0^2xX + \hat{p}^2/2\mu + u(x) \\ + (\mu^2/2M_0)\omega_0^2x^2 - \sum_k \gamma_k q_k x + H_L \end{aligned} \quad (106)$$

with

$$\omega^2 = \omega_0^2M_0/M$$

where, for the sake of simplicity, the interaction is chosen to be linear in the relative coordinate x which corresponds to the interaction of the medium with the dipolar momentum created by the

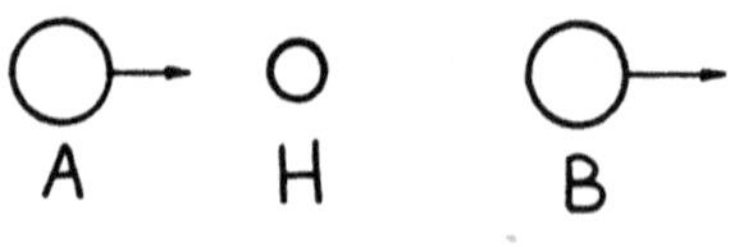

Figure 9. Linear reaction complex for the atom transfer reaction.

shift of the proton with respect to the atom. This approximation is not of fundamental importance and is chosen only for the sake of simplicity.

Equation (106) shows that the interaction of the proton with the motion of the center of mass, described by the terms proportional to μ, is formally of the same form as the interaction with the medium atoms, and the first three terms in the Hamiltonian in Eq. (106) are equivalent to addition of one more degree of freedom to the vibrational subsystem. Thus, this problem does not differ from that for the process of tunnel transfer of the particles stimulated by the vibrations which were discussed in Section IV. So we may use directly the expressions obtained previously with substitution of the appropriate parameters.

First, we will consider the symmetric transition and will assume that proton transfer occurs between unexcited vibrational states, the other part of the vibrational subsystem being described by classical mechanics. Then we obtain[67]

$$W = V^2[\pi/\hbar^2 kT(E_r^L + E_r^{\rm cm})]^{1/2} \exp[-(E_r^L + E_r^{\rm cm})/4kT] \quad (107)$$

where E_r^L is the reorganization energy of the medium, and $E_r^{\rm cm}$ is the reorganization energy of the degree of freedom describing the motion of the center of mass:

$$E_r^{\rm cm} = \tfrac{1}{2}M\omega^2(X_{0i} - X_{0f})^2 \quad (108)$$

where

$$X_{0i} = (\mu/M_0)\langle x\rangle_i; \qquad x_{0f} = (\mu/M_0)\langle x\rangle_f \quad (109)$$

Here $\langle x\rangle_i$ and $\langle x\rangle_f$ denote the mean values of the relative coordinate x over the states of the proton in the first and second potential wells, respectively. Equation (107) shows that the inertia effects lead to a decrease of the activation factor in the transition probability due to an increase of the reorganization energy. The greater the mass, m of the tunneling particle and the frequency of the vibrations of the atom, ω_0, the greater is this effect. The above result corresponds to the conclusion drawn in Ref. 66.

However, the inertia leads to one more effect which was not considered in the earlier theory. Equation (106) shows that the inertia results in the appearance of an additional term in the potential energy of the proton, which leads to a change of the proton resonance integral.[67] To estimate this effect, we will consider a model potential for $u(x)$ of the type

$$u(x) = \begin{cases} \frac{1}{2}\mu\Omega^2(x - x_0)^2, & x \geq 0 \\ \frac{1}{2}\mu\Omega^2(x + x_0)^2, & x < 0 \end{cases} \tag{110}$$

Using the wave functions of the harmonic oscillator in each potential well of the proton, we can estimate the total effect of the inertia on the transition probability in the high-temperature approximation for the medium[67]:

$$W/W_0 \simeq \exp[(2m\Omega/\hbar)(1 - M_0\Omega^3/M\tilde{\Omega}^3)x_0^2 - (\mu^2/MM_0)(M\omega_0^2x_0^2/2)/kT] \tag{111}$$

At $4kT\hbar\Omega > (\hbar\omega_0)^2$, the first term in the exponent is greater than the absolute value of the second term and the inertia effect leads to an increase of the transition probability.

If the motion of the center of mass is of quantum character (low temperatures), the inertia effect leads only to the renormalization of the resonance integral,

$$V^2 = V_0^2\exp[(2m\Omega/\hbar)(1 - M_0\Omega^3/M\tilde{\Omega}^3)x_0^2 - E_r^{cm}/\hbar\omega] \tag{112}$$

and at $(M/M_0)^{3/2}(\Omega/\omega_0) > 1$, this leads also to an increase of the transition probability.

At arbitrary temperatures, taking into account the contribution from the excited vibrational states of the tunneling particle, we have approximately,[67]

$$W = W_0\exp\left[-2(\mu\tilde{\Omega}/\hbar)\tilde{x}_0^2\tanh(\beta\hbar\tilde{\Omega}/4) + 2(m\Omega/\hbar)x_0^2\tanh\left(\frac{\beta\hbar\Omega}{4}\right) - 2(M\omega/\hbar)(\mu^2/M_0^2)x_0^2\tanh(\beta\hbar\omega/4)\right] \tag{113}$$

where

$$\tilde{\Omega}^2 = \Omega^2 + \omega_0^2\mu/M_0$$

$$\tilde{x}_0 = x_0\Omega^2/\tilde{\Omega}^2$$

$$W_0 = V_0^2\left[\pi/\hbar^2\left(\sum_k \hbar\omega_k^L E_{rk}^L/2\sinh(\beta\hbar\omega_k^L/2) + \hbar\Omega m\Omega^2 x_0^2/\sinh(\beta\hbar\Omega/2)\right)\right]^{1/2} \times \exp\left[-\sum_k (E_{rk}^L/\hbar\omega_k^L)\tanh(\beta\hbar\omega_k^L/4) - 2(m\Omega/\hbar)x_0^2\tanh(\beta\hbar\Omega/4)\right] \quad (114)$$

Thus, the inertia of the tunneling particle leads to two opposite effects: a decrease of the transition probability due to the reorganization along the coordinate of the center of mass and an increase of the transition probability due to the increase of the Franck–Condon factor of the tunneling particle. Unlike the result in Ref. 66, it is found in Ref. 67 that for ordinary relationships between the physical parameters, the inertia leads to an increase of the transition probability.

3. Nonadiabaticity Effects in Processes Involving Transfer of Atoms and Atomic Groups

The effects of deviations from the Born–Oppenheimer approximation (BOA) due to the interaction of the electron in the sub-barrier region with the local vibrations of the donor or the acceptor were considered for electron transfer processes in Ref. 68. It was shown that these effects are of importance for long-distance electron transfer since in this case the time when the electron is in the sub-barrier region may be long as compared to the period of the local vibration.[68] A similar approach has been used in Ref. 65 to treat nonadiabatic effects in the sub-barrier region in atom transfer processes. However, nonadiabatic effects in the classically attainable region may also be of importance in atom transfer processes. In the harmonic approximation, when these effects are taken into account exactly, they manifest themselves in the noncoincidence of the normal coordinates with the coordinates of individual particles.[44] However, in reactions involving the breaking of chemical bonds, the anharmonicity of the vibrations of atoms may be of importance. A model has been suggested in Ref. 69 which enables us to calculate the probability of partially nonadiabatic transitions taking account of the effects of nonadiabaticity both in the sub-barrier region and in the classically attainable one.

According to Ref. 69, we consider potential energy surfaces of the type

$$u_i = M\omega^2R^2/2 + m\Omega^2x^2/2 - V_i(x, R) + \tfrac{1}{2}\sum_k \hbar\omega_k(q_k - q_{k0i})^2 \quad (115)$$

$$u_f = M\omega^2(R - R_0)^2/2 + m\Omega^2(x - x_0)^2/2 - V_f(x, R) + \tfrac{1}{2}\sum_k \hbar\omega_k(q_k - q_{k0f})^2 \quad (116)$$

where V_i and V_f describe the interaction of the motion of the tunneling particle (x) with an atom (R), the motion of the latter being considered to be classical, and the q_k are the dimensionless coordinates of the other atoms of the reactants and the medium which do not interact directly with the motion along the coordinate x.

It is assumed that the interactions V_i and V_f are linear in x:

$$V_i(x, R) = xf_i(R) = m\Omega^2xx_i^0(R) \quad (117)$$

$$V_f(x, R) = (x - x_0)f_f(R) = m\Omega^2(x - x_0)[x_f^0(R) - x_0] \quad (118)$$

The original expression in Eq. (116) for the transition probability is transformed to the form

$$W \simeq A\int d\theta \int dx\, dR\, F_q(\theta)\rho_i(x, R, x, R; 1 - \theta)\rho_f(x, R, x, R; \theta) \quad (119)$$

where $F_q(\theta)$ is the known expression for the generating function for the vibrational subsystem $\{q\}$.

The expressions for the density matrices ρ_i and ρ_f are written through the Feynman path integrals over $x(t)$ and $R(t)$.[70] The path integral over $x(t)$ can be calculated exactly for the arbitrary form of the functions $f_i(R)$ and $f_f(R)$. The path integrals over $R(t)$ cannot, in general, be calculated exactly. However, taking into account that the major contribution to the integral over R in Eq. (119) comes from a small region in the vicinity of a point R^*, approximate expressions may be used for the functions $f_i(R)$ and $f_f(R)$:

$$f_i(R) = m\Omega^2 x_i^0(R^*) + b_i y \tag{120}$$

$$f_f(R) = m\Omega^2[x_f^0(R^*) - x_0] + b_f y \tag{121}$$

where $y = R - R^*$.

Then the actions $S_\gamma[y(t)](\gamma = i, f)$ in the path integrals are quadratic functions of $y(t)$ and the path integrals over $y(t)$ can be calculated using standard methods,[70] viz., by introducing new variables $y(t) = z(t) + s(t)$ where $z(t)$ is the trajectory minimizing the action. The insertion of the density matrices ρ_i and ρ_f calculated in this way in Eq. (119) leads to Gaussian integrals over x and R, which can be easily calculated. The remaining integral over θ is calculated using the saddle point method. The final expression for W is rather cumbersome but it takes into account, in a rigorous manner, the interaction of the tunneling particle with the motion along the other degrees of freedom (R).[69]

Below we will present the result for the symmetric transition $[\theta^* = \frac{1}{2},\ R^* = R/2,\ b_i = b_f = b > 0,\ x_i^0(R^*) = x_0 - x_f^0(R^*)]$ with due account of the effects in the lowest order in the quantity b[69]:

$$W = A_0 \exp(-F_a/kT - \sigma) \tag{122}$$

where

$$F_a = F_a^{\mathrm{B0}} + \delta E_a; \qquad \sigma = \sigma_{\mathrm{B0}} + \delta\sigma \tag{123}$$

$$F_a^{\mathrm{B0}} = F_a^0 - m\Omega^2[x_i^0(R^*)]^2/2; \qquad F_a^0 = (E_r + E_R)/4 \tag{124}$$

$$E_r = \tfrac{1}{2}\sum_k \hbar\omega_k(q_{k0i} - q_{k0f})^2; \qquad E_R = M\omega^2 R_0^2/2 \tag{125}$$

$$\begin{aligned}\sigma_{\mathrm{B0}} &= \sigma_0 + (m\Omega/2\hbar)\{[x_f^0(R^*) - x_i^0(R^*)]^2 - x_0^2\}\\ &= \sigma_0 + (m\Omega/2\hbar)\{[x_0 - 2x_i^0(R^*)]^2 - x_0^2\};\\ \sigma_0 &= m\Omega x_0^2/2\hbar\end{aligned} \tag{126}$$

The quantities F_a^0 and σ_0 describe the activation free energy and the tunneling factor in the absence of interaction between the motion of the tunneling particle and that along the coordinate R (i.e., at $b = 0$). The quantities F_a^{BO} and σ_{BO} are the activation free energy and the tunneling factor in the Born–Oppenheimer approximation (i.e., in the approximation when the motion along the x-coordinate in the initial and final states is adiabatically separated from the motion along the R-coordinate). The terms δE_a and $\delta\sigma$ give the corrections for the effects of nonadiabaticity:

$$\delta E_a = (\omega^2/\Omega^2) b \Delta x \cdot R_0/4 \tag{127}$$

$$\delta\sigma = -(\omega^2/\Omega^2) b \Delta x \cdot R_0/\hbar\Omega$$

with

$$\Delta x = x_f^0(R^*) - x_i^0(R^*) \tag{128}$$

Equations (124)–(126) show that even in the BOA the interaction between the motion of the tunneling particle and the motion along the coordinate R affects significantly the kinetic parameters, leading to a decrease of the activation free energy and of the tunneling factor σ. This effect is in accordance with the lability principle.[26] In this case, the lability of the transferable particle is determined by the value of the shift of its equilibrium position $x^0(R)$ with the change of the coordinate R. The greater this shift, the smaller is the tunneling distance for the particle in the transition configuration (i.e., at $R = R^*$) and the smaller is σ. Note that we refer here to the change of the equilibrium lengths of the chemical bonds and valence angles with the change of the nuclear configuration of the molecule rather than to the trivial decrease of the transfer distance due to reactants approaching each other. For example, the change may be in the H—C—H valence angles in ligand substitution reactions of alkyl halides, or it may be the elongation of the chemical bond of the proton with the change in configuration of the reactants in intramolecular proton transfer reactions. The lability of the transferable particle leads to a decrease of its energy by the value $m\Omega^2[x_i^0(R^*)]^2/2$. The activation energy is decreased by the same quantity.

In the process considered above, the corrections due to nonadiabaticity effects lead to an increase of F_a and to a decrease of σ. However, the first effect is greater than the second, and the joint effect of the nonadiabaticity is a decrease of the transition probability.

4. Ligand Substitutions in Alkyl Halides

Ligand substitutions in alkyl halides,

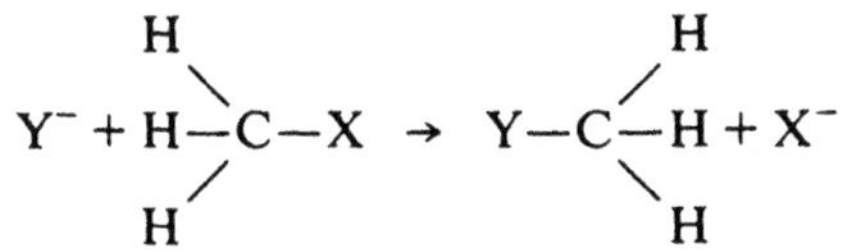

represent an example which enables us to illustrate the effects of interrelation of the motions of individual atoms and some additional effects due to interactions with the medium polarization.

These reactions were considered in Ref. 71 in the model of the linear complex, and the motion of all the atoms was described with the use of classical mechanics. However, the frequencies of the intramolecular vibrations are rather high (500–1000 cm^{-1}). Therefore, a model has been suggested[72] which enables us to take into account possible quantum effects. In this model,[72] it is assumed that the reactants are in a "cage" formed by the solvent molecules, and the interaction potentials between the atoms Y, C, and X and the nearest medium molecules are replaced by the effective ones V_{YC} and V_{CX}, depending only on the distances r_{CY} and r_{CX}, respectively. Thus, a linear complex Y···C···X is considered whose center of mass is assumed to be fixed. In the initial state the motion of the ion Y^- relative to C—X is a low-frequency one ($\omega \sim$ 40–100 cm^{-1}). In the final state, the motion of the ion X^- relative to CY is a low-frequency one. The reaction is assumed to be nonadiabatic since it leads to a considerable redistribution of the electron density. The effective charge is transferred from Y to X. The linear complex Y···C···X is characterized by the set of normal vibrations. In this model three normal vibrations are considered: one antisymmetric vibration (q_H) describing the motion of the protons in the CH_3 group, and two normal vibrations (q_1 and q_2) describing the relative motions of the atoms Y, C, and X. It is important that the dimensionless normal coordinates q_1 and q_2 in the initial and final states are different (the effect of mixing of the normal coordinates):

$$q_1^i \simeq [\omega_1^i m_C m_X/\hbar(m_C + m_X)]^{1/2} r_{CX}$$

$$q_2^i \simeq [\omega_2^i m_Y(m_C + m_X)/\hbar(m_C + m_X + m_Y)]^{1/2} \times \left(r_{YC} + \frac{m_X}{m_C + m_X} r_{CX}\right) \tag{129}$$

$$q_1^f \simeq [\omega_1^f m_X(m_C + m_Y)/\hbar(m_C + m_X + m_Y)]^{1/2} \times \left(r_{CX} + \frac{m_Y}{m_C + m_Y} r_{YC}\right) \tag{130}$$

$$q_2^f \simeq [\omega_2^f m_C m_Y/\hbar(m_C + m_Y)]^{1/2} r_{YC}$$

where ω_1^i, $\omega_2^f \gg \omega_2^i$, ω_1^f.

The normal vibrations q_1^i and q_2^f are related to the shifts of the ions Y^- and X^-. The low-frequency part of the inertial polarization of the medium, $\xi_k(\omega_k \ll \omega_1^i, \omega_2^f)$, cannot follow these shifts. The high-frequency part of the inertial polarization, $\zeta_l(\omega_l \gg \omega_1^i, \omega_2^f)$, adiabatically follows the shifts of the ions Y^- and X^-, and the equilibrium coordinates of the effective oscillators describing this part of the polarization depend on the normal coordinates of the corresponding normal vibrations, viz. $\zeta_{l0i}(q_1^i)$, $\zeta_{l0f}(q_2^f)$.

The calculation is performed using Eq. (16) and the model potential energy surfaces

$$\begin{aligned} U_i = {} & \tfrac{1}{2}\sum_k \hbar\omega_k(\xi_k - \xi_{k0i})^2 + \tfrac{1}{2}\sum_l \hbar\omega_l[\zeta_l - \zeta_{l0i}(q_1^i)]^2 \\ & + \tfrac{1}{2}\hbar\Omega(q_H - q_{H0i})^2 + \tfrac{1}{2}\hbar\omega_1^i(q_1^i - q_{10i}^i)^2 \\ & + \tfrac{1}{2}\hbar\omega_2^i(q_2^i - q_{20i}^i)^2 + J_i \\ U_f = {} & \tfrac{1}{2}\sum_k \hbar\omega_k(\xi_k - \xi_{k0f})^2 + \tfrac{1}{2}\sum_l \hbar\omega_l[\zeta_l - \zeta_{l0f}(q_2^f)]^2 \\ & + \tfrac{1}{2}\hbar\Omega(q_H - q_{H0f})^2 + \tfrac{1}{2}\hbar\omega_1^f(q_1^f - q_{10f}^f)^2 \\ & + \tfrac{1}{2}\hbar\omega_2^f(q_2^f - q_{20f}^f)^2 + J_f \end{aligned} \tag{131}$$

The resulting expression for the transition probability has the form[72]

$$W = (V_i^2/\hbar)\exp[-(\Delta q_H^0)^2/2](2\pi kT/|\partial^2 H(\theta)/\partial\theta^2|_{\theta^*})^{1/2} g(\theta^*) \times \exp[-H(\theta^*)/kT] \tag{132}$$

where $g(\theta^*)$ is a constant, θ^* is determined by the equation $\partial H(\theta)/\partial\theta = 0$, and

$$H(\theta) = \theta \cdot \Delta F + F_{\text{outer}}(\theta) + F_{\text{inner}}(\theta) \tag{133}$$

Neglecting the quantum tail of the inertial polarization of the medium, we can write the outer-sphere contribution to $H(\theta)$ in the form

$$F_{\text{outer}}(\theta) = \theta(1-\theta)E_s \tag{134}$$

where

$$E_s = \tfrac{1}{2}\sum_k \hbar\omega_k(\xi_{k0i} - \xi_{k0f})^2 + \tfrac{1}{2}\sum_l \hbar\omega_l[\zeta_{l0i}(q_1^{i*}) - \zeta_{l0f}(q_2^{f*})]^2 \tag{135}$$

The quantity Δq_{H}^0 in Eq. (132) is the shift of the equilibrium positions of the protons (at fixed transitional configurations of the other nuclei).

The contribution to $H(\theta)$ from the reorganization of the complex has the form[72]

$$F_{\text{inner}}(\theta)/kT = \phi_{11}(\Delta q_{10}^f)^2 + \phi_{22}(\Delta q_{20}^f)^2 + 2\phi_{12}\Delta q_{10}^f \Delta q_{20}^f \tag{136}$$

where

$$\begin{aligned}
\phi_{11} &= (b_1/Q)(C_1C_2^2a_1b_2x_1y_2 + C_1a_2b_2x_2y_2 + a_1a_2x_1x_2)\\
\phi_{22} &= (b_2/Q)(C_1a_1b_1x_1y_1 + C_1C_2^2a_2b_1x_2y_1 + a_1a_2x_1x_2)\\
\phi_{12} &= -(C_1C_2b_1b_2/Q)(a_1x_1 - a_2x_2)(y_1y_2)^{1/2}\\
Q &= b_1b_2y_1y_2 + C_1a_1b_1x_1y_1 + C_1C_2^2a_1b_2x_1y_2\\
&\quad + C_1C_2^2a_2b_1x_2y_1 + C_1a_2b_2x_2y_2 + a_1a_2x_1x_2
\end{aligned} \tag{137}$$

with

$$a_k = \tanh[x_k(1-\theta)]; \qquad b_k = \tanh(y_k\theta);$$
$$x_k = \hbar\omega_k^i/2kT; \qquad y_k = \hbar\omega_k^f/2kT$$
$$C_1 = m_{\text{X}}m_{\text{Y}}/(m_{\text{X}} + m_{\text{C}})(m_{\text{Y}} + m_{\text{C}});$$
$$C_2 = [(m_{\text{X}} + m_{\text{Y}} + m_{\text{C}})m_{\text{C}}/m_{\text{X}}m_{\text{Y}}]^{1/2}$$

The shifts of the normal coordinates Δq_{10}^f and Δq_{20}^f are determined from Eqs. (130) using the equilibrium values $r_{0\text{CX}}^{i,f}$ and $r_{0\text{YC}}^{i,f}$.

The expressions obtained take into account the mixing of the normal coordinates and possible quantum effects due to the reorganization of the complex and the different character of the interactions of various normal vibrations with the medium polarization. Estimations using Eqs. (132)-(135) for various reaction pairs in which F, Cl, Br, and I serve as the ligands Y and X gave satisfactory agreement with experiment except in the case of the system CH_3Cl/Br.[72]

VI. DYNAMIC AND STOCHASTIC APPROACHES TO THE DESCRIPTION OF THE PROCESSES OF CHARGE TRANSFER

It is known that the interaction of the reactants with the medium plays an important role in the processes occurring in the condensed phase. This interaction may be separated into two parts: (1) the interaction with the degrees of freedom of the medium which, together with the intramolecular degrees of freedom, represent the reactive modes of the system, and (2) the interaction between the reactive and nonreactive modes. The latter play the role of the thermal bath. The interaction with the thermal bath leads to the relaxation of the energy in the reaction system. Furthermore, as a result of this interaction, the motion along the reactive modes is a complicated function of time and, on average, has stochastic character.

Recently, much attention has been paid to the investigation of the role of this interaction in relation to the calculations for adiabatic reactions. For steady-state nonadiabatic reactions where the initial thermal equilibrium is not disturbed by the reaction, the coupling constants describing the interaction with the thermal bath do not enter explicitly into the expressions for the transition probabilities. The role of the thermal bath in this case is reduced to that the activation factor is determined by the free energy in the transitional configuration, and for the calculation of the transition probabilities, it is sufficient to know the free energy surfaces of the system as functions of the coordinates of the reactive modes.

Two different approaches are used at present in the theory of the processes of charge transfer in polar media. One of them is based on the dynamic description of the polar medium and intramolecular vibrations. The other one originates from the works of Kramers[73] and is based on the stochastic description of the medium and some local vibrations.

1. Dynamic and Fluctuational Subsystems

The problem of the relaxation of the energy in various subsystems is of great importance in the discussion of the kinetics of a transition. Strictly speaking, motion with constant energy is possible only in the isolated system. The relaxation of the energy occurs in any subsystem which is a part of a large system. It is clear qualitatively that if the characteristic time of the energy relaxation τ_E of a subsystem is long as compared to the characteristic time of the transition τ_e, such a subsystem may be considered as a dynamic one. For systems for which the opposite inequality is valid ($\tau_E \ll \tau_e$), the exchange of energy between both parts of the system will take place, and the motion in such a subsystem will be, to some extent, a fluctuational one.

In the condensed phase for the system in which the interaction with the nonreactive modes is of short-range character, τ_E increases with an increase in the size of the subsystem. However, the larger the subsystem, the more difficult it is to describe it in the framework of the dynamical approach. Thus, the separation of the whole system into the dynamic and fluctuational subsystems is not always unambiguous and depends on the possibilities of dynamic or stochastic descriptions of various subsystems. For example, at low temperatures, where the harmonic approximation is a good one, the whole subsystem may be described, in principle, in a dynamic way. At high temperatures, high-frequency (quantum) and some low-frequency intramolecular vibrations with long relaxation times may also be described in this way. The behavior of individual degrees of freedom describing the collective state of the solvent is of a stochastic (fluctuational) character.

In the stochastic approach, the Markovian random process is usually used for the description of the solvent, and it is assumed that the velocity relaxation is much faster than the coordinate relaxation.[74] Such a description is applicable at long time intervals which considerably exceed the characteristic times of the electron motion. In general, it is inapplicable in the nonadiabaticity regions where considerable rearrangement of the electron wave function occurs.

The pair correlation function of the velocities and the pair correlation functions of some time derivatives of the velocity are sometimes taken into account.[75] However, the validity of this description in the nonadiabaticity regions also has to be proved. The dynamic description or the description using the differentiable random process is more rigorous in this region.[76]

2. Transition Probability and Master Equations

In general, the equations for the density operator should be solved to describe the kinetics of the process. However, if the nondiagonal matrix elements of the density operator (with respect to electron states) do not play an essential role (or if they may be expressed through the diagonal matrix elements), the problem is reduced to the solution of the master equations for the diagonal matrix elements. Equations of two types may be considered. One of them is the equation for the reduced density matrix which is obtained after the calculation of the trace over the states of the nuclear subsystem. We will consider the other type of equation, which describes the change with time of the densities of the probability to find the system in a given electron state as a function of the coordinates of heavy particles $P_i(R, q, Q, s, \ldots)$ and $P_f(R, q, Q, s, \ldots)$.[74,77-80]

Let us assume that all the nuclear subsystems may be separated into several subsystems $(R, q, Q, s, \ldots)$ characterized by different times of motion, for example, low-frequency vibrations of the polarization or the density of the medium (q), intramolecular vibrations, etc. Let (r) be the fastest classical subsystem, for which the concept of the transition probability per unit time $W_{if}(q, Q, s)$ at fixed values of the coordinates of slower subsystems (q, Q, s) may be introduced.

Then the master equations for the densities of the probabilities to find the system in the state i $[P_i(q, Q, s)]$ or in the state f $[P_f(q, Q, s)]$ at given values of the coordinates q, Q, s have the form

$$\begin{aligned} \partial P_i/\partial t &= \hat{L}_i P_i - W_{if} P_i + W_{fi} P_f \\ \partial P_f/\partial t &= \hat{L}_f P_f - W_{fi} P_f + W_{if} P_i \end{aligned} \tag{138}$$

where $\hat{L}_i$ and $\hat{L}_f$ are the operators describing the evolution of the subsystems (q, Q, s) in the initial and final electron states.

We may also introduce the transition probability per unit time at fixed values of the coordinates of slower subsystems, $\bar{W}_{if}(q, Q)$ and $\bar{W}_{fi}(q, Q)$, and consider the master equations for the corresponding probability densities $R_i(q, Q)$ and $R_f(q, Q)$, etc.

The form of the operators of evolution involved in these equations depends on the way in which they are described. The solution of the master equations enables us, in principle, to find the average rate of transition for both small and large values of the transition probabilities W_{if}, $\bar{W}_{if}$ and W_{fi}, $\bar{W}_{fi}$.

3. Frequency Factor in the Transition Probability

The value of the frequency factor (the pre-exponential factor), A, in the expression for the average transition probability

$$w = A \exp(-E_a/kT) \tag{139}$$

is usually of interest when one considers the problem of the role of relaxation and friction in the kinetics of the transition.

For entirely nonadiabatic transitions, the transition probabilities are so small that the reaction does not disturb the equilibrium distribution in the nuclear subsystems (q, Q, s), and the calculation of the mean transition probability is reduced to averaging the corresponding local transition probability over the equilibrium distribution of the coordinates q, Q, s:

$$W_{\text{n.ad.}} = \langle W_{if}^{\text{n.ad.}}(q, Q, s)\rangle_{q,Q,s} \tag{140}$$

In this case, A is independent of the frequency characteristics of the nuclear subsystems.

As for the adiabatic transitions, various situations are possible when the system involves several subsystems (e.g., r, s, and q).

(i) *Adiabatic transition in the subsystem* (r). If at a fixed value of the coordinates q and s, the transition is the adiabatic one, the frequency factor in the transition probability $W_{if}(q, s)$ has the form[3]

$$A_{\text{ad.}}^r = \omega_{\text{eff}}^r/2\pi \tag{141}$$

where ω_{eff}^r is determined by the dynamical behavior of subsystem (r) and depends on the profiles of the corresponding potential energy surfaces. In particular, in the model of harmonic vibrations, ω_{eff}^{r} has the form[3]

$$(\omega_{\text{eff}}^{r})^2 = \sum_k \omega_k^2 E_{rk}^{(r)} / \sum_k E_{rk}^{(r)} \tag{142}$$

where $E_{rk}^{(r)}$ is the energy of the reorganization of the kth degree of freedom in subsystem (r).

The frequency factor A^s for the transition probability $\bar{W}_{if}(q)$ in subsystem (s) depends on the values of the probabilities $W_{if}(q, s)$ and $W_{fi}(q, s)$. If the latter are sufficiently small, then we have

$$\bar{W}_{if}(q) = \langle W_{if}(q, s) \rangle_s \tag{143}$$

and in this case, the frequency factor A^s is still determined mainly by the frequency characteristics of subsystem (r).

If the probabilities $W_{if}(q, s)$ and $W_{fi}(q, s)$ are large, the frequency factor A^s depends on the frequency characteristics of subsystem (s)

$$A^s \sim \omega_{\text{eff}}^{r,s} / 2\pi \tag{144}$$

The expression for the frequency factor A in the mean transition probability depends on the values of $\bar{W}_{if}(q)$ and $\bar{W}_{fi}(q)$. If they are sufficiently small so that the conditions

$$\bar{W}_{if(fi)}(q^*) < \tau_q^{-1}$$

are fulfilled[78] [where q^* is the value of the coordinate at which the transition is most effective, and τ_q is the relaxation time in the subsystem (q)], the total transition probability may be obtained by averaging

$$w = \langle \bar{W}_{if}(q) \rangle_q \tag{145}$$

If $\bar{W}_{if}(q)$ is described by Eq. (143), the frequency factor A is still determined by the frequency characteristics of the fastest subsystem (r). If Eq. (144) holds, A depends on the frequency characteristics of the subsystems (r, s). Finally, if the probabilities $\bar{W}_{if}(q)$ and $\bar{W}_{fi}(q)$ are sufficiently large, the frequency factor A depends on the frequency (or relaxational) characteristics of subsystem (q).

(ii) *Nonadiabatic transition in subsystem* (r), *adiabatic transition in subsystem* (s). In this case, the result is independent of the frequency characteristics of subsystem (r). The frequency factor in the probability $\bar{W}_{if}(q)$ is determined by the frequency characteristics of subsystem (s), and the frequency factor in the mean transition probability depends on the values of the probabilities $\bar{W}_{if}(q)$ and $\bar{W}_{fi}(q)$ (see above).

(iii) *Nonadiabatic transition in the subsystems* (r, s). In this case, the frequency factor in the total rate of the adiabatic transition, $A_{\text{ad.}}$, is determined by the frequency (or relaxational) characteristics of subsystem (q).

Thus, if the system involves several nuclear subsystems, the frequency factor is not necessarily determined by the relaxational characteristics of the slowest subsystem. Under some conditions, it may be determined by the frequency characteristics of faster subsystems (including the dynamic ones).

4. Effect of Relaxation on the Probability of the Adiabatic Transition: A Dynamic Approach in the Classical Limit

A dynamic description of the effect of relaxation on the probability of the adiabatic transition may be performed using various methods, e.g., a Feynman path integral approach similar to that presented in Section III (see also Refs. 81-84). Here we shall present the results for a simple model obtained by another method.[85]

Let us consider the one-dimensional motion of a classical particle (q-oscillator) in a potential well of the type

$$U(q) = \begin{cases} \frac{1}{2}\hbar\omega q^2 & q < q_0/2 \\ \frac{1}{2}\hbar\omega(q - q_0)^2 & q_0 > q > q_0/2, \\ 0 & q > q_0 \end{cases}$$

taking account of its interaction with the set of classical oscillators (Q-oscillators) modeling the thermal bath.

This model enables us to investigate the character of the motion of the system in the course of the transition in various limits and to analyze under what conditions the stochastic description is applicable.

In the second quantization representation, the Hamiltonian H_L describing the motion of the reactive q-oscillator in the left potential well has the form

$$H_L = \hbar\omega a^+ a + \sum_m \hbar\Omega_m b_m^+ b_m + \hbar \sum_m \lambda_m (a^+ b_m + b_m^+ a) \tag{146}$$

where a^+ and a are the creation and annihilation operators for the reactive q-oscillator, b_m^+ and b_m are the creation and annihilation operators for the oscillators of the thermal bath (Q-oscillators), and the last term describes the interaction V_{qQ} of the reactive oscillator with the oscillators of the thermal bath.

We introduce the eigenstates and eigenvalues for the creation and annihilation operators (coherent states[86]):

$$a|\alpha\rangle = \alpha|\alpha\rangle; \qquad \langle\alpha|a^+ = \alpha^*\langle\alpha|$$
$$b_m|\beta_m\rangle = \beta_m|\beta_m\rangle; \qquad \langle\beta_m|b^+ = \beta_m^*\langle\beta_m| \tag{147}$$

The real and imaginary parts of the eigenvalues $\alpha(t)$ and $\beta(t)$ are related to the expected values of the coordinates and momenta of the corresponding oscillators. In particular,

$$\text{Re}\,\alpha(t) = q(t)/\sqrt{2} \tag{148}$$

It is known that if the system described by the Hamiltonian in Eq. (146) has been in a coherent state $\alpha(0) = \alpha$ and $\beta_m(0) = \beta_m$ at the initial moment of time, then its state will also be a coherent one at subsequent moments of time, and the dependence of the eigenvalues on time will be determined by the coupled equations

$$d\alpha(t)/dt = -i\omega\alpha(t) - i\sum_m \lambda_m\beta_m(t)$$
$$d\beta_m(t)/dt = -i\Omega_m\beta_m(t) - i\lambda_m\alpha(t) \tag{149}$$

with the initial conditions

$$\alpha(0) = \alpha; \qquad \beta_m(0) = \beta_m \tag{150}$$

We will find the probability $P(t)$ for the system to pass the point $q^* = q_0/2$ up to the moment of time t. This probability gives the upper estimate for the transition probability since, in principle, there are trajectories for which the system goes back to the left potential well after crossing the top of the potential barrier. However, if the contribution of these trajectories is small, as is the case for not too strong an interaction with the thermal bath at large narrow barriers, $P(t)$ is close to the exact value of the transition probability.

The probability $P(t)$ may be found by integration of the distribution function $\phi(\alpha, \{\beta_m\})$ over all possible initial values α

and β_m provided the condition

$$\operatorname{Re}\alpha(t) \geq q^*/\sqrt{2} \tag{151}$$

is fulfilled. Thus, we have for $P(t)$[85]

$$P(t) = \int \phi(\alpha_1\{\beta_m\}) \prod_m d^2\alpha d^2\beta_m \tag{152}$$

where the integration is performed over the real and imaginary parts of the initial eigenvalues α and β_m, which corresponds to integration over all possible initial values of the coordinates and momenta of the reactive q-oscillator and the oscillators of the thermal bath (Q-oscillators).

The transition probability per unit time is determined as

$$w = dP(t)/dt \tag{153}$$

To calculate the probability $P(t)$, we must know the form of the distribution function $\phi(\alpha, \{\beta_m\})$.

(i) The Transition from the Equilibrium State

First, we shall consider the case when, at the initial moment of time, the distribution of the states in the thermal bath and for the reactive oscillator is an equilibrium one, i.e.,

$$\phi(\alpha, \{\beta_m\}) = Z_{q,Q}^{-1} \prod_m \exp\{-[\hbar\Omega_m|\beta_m|^2 + \hbar\omega|\alpha|^2 + \hbar\lambda_m(\alpha^*\beta_m + \beta_m^*\alpha)]/kT\} \tag{154}$$

where $Z_{q,Q}$ is the partition function of the whole system.

In this case, to calculate the integral in Eq. (152), we have to take into account that in addition to the condition given by Eq. (151), there is also a restrictive condition for the initial eigenvalues.

$$\operatorname{Re}\alpha \leq q^*/\sqrt{2} \tag{155}$$

which means that at time $t = 0$, the particle is located in the left potential well $\langle L\rangle$. The exact solution for $\alpha(t)$ in the region $\langle L\rangle$ may be written formally for an arbitrary type of frequency spectrum of the thermal bath. We will accept the frequently used approximation in which the spectrum of the oscillators of the thermal bath near the frequency ω is dense enough so that the solution for $\alpha(t)$ has the form

$$\alpha(t) \simeq \alpha \exp(-i\bar{\omega}t) + \sum_m \lambda_m \beta_m [\exp(-i\bar{\omega}t) - \exp(-i\Omega_m t)]/(\bar{\omega} - \Omega_m) \tag{156}$$

where

$$\bar{\omega} = \omega + \delta\omega - i\Gamma \tag{157}$$

The damping constant Γ and the frequency shift $\delta\omega$ are expressed through the coupling constants λ_m for the interaction of the oscillator with the thermal bath and through the frequency characteristics of the latter.[86] The frequency shift will be neglected in what follows for the sake of simplicity.

At short times ($t < \omega^{-1}$), the condition in Eq. (151) takes the form

$$\text{Re}\,\alpha + \omega t\, \text{Im}\,\alpha + t \sum_m \lambda_m\, \text{Im}\,\beta_m \geq q^*/\sqrt{2} \tag{158}$$

It may be easily seen that the major contribution to the integral over $\text{Re}\,\alpha$ and $\text{Im}\,\alpha$ comes from a small region near the point ($\text{Re}\,\alpha = q^*/\sqrt{2}$; $\text{Im}\,\alpha = 0$). Therefore, the term V_{qQ} in the exponent in Eq. (154) describing the interaction of the reactive oscillator with the thermal bath may be considered to be approximately equal, in this region, to $V_{qQ}(\text{Re}\,\alpha, \text{Im}\,\alpha, \beta_m) \simeq V_{qQ}(q^*/\sqrt{2}, 0, \beta_m)$. Then, calculating the integral in Eq. (152) using the distribution function in Eq. (154), we obtain

$$P(t) = Z_{q,Q}^{-1} \prod_m \int d^2\gamma_m (\omega_{\text{eff}} tkT/2\hbar\omega) \times \exp\{-[\hbar\omega q^{*2}/2 + \hbar\Omega_m|\gamma_m|^2 + V_{q,Q}(q^*/\sqrt{2}, 0, \gamma_m)]/kT\} \tag{159}$$

Taking into account that the major contribution to the integrals over $\text{Re}\,\alpha$ and $\text{Im}\,\alpha$ in $Z_{q,Q}$ comes from a small region near the point $\text{Re}\,\alpha = \text{Im}\,\alpha = 0$, $Z_{q,Q}$ may be written in the form

$$Z_{q,Q} = \pi(kT/\hbar\omega) \exp[-F_i(0)/kT] \tag{160}$$

where $F_i(0) = F_i(\text{Re}\,\alpha = 0; \text{Im}\,\alpha = 0)$ is the free energy of the system in the initial equilibrium configuration for the reactive oscillator ($q = 0, dq/dt = 0$).

With the aid of Eq. (160), the transition probability $P(t)$ may be finally written in the form

$$P(t) = (\omega_{\text{eff}} t/2\pi) \exp(-F_a/kT) = wt \tag{161}$$

where

$$\begin{aligned} F_a &= F(\text{Re}\,\alpha = q^*/\sqrt{2}; \text{Im}\,\alpha = 0) - F_i(0) \\ \omega_{\text{eff}}^2 &= \omega^2 + \sum_m \lambda_m^2 \omega/\Omega_m \end{aligned} \tag{162}$$

The quantity $F(\text{Re}\,\alpha = q^*/\sqrt{2}; \text{Im}\,\alpha = 0)$ is the free energy of the system in the transitional configuration, i.e., at the value of the coordinate of the reactive oscillator $q^* = q_0/2$.

Thus, in this limit, $P(t)$ increases linearly with t and the concept of the transition probability per unit time w may be introduced. The calculation for long times leads to w decreasing with an increase of t.

(ii) The Transition from the Nonequilibrium State

The above method enables us to calculate the transition probability at various initial nonequilibrium conditions. As an example, we will consider the transition from the state in which the initial values of the coordinate and velocity of the reactive oscillator are equal to zero.[85] In this case, the normalized distribution function has the form

$$\phi(\alpha, \{\beta_m\}) = \delta^{(2)}(\alpha) \prod_m \exp(-|\beta_m|^2/\langle n_m \rangle)/\pi\langle n_m \rangle \tag{163}$$

where $\delta^{(2)}(\alpha) = \delta(\text{Re}\,\alpha)\delta(\text{Im}\,\alpha)$, and $\langle n_m \rangle$ and $\langle n \rangle$ (which appears below in eq. (169)) are the average occupation numbers for the oscillators:

$$\begin{aligned} \langle n \rangle &= [\exp(\hbar\omega/kT) - 1]^{-1} \simeq kT/\hbar\omega \\ \langle n_m \rangle &= [\exp(\hbar\Omega_m/kT) - 1]^{-1} \simeq kT/\hbar\Omega_m. \end{aligned} \tag{164}$$

For $P(q, t)$ we obtain from Eq. (152)[85]

$$P(q, t) = \int \prod_m (d^2\gamma_m/\pi) \exp(-|\gamma_m|^2) \tag{165}$$

where $\gamma_m = \beta_m/\sqrt{\langle n_m \rangle}$.

For $\alpha(t)$ in the region of the left potential well, $\langle L \rangle$, in this case we have

$$\alpha(t) = \sum_m (\langle\langle n_m \rangle\rangle)^{1/2} \gamma_m v_m(t) = \sum_m \lambda_m (\langle\langle n_m \rangle\rangle)^{1/2} \gamma_m \times [\exp(-i\omega t - \Gamma t) - \exp(-i\Omega_m t)]/(\omega - \Omega_m - i\Gamma) \quad (166)$$

Integration in Eq. (165) is restricted by the condition of the type in Eq. (151). The rotation of the coordinate system and subsequent integration transform the expression for $P(q, t)$ to the form

$$p(q, t) = \int_{1/a(t)}^{\infty} \exp(-s^2)\, ds/\sqrt{\pi} \quad (167)$$

where

$$a(t) = (\sqrt{2}/q)\left\{\sum_m \langle n_m \rangle |v_m(t)|^2\right\}^{1/2} \quad (168)$$

The last expression may be transformed to the form[86]

$$a(t) \simeq (\sqrt{2}/q)\{\langle n \rangle [1 - \exp(-2\Gamma t)]\}^{1/2}$$

Differentiating Eq. (167) with respect to q and t we obtain

$$\partial P/\partial t = D[\partial^2 P/\partial q^2 + (\hbar\omega/kT) qP] \quad (169)$$

where

$$D = \Gamma kT/\hbar\omega \quad (170)$$

The solution of this equation gives the transition probability

$$\omega = (2\tau)^{-1}(E_a/4\pi kT)^{1/2} \exp(-E_a/kT) \quad (171)$$

where

$$\tau^{-1} = 2\Gamma; \qquad E_a = \hbar\omega(q^*)^2/2 \quad (172)$$

Thus unlike the previous case where the transition probability per unit time exists at some small time and is determined by the frequency characteristics of the reactive oscillator, here the concept of the transition probability per unit time exists only at some sufficiently long time. Note two more differences between the formulas (161)-(162) and (171)-(172). In the first case the frequency factor ω_{eff} in the transition probability (i.e., preexponential factor) is determined mainly by the frequency of the reactive oscillator ω. In the second case it depends on the inverse relaxation time $\tau^{-1} = 2\Gamma$ determined by the interaction of the reactive oscillator with the thermal bath.

The activation factor in the first case is determined by the free energy of the system in the transitional configuration F_a, whereas in the second case it involves the energy of the reactive oscillator $U(q^*) = (1/2)\hbar\omega q^{*2}$ in the transitional configuration. The contrast due to the fact that in the first case the transition probability is determined by the equilibrium probability of finding the system in the transitional configuration, whereas in the second case the process is essentially a nonequilibrium one, and a Newtonian motion of the reactive oscillator in the field of external random forces in the potential $U(q)$ from the point $q = 0$ to the point q^* takes place. The result in Eqs. (171) and (172) corresponds to that obtained from Kramers' theory[73] in the case of small friction ($\Gamma \to 0$) but differs from the latter in the initial conditions.

5. Stochastic Equations

In recent times, a great deal of interest has been devoted to the description of the reactions in condensed media using a stochastic approach such as that of Kramers.[73] As has been noted by Frauenfelder and Wolynes,[87] publications in this area are voluminous and it is impossible even to give a complete list of references. The papers cited in the reference list[74,75,77,78,81,82,88-101] are only a small part of what exists at present in the literature. Additional references may also be found in Ref. 87 and in papers cited therein. Langevin equations for the reactive modes or Focker–Planck equations for the distribution functions for the coordinates and velocities of the reactive modes are usually the starting points when the classical description of the nuclear motion is used. An approximation is often used in which it is assumed that the relaxation of the velocities occurs much faster than that of the coordinates. In this case, the equations for the distribution functions are reduced to equations of the diffusion type. For electron transfer reactions, these equations have the form[77,78]

$$\frac{\partial P_i}{\partial t} = \frac{\Delta^2}{\tau_q}\left[\frac{\partial^2 P_i}{\partial q^2} + \frac{1}{kT}\frac{\partial}{\partial q}\left(\frac{\partial U_i}{\partial q}P_i\right)\right] - \bar{\omega}_{if}P_i + \bar{\omega}_{fi}P_f$$

$$\frac{\partial P_f}{\partial t} = \frac{\Delta^2}{\tau_q}\left[\frac{\partial^2 P_f}{\partial q^2} + \frac{1}{kT}\frac{\partial}{\partial q}\left(\frac{\partial U_f}{\partial q}P_f\right)\right] - \bar{\omega}_{fi}P_f + \bar{\omega}_{if}P_i \qquad (173)$$

where P_i and P_f are the densities of the probabilities to find the system in the initial and final electron states, respectively, at a given value of the relaxational coordinate q, $\bar{\omega}_{if}$ and $\bar{\omega}_{fi}$ are the transition probabilities per unit time between these electron states at a given value of q, τ_q is the relaxation time, $\Delta^2 = K(0)$, with $K(t) = \langle q(t)q(0)\rangle_i = \Delta^2 \exp(-|t|/\tau_q)$ the correlation function of the random process $q(t)$, and $U_i(q)$ and $U_f(q)$ are the initial and final potential energies, respectively, as functions of the reactive coordinate q.

Using equations of this type, the expressions for the average transition probability at an arbitrary value of the electron resonance integral V were obtained.[77] For the symmetric transition, W_{if} has the form

$$W_{if} = \frac{1}{\hbar}\frac{2\pi V^2 \exp(-E_a/kT)}{(1 + 2\pi V^2\tau_q/2\hbar E_a)(4\pi E_r kT)^{1/2}} \tag{174}$$

For the nonadiabatic transition,

$$2(2\pi V^2\tau_q/\hbar E_a) \ll 1 \tag{175}$$

Eq. (174) gives the well-known expression for the transition probability [see Eqs. (9) and (10)]. If the condition opposite to Eq. (175) holds, the transition probability for the adiabatic process takes the form

$$W_{if}^{\text{ad.}} = \frac{1}{2\tau_q^{-1}}(\pi kT/E_a)^{-1/2} \exp(-E_a/kT) \tag{176}$$

Thus, in this case the pre-exponential factor (the frequency factor) in the transition probability depends on the relaxation time τ_q. Various refinements of this simple model were made in many papers by considering the change of the intramolecular structure of the reactants,[79,80] cases of several relaxation times,[88] etc.

An approach of this type has an advantage in that it is based only on rather general characteristics of the random process describing the motion along the reaction coordinate. However, it should be noted that equations of the type (173) describing a nondifferentiable random Markovian process are, strictly speaking, only valid outside the nonadiabaticity regions. In the nonadiabaticity regions, where an essential rearrangement of the electron wave functions takes place, we have to use the dynamic description or the differentiable random process since the electron "sees" the dynamic motion of the nuclear subsystem. Stochastic equations are more appropriate for the description of processes where the effects of the nonadiabaticity are unimportant.

We note, however, that one more disadvantage is inherent to the stochastic description. The stochastic approach assumes the averaging of all the physical values over a time interval Δt which exceeds considerably the time of "free" motion $\tau(\Delta t \gg \tau)$ (τ is the time during which the motion along the coordinate q may be considered as a dynamic motion in the corresponding potential field). This means that Δt is the smallest physical time unit and all the results have the corresponding uncertainty. In particular, for high, narrow potential barriers, the uncertainty in the activation energy in the stochastic approach may exceed kT.

The influence of the fluctuational motion along the reaction coordinates on the probability of the electron transition has been considered recently in the framework of the Landau-Zener method.[102] A Hamiltonian of the form

$$H(t) = \tfrac{1}{2}[vt + f(t)](|1\rangle\langle 1| - |2\rangle\langle 2|) + J(|1\rangle\langle 2| + |2\rangle\langle 1|) \tag{177}$$

was used where v is a constant velocity, J is the resonance integral leading to transitions between the states $|1\rangle$ and $|2\rangle$, and $f(t)$ is a random function of time.

It was, however, assumed[102] that $f(t)$ is a random Gaussian process. The expressions for the probability P to find the system in the state $|2\rangle$ at $t \to \infty$ if at $t \to -\infty$ it was in the state $|1\rangle$ was obtained in two limiting cases:

1. The case of slow fluctuations corresponds to the inequalities

$$\tau_{tr} \ll \tau_c \tag{178}$$

where

$$\tau_{tr} = \max(J/|v|, D/|v|); \qquad \tau_c = 1/\gamma \tag{179}$$

and D and γ are the characteristics of the random process

$$\langle f(t)f(t')\rangle = 2D^2 \exp(-\gamma|t - t'|) \tag{180}$$

In this case, P is equal to the Landau-Zener probability P_{LZ}.

2. The case of fast fluctuations corresponds to the relationships

$$\tau_{tr} \gg \tau_c; \qquad D/\gamma \ll 1 \tag{181}$$

$$\langle f(t)f(t')\rangle \simeq (4D^2/\gamma)\delta(t-t') \tag{182}$$

In this case, the expression for P has the form[102]

$$P = [1 - \exp(-4\pi J^2/|v|)]/2 \tag{183}$$

The Landau–Zener formula in the limit $J^2/|v| \to \infty$ gives $P_{LZ} = 1$ whereas Eq. (183) for the case of fast fluctuations in this limit gives $P = \frac{1}{2}$. At small values of $J^2/|v|$, both formulas give the same result, $P = 2\pi J^2/|v|$.

6. Effect of Dissipation on Tunneling

A number of papers are devoted to the effect of dissipation on tunneling.[81–83,103,104] Wolynes[81] was one of the first to consider this problem using the Feynman path integral approach to calculate the correlation function of the reactive flux involved in the expression for the rate constant,

$$k = (1/\beta\hbar) \lim_{t\to\infty} \int_0^{\beta\hbar} d\eta \langle \zeta(0)\dot{\zeta}(t+i\eta)\rangle / \langle \delta\zeta(0)\rangle^2 \tag{184}$$

where ζ is the unit stepwise Heaviside function determining the occupation of one side of the double well. One of the main results of the paper[81] consists in the fact that the dissipation leads to a decrease of the tunnel effects.

A rather general method of the calculation of the tunneling taking account of the dissipation was given in Ref. 82. The cases of rather strong dissipation were considered in Refs. 81 and 82, where it was assumed that a thermodynamical equilibrium in the initial potential well exists. The case of extremely weak friction has been considered using the equations for the density matrix in Ref. 83. A quantum analogue of the Focker–Planck equation for the adiabatic and nonadiabatic processes in condensed media was obtained in Refs. 105 and 106.

VII. CONCLUSION

The brief review of the newest results in the theory of elementary chemical processes in the condensed phase given in this chapter shows that great progress has been achieved in this field during recent years, concerning the description of both the interaction of electrons with the polar medium and with the intramolecular vibrations and the interaction of the intramolecular vibrations and other reactive modes with each other and with the dissipative subsystem (thermal bath). The rapid development of the theory of the adiabatic reactions of the transfer of heavy particles with due account of the fluctuational character of the motion of the medium in the framework of both dynamic and stochastic approaches should be mentioned. The stochastic approach is described only briefly in this chapter. The number of papers in this field is so great that their detailed review would require a separate article.

It should be noted that recent work has not been devoted only to the application of the theory to new processes and phenomena but has also been concerned with the basis of the theory. Therefore, new important results have been obtained also for processes which have been under theoretical investigation for many years, in particular, for electron and proton transfer reactions. These results open new directions for further investigations.

REFERENCES

1. R. R. Dogonadze and A. M. Kuznetsov, *Prog. Surf. Sci.* **6** (1975) 1.
2. P. P. Schmidt, *Specialist Periodical Report, Electrochemistry*, Vol. 5, The Chemical Society, London, 1975, p. 21.
3. R. R. Dogonadze and A. M. Kuznetsov, *Itogi Nauki i Tekhniki, Ser. Kinetika i Katalis*, Vol. 5, VINITI, Moscow, 1978, p. 2.
4. J. Ulstrup, *Charge Transfer Processes in Condensed Media*, Springer-Verlag, Berlin, 1979.
5. R. R. Dogonadze and A. M. Kuznetsov, *Itogi Nauki i Tekhniki, Ser. Fizicheskaya Khimiya, Kinetika*, Vol. 2, VINITI, Moscow, 1973, p. 3.
6. A. M. Kuznetsov, *Nouv. J. Chimie* **5** (1981) 427.
7. Sh. Efrima and M. Bixon, *J. Chem. Phys.* **64** (1976) 3639.
8. R. A. Marcus, *J. Phys. Chem.* **24** (1956) 966, 979.
9. M. Bixon and J. J. Jortner, *Faraday Disc. Chem. Soc.* **74** (1982) 17.
10. A. A. Ovchinnikov and V. A. Benderskii, *Fizicheskaya Khimiya. Sovremennye Problemy*, Ed. by Ya. M. Kolotyrkin, Khimiya, Moscow, 1980, p. 159.
11. A. M. Kuznetsov, *Elektrokhimiya* **17** (1981) 84.

[12] R. R. Dogonadze and A. M. Kuznetsov, *J. Res. Inst. Catal., Hokkaido Univ.* **22** (1974) 93.
[13] G. A. Bogdanchikov, A. I. Burshtein, and A. A. Zharikov, *Chem. Phys.* **86** (1984) 9.
[14] A. M. Kuznetsov, *J. Electroanal. Chem.* **159** (1983) 241.
[15] A. K. Churg, R. M. Weiss, A. Warshel, and T. Takano, *J. Phys. Chem.* **87** (1983) 1683.
[16] A. M. Kuznetsov and J. Ulstrup, *Phys. Status Solidi B* **114** (1982) 673.
[17] A. M. Kuznetsov, Ph.D. thesis, The Moscow Engineering Physics Institute, Moscow 1964.
[18] M. D. Newton, *Faraday Disc. Chem. Soc.* **74** (1982) 95.
[19] V. A. Zasukha and S. V. Volkov, *Teor. Eksp. Khim.* **18** (1982) 392.
[20] S. I. Pekar, *Untersuchungen über die Elektronentheorie der Kristalle*, Akademie-Verlag, Berlin, 1954.
[21] A. M. Kuznetsov and J. Ulstrup, *Faraday Disc. Chem. Soc.* **74** (1982) 31.
[22] A. M. Kuznetsov, *Poverkhnost* **9** (1982) 119.
[23] A. M. Kuznetsov, *Chem. Phys. Lett.* **91** (1982) 34.
[24] A. M. Kuznetsov, *Khim. Fiz.*, **1** (1982) 1496.
[25] A. I. Burshtein, G. K. Ivanov, and M. A. Kozhushner, *Khim. Fiz.*, **1** (1982) 195.
[26] A. M. Kuznetsov, *Elektrokhimiya* **19** (1983) 1596.
[27] W. P. Jenks, *Catalysis in Chemistry and Biochemistry*, McGraw-Hill, New York, 1969.
[28] C. K. Ingold, *Structure and Mechanism in Organic Chemistry*, 2nd Ed. Bell, London, 1969.
[29] B. S. Brunschwig, C. Creutz, D. H. Macartney, T.-K. Sham, and N. Sutin, *Faraday Disc. Chem. Soc.* **74** (1982) 113.
[30] S. U. M. Khan and J. O'M. Bockris, *J. Res. Inst. Catal., Hokkaido Univ.* **31** (1983) 35.
[31] S. U. M. Khan and J. O'M. Bockris, *J. Phys. Chem.* **87** (1983) 4012.
[32] E. D. German and A. M. Kuznetsov, *Itogi Nauki i Tekhniki, Ser. Kinetika i Katalis*, Vol. 10, VINITI, Moscow, 1982.
[33] M. Bixon and J. Jortner, *Faraday Disc. Chem. Soc.* **74** (1982) 17.
[34] P. Siders and R. A. Marcus, *J. Am. Chem. Soc.* **103** (1981) 748.
[35] A. M. Kuznetsov and J. Ulstrup (in preparation).
[36] R. R. Dogonadze, A. M. Kuznetsov, and V. G. Levich, *Electrochim. Acta* **13** (1968) 1025.
[37] B. Fain, *Theory of Rate Processes in Condensed Media*, Springer-Verlag, Berlin 1980.
[38] J. O'M. Bockris and S. U. M. Khan, *Quantum Electrochemistry*, Plenum Press, New York, 1979.
[39] J. O'M. Bockris, S. U. M. Khan, and D. B. Matthews, *J. Res. Inst. Catal., Hokkaido Univ.* **22** (1974) 1.
[40] R. R. Dogonadze, E. D. German, and A. M. Kuznetsov, *J. Chem. Soc., Faraday Trans. 2* **76** (1980) 1128.
[41] R. R. Dogonadze, G. M. Chonishvili, and T. A. Marsagishvili, *J. Chem. Soc., Faraday Trans. 2* **80** (1984) 355.
[42] E. D. German, R. R. Dogonadze, A. M. Kuznetsov, V. G. Levich, and Yu. I. Kharkats, *Elektrokhimiya* **6** (1970) 350.
[43] R. P. Bell, *J. Chem. Soc., Faraday Trans. 2* **76** (1980) 954.
[44] R. R. Dogonadze, A. M. Kuznetsov, and M. A. Vorotyntsev, *Phys. Status Solidi B* **54** (1972) 125, 425.

[45] E. D. German and A. M. Kuznetsov, *J. Chem. Soc., Faraday Trans. 2* **77** (1981) 2203.
[46] J. Sühnel and K. Gustav, *Chem. Phys.* **70** (1982) 109.
[47] V. L. Klochikhin, S. Ya. Pshezhetskii, and L. I. Trakhtenberg, *Dokl. Akad. Nauk SSSR* **239** (1978) 879; L. I. Trakhtenberg, *Khim. Fiz.* **1** (1982) 53.
[48] A. M. Kuznetsov, *Elektrokhimiya* **22** (1986) 240.
[49] M. M. Kreevoy, T. M. Liang, and K. C. Chang, *J. Am. Chem. Soc.* **99** (1977) 5207.
[50] Yu. I. Kharkats and J. Ulstrup, *J. Electroanal. Chem.* **65** (1975) 555.
[51] A. A. Ovchinnikov and V. A. Benderskii, *J. Electroanal. Chem.* **100** (1979) 563.
[52] A. A. Ovchinnikov, V. A. Benderskii, S. D. Babenko, and A. G. Krivenko, *J. Electroanal. Chem.* **91** (1978) 321.
[53] A. M. Kuznetsov, *J. Electroanal. Chem.* **151** (1983) 227.
[54] A. M. Kuznetsov, *J. Electroanal. Chem.* **180** (1984) 121.
[55] A. M. Kuznetsov, *J. Electroanal. Chem.* **159** (1983) 241.
[56] J. P. Muscat and D. M. Newns, *Prog. Surf. Sci.* **9** (1978) 1.
[57] J. R. Smith, Ed., *Theory of Chemisorption*, Springer-Verlag, Berlin, 1980.
[58] L. I. Krishtalik, *Charge Transfer Reactions. Electrochemical and Chemical Processes*, Plenum Press, New York, 1984.
[59] A. M. Kuznetsov and J. Ulstrup (in preparation).
[60] I. G. Medvedev, *Elektrokhimiya* **15** (1979) 713, 886.
[61] P. P. Schmidt, *J. Chem. Soc. Faraday Trans. 2* **80** (1984) 157, 181.
[62] C. P. Flynn and A. M. Stoneham, *Phys. Rev. B* **1** (1970) 3966.
[63] H. Teichler, *Phys. Status Solidi B* **104** (1981) 239.
[64] Yu. Kagan and M. I. Klinger, *Sov. Phys.—J. Exp. Theor. Phys.* **43** (1976) 132.
[65] G. K. Ivanov and M. A. Kozhushner, *Khim. Fiz.* **2** (1983) 1299.
[66] W. Kuhn and M. Wagner, *Phys. Rev. B* **23** (1981) 685.
[67] A. M. Kuznetsov, *Elektrokhimiya* **21** (1985) 836.
[68] G. K. Ivanov and M. A. Kozhushner, *Fiz. Tverd. Tela* **20** (1978) 9.
[69] A. M. Kuznetsov, *Elektrokhimiya* **22** (1986) 291.
[70] R. P. Feynman and A. R. Hibbs, *Quantum Mechanics and Path Integrals*, McGraw-Hill, New York, 1965.
[71] E. D. German and R. R. Dogonadze, *Dokl. Akad. Nauk SSSR* **210** (1973) 377.
[72] E. D. German and A. M. Kuznetsov *J. Chem. Soc. Faraday Trans. 2* **82** (1986) 1885.
[73] H. A. Kramers, *Physica* **7** (1940) 284.
[74] I. V. Alexandrov, *Teor. Eksp. Khim.* **16** (1980) 435.
[75] A. I. Shushin, *Teor. Eksp. Khim.* **17** (1981) 3.
[76] A. M. Kuznetsov, *Elektrokhimiya* **7** (1971) 1067.
[77] L. D. Zusman, *Teor. Eksp. Khim.* **15** (1979) 227.
[78] M. Ya. Ovchinnikova, *Teor. Eksp. Khim.* **17** (1981) 651.
[79] V. K. Bykhovskii, E. E. Nikitin, and M. Ya. Ovchinnikova, *Sov. Phys.—J. Exp. Theor. Phys.* **47** (1964) 750.
[80] M. A. Vorotyntsev, R. R. Dogonadze, and A. M. Kuznetsov, *Vestn. Mosk. Gos. Univ., Ser. Phys.*, no. 2 (1973) 224.
[81] P. G. Wolynes, *Phys. Rev. Lett.* **47** (1981) 968.
[82] A. I. Larkin and Yu. N. Ovchinnikov, *Pis'ma Zh. Eksp. Teor. Fiz.* **37** (1983) 322.
[83] V. Melnikov and A. Süto, *J. Phys. C: Solid State Phys.* **17** (1984) L207.
[84] T. P. Sethna, *Phys. Rev. B* **24** (1981) 698; **25** (1982) 5050.
[85] A. M. Kuznetsov, *Elektrokhimiya* **20** (1984) 1233.
[86] R. J. Glauber, in *Coherent States in Quantum Theory*, Mir, Moscow, 1972, p. 26.
[87] H. Frauenfelder and P. G. Wolynes, *Science* **229** (1985) 337.

[88] L. D. Zusman, *Chem. Phys.* **80** (1983) 29.
[89] L. D. Zusman, *Chem. Phys.* **49** (1980) 295.
[90] A. B. Helman, *Chem. Phys.* **65** (1982) 271.
[91] I. V. Alexandrov and V. I. Goldanskii, *Khim. Fiz.* **3** (1984) 185.
[92] B. L. Tembe, H. L. Friedman, and M. D. Newton, *J. Chem. Phys.* **76** (1982) 1490.
[93] T. Fonseca, J. A. N. F. Gomes, P. Grigolini, and F. Marchesoni, *J. Chem. Phys.* **80** (1984) 1826.
[94] E. Marechal and M. Moreau, *Mol. Phys.* **51** (1984) 133.
[95] K. Schulten, Z. Schulten, and A. Szabo, *J. Chem. Phys.* **74** (1981) 4426.
[96] B. Carmeli and A. Nitzan, *J. Chem. Phys.* **80** (1984) 3596.
[97] R. F. Grote and J. T. Hynes, *J. Chem. Phys.* **74** (1981) 4465; **75** (1981) 2191; **76** (1981) 2715; **77** (1982) 3736.
[98] J. L. Skinner and P. G. Wolynes, *J. Chem. Phys.* **69** (1978) 2143.
[99] G. van der Zwan and J. T. Hynes, *J. Chem. Phys.* **76** (1982) 2993; **77** (1982) 1295.
[100] D. P. Ali and W. H. Miller, *Chem. Phys. Lett.* **105** (1984) 501.
[101] B. Carmeli and A. Nitzan, *Phys. Rev. A* **29** (1984) 1481.
[102] Y. Kayanuma, *J. Phys. Soc. Jpn.* **53** (1984) 108.
[103] M. V. Basilevsky and V. M. Ryaboy, *Mol. Phys.* **44** (1981) 785.
[104] J. Brickmann, *Ber. Bunsenges., Phys. Chem.* **85** (1981) 106.
[105] W. A. Wassam, Jr. and J. H. Freed, *J. Chem. Phys.* **76** (1982) 6133.
[106] W. A. Wassam, Jr. and J. H. Freed, *J. Chem. Phys.* **76** (1982) 6150.

3

Recent Developments in Faradaic Rectification Studies

H. P. Agarwal

Department of Chemistry, M.A. College of Technology, Bhopal, India

I. INTRODUCTION

The mid-twentieth century witnessed the discovery of faradaic rectification by Doss and Agarwal.[1-4] In the late 1950s when Oldham,[5,32] Vdovin,[6] Barker,[7] and Rangarajan[33] each independently worked out the theoretical formulations which corroborated the results reported earlier,[3,4] the effect came into the limelight. After the pioneering work of Barker[7,8] and Delahay and coworkers[9-14] in the early 1960s, its potential and applicability in the study of fast electrode kinetics were recognized. The method finds mention in books on electrode processes[15-17] and in reviews on relaxation methods.[17,18] Earlier reviews on the subject deal with preliminary details and development in instrumentation techniques.[19-24] Soon, it made an impact and led to the development of other related nonlinear phenomena such as radio frequency polarography,[7,8,25,26] second-harmonic polarography,[27,30] intermodulation polarography,[9,26] and high-level faradaic rectification[26] for the study of fast electrode reactions. The first comprehensive review by the author on the subject appeared in *Electroanalytical Chemistry*, Vol. 7. A later review article describes the work done in the field up to early 1972.[40] Since then, no voluminous work has been reported, yet even the little progress made is significant as it has opened up new frontiers in electroanalytical chemistry and in electrodics in particular. The electrodics can be linked directly to problems of technological importance.

Relaxation methods for the study of fast electrode processes are recent developments but their origin, except in the case of faradaic rectification, can be traced to older work. The other relaxation methods are subject to errors related directly or indirectly to the internal resistance of the cell and the double-layer capacity of the test electrode. These errors tend to increase as the reaction becomes more and more reversible. None of these methods is suitable for the accurate determination of rate constants larger than 1.0 cm/s. Such errors are eliminated with faradaic rectification, because this method takes advantage of complete linearity of cell resistance and the slight nonlinearity of double-layer capacity. The potentialities of the faradaic rectification method for measurement of rate constants of the order of 10 cm/s are well recognized, and it is hoped that by suitably developing the technique for measurement at frequencies above 20 MHz, it should be possible to measure rate constants even of the order of 100 cm/s.

The present chapter will cover detailed studies of kinetic parameters of several reversible, quasi-reversible, and irreversible reactions accompanied by either single-electron charge transfer or multiple-electrons charge transfer. To evaluate the kinetic parameters for each step of electron charge transfer in any multistep reaction, the suitably developed and modified theory of faradaic rectification will be discussed. The results reported relate to the reactions at redox couple/metal, metal ion/metal, and metal ion/mercury interfaces in the audio and higher frequency ranges. The zero-point method has also been applied to some multiple-electron charge transfer reactions and, wheresoever possible, these results have been incorporated. Other related methods and applications will also be treated.

II. THEORETICAL ASPECTS

Rectification effects are due to the asymmetry of a current–voltage curve of an electrode system. The asymmetry of these curves may arise from the intrinsic asymmetry of the charge transfer reaction or from the extraneous asymmetry produced by inequalities in mass transfer rates of oxidant and reductant. Devanathan[31] has rightly categorized the former effect as faradaic rectification and the latter as a redoxokinetic effect. However, the term faradaic rectification has generally been used to cover both effects. By using the basic equation of a redox electrode superimposed by an alternating voltage, being a function of cos ωt, Devanathan[31] has derived the expression for the faradaic impedance, faradaic distortion, and faradaic rectification using only elementary mathematics. He has established that faradaic rectification results because of faradaic distortion and that these two phenomena are interdependent.

1. Single-Electron Charge Transfer Reactions

Devanathan[31] obtained the equation for the total rectification current due to mass transfer (I_{MTR}) as

$$I_{\text{MTR}} = -I_0(\alpha_c q_O + \alpha_a q_R) \cdot z \frac{(\cos\theta + \sin\theta)}{2 \cdot 2^{1/2}} \tag{1}$$

where I_0 is the exchange current density, given by

$$I_0 = nFk^0(C_O^0)^{\alpha_a}(C_R^0)^{\alpha_c}$$

and

$$q_O = -\frac{I_\omega}{nFC_O^0 D_O^{1/2}\omega^{1/2}}, \qquad q_R = \frac{I_\omega}{nFC_R^0 \omega^{1/2} D_R^{1/2}}, \qquad z \equiv \frac{nF}{RT} V_A < 1$$

and θ is the phase difference between the alternating current and applied alternating voltage, $\omega = 2\pi f$ (f is the frequency of the alternating current), n is the number of electrons involved in the charge transfer reaction, I_ω is the amplitude of the current corresponding to the fundamental frequencies, k^0 is the rate constant, and F, R, and T have their usual meanings.

The total rectification current, I_{FR}, considering charge transfer and mass transfer is given by

$$I_{\text{FR}} = I_0(\alpha_c^2 - \alpha_a^2)\frac{z^2}{4} - I_0(\alpha_c q_O + \alpha_a q_R)z \frac{(\cos\theta + \sin\theta)}{2 \cdot 2^{1/2}} \tag{2}$$

Charge transfer Mass transfer

Which can be written as[31]

$$I_{FR} = I_0 \frac{z^2}{4}(\alpha_c^2 - \alpha_a^2) + \frac{2RT(\cos\theta + \sin\theta)}{n^2F^2Z_\omega 2^{1/2}(\alpha_c + \alpha_a)\omega^{1/2}} \times \left(\frac{\alpha_a}{C_R^0 D_R^{1/2}} - \frac{\alpha_c}{C_O^0 D_O^{1/2}}\right) \tag{3}$$

where Z_ω is the faradaic impedance[31] (noting the convention that the cathodic current is taken as positive when V_A is negative), $Z_\omega = [(-R_r + \sigma\omega^{-1/2})^2 + \sigma^2\omega^{-1}]^{1/2}$, with R_r the reaction resistance and σ the Warburg coefficient. It can be shown that

$$\frac{2RT(\cos\theta + \sin\theta)}{\omega^{1/2}n^2F^2Z_\omega 2^{1/2}(\alpha_c + \alpha_a)}\left(\frac{\alpha_a}{C_R^0 D_R^{1/2}} - \frac{\alpha_c}{C_O^0 D_O^{1/2}}\right) = 2\left(\frac{1 + \cot\theta}{1 + \cot^2\theta}\right)\left(\frac{\alpha_a C_O^0 D_O^{1/2} - \alpha_c C_R^0 D_R^{1/2}}{C_O^0 D_O^{1/2} + C_R^0 D_R^{1/2}}\right) \tag{4}$$

Hence,

$$I_{FR} = I_0 \frac{z^2}{4}(\alpha_c^2 - \alpha_a^2) + 2\left(\frac{\alpha_a C_O^0 D_O^{1/2} - \alpha_c C_R^0 D_R^{1/2}}{C_O^0 D_O^{1/2} + C_R^0 D_R^{1/2}}\right)\frac{1 + \cot\theta}{1 + \cot^2\theta} \tag{5}$$

The rectification ratio is obtained by multiplying I_{FR} by the reaction resistance R_r:

$$\frac{RT\Delta E_\infty}{nFV_A^2} = -\left(\frac{\alpha_c - \alpha_a}{4}\right) - \frac{1}{2(\alpha_c + \alpha_a)} \times \left(\frac{\alpha_a C_O^0 D_O^{1/2} - \alpha_c C_R^0 D_R^{1/2}}{C_O^0 D_O^{1/2} + C_R^0 D_R^{1/2}}\right)\frac{1 + \cot\theta}{1 + \cot^2\theta} \tag{6}$$

where ΔE_∞ is the shift in mean equilibrium potential. At higher frequencies and for moderately fast reactions

$$\frac{1 + \cot\theta}{1 + \cot^2\theta} = \frac{1}{\cot\theta} = \frac{1}{p} \tag{7}$$

where

$$p = \frac{2^{1/2}\omega^{1/2}C_O^{0\alpha}C_R^{0(1-\alpha)}D_O^{1/2}D_R^{1/2}}{k^0(C_O^0D_O^{1/2} + C_R^0D_R^{1/2})} \tag{8}$$

Substituting $\alpha_C + \alpha_a = 1$, for single-electron charge transfer reactions, the above expression reduces to that of Delahay *et al.*[11]

$$\frac{\Delta E_\infty}{V_A^2} = \frac{nF}{4RT}\left[(2\alpha - 1) - \left(\alpha - \frac{C_O^0D_O^{1/2}}{C_O^0D_O^{1/2} + C_R^0D_R^{1/2}}\right)\frac{2}{p}\right]$$

By assuming $D_O = D_R$ (as a first approximation for the sake of simplicity), the equation can be written as

$$\frac{4RT}{nF}\frac{\Delta E_\infty}{V_A^2} = (2\alpha - 1) + [C_O^0(1-\alpha) - \alpha C_R^0] \times \frac{2^{1/2}k^0}{\omega^{1/2}(C_O^0)^\alpha(C_R^0)^{(1-\alpha)}D_O^{1/2}} \tag{9}$$

In any fast multielectron transfer reaction all the electrons cannot be transferred in one step but only by a succession of single-electron transfer steps, whereas Eq. (6) was arrived at by Devanathan[31] for the simple case in which it is assumed that the same step is rate determining in both directions, irrespective of the number of electrons involved in the reaction.

If a two-electron charge transfer reaction takes place in two separate steps, each being accompanied by transfer of a single electron, the mathematical expression for the determination of kinetic parameters becomes more involved and complicated.

Recently, Rangarajan has worked out a comprehensive theory[35,36] of the electrode/electrolyte interface using the partial derivative formalism introduced by Grahame[34] and considering four phenomenological components: (1) charge separation, (2) adsorption–desorption (3) charge transfer at the electrode, and (4) mass transfer with or without volume sources/sinks. It may be mentioned that the Rangarajan treatment is restricted to the four phenomenological components, and it applies only to certain aspects of the general nonlinear case,[37] i.e., regime of a generalized Randles–Ershler scheme. Rangarajan's expressions[35,36] are highly complicated, very involved, and difficult to apply in the determination of kinetic parameters of each step of a multiple-electron charge transfer reaction.

It was therefore thought appropriate to suitably modify and develop the faradaic rectification theory for the study of multiple-electron charge transfer reactions.

2. Two-Electron Charge Transfer Reactions

In any two-electron charge transfer reaction, the two steps can be represented as follows:

$$\text{(a)}\quad \underset{C_O^0}{M^{2+}} + e^- \underset{}{\overset{k_1^0}{\rightleftharpoons}} \underset{C_{R_I}^0}{M^+}$$

$$\text{(b)}\quad \underset{C_{R_I}^0}{M^+} + e^- \overset{k_2^0}{\rightleftharpoons} \underset{C_R^0}{M}$$

where C_O^0 is the concentration of a bivalent oxidant M^{2+}, $C_{R_I}^0$ is the concentration of an intermediate species formed intermittently, C_R^0 is the concentration of reductant present in solution, and k_1^0 and k_2^0 are the rate constants for the two respective reactions.

Physical Picture

When a bivalent metallic ion or a redox couple differing by a bivalent charge is in solution, it is difficult to assign any particular boundary of separation between a bivalent ion and a univalent ion formed intermittently. It would be very hypothetical to assign separate interfacial potentials for $M^{2+} \overset{e^-}{\rightleftharpoons} M^+$ and $M^+ \overset{e^-}{\rightleftharpoons} M$. It is difficult to isolate the stage when the bivalent metal ion is converted into a univalent ion which is subsequently converted into a final reduced state. The only thing that is understood is that in the Helmholtz layer, the bimetallic ion is first converted to a univalent ion state before being finally reduced at the electrode surface. It would be an ideal situation if one could take into account, for separate transfer resistances, transfer coefficients and the amplitude V_A corresponding to each step of electron charge transfer in any multistep electron charge transfer theory. In such a case, a very complicated and highly involved mathematical expression would be obtained which may be impossible to solve. For purposes of simplification, the expression has been worked out under the following assumptions:

1. The value of α is taken to be the same in both of the steps of electron charge transfer and its value is assumed to be close to 0.5.

2. $D_O = D_R$.

3. V_A, the amplitude of the interfacial potential, is considered for the overall reaction.

If $\Delta E_{\infty_{\rm I}}$ is the rectification potential due to the first step of the reaction and $\Delta E_{\infty_{\rm II}}$ is the rectification potential contribution due to the second step, then the total rectified potential should be the sum of the rectified potentials for each individual step, i.e., $\Delta E_{\infty_{\rm I}} + \Delta E_{\infty_{\rm II}}$. The combined faradaic rectification change for both the steps of electron charge transfer can be represented as[38]

$$\frac{4RT}{F}\frac{(\Delta E_{\infty_{\rm I}} + \Delta E_{\infty_{\rm II}})}{V_A^2} = (2\alpha - 1) + \frac{[C_O^0(1-\alpha) - \alpha C_{R_{\rm I}}^0]2^{1/2}k_1^0}{\omega^{1/2}(C_O^0)^{\alpha}(C_{R_{\rm I}}^0)^{(1-\alpha)}D_O^{1/2}}$$
$$+ \frac{[C_{R_{\rm I}}^0(1-\alpha) - \alpha C_R^0]2^{1/2}k_2^0}{\omega^{1/2}(C_{R_{\rm I}}^0)^{\alpha}(C_R^0)^{(1-\alpha)}D_O^{1/2}} \quad (10)$$

Assuming that α practically remains constant in both the steps of electron charge transfer and putting $\Delta E_{\infty_{\rm I}} + \Delta E_{\infty_{\rm II}} = \Delta E_\infty$,

$$\frac{4RT}{F}\frac{\Delta E_\infty}{V_A^2} = (2\alpha - 1) + \frac{[C_O^0(1-\alpha) - \alpha C_{R_{\rm I}}^0]2^{1/2}k_1^0}{\omega^{1/2}(C_O^0)^{\alpha}(C_{R_{\rm I}}^0)^{(1-\alpha)}D_O^{1/2}}$$
$$+ \frac{[C_{R_{\rm I}}^0(1-\alpha) - \alpha C_R^0]2^{1/2}k_2^0}{\omega^{1/2}(C_{R_{\rm I}}^0)^{\alpha}(C_R^0)^{(1-\alpha)}D_O^{1/2}} \quad (11)$$

In order to simplify the expression, it can be assumed that $(C_{R_{\rm I}}^0)^{\alpha} \approx (C_{R_{\rm I}}^0)^{(1-\alpha)} \approx (C_{R_{\rm I}}^0)^{1/2}$, a condition that is met when α can be approximated as 0.5. Equation (11) then becomes

$$\left[\frac{4RT}{F}\frac{\Delta E_\infty}{V_A^2} - (2\alpha - 1)\right]\frac{\omega^{1/2}(C_{R_{\rm I}}^0)^{1/2}D_O^{1/2}}{2^{1/2}}$$
$$= [C_O^0(1-\alpha) - \alpha C_{R_{\rm I}}^0]\frac{k_1^0}{(C_O^0)^{\alpha}}$$
$$+ \left[C_{R_{\rm I}}^0(1-\alpha) - \alpha C_R^0\right]\frac{k_2^0}{(C_R^0)^{(1-\alpha)}} \quad (12)$$

As the expression involves three unknowns, k_1^0, k_2^0, and $C_{R_I}^0$, it is necessary to carry out the experiment at three different concentrations of redox couples. This can be achieved in two ways:

1. By keeping the concentration of the reductant constant while varying that of the oxidant.
2. By keeping the concentration of the oxidant constant while varying that of the reductant.

The separate mathematical expressions for the above two cases have been obtained, and they are given in Appendix A. Whichever experimental condition is chosen, the corresponding expressions will be used for the determination of the values of $C_{R_I}^0$, k_1^0, k_2^0.

3. Three-Electron Charge Transfer Reactions

In any three-electron charge transfer reaction, the three steps can be represented as follows. Considering a trivalent metal ion reduction,

(a) $\underset{C_O^0}{M^{3+}} + e^- \overset{k_1^0}{\rightleftharpoons} \underset{C_{R_{II}}^0}{M^{2+}}$ (b) $\underset{C_{R_{II}}^0}{M^{2+}} + e^- \overset{k_2^0}{\rightleftharpoons} \underset{C_{R_I}^0}{M^{+}}$

(c) $\underset{C_{R_I}^0}{M^{+}} + e^- \overset{k_3^0}{\rightleftharpoons} \underset{C_R^0}{M}$

where C_O^0 is the concentration of a trivalent ion M^{3+}, $C_{R_{II}}^0$ and $C_{R_I}^0$ are the concentrations of the two intermediate species, C_R^0 is the concentration of the final reduced state, and k_1^0, k_2^0, and k_3^0 are the three rate constants.

Assuming that α practically remains constant for all three steps of electron charge transfer, the expression for the total rectified potential can suitably be represented,[39] using Eq. (9) as

$$\frac{4RT}{F}\left[\frac{\Delta E_{\infty_I} + \Delta E_{\infty_{II}} + \Delta E_{\infty_{III}}}{V_A^2}\right] = (2\alpha - 1) + \frac{[C_O^0(1-\alpha) - C_{R_{II}}^0]2^{1/2}k_1^0}{\omega^{1/2}(C_O^0)^\alpha (C_{R_{II}}^0)^{1-\alpha} D_O^{1/2}} + \frac{[C_{R_{II}}^0(1-\alpha) - \alpha C_{R_I}^0]2^{1/2}k_2^0}{\omega^{1/2}(C_{R_{II}}^0)^\alpha (C_{R_I}^0)^{1-\alpha} D_O^{1/2}} + \frac{[C_{R_I}^0(1-\alpha) - \alpha C_R^0]2^{1/2}k_3^0}{\omega^{1/2}(C_{R_I}^0)^\alpha (C_R^0)^{1-\alpha} D_O^{1/2}} \tag{13}$$

The observed rectified potential ΔE_∞ may be taken as equal to the total sum of the rectified potentials of the individual steps, ΔE_{∞_I}, $\Delta E_{\infty_{II}}$, and $\Delta E_{\infty_{III}}$, in Eq. (13). Then rearranging terms,

$$\left[\frac{4RT}{F}\frac{\Delta E_\infty}{V_A^2} - (2\alpha - 1)\right]\frac{\omega^{1/2}D_O^{1/2}}{2^{1/2}}$$
$$= \frac{[C_O^0(1-\alpha) - \alpha C_{R_{II}}^0]k_1^0}{(C_O^0)^\alpha (C_{R_{II}}^0)^{(1-\alpha)}} + \frac{[C_{R_{II}}^0(1-\alpha) - \alpha C_{R_I}^0]k_2^0}{(C_{R_{II}}^0)^\alpha (C_{R_I}^0)^{(1-\alpha)}}$$
$$+ \frac{[C_{R_I}^0(1-\alpha) - \alpha C_R^0]k_3^0}{(C_{R_I}^0)^\alpha (C_R^0)^{(1-\alpha)}} \tag{14}$$

If C_R^0 is kept constant, then the expression involves five unknowns, k_1^0, k_2^0, k_3^0, $C_{R_{II}}^0$, and $C_{R_I}^0$. Hence, it is necessary to carry out experiments at five different redox concentration ratios. This has been done by using five oxidant concentrations, C_O^0, $C_O^{0'}$, $C_O^{0''}$, $C_O^{0'''}$, and $C_O^{0^{IV}}$, keeping the concentration of the reductant constant, i.e., C_R^0.

In Appendix B, the mathematical derivation is given for obtaining experimentally the values of $C_{R_{II}}^0$, $C_{R_I}^0$, k_1^0, k_2^0, and k_3^0 in the case of any three-electron charge transfer reaction.

From the derivations in Appendix B, it is evident that the present faradaic rectification formulations for multiple-electron charge transfer not only enable the determination of kinetic parameters for each step of three-electron charge transfer processes but may also be extended to charge transfer processes involving a higher number of electrons. However, the calculations become highly involved and complicated.

4. Zero-Point Method

(*i*) *Single-Electron Charge Transfer Reactions*

Kinetic parameters can also be obtained by using the zero-point method as described earlier.[40] The advantage of this method is that the values of α and k^0 can be deduced independent of the determination of values of the double-layer capacitance, electrode impedance, and potential difference across the electrode/solution interface. The expression used is[40]

$$\frac{p'^2\omega + p'\omega^{1/2}}{p'\omega^{1/2} + 2} = -\frac{C_O^0 D_O^{1/2} - C_R^0 D_R^{1/2}}{(C_O^0 D_O^{1/2} + C_R^0 D_R^{1/2})(2\alpha - 1)} \tag{15}$$

where

$$p' = \frac{p}{\omega^{1/2}} \frac{2^{1/2}(C_O^0)^\alpha (C_R^0)^{(1-\alpha)} D_O^{1/2} D_R^{1/2}}{k^0(C_O^0 D_O^{1/2} + C_R^0 D_R^{1/2})}$$

In order to apply the zero-point method, it is necessary to know the value of the frequency at which the rectification voltage tends to zero. The experimental determination of the zero-point frequency has some practical difficulties, because over a small frequency range, the rectification signal is indistinguishable. Hence, the zero-point frequencies have been determined by extrapolating plots of $\Delta E_\infty / V_A^2$ versus $\omega^{-1/2}$ for those redox concentration ratios which intercept the abscissa (i.e., when $\Delta E_\infty = 0$).

(ii) Two-Electron Charge Transfer Reactions

(a) Case I: C_R^0 is kept constant and C_O^0 varies

The kinetic parameters can also be obtained using the values of ω corresponding to zero-point frequencies (ω, ω_1, and ω_2, i.e., the frequencies at the respective concentrations). On substituting these values in Eq. (12) and Eqs. (a) and (b) of Appendix A, the expressions reduce to[39]

$$\begin{aligned} -(2\alpha - 1)\omega^{1/2}(C_{R_1}^0)^{1/2} D_O^{1/2}/2^{1/2} &= [C_O^0(1-\alpha) - \alpha C_{R_1}^0]\frac{k_1^0}{(C_O^0)^\alpha} \\ &\quad + [C_{R_1}^0(1-\alpha) - \alpha C_R^0]\frac{k_2^0}{(C_R^0)^{(1-\alpha)}} \end{aligned} \tag{16}$$

$$\begin{aligned} -(2\alpha - 1)\omega^{1/2} a^{1/2}(C_{R_1}^0)^{1/2} D_O^{1/2}/2^{1/2} &= [C_O^{0\prime}(1-\alpha) - \alpha a C_{R_1}^0]\frac{k_1^0}{(C_O^{0\prime})^\alpha} \\ &\quad + [a C_{R_1}^0(1-\alpha) - \alpha C_R^0]\frac{k_2^0}{(C_R^0)^{(1-\alpha)}} \end{aligned} \tag{17}$$

$$
\begin{aligned}
&- (2\alpha - 1)\omega_2^{1/2} b^{1/2} (C^0_{R_1})^{1/2} D_O^{1/2}/2^{1/2} \\
&= [C_O^{0''}(1-\alpha) - b\alpha C^0_{R_1}] \frac{k_1^0}{(C_O^{0''})^{\alpha}} \\
&+ [bC^0_{R_1}(1-\alpha) - \alpha C^0_R] \frac{k_2^0}{(C^0_R)^{(1-\alpha)}}
\end{aligned} \tag{18}
$$

For simplification, the second term on the right-hand side in each of the above three equations may be taken as approximately equal. On subtracting Eq. (17) from Eq. (16) and Eq. (18) from Eq. (16), one obtains

$$
\begin{aligned}
&- (2\alpha - 1)(C^0_{R_1})^{1/2} D_O^{1/2} (\omega^{1/2} - a^{1/2}\omega_1^{1/2})/2^{1/2} \\
&= k_1^0\{[C^0_O(1-\alpha) - \alpha C^0_{R_1}]/(C^0_O)^{\alpha} \\
&- [C_O^{0'}(1-\alpha) - a\alpha C^{0'}_{R_1}]/(C_O^{0'})^{\alpha}\}
\end{aligned} \tag{19}
$$

$$
\begin{aligned}
&- (2\alpha - 1)(C^0_{R_1})^{1/2} D_O^{1/2} (\omega^{1/2} - b^{1/2}\omega_2^{1/2})/2^{1/2} \\
&= k_1^0\{[C^0_O(1-\alpha) - \alpha C^0_{R_1}]/(C^0_O)^{\alpha} \\
&- [C_O^{0''}(1-\alpha) - b\alpha C^0_{R_1}]/(C_O^{0''})^{\alpha}\}
\end{aligned} \tag{20}
$$

On dividing Eq. (19) by Eq. (20),

$$
\begin{aligned}
\frac{\omega^{1/2} - a^{1/2}\omega_1^{1/2}}{\omega^{1/2} - \omega_2^{1/2} b^{1/2}} &= (C_O^{0''})^{\alpha}\{(C_O^{0'})^{\alpha}[C^0_O(1-\alpha) - \alpha C^0_{R_1}] \\
&- (C^0_O)^{\alpha}[C_O^{0'}(1-\alpha) - a\alpha C^0_{R_1}]\}/ \\
&(C_O^{0'})^{\alpha}\{(C_O^{0''})^{\alpha}[C^0_O(1-\alpha) - \alpha C^0_{R_1}] \\
&- (C^0_O)^{\alpha}[C_O^{0''}(1-\alpha) - b\alpha C^0_{R_1}]\}
\end{aligned}
$$

or

$$
\begin{aligned}
\frac{(C_O^{0'})^{\alpha}[\omega^{1/2} - \omega_1^{1/2}a^{1/2}]}{(C_O^{0''})^{\alpha}[\omega^{1/2} - \omega_2^{1/2}b^{1/2}]} &= (C_O^{0'})^{\alpha} C^0_O(1-\alpha) - \alpha (C_O^{0'})^{\alpha} C^0_{R_1} \\
&- (C^0_O)^{\alpha} C_O^{0'}(1-\alpha) + a\alpha C^0_{R_1}(C^0_O)^{\alpha}/ \\
&(C_O^{0''})^{\alpha} C^0_O(1-\alpha) - \alpha (C_O^{0''})^{\alpha} C^0_{R_1} \\
&- (C^0_O)^{\alpha} C_O^{0''}(1-\alpha) + b\alpha C^0_{R_1}(C^0_O)^{\alpha}
\end{aligned}
$$

On rearranging the terms in the above equation,

$$C^0_{R_1} = (1-\alpha)\left\{[(C^{0\prime}_O)^\alpha C^0_O - (C^0_O)^\alpha C^{0\prime}_O] - \frac{(C^{0\prime}_O)^\alpha}{(C^{0\prime\prime}_O)^\alpha}\left(\frac{\omega^{1/2} - a^{1/2}\omega_1^{1/2}}{\omega^{1/2} - b^{1/2}\omega_2^{1/2}}\right)\right.$$

$$\left.\times [(C^{0\prime\prime}_O)^\alpha C^0_O - (C^0_O)^\alpha C^{0\prime\prime}_O]\right\} \Big/ -\alpha\left\{[(C^{0\prime}_O)^\alpha - a(C^0_O)^\alpha]\right.$$

$$\left. + \frac{(C^{0\prime}_O)^\alpha}{(C^{0\prime\prime}_O)^\alpha}\left(\frac{\omega^{1/2} - \omega_1^{1/2}a^{1/2}}{\omega^{1/2} - b^{1/2}\omega_2^{1/2}}\right)[C^{0\prime\prime}_O - b(C^0_O)^\alpha]\right\} \tag{21}$$

Substituting $\omega^{-1/2}$, $\omega_1^{-1/2}$, $\omega_2^{-1/2}$, and all other terms given earlier, $C^0_{R_1}$ can be obtained. The value of k_1^0 can be obtained on substituting the value of $C^0_{R_1}$ and all other terms in Eq. (19). On substituting the values of $C^0_{R_1}$, k_1^0, and all other terms in Eq. (16), the value of k_2^0 can be obtained.

(*b*) *Case II*: C^0_O *is kept constant and* C^0_R *varies*

Let the values of $\omega^{-1/2}$ at the zero-point frequencies be $\omega^{-1/2}$, $\omega_1^{-1/2}$, and $\omega_2^{-1/2}$ at three oxidant/reductant ratios. Referring to Eqs. (f), (g), and (h) in Appendix A and substituting $\Delta E_\infty = 0$, the expressions obtained are[39]

$$-(2\alpha - 1)\omega^{1/2}(C^0_{R_1})^{1/2}D_O^{1/2}2^{-1/2}$$
$$= [C^0_O(1-\alpha) - \alpha C^0_{R_1}]k_1^0/(C^0_O)^\alpha$$
$$+ [C^0_{R_1}(1-\alpha) - \alpha C^0_R]k_2^0/(C^0_R)^{(1-\alpha)} \tag{22}$$

$$-(2\alpha - 1)\omega^{1/2}a^{1/2}(C^0_{R_1})^{1/2}D_O^{1/2}2^{-1/2}$$
$$= [C^0_O(1-\alpha) - a\alpha C^0_{R_1}]k_1^0/(C^0_O)^\alpha$$
$$+ [aC^0_{R_1}(1-\alpha) - \alpha C^{0\prime}_R]k_2^0/(C^{0\prime}_R)^{(1-\alpha)} \tag{23}$$

$$-(2\alpha - 1)\omega^{1/2}b^{1/2}(C^0_{R_1})^{1/2}D_O^{1/2}2^{-1/2}$$
$$= [C^0_O(1-\alpha) - b\alpha C^0_{R_1}]k_1^0/(C^0_O)^\alpha$$
$$+ [bC^0_{R_1}(1-\alpha) - \alpha C^{0\prime\prime}_R]k_2^0/(C^{0\prime\prime}_R)^{(1-\alpha)} \tag{24}$$

The first term on the right-hand side in each of the above three equations can be approximately taken as equal so as to simplify the expressions. Now, on subtracting Eq. (23) from Eq. (22) and Eq. (24) from Eq. (23), we obtain

$$(2\alpha - 1)(C_{R_1}^0)^{1/2} D_0^{1/2} 2^{-1/2} [\omega_1^{1/2} a^{1/2} - \omega^{1/2}]$$
$$= k_2^0 \{[C_{R_1}^0(1-\alpha) - \alpha C_R^0]/(C_R^0)^{(1-\alpha)}$$
$$- [aC_{R_1}^0(1-\alpha) - \alpha C_R^{0\prime}]/(C_R^{0\prime})^{(1-\alpha)}\} \tag{25}$$

$$(2\alpha - 1)(C_{R_1}^0)^{1/2} D_O^{1/2} 2^{-1/2} [\omega_2^{1/2} b^{1/2} - \omega_1^{1/2} a^{1/2}]$$
$$= k_2^0 \{[C_{R_1}^0 a(1-\alpha) - \alpha C_R^{0\prime}]/(C_R^{0\prime})^{(1-\alpha)}$$
$$- [bC_{R_1}^0(1-\alpha) - \alpha C_R^{0\prime\prime}]/(C_R^{0\prime\prime})^{(1-\alpha)}\} \tag{26}$$

On dividing Eq. (25) by Eq. (26) and rearranging,

$$\frac{\omega_1^{1/2} a^{1/2} - \omega^{1/2}}{\omega_2^{1/2} b^{1/2} - \omega_1^{1/2} a^{1/2}}$$
$$= (C_R^{0\prime\prime})^{(1-\alpha)} [(C_R^{0\prime})^{(1-\alpha)} C_{R_1}^0 (1-\alpha) - \alpha C_R^0 (C_R^{0\prime})^{(1-\alpha)}$$
$$- a(C_R^0)^{(1-\alpha)} C_{R_1}^0 (1-\alpha) + \alpha C_R^{0\prime} (C_R^0)^{(1-\alpha)}]/$$
$$(C_R^0)^{(1-\alpha)} [aC_{R_1}^0 (1-\alpha)(C_R^{0\prime\prime})^{(1-\alpha)} - \alpha C_R^{0\prime} (C_R^{0\prime\prime})^{(1-\alpha)}$$
$$- bC_{R_1}^0 (1-\alpha)(C_R^{0\prime})^{(1-\alpha)} + \alpha C_R^{0\prime\prime} (C_R^{0\prime})^{(1-\alpha)}]$$

$$\frac{(C_R^0)^{(1-\alpha)}}{(C_R^{0\prime\prime})^{(1-\alpha)}} \frac{[\omega_1^{1/2} a^{1/2} - \omega^{1/2}]}{[\omega_2^{1/2} b^{1/2} - \omega_1^{1/2} a^{1/2}]}$$
$$= (C_R^{0\prime})^{(1-\alpha)} C_{R_1}^0 (1-\alpha) - \alpha C_R^0 (C_R^{0\prime})^{(1-\alpha)}$$
$$- a(C_R^0)^{(1-\alpha)} C_{R_1}^0 (1-\alpha) + \alpha C_R^{0\prime} (C_R^0)^{(1-\alpha)}/$$
$$a(C_R^{0\prime\prime})^{(1-\alpha)} C_{R_1}^0 (1-\alpha) - \alpha C_R^{0\prime} (C_R^{0\prime\prime})^{(1-\alpha)}$$
$$- b(C_R^{0\prime})^{(1-\alpha)} C_{R_1}^0 (1-\alpha) + \alpha C_R^{0\prime\prime} (C_R^{0\prime})^{(1-\alpha)}$$
$$= C_{R_1}^0 (1-\alpha)[(C_R^{0\prime})^{(1-\alpha)} - a(C_R^0)^{(1-\alpha)}]$$
$$- \alpha[C_R^0 (C_R^{0\prime})^{(1-\alpha)} - C_R^{0\prime} (C_R^0)^{(1-\alpha)}]/$$
$$C_{R_1}^0 (1-\alpha)[(C_R^{0\prime\prime})^{(1-\alpha)} - b(C_R^{0\prime})^{(1-\alpha)}]$$
$$- \alpha[C_R^{0\prime} (C_R^{0\prime\prime})^{(1-\alpha)} - (C_R^{0\prime})^{(1-\alpha)} C_R^{0\prime\prime}] \tag{27}$$

On rearranging the terms in the above equation,

$$\begin{aligned} C^0_{R_1} = \alpha\{[(C^0_R)^{(1-\alpha)}(\omega_1^{1/2}a^{1/2} - \omega^{1/2})/(C^{0''}_R)^{(1-\alpha)} \\ \times (\omega_2^{1/2}b^{1/2} - \omega_1^{1/2}a^{1/2})][C^{0'}_R(C^{0''}_R)^{(1-\alpha)} - (C^{0'}_R)^{(1-\alpha)}C^{0''}_R] \\ - [C^0_R(C^{0'}_R)^{(1-\alpha)} - C^{0'}_R(C^0_R)^{(1-\alpha)}]\}/(1-\alpha)\{[(C^0_R)^{(1-\alpha)} \\ \times (\omega_1^{1/2}a^{1/2} - \omega^{1/2})/(C^{0''}_R)^{(1-\alpha)}(\omega_2^{1/2}b^{1/2} - \omega_1^{1/2}a^{1/2})] \\ \times [(C^{0''}_R)^{(1-\alpha)}a - b(C^{0'}_R)^{(1-\alpha)}] \\ - [(C^{0'}_R)^{(1-\alpha)} - a(C^0_R)^{(1-\alpha)}]\} \end{aligned} \tag{28}$$

The value of k_2^0 can be obtained on substituting the value of $C^0_{R_1}$ and all other terms in Eq. (25). On substituting the values of $C^0_{R_1}$, k_2^0, and all other terms in Eq. (22), the value of k_1^0 can be obtained.

III. INSTRUMENTATION AND RESULTS

1. Faradaic Rectification Studies at Metal Ion/Metal(s) Interfaces

(i) *Experimental Techniques*

The study of metal ion/metal(s) interfaces has been limited because of the excessive adsorption of the reactants and impurities at the electrode surface and due to the inseparability of the faradaic and nonfaradaic impedances. For obtaining reproducible results with solid electrodes, the important factors to be considered are the fabrication, the smoothness of the surface (by polishing), and the pretreatment of the electrodes, the treatment of the solution with activated charcoal, the use of an inert atmosphere, and the constancy of the equilibrium potential for the duration of the experiment. It is appropriate to deal with some of these details from a practical point of view.

(*a*) *Fabrication of the electrodes*

All the three electrodes (test electrode, counter electrode, and reference electrode) are made from the same smooth, bright, polished metal foil (A.R.). The metal foil is cut in rectangular shapes with continuity provided by a thin contact strip. The other end of the strip is attached to a platinum wire using epoxy silver paste. The other end of the platinum wire is connected to silver wire and the platinum wire is sealed into a glass tube. The size of the test electrode is 20 times smaller than that of the reference electrode. The counter electrode and reference electrode are of the same size.

The details of the pretreatment of the electrodes and purification of the charcoal and the solution have already been described in an earlier publication[43] and review.[40]

(*b*) *Measurement of AC and rectified signals*

The cell and the circuit diagram are shown in Fig. 1. The cell consists of a test electrode, E_1, reference electrode, R, and counter electrode, E_2. The ac potential between the test electrode E_1 and the reference electrode R is measured by connecting them to a sensitive ac millivoltmeter through the contact key (by connecting point a to point b). The rectified voltage between E_1 and R is measured across a dc microvoltmeter (sensitivity, 1 μV/smallest

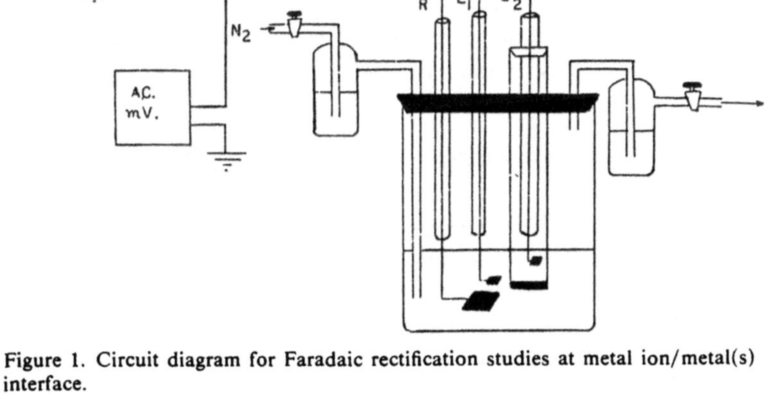

Figure 1. Circuit diagram for Faradaic rectification studies at metal ion/metal(s) interface.

division) after filtering the ac through a low-pass filter F when point a is connected to point c by the key K. After switching off the ac the initial dc potential, if any is again measured subtracted from the former to obtain the actual magnitude of the rectified voltage. All measurements are made below 5 mV, with the applied ac varying at frequencies between 50 Hz and 15 kHz.

Before starting any experiment, the potential of the test electrode E_1 is measured with reference to a saturated calomel electrode which is connected to the experimental cell through a bridge containing the same supporting electrolyte solution. Such measurements are taken whenever the concentration of the metal ion is changed. The cell is kept immersed in a thermostated bath maintained at a known temperature.

(*c*) *Diffusion coefficient*

For obtaining the value of the rate constant, it is desirable to determine the value of the diffusion coefficient of the metal ions or of one of the reactants (in the case of a redox couple) in the supporting electrolyte at an appropriate temperature. The value of the diffusion coefficient is experimentally determined using a McBain–Dowson cell and the King–Cathard equation, as described earlier.[40]

(*ii*) *Reactions Occurring through a Single-Electron Charge Transfer*

$Ag^+ + e^- \rightleftharpoons Ag$ is the simplest reaction which can be studied easily in a potassium nitrate solution. In any metal ions metal reaction, $C_R^0 \gg C_O^0$, and the theoretical expression in Eq. (9) reduces to[43]

$$\Delta E_\infty / V_A^2 = nF/4RT(2\alpha - 1) - \frac{1}{RT}\left(\frac{\alpha I_a^0}{2^{3/2} C_O^0 D_O^{1/2}}\right)\omega^{-1/2}$$

Thus, it is evident that the linear portion of a $\Delta E_\infty / V_A^2$ versus $\omega^{-1/2}$ plot can be used for determining the value of α from its intercept with the ordinate, while the slope of the plot will give the value of I_a^0, provided the diffusion coefficient of the metal ion is known. The plots for two concentrations of Ag^+ (1.0 mM and 2.0 mM in 1.0 *M* KNO_3) are given in Fig. 2. It can be seen that the two plots,

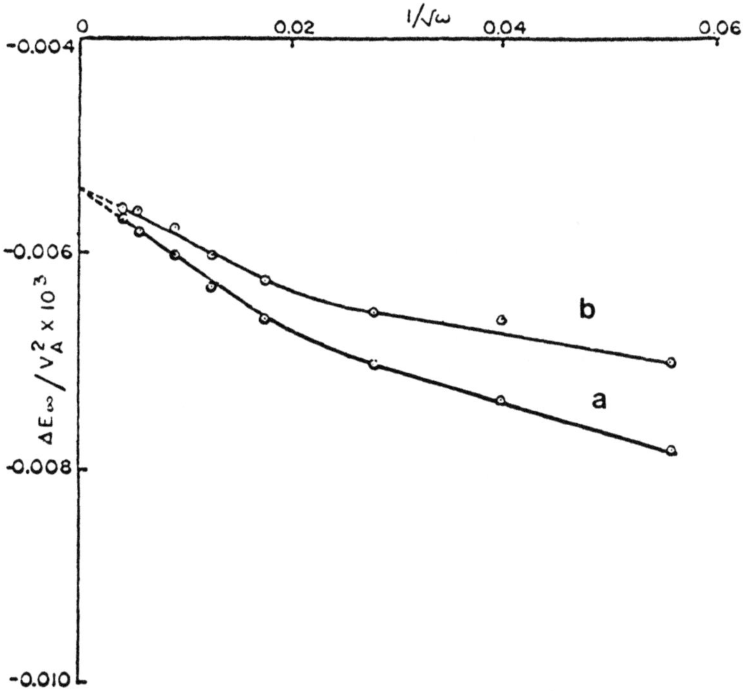

Figure 2. $\Delta E_\infty / V_A^2$ versus $\omega^{-1/2}$ plots for Ag^+, 1.0 N KNO_3/Ag electrode: (a) 1.0 mM, (b) 2.0 mM. (From Ref. 43, courtesy Pergamon Press.)

on extrapolation in the high-frequency region, meet at a point on the ordinate. The kinetic parameters obtained are $\alpha = 0.22$, $k_a^0 = 0.30 \times 10^{-2}$ cm/s, and $I_a^0 = 7.3$ mA/cm^2.[43]

(*iii*) *Reactions Involving Two-Electron Charge Transfer*

Very few references are available on the determination of the rate constant for each step of electron charge transfer in the reaction $M^{2+} + 2e^- \rightarrow M(s)$, i.e., $M^{2+} + e^- \rightarrow M^+$, $M^+ + e^- \rightarrow M(s)$. Earlier studies are mostly related to two-electron charge transfer reactions either at M^{2+}/Hg(dme), M^{2+}/metal amalgam, or redox couple/Pt interfaces. Even in these studies, the kinetic parameters have been determined assuming that one of the two steps of the reaction is much slower and is in overall control of the rate of reaction in both

Table 1
Kinetic Parameters of Cu(II)/Cu and Cd(II)/Cd in Various Supporting Electrolytes[a]

Reaction	Supporting electrolyte	$D_O \times 10^6$ (meas.) (cm^2/s)	α (Cathodic meas.)	Eliminating C_R^0: $k_1^0 \times 10$ (cm/s)	Eliminating C_R^0: $k_2^0 \times 10$ (cm/s)	Taking $C_R^0 = 1$: $k_1^0 \times 10^2$ (cm/s)	Taking $C_R^0 = 1$: $k_2^0 \times 10^5$ (cm/s)	$k^0 \times 10^5$ [b] (cm/s)	References
Cu(II)/Cu	1.0 *N* KCl	4.85	0.51	0.32	7.3	12.0	2.3	4.6	39, 42
	1.0 *N* K_2SO_4	4.00	0.50	0.18	0.78	0.72	3.7	2.9	
	1.0 *N* KNO_3	4.12	0.51	0.30	0.54	1.7	7.3	6.0	
	1.0 *N* $NaClO_4$	4.46	0.51	0.27	1.7	0.28	1.2	2.4	
Cd(II)/Cd	1.0 *N* KCl	7.4	0.50	—	0.20	5.7	3.4	22.0	38, 39
	1.0 *N* K_2SO_4	6.2	0.51	0.79	4.97	1.4	1.4	8.0	
	1.0 *N* $NaClO_4$	5.8	0.50	—	0.14	39.0	1.70	6.4	
	1.0 *N* KI	11.3	0.50	—	0.32	32.0	3.80	3.0	
	1.0 *N* KNO_3	6.4	0.52	0.80	1.42	1.1	7.2	10.0	

[a] Temperature = 27°C.
[b] k^0 is the rate constant obtained by considering two-electron charge transfer to occur in a single step or the slower step to be in overall control of the rate process.

Table 2
Kinetic Parameters of Ni(II)/Ni and Zn(II)/Zn in Various Supporting Electrolytes[a]

Reaction	Supporting electrolyte	$D_0 \times 10^6$ (meas.) (cm^2/s)	α (Cathodic meas.)	Eliminating C_R^0: $k_1^0 \times 10$ (cm/s)	Eliminating C_R^0: $k_2^0 \times 10$ (cm/s)	Taking $C_R^0 = 1$: $k_1^0 \times 10^2$ (cm/s)	Taking $C_R^0 = 1$: $k_2^0 \times 10^3$ (cm/s)	$k^0 \times 10^4$ [b] (cm/s)	References
Ni(II)/Ni	1.0 *N* KCl	10.4	0.49	0.22	0.30	2.96	0.49	0.22	39, 51
	1.0 *N* KNO_3	8.6	0.49	0.11	0.29	7.01	0.20	0.94	
	1.0 *N* KI	12.4	0.48	—	0.15	1.32	1.21	2.5	
	1.0 *N* $NaClO_4$	6.6	0.49	—	0.11	7.65	0.24	0.99	
	1.0 *N* K_2SO_4	10.0	0.44	0.94	1.70	52.5	3.07	9.1	
Zn(II)/Zn	1.0 *N* KCl	8.4	0.58	0.13	0.17	3.85	0.25	0.33	39, 42
	1.0 *N* K_2SO_4	4.8	0.51	—	0.13	3.33	0.05	1.14	
	1.0 *N* $NaClO_4$	4.6	0.61	0.06	0.11	2.50	0.02	0.22	
	1.0 *N* KNO_3	7.8	0.51	0.67	13.7	3.2	0.09	0.91	

[a] Temperature = 27°C.

[b] k^0 is the rate constant obtained by considering two-electron charge transfer to occur in a single step or the slower step to be in overall control of the rate process.

directions. The kinetic parameters reported earlier for two-electron charge transfer reactions, even those obtained by the faradaic rectification method, are based on this assumption. In discussing each individual reaction in detail below, the results obtained by applying the recently developed theory of faradaic rectification for multiple-electron charge transfer reactions will be compared with those reported in the literature.

Some of the two-electron charge transfer reactions which have recently been studied are Cu(II)/Cu(s), Ni(II)/Ni(s), Cd(II)/Cd(s), and Zn(II)/Zn(s). Their kinetic parameters in different supporting electrolytes are given in Tables 1 and 2.

(a) Cu(II)/Cu(s)

The Cu(II)/Cu reaction has been extensively studied[44-50] using uncomplexed Cu(II) ion species. Hampson and Latham[50] have studied this reaction in an aqueous nitrate electrolyte medium using faradaic impedance and galvanostatic pulse methods at varying temperatures. From their studies, it is concluded that the reduction of Cu(II) → Cu(I) is a slow process and that the second step, Cu(I) → Cu, is fast. This reaction has been studied earlier in 1 *N* KNO_3 by the faradaic rectification method[43] assuming that $k_1^0 \ll k_2^0$ and the slowest step is in overall control of the rate of reaction in both directions. The value of α reported is 0.45 and $k_a^0 = 0.11 \times 10^2$ cm/s.

Recently, the kinetic parameters for each step of this reaction in different supporting electrolytes have been obtained[39,42] by applying the faradaic rectification theory as extended to multiple-electron charge transfer reactions. The kinetic parameters are listed in Table 1.

The $\Delta E_\infty / V_A^2$ versus $\omega^{-1/2}$ plots in 1.0*N* KCl at three different concentrations of Cu(II) are given in Fig. 3. All three plots tend to be linear in the high-frequency region, and on extension, they meet on the ordinate. From the intercept on the ordinate, the value of α is obtained from Eq. (12), which reduces to

$$\Delta E_\infty / V_A^2 = \frac{nF}{4RT}(2\alpha - 1)$$

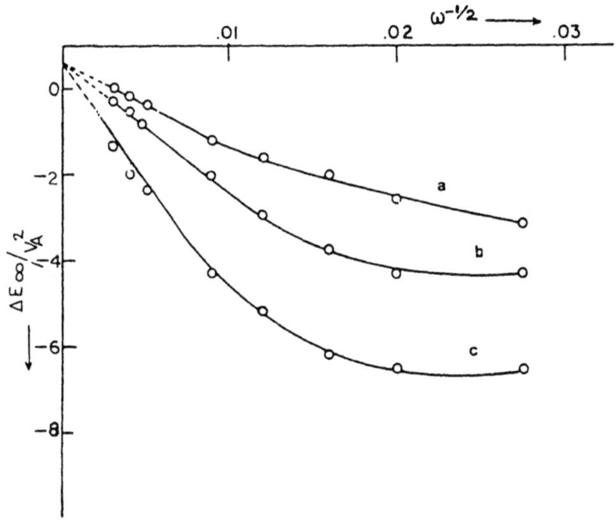

Figure 3. $\Delta E_\infty / V_A^2$ versus $\omega^{-1/2}$ plots for Cu^{2+}, 1.0 N KCl/Cu. Concentration of Cu^{2+}: (a) 2.0 mM; (b) 1.0 mM; (c) 0.5 mM.

Knowing the value of α, the value of $C_{R_1}^0$ is obtained from Eq. (e) in Appendix A as all other terms corresponding to the three concentrations of Cu(II) are known. On further substituting the values of α and $C_{R_1}^0$ in Eq. (c) Appendix A, the value of k_1^0 can be obtained. On substituting the values of α, $C_{R_1}^0$, and k_1^0 in Eq. (a) of Appendix A, only, two unknowns, i.e., k_2^0 and C_R^0, are left. Hence, using Eq. (a), at any two concentrations of Cu(II), the C_R^0 can be eliminated and the value of k_2^0 is obtained independently of C_R^0. It is very interesting to note that the value of k_2^0 thus determined is of the order of 10^{-1} and that the second step of the reaction is faster than the first, as has been observed earlier. It should, however, be noted that the value of k_2^0 in KCl is exceptionally high. It is 10 times higher than that in a potassium sulfate medium and is about 4 times higher than the value obtained in sodium perchlorate. The value of k_2^0 varies in different supporting electrolytes in the order $Cl^- > ClO_4^- > SO_4^{2-} > NO_3^-$, whereas k_1^0 in all supporting electrolytes remains practically the same and is much lower than k_2^0, particularly in a KCl medium. From the values of the kinetic parameters listed in Table 1, it becomes evident why there is so much variance in the values of kinetic data reported in the literature: they hold only for the specific experimental conditions and methods followed by the worker.

If the rate constant is obtained by assuming that only one of the steps is in overall control of the rate of reaction, then the rate constant obtained for the slowest reaction is of the order of 10^{-5} cm/s, which is the same order of magnitude as obtained when $C_R^0 = 1$.

Considering the activity of the finally reduced species in the second step of reduction, i.e., $M^+ \rightarrow M(s)$, as 1.0, the value of k_1^0 is invariably found to be much higher than that of k_2^0 in all supporting electrolytes. For the sake of comparison, such values are also included in Table 1.

(*b*) *Ni(II)/Ni(s)*

The $\Delta E_\infty / V_A^2$ versus $\omega^{-1/2}$ plots for the Ni(II)/Ni(s) reaction in 1.0 *N* KI are shown in Fig. 4. The values of α in different

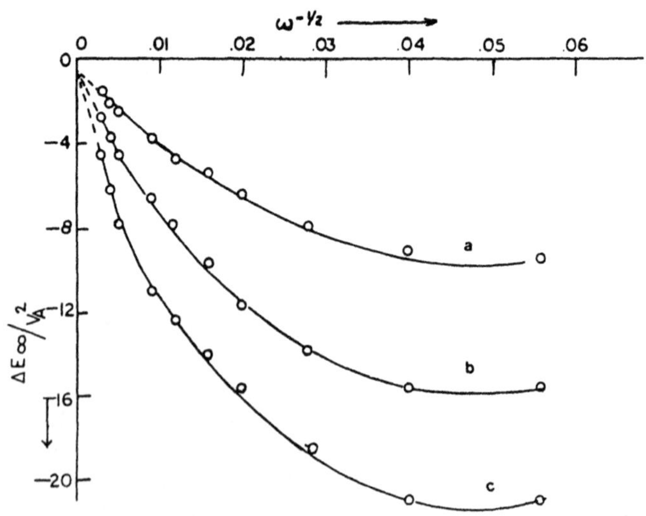

Figure 4. $\Delta E_\infty / V_A^2$ versus $\omega^{-1/2}$ plots for Ni^{2+}, 1.0 *N* KI/Ni. Concentration of Ni^{2+}: (a) 2.0 mM; (b) 1.0 mM; (c) 0.5 mM.

supporting electrolytes are obtained from the intercepts of the plots on the ordinate. Kinetic parameters for this reaction have been obtained by the faradaic rectification method under the following conditions.

1. By using Delahay's equation, assuming that both electron charge transfers occur in a single step. In this case, the rate constant obtained for the slowest reaction is of the order of 10^{-5} cm/s (Table 2).

2. By applying the recently developed theory of faradaic rectification as applied to multiple-electron charge transfer reactions under the condition that $k_1^0 \gg k_2^0$ and $C_R^0 = 1$. Kinetic parameters are obtained for each step of the electron charge transfer. The value of k_2^0 reported is of the order of 10^{-6} to 10^{-9} cm/s whereas that of k_1^0 is of the order of 10^{-3} cm/s in different supporting electrolytes.[51]

3. By applying the faradaic rectification theory of multiple-electron charge transfer reactions under the condition that $C_R^0 = 1$. The rate constant for the first step of electron charge transfer is found to be 100 to 1000 times higher than that for the second step.[39] The kinetic parameters determined in different supporting electrolytes[42] are given in Table 2.

4. By applying the theory of multiple-electron charge transfer reactions and eliminating C_R^0 by carrying out the experiment at varying concentrations of Ni(II). It is interesting to find that the values of the rate constant for the second step of charge transfer are invariably higher than those for the first step of the reaction in all supporting electrolytes. The kinetic parameters are given in Table 2. Considering the k_2^0 values, it can be seen that they vary in different supporting electrolytes in the following order:

$$SO_4^{2-} > Cl^- > NO_3^- > I^- > ClO_4^-$$

The reaction is almost 10 times faster in sulfate media than in other anionic media, which may perhaps be due to preferential adsorption of sulfate anions at the electrode surface. It may be pointed out that the second step of the reaction is generally expected to be fast because of the speed with which the intermediate species formed is discharged at the electrode surface. This finds support in other studies carried out at different metal ion/metal(s) interfaces.

(c) *Cd(II)/Cd(s)*

The discharge of Cd(II) at dropping mercury or cadmium amalgam (dropping or hanging) electrodes has been extensively studied using polarographic as well as relaxation techniques. Because of complications arising due to preferential adsorption of metal ions at the solid electrode surface, studies at metal ion/metal(s) interfaces have generally been avoided. Earlier studies were mostly confined to the study of the overall rate of reaction rather than determination of kinetic parameters for each step of electron charge transfer. More recently, Cd(II)/Cd(s) studies have been made in different electrolytes for the first time,[38] and the kinetic parameters have been obtained for each step of electron charge transfer by the faradaic rectification method. As in the case of the Ni(II)/Ni(s) system, the kinetic parameters for this system are also reported under the conditions when $C_R^0 = 1$, when C_R^0 is eliminated, and when both electrons are taken to be transferred in a single step and $C_R^0 = 1$.

The nature of $\Delta E_\infty / V_A^2$ versus $\omega^{-1/2}$ plots is similar for all supporting electrolytes to those shown in Fig. 5 for a KNO_3 medium. The kinetic parameters are given in Table 1.

As has already been explained, when $C_R^0 = 1$, the rate constant k_1^0 is 100 times higher than k_2^0 in all supporting electrolytes and the k_2^0 values are comparable to those obtained from Delahay's expression, assuming that the slowest step is in overall control of the rate of reaction.[38] On determining the kinetic parameters independent of C_R^0 (i.e., by elimination of C_R^0), it is interesting to note that k_2^0 is invariably found to be higher than k_1^0 and is in the range of 10^{-2} to 10^{-1} cm/s. The value of k_2^0 in different electrolytes varies in the order $SO_4^{2-} > NO_3^- > I^- > Cl^- > ClO_4^-$.

(d) *Zn(II)/Zn(s)*

Gaiser and Heusler[53] have shown that the electrode reaction $Zn^{2+} + 2e^- \rightarrow Zn$ proceeds in two steps: $Zn^{2+} + e^- \rightarrow Zn^+$ and $Zn^+ + e^- \rightarrow Zn(s)$. Van Der Pol *et al.*,[54] using ac coupled with the faradaic rectification polarography method, also concluded that this reaction is a multistep reaction. Hurlen and Fischer[55] have studied this reaction in an acid solution of potassium chloride and

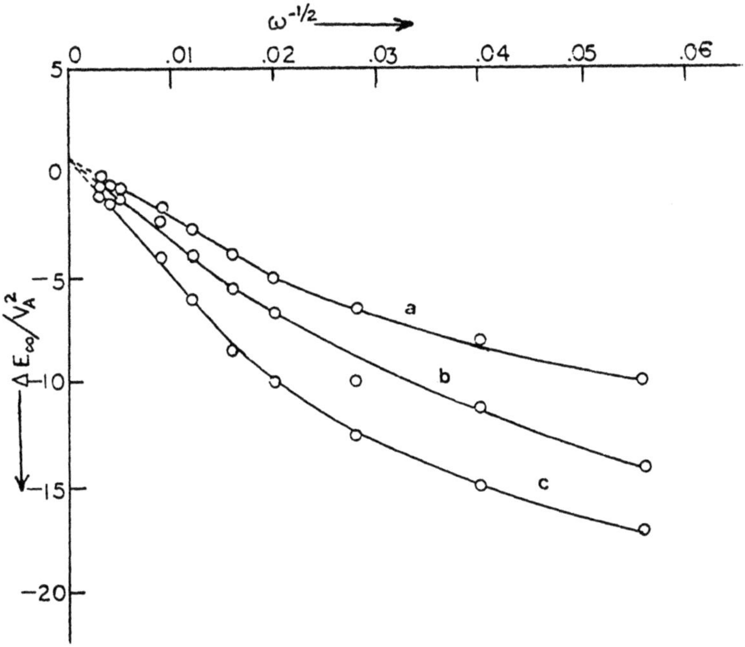

Figure 5. $\Delta E_\infty / V_A^2$ versus $\omega^{-1/2}$ plots for Cd^{2+}, 1.0 N KNO_3/Cd. Concentration of Cd^{2+}: (a) 2.0 mM; (b) 1.0 mM; (c) 0.5 mM.

concluded that of the two consecutive charge transfer steps, the ion transfer step Zn(I)/Zn(s) is a fast reaction. In all earlier studies, the exchange current density reported is for the overall reaction.

The kinetic parameters for each of the two steps of this reaction have been obtained in different supporting electrolytes by the faradaic rectification method and are given in Table 2. The $\Delta E_\infty / V_A^2$ versus $\omega^{-1/2}$ plots in 1.0 N $NaClO_4$ are given in Fig. 6. Similar curves are obtained in other electrolytes as well. On extending the three plots, they meet at a point on the ordinate. From the intercept at the ordinate, the value of the transfer coefficient determined in different supporting electrolytes varies in the range 0.51 to 0.61. Also, in this system, k_1^0 is about 1000 times higher than k_2^0 in practically all the media if C_R^0 is taken to be 1.0. The k_2^0 values are comparable to those obtained from Delahay's expression considering that both electrons are transferred in a single step and with the value of C_R^0 taken to be 1.

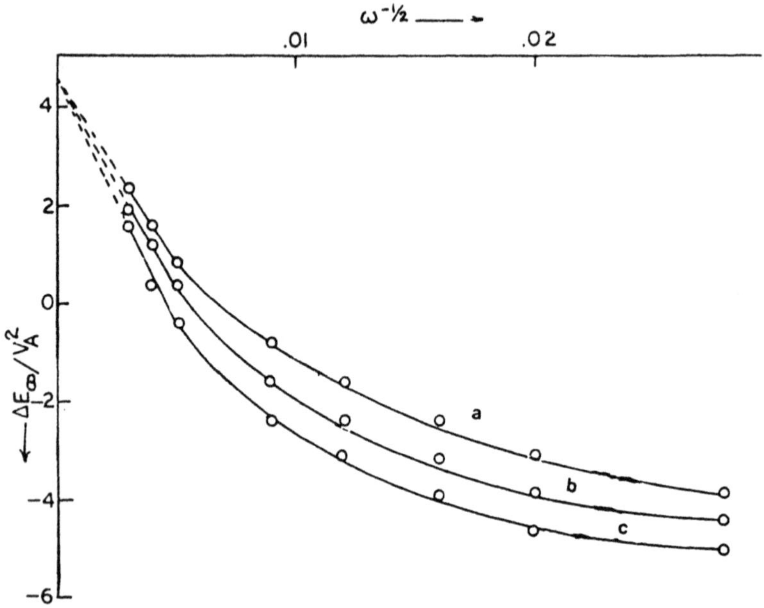

Figure 6. $\Delta E_{\infty}/V_A^2$ versus $\omega^{-1/2}$ plots for Zn^{2+}, 1.0 N $NaClO_4/Zn$. Concentration of Zn^{2+}: (a) 2.0 mM; (b) 1.0 mM; (c) 0.5 mM.

When the kinetic parameters are obtained by eliminating C_R^0, as has been done in the previous cases, k_2^0 is always higher than k_1^0, in accordance with earlier observations.[55] It is interesting to note that the rate constant in a potassium nitrate medium is very fast (100 times higher) as compared to that in other media. The k_2^0 values are, in general, of the order of 10^{-2} cm/s except in potassium nitrate, for which the rate constant is 1.37 cm/s. In general, k_1^0 is also of the order of 10^{-2} cm/s but lesser in magnitude than k_2^0.

(*iv*) *Reactions Involving Three-Electron Charge Transfer*

Very few references are available relating to the study of the Al(III)/Al(s) reaction. Most of the earlier studies have been made in molten cryolite.[56,57] Armalis and Levinskas[58] have reported the overall exchange current density of the reaction under a steady state of deposition and showed that it is a moderately fast reaction.

This reaction is found to be stable in sodium acetate and acetic acid buffer (pH 4.65), and so it has only been studied in this medium. The faradaic rectification theory becomes highly complicated when extended to three-electron charge transfer reactions due to the formation of the two intermediate species Al(II) and Al(I). In order to determine the three rate constants and the two unknown concentration terms, $C^0_{R_I}$ and $C^0_{R_{II}}$, corresponding to the two intermediate species formed, it becomes necessary to carry out the experiment at five different concentrations of aluminum ion, each below 2.00 mM.

The $\Delta E_\infty / V_A^2$ versus $\omega^{-1/2}$ plots at the five different concentrations of Al(III) meet on the ordinate at a point (Fig. 7), and from their intercept on the ordinate, the value of α obtained is 0.52.

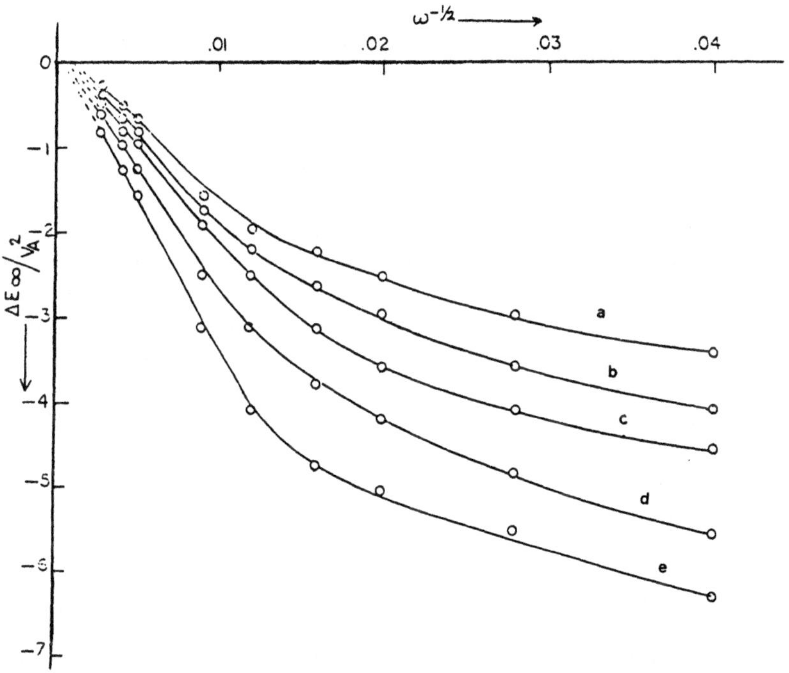

Figure 7. $\Delta E_\infty / V_A^2$ versus $\omega^{-1/2}$ plots for Al^{3+} in sodium acetate-acetic acid buffer (pH 4.65)/Al. Concentration of Al^{3+}: (a) 2.0 mM; (b) 1.5 mM; (c) 1.0 mM; (d) 0.75 mM; (e) 0.5 mM.

Having determined the value of α, the value of $C^0_{R_{II}}$ is obtained from Eqs. (o) and (p) of Appendix B. On substituting the value of α and $C^0_{R_{II}}$ in Eq. (m) of Appendix B, the value of $C^0_{R_I}$ can be obtained. Knowing these parameters, the value of k^0_1 can be determined from Eq. (j) of Appendix B, taking $C^0_R = 1$. To determine the value of k^0_3, Eq. (b) of Appendix B is used. On substituting the values of k^0_1, k^0_3, $C^0_{R_I}$, $C^0_{R_{II}}$, and all other terms in Eq. (a) of Appendix B, the value of k^0_2 is determined.

The three rate constants thus obtained are $k^0_1 = 3.68 \times 10^{-5}$, $k^0_2 = 2.70 \times 10^{-4}$, and $k^0_3 = 5.36 \times 10^{-4}$ cm/s, respectively. All three steps of the reaction are irreversible. However, on comparing the three rate constants, they vary in the order given below:

$$k^0_3 > k^0_2 > k^0_1$$

2. Faradaic Rectification Studies at Redox Couple/Inert Metal(s) Interfaces

Faradaic rectification studies in earlier stages were mostly confined to the commonly known redox couples and platinum electrode interfaces. Results relating to Fe^{2+}, Fe^{3+} in 1.0 N H_2SO_4 and $Fe(CN)_6^{4-}$, $Fe(CN)_3^{3-}$ in 1.0 N KNO_3 have already been included in an earlier review of the subject.[40] Some studies were also carried out with redox couples involving two-electron charge transfer either at a platinum interface or at a dropping mercury electrode, assuming that both the electrons are transferred in a single step and that the slow reaction is in overall control of the rate process. The kinetic parameters obtained have already been included in the earlier review. Those systems for which results were published after 1972 will be discussed in this section. The results for single-electron and two-electron charge transfer reactions will be given separately.

(i) *Redox Couples Accompanied by Single-Electron Charge Transfer*

A few reversible redox couples are known for their stability and reproducibility such as $Fe(CN)_6^{3-}$, $Fe(CN)_6^{4-}$; Cr^{3+}, Cr^{2+}; Ti^{4+}, Ti^{3+}; Ce^{4+}; Ce^{3+}; and Cu^{2+}, Cu^+. For all of these reactions, studies have been carried out at varying redox concentration ratios, and for some of these reactions, the zero-point method has also been applied. Each of the individual reactions will be discussed in detail.

(*i*) $Fe(CN)_6^{3-}$, $Fe(CN)_6^{4-}/Pt(s)$

This reaction is fairly stable in a 1.0 *N* KNO_3 medium. Earlier studies have been made using equimolar concentrations of the oxidant and reductant.[59] This reaction has been studied again by varying the redox concentration ratios.[60] The plot of $\Delta E_\infty/V_A^2$ versus $\omega^{-1/2}$ is shown in Fig. 8. Although the value of α remains unaltered, the rate constant is found to be of the order of 10^{-1} cm/s and is

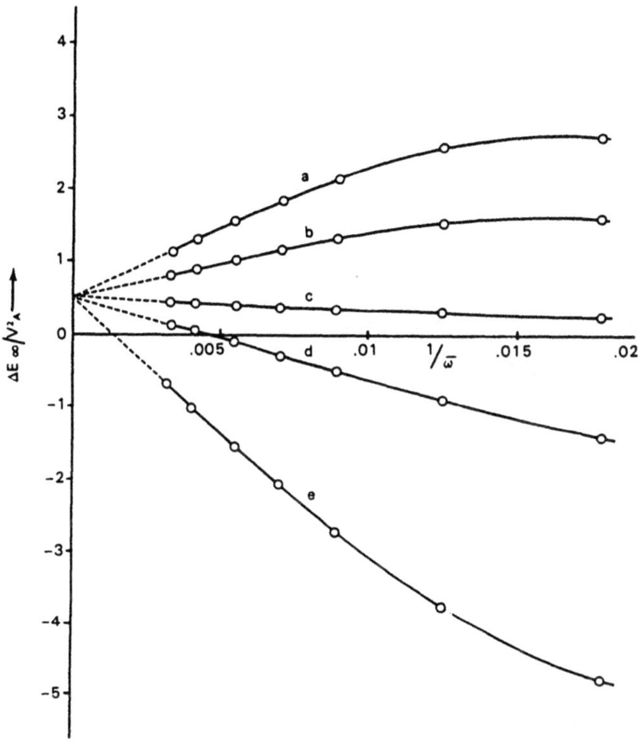

Figure 8. $\Delta E_\infty/V_A^2$ versus $\omega^{-1/2}$ plots for $Fe(CN)_6^{3-}$, $Fe(CN)_6^{4-}$ in 1.0 *N* KNO_3/Pt: (a) 4:1; (b) 2:1; (c) 1:1; (d) 1:2; (e) 1:4. (From Ref. 60, courtesy Elsevier.)

Table 3
Kinetic Parameters of Some Single-Electron Charge Transfer Redox Couples at a Platnium Interface[a]

Redox couple	Supporting electrolyte	Concentration of redox couple (oxid : red)	$D_O \times 10^6$ (cm^2/s)	α (meas.)	k_a^0 (cm/s)	Zero-point method α	Zero-point method k_a^0	References
Ce^{4+}, Ce^{3+}/Pt	1.0 *N* HNO_3	0.5 : 2.0	6.8	0.51	0.04	0.51	0.01	39
		1.0 : 2.0			0.06			
		1.0 : 1.0			0.05			
		2.0 : 1.0			0.06			
		2.0 : 0.5			0.03			
	1 *N* H_2SO_4	0.5 : 2.0	1.2	0.60	0.04	0.53	0.04	39
		1.0 : 2.0			0.05			
		1.0 : 1.0			0.31			
		2.0 : 1.0			0.16			
	1 *N* HCl	0.5 : 2.0	4.6	0.70	0.23	0.51	0.04	39
		1.0 : 2.0			0.33			
		1.0 : 1.0			0.76			
Cu^{2+}, Cu^{+}/Pt	4.5 *N* HCl + 1.0 *N* sodium citrate	0.5 : 2.0	7.16	0.495	0.03	0.51	1.12×10^{-2}	39
		1.0 : 2.0			0.06			
$Fe(CN)_6^{3-}$, $Fe(CN)_6^{4-}$/Pt	1.0 *N* KNO_3	2.0 : 0.5		0.49	10.5×10^{-2}			60, 67
		2.0 : 1.0			12.6×10^{-2}			
		1.0 : 2.0			14.2×10^{-2}			
		0.5 : 2.0			11.5×10^{-2}			
		1.0 : 1.0	12.2	0.49	6.6×10^{-2}			59
Cr^{3+}, Cr^{2+}/Pt	1.0 *N* H_2SO_4	1.0 : 5.0	5.16	0.47	1.98×10^{-3}			60, 67
		0.5 : 5.0			2.00×10^{-3}			
Ti^{4+}, Ti^{3+}/Pt	1.0 *N* HCl	0.5 : 5.0	7.14	0.49	5.56×10^{-4}			

[a] Temperature = 27°C.

almost twice the magnitude of the value reported earlier. The kinetic parameters are given in Table 3.

(*ii*) Cr^{3+}, $Cr^{2+}/Pt(s)$

The chromous-chromic reaction is highly unstable because of the high instability of the chromous salt. This redox couple is only stable when the concentration of the chromous salt is 5 to 10 times higher than that of the chromic salt. This reaction has been studied[60] in 1.0 N H_2SO_4, and the plots of $\Delta E_\infty / V_A^2$ versus $\omega^{-1/2}$ are given in Fig. 9. The value of α is 0.47 and $k_a^0 = 2 \times 10^{-3}$ cm/s. The results (Table 3) are comparable to those obtained by direct current polarography.

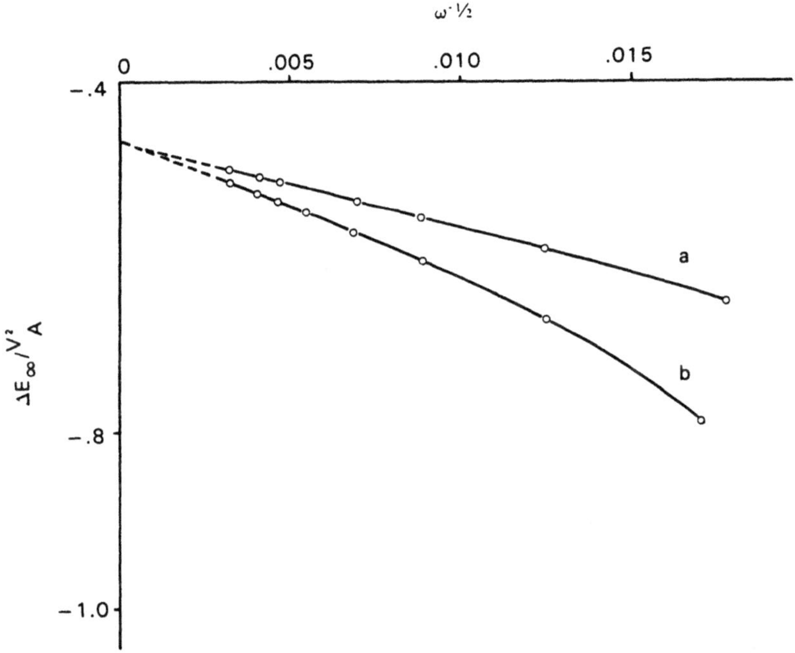

Figure 9. $\Delta E_\infty / V_A^2$ versus $\omega^{-1/2}$ plots for Cr^{3+}, Cr^{2+}, 1.0 M H_2SO_4/Pt. Cr^{3+}(mM)/Cr^{2+}(mM): (a) 1.0:5.0; (b) 0.5:5.0. (From Ref. 60, courtesy Elsevier.)

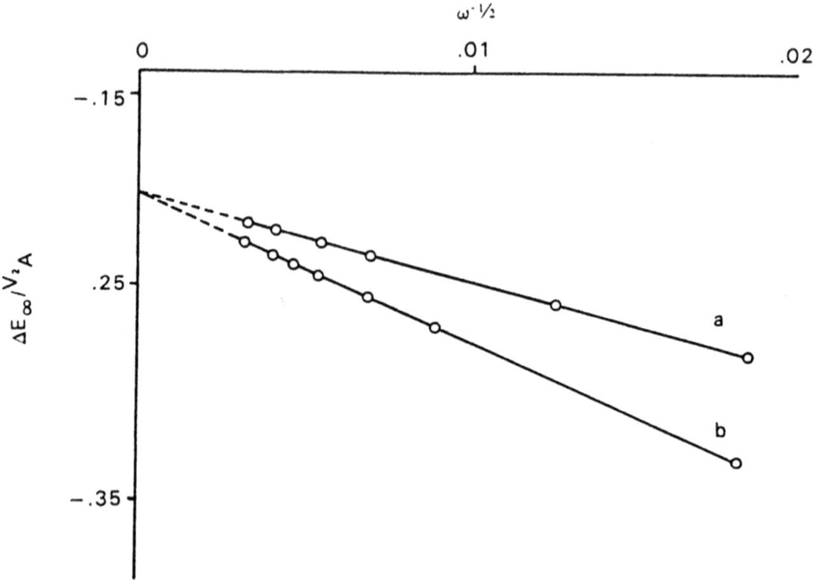

Figure 10. $\Delta E_\infty/V_A^2$ versus $\omega^{-1/2}$ plots for Ti^{4+}, Ti^{3+}, 1.0 N HCl/Pt. Ti^{4+} (mM)/Ti^{3+}(mM): (a) 0.5:5.0; (b) 0.25:5.0. (From Ref. 60, courtesy Elsevier.)

(c) Ti^{4+}, $Ti^{3+}/Pt(s)$

This redox couple has been studied in H_2SO_4 and tartaric acid at the dropping mercury interface by Delahay *et al.*[11] They only reported the value of α for the reaction. This system is only stable when the concentration of Ti^{3+} is 10 to 20 times higher than that of Ti^{4+}. The $\Delta E_\infty/V_A^2$ versus $\omega^{-1/2}$ plots for this reaction in 1.0 N HCl are shown in Fig. 10 and the kinetic parameters[60] are given in Table 3. The value of α is 0.49 and $k_a^0 = 5.56 \times 10^{-4}$ cm/s. The reaction appears to be irreversible.

(d) Ce^{4+}, $Ce^{3+}/Pt(s)$

Galus and Adams[61] reported that the rate constant for this redox couple is of the order of 10^{-4} cm/s. Some studies have been carried out for the cerous, ceric redox couple in 1.0 M H_2SO_4 using a rotating tungsten electrode[62] and also in an HNO_3 medium using a platinum electrode.[63] A value of α slightly less than 0.5 was reported in the latter study. This redox couple has been studied at varying redox concentration ratios in mineral acids, and the kinetic parameters have been obtained by the faradaic rectification method.[39,42] Kinetic parameters have also been determined by the zero-point method.[120]

The $\Delta E_\infty / V_A^2$ versus $\omega^{-1/2}$ plots in 1.0 N H_2SO_4 are shown in Fig. 11 and the kinetic parameters are given in Table 3. The value of α is 0.70 in HCl, 0.60 in H_2SO_4, and 0.51 in HNO_3. The rate constants in the three acids vary in the order given below:

$$HCl > H_2SO_4 > HNO_3$$

However, the rate constant determined by the zero-point method is 0.04 cm/s in HCl and H_2SO_4 media and in HNO_3 it is

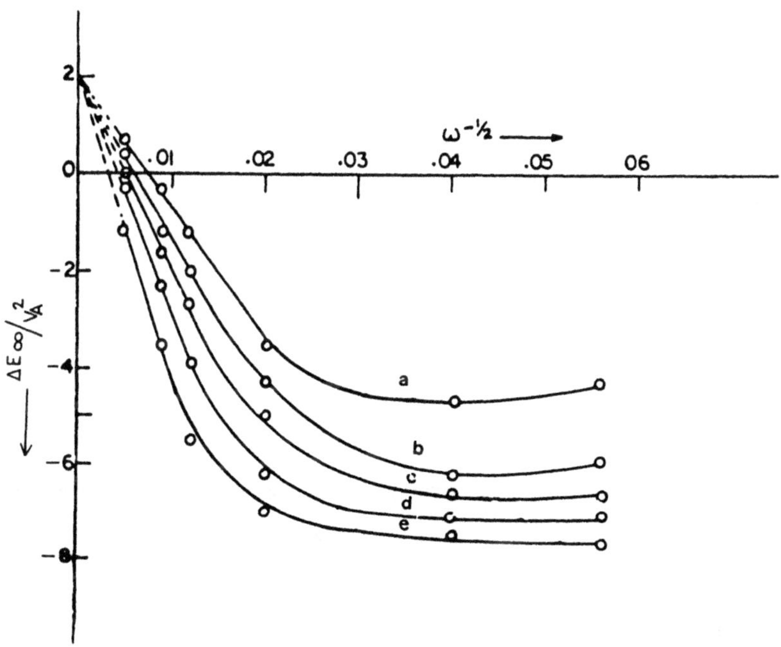

Figure 11. $\Delta E_\infty / V_A^2$ versus $\omega^{-1/2}$ plots for Ce^{4+}, Ce^{3+}, 1.0 N H_2SO_4/Pt. [Ce^{4+} (mM)/Ce^{3+}(mM)]; (a) 0.5:2.0; (b) 1.0:2.0; (c) 1.0:1.0; (d) 2.0:1.0; (e) 2.0:0.5.

only 0.01 cm/s. This shows that the reaction is much slower in HNO_3 than in the other two acids.

(e) Cu^{2+}, $Cu^{+}/Pt(s)$

Very few references are available regarding the study of this redox couple. Gorbachev and co-workers[64,65] reported from polarization curves that in citrate and chloride complex electrolytes, the oxidant and reductant reduce to the metal state.

The $\Delta E_{\infty}/V_A^2$ versus $\omega^{-1/2}$ plots for Cu^{2+}, Cu^{+} in HCl and sodium citrate media are shown in Fig. 12 and the kinetic

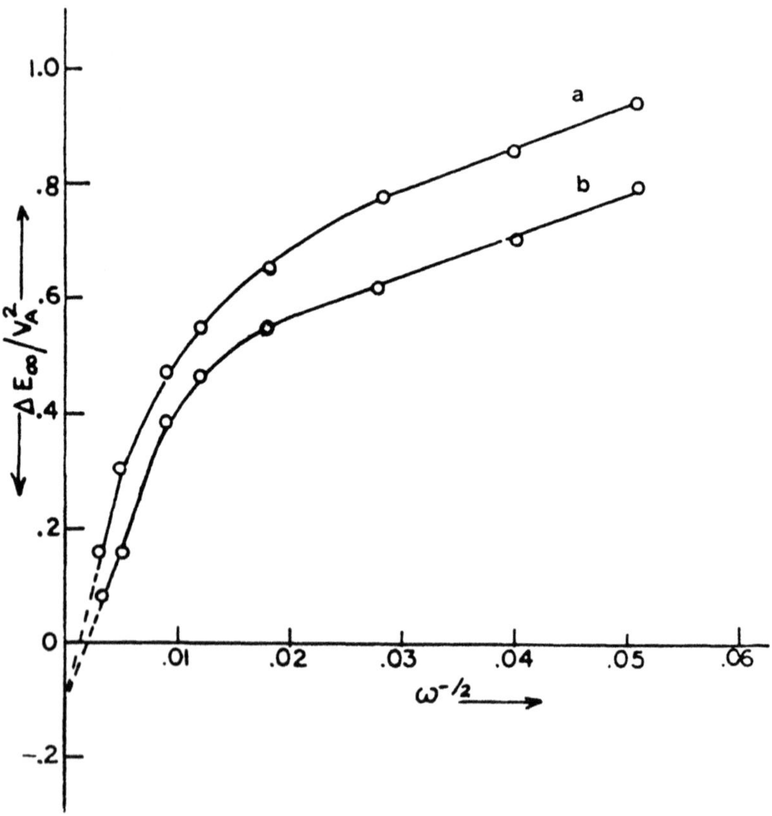

Figure 12. $\Delta E_{\infty}/V_A^2$ versus $\omega^{-1/2}$ plots for Cu^{2+}, Cu^{+} (buffer)/Pt. Cu^{2+} (mM)/Cu^{+}(mM): (a) 0.5:2.0; (b) 1.0:2.0.

parameters[39,42,120] are given shown in Table 3. In 4.5 N HCl, the value of α is 0.495 and the rate constant is 0.03 cm/s. On comparing the latter value with that obtained from the zero-point method, it is found to be 3 times higher.

(ii) *Redox Reactions Involving Two-Electron Charge Transfer*

The earlier studies by the faradaic rectification method of redox reactions involving two-electron charge transfer were done assuming that both electron charge transfers occur in a single step. This was so because the earlier theoretical formulations were only applicable to the study of single-electron charge transfer reactions. Some of the redox couples which have been studied[66,67] are I_2, I^-/Pt(s); QH_2, Q/Pt(s); Sn^{4+}, Sn^{2+}/Pt(s); and Tl^{3+}, Tl^+/Pt(s). Each of these reactions is discussed separately below.

(a) I_2, $I^-/Pt(s)$

The mechanism for this reaction has been given as[66]

$$I^- - e^- \rightleftharpoons I_{ads};\ I_{ads} + I^- \rightleftharpoons I_{2ads} + e^-$$

The absorption in the form of an adatom has already been reported by Tyagai and Kolbasov.[68] The second-step charge transfer reaction appears to be rate determining in both directions and is in overall control of the reaction. The $\Delta E_\infty/V_A^2$ versus $\omega^{-1/2}$ plots for this reaction in 1.0 N KNO_3 are presented in Fig. 13. On extrapolating the curves in the high-frequency region, they meet at a point on the ordinate. The value of α is determined from the intercept of the ordinate and that of k_a^0 from the slope of the curve. The value of α is 0.49 and $k_a^0 = 0.55 \times 10^{-2}$ cm/s (Table 4). These parameters are comparable to those obtained in KCl.[68]

(b) Sn^{4+}, $Sn^{2+}/Pt(s)$

Very few references are available for the kinetic parameters of the Sn(IV)/Sn(II) reaction except for the Sn(II)/Sn(Hg) reaction in chloride and perchlorate media.[69,70] The $\Delta E_\infty/V_A^2$ versus $\omega^{-1/2}$ plots for this reaction in 1.0 N HCl are given in Fig. 14. The transfer coefficient is obtained from the intercept of the extrapolated curves

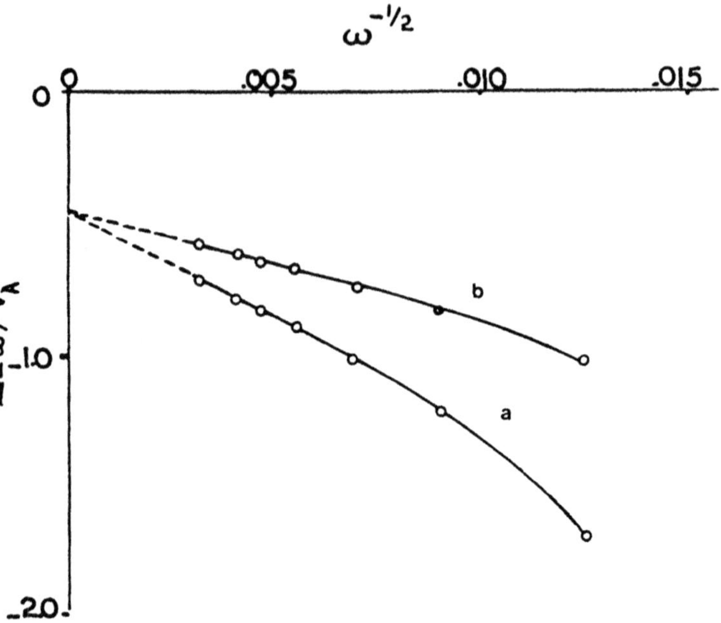

Figure 13. $\Delta E_{\infty}/V_A^2$ versus $\omega^{-1/2}$ plots for I_2, I^-, 1.0 N KNO_3/Pt. I_2 (mM)/I^- (mM): (a) 0.5:5.0; (b) 1.0:5. (From Ref. 66, courtesy Pergamon Press.)

at a point on the ordinate. Agarwal and Qureshi,[66] in their earlier studies, have reported values of $\alpha = 0.48$ and $k_a^0 = 2.1 \times 10^{-3}$ cm/s, assuming that the slower step is in overall control of the rate of the reaction. Recently, the kinetic parameters have been determined for each step of electron charge transfer involved in the reaction and the values of the two rate constants obtained are $k_1^0 = 5.47 \times 10^{-2}$ cm/s and $k_2^0 = 1.15 \times 10^{-1}$ cm/s (Table 4).

(c) Tl^{3+}, $Tl^+/Pt(s)$

Vetter and Thiemke[71] studied Tl(III), Tl(I)/Pt(s) and tried to explain the mechanism of the reaction from a Tafel plot. The kinetic parameters of this reaction have been determined by Toshima *et al.*[72] using a potential step method, by Fasco *et al.*[73] using an electrode vibration technique, and by Agarwal and Qureshi[67,74] using the

Table 4
Kinetic Parameters of Some Two-Electron Charge Transfer Reactions at a Platinum Interface

Reaction	Supporting electrolyte	Concentration of redox couple (mM) (oxid : red)	T (°C)	Diffusion coefficient, $D_O \times 10^6$ (meas.) (cm^2/s)	α_c (Cathodic meas.)	α_a (Anodic meas.)	k_a^0 (cm/s) (meas.)	References
Iodine, Iodide/Pt	1.0 M KI	1.0 : 1000	27	6.81	0.5	0.5	5.10×10^{-5}	66, 67
		2.0 : 1000			0.5	0.5	4.94×10^{-5}	
	1.0 M KNO_3	0.5 : 5.0	25		0.49	0.51	5.59×10^{-3}	
		1.0 : 5.0			0.49	0.51	4.44×10^{-3}	
				$C_R^0 > C_O^0$		Zero-point method; $C_R^0 > C_O^0$		
		$D_O \times 10^5$ (cm^2/s) (meas.)	α (Cathodic meas.)	$k_1^0 \times 10$ (cm/s) (meas.)	$k_2^0 \times 10$ (cm/s) (meas.)	$k_1^0 \times 10^3$ (cm/s) (meas.)	$k_2^0 \times 10^2$ (cm/s) (meas.)	
Tl^{3+}, Tl^{+}/Pt	1.0 N KNO_3	1.84	0.52	4.52	5.99	113	68.2	39, 52
	1.0 N K_2SO_4	1.76	0.59	1.26	8.55	8.14	2.38	
	1.0 N $NaClO_4$	1.21	0.51	1.58	1.76	6.88	1.92	
Sn^{4+}, Sn^{2+}/Pt	1.0 N HCl	0.68	0.48	0.55	1.15	—	—	

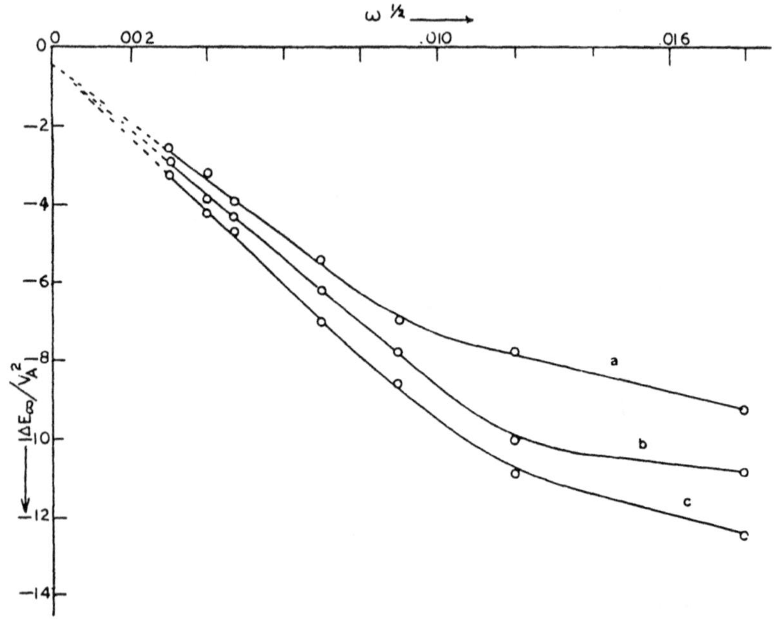

Figure 14. $\Delta E_{\infty}/V_A^2$ versus $\omega^{-1/2}$ plots for Sn^{4+}, Sn^{2+}, 1.0 *N* HCl/Pt. Sn^{4+}(mM)/Sn^{2+} (mM): (a) 2.0 : 5.0; (b) 1.0 : 5.0; (c) 0.5 : 5.0. (From Ref. 66, courtesy Pergamon Press.)

faradaic rectification method. In all the earlier studies, the kinetic parameters of the overall reaction were determined assuming that the same step is rate determining in both directions. Values of $\alpha = 0.48$ and $k_a^0 = 0.12$ cm/s in 1.0 *N* $HClO_4$ have been reported.

The $\Delta E_{\infty}/V_A^2$ versus $\omega^{-1/2}$ plots in 1.0 *N* $NaCiO_4$ are shown in Fig. 15 and the kinetic parameters obtained from extrapolation of these plots and using the zero-point method are given in Table 4. It may be pointed out that when C_R^0 is kept constant and C_O^0 is varied, the value of $C_{R_1}^0$ is obtained from Eq. (e) of Appendix A, that of k_1^0 from Eq. (c) of Appendix A, and that of k_2^0 from Eq. (12). For determining the value of the two rate constants by the zero-point method, the theoretical formulations for multiple-electron charge transfer have suitably been modified, and corresponding expressions for k_2^0, k_1^0, and $C_{R_1}^0$ have been deduced from Eqs. (16), (19), and (21).

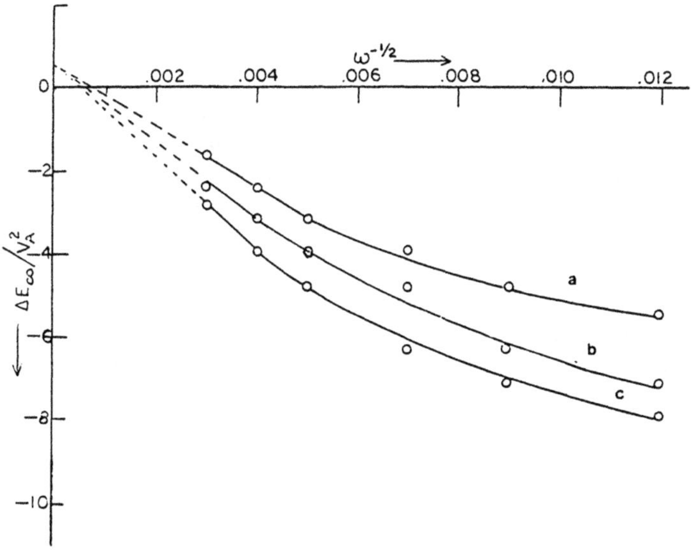

Figure 15. $\Delta E_\infty / V_A^2$ versus $\omega^{-1/2}$ plots for Tl(III), Tl(I) in 1.0 N $NaClO_4$/Pt. Tl^{3+}(mM)/Tl^{+}(mM): (a) 1.0 : 2.0; (b) 0.75 : 2.0; (c) 0.5 : 2.0.

The two rate constants obtained in different supporting electrolytes[52] are of the order of 10^{-1} cm/s, indicating that the reaction is fast. It can also be seen that k_2^0 is invariably higher than k_1^0 in all media but varies less. Thus, it is evident that in such reactions one cannot assume either of the two reactions as slow and in overall control of the charge transfer reaction. On comparing the values of k_2^0 in different supporting electrolytes, they are found to vary in the following order:

$$SO_4^{2-} > NO_3^- > ClO_4^-$$

It is interesting to note that the influence of these supporting electrolytes on k_1^0 is in a different order, i.e.,

$$NO_3^- > ClO_4^- > SO_4^{2-}$$

The rate constants k_1^0 and k_2^0 determined by the zero-point method are generally 10^{-1} to 10^{-2} times lower than those obtained by extrapolation techniques. The influence of different electrolytes on both rate constants is as given below:

$$NO_3^- > SO_4^{2-} > ClO_4^-$$

It is interesting to observe that in nitrate media both rate constants are 100 times higher than in the other two media. The high values of rate constants in NO_3^- and SO_4^{2-} media are due to their preferential adsorption at the electrode/solution interface.[7,75-77]

(d) QH_2, $Q/Pt(s)$

The most widely studied and stable organic redox couple is quinone–hydroquinone, and this reaction can be studied at varying pH's up to 7.5. The earlier faradaic rectification studies involved the determination of the transfer coefficient at varying pH's using equimolar concentrations of quinone and hydroquinone. Agarwal and Qureshi[67] determined the kinetic parameters of this redox couple using varying redox concentrations but assuming that the first step of electron charge transfer is in overall control of the reaction. More recently, the kinetic parameters for each step of electron charge transfer have been determined for the first time at varying pH's by the usual extrapolation method as well as by the zero-point method. The $\Delta E_\infty/V_A^2$ versus $\omega^{-1/2}$ plots at pH 7.4 are presented in Fig. 16 and the kinetic parameters are given[119] in Table 5.

The studies have been carried out with C_O^0 kept constant and larger than C_R^0 and vice versa, maintaining a total redox concentration not exceeding 4.0 mM. It is generally observed that when the oxidant/reductant concentration ratio is increased, at any frequency the rectification potential becomes more positive. Conversely, on increasing the redox concentration ratio, the rectification potential tends to become more negative at any frequency. On extrapolating the plots at high frequencies (Fig. 16) at all redox concentration ratios, they meet on the ordinate at a point and the value of α determined at all pH's is 0.475 to 0.480. When $C_O^0 > C_R^0$, the $\Delta E_\infty/V_A^2$ versus $\omega^{-1/2}$ plots on extrapolation intersect the abscissa, and it is possible to determine the two rate constants by the zero-point method in such cases.

The values of $C_{R_1}^0$ are obtained from Eq. (j) of Appendix A and those of k_2^0 and k_1^0 are obtained from Eqs. (f) and (g) and Eq.

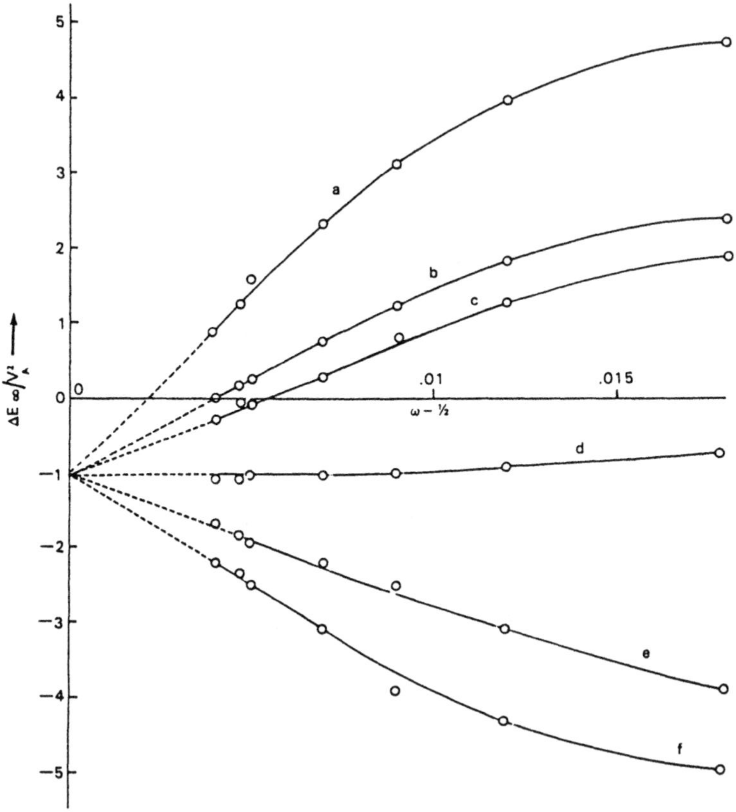

Figure 16. $\Delta E_\infty / V_A^2$ versus $\omega^{-1/2}$ plots for Q, QH_2 (buffer of pH 7.4)/Pt. Q (mM)/QH_2 (mM) = (a) 2:0.5; (b) 2:1.0; (c) 2:1.5; (d) 1.5:2; (e) 1:2; (f) 0.5:2.

(f) of Appendix A, respectively. For obtaining the value of $C_{R_1}^0$ by the zero-point method, the values of zero-point frequencies (when $\Delta E_\infty = 0$) are substituted along with other terms in Eq. (28). Knowing $C_{R_1}^0$, the value of k_2^0 is determined from Eq. (25) . Finally, the value of k_1^0 is obtained from Eq. (22).

It is interesting to note that at all pH's, when $C_O^0 > C_R^0$, the value of k_1^0 is invariably greater than that of k_2^0, but on the other hand, when $C_R^0 > C_O^0$, k_2^0 is always found to be higher than k_1^0. These results[119] explain clearly for the first time why the results

Table 5
Kinetic Parameters of Quinone, Hydroquinone Redox Couple at a Platinum Interface[a]

pH	$D_O \times 10^5$ (meas.) (cm^2/s)	α (Cathodic meas.)	$C_O^0 > C_R^0$: $k_1^0 \times 10^2$ (cm/s)	$C_O^0 > C_R^0$: $k_2^0 \times 10^2$ (cm/s)	$C_R^0 > C_O^0$: $k_1^0 \times 10$ (cm/s)	$C_R^0 > C_O^0$: $k_2^0 \times 10$ (cm/s)	Zero-point method; $C_O^0 > C_R^0$: $k_1^0 \times 10$ (cm/s)	Zero-point method; $C_O^0 > C_R^0$: $k_2^0 \times 10^3$ (cm/s)	References
4.6	0.16	0.487	0.14	0.09	3.8	6.59	—	—	39, 119
7.0	6.9	0.475	28.6	3.05	1.26	2.33	0.47	7.4	
7.4	6.0	0.475	3.19	1.85	2.67	0.56	5.7	2.1	

[a] Temperature = 27°C.

reported earlier using different redox concentration ratios do not tally. The kinetic parameters obtained by the zero-point method are generally lower in magnitude than those obtained by the extrapolation method. k_2^0 is invariably much less than k_1^0 when $C_O^0 > C_R^0$.

IV. FARADAIC RECTIFICATION POLAROGRAPHY AND ITS APPLICATIONS

The merits of the method in which a modulated high-frequency signal is superimposed on a varying dc polarizing potential were investigated by Barker,[7] Delahay and co-workers,[11,13] and Agarwal.[40,41] Barker initially reported the technique as low level faradaic rectification (LLFR) or radio frequency polarography. Later, Van Der Pol, Sluyters-Rehbach, and Sluyters[78,79] extended the method and carried out a more systematic study, renaming the technique faradaic rectification polarography.[78] Agarwal and Saxena[80,82] superimposed an unmodulated audio frequency alternating current on a polarizing dc potential and recorded the rectified current at varying frequencies. On plotting the rectified current versus dc potential, the polarograms (FR polarograms) obtained are somewhat similar to ac polarograms. Using the faradaic rectification polarographic technique, the kinetics of reduction of several metal ions in different supporting electrolytes and of some organic depolarizers at varying pH's have been studied so as to examine the potential of the method.

1. Studies Using Inorganic Ions as Depolarizers

Faradaic rectification polarographic studies have been carried out for a mixture containing several metal ions together and also for individual inorganic depolarizers so as to explore the applicability and limitations of the method and to determine kinetic parameters for some of them. For comparison, some of the dc and ac polarograms have also been recorded simultaneously. In the following, the details of the experimental technique used will be described and the potentiality of the technique in qualitative and quantitative analysis will be examined. The applicability of the method in the determination of kinetic parameters of individual depolarizers will also be described.

(i) *Circuit and Measurement Technique*

The circuit diagram for ac polarography and faradaic rectification polarography[81] is shown in Fig. 17. The output from an audio oscillator is made incident through a two-way key on a precision potentiometer. The negative end of the potentiometer is connected to the dropping mercury electrode (d.m.e.). The pool is grounded through a standard precision resistance of 100 Ω. The ac and the rectified current are measured across the precision resistance. For measuring ac the key K_2 is disconnected and key K_1 is pressed. The ac signal is amplified by the transistorized mixer preamplifier. The amplified ac is measured on a Solartron double-beam oscillograph. For some measurements, an ac multivoltmeter has also been used.

The rectified dc is measured by pressing key K_2 and leaving K_1 open, thus enabling filtration of ac through a low-pass filter. The output of the filter is connected to the dc microvoltmeter so as to measure the rectified voltage. The residual dc potential, if any, is also measured by switching off the ac using the two-way

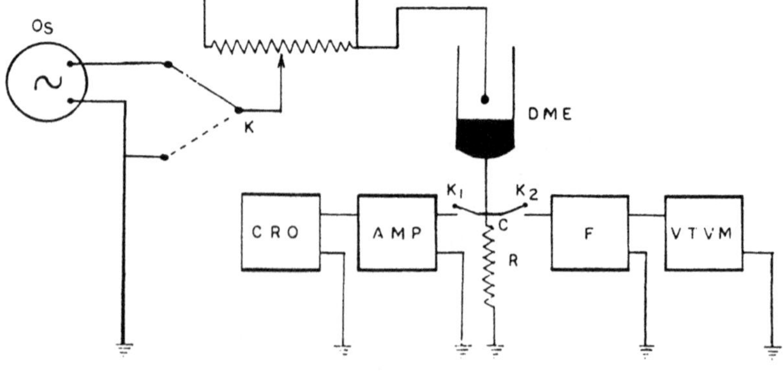

Figure 17. Circuit diagram for FR polarography.

key K. This is subtracted from the measured rectified voltage so as to obtain the actual magnitude of the rectified voltage. Throughout the measurements, the interfacial potential is kept below 15 mV.

The dropping mercury electrode has the following characteristics: $m = 0.64$ mg/s; $t = 8.0$ s in distilled water in an open circuit.

All solutions are made in bidistilled water by dissolving the required quantities of analytical-grade reagents. Pure nitrogen gas is bubbled through the solutions to get rid of dissolved oxygen.

The diffusion coefficient of the depolarizer in any supporting electrolyte is determined using a McBain–Dowson cell and the King-Cathard equation.

Faradaic rectification polarographic studies have been carried out in a mixture[82] containing seven metal ions—Pb(II), Tl(I), In(III), Cd(II), Ni(II), Zn(II), and Co(II), the concentration of each being as low as 1.0 mM in 1.0 *M* KCl and the half-wave potential of some of them differing by less than 20 mV—to examine the applicability of the method. Simultaneously, the dc steps and ac polarograms have also been obtained so as to compare the limitations of the three methods. The FR polarograms at varying ac frequencies have been recorded. For the purpose of determining the kinetic parameters of each individual depolarizer in different supporting electrolytes, the FR polarogram of each of the depolarizers has been obtained. The results are given in Figs. 18–20.

The dc polarogram of the mixture containing the above-mentioned metal ions shows distinct steps corresponding to Tl(I), Cd(II), Zn(II), and Co(II) ions only (Fig. 18) but does not provide information about the presence of the other three ions, viz., lead, indium, and nickel, in the mixture. Further, the limiting current corresponding to each step seems to be due to a combined concentration effect of lead and thallium ions, cadmium and indium ions, and zinc and nickel ions.

The ac polarograms of the mixture at varying frequencies (Fig. 19) shows four ac summit peaks corresponding to reduction of Tl(I), In(III), Cd(II), and Zn(II). The summit peaks for In(III) and Cd(II) are very close and so their ac waves are not very sharp. The first summit peak corresponding to Tl(I) appears to be due to the combined reduction of lead and thallium ions, as is evident from the summit peak height. Hence, ac polarographic analysis only enables the identification of four metal ions out of seven and

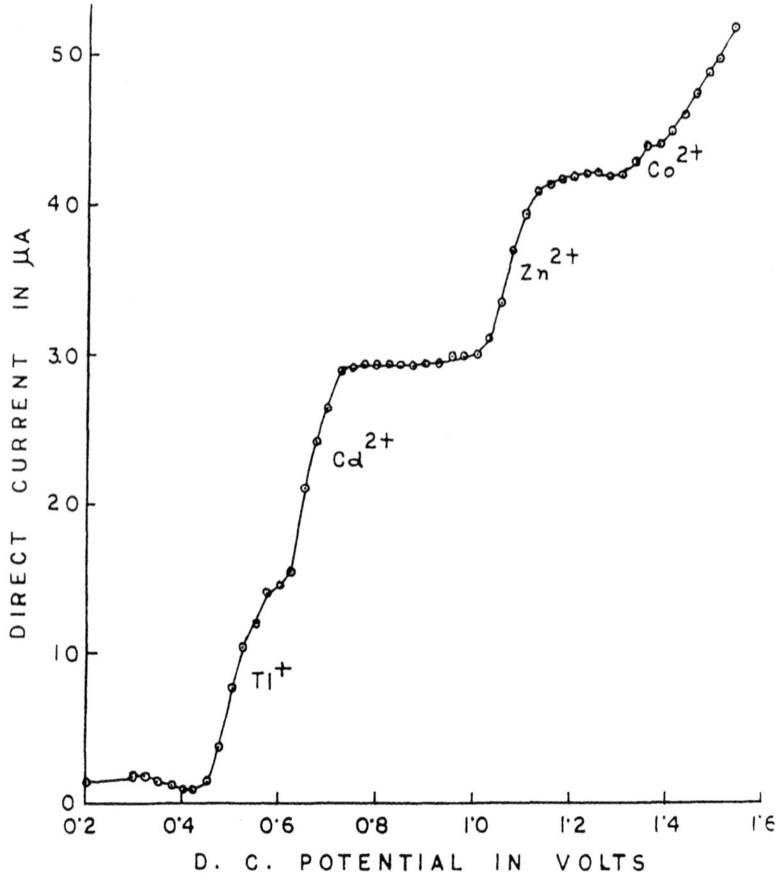

Figure 18. DC polarogram of a mixture containing 1.0 mM concentrations of Pb^{2+}, Tl^{+}, In^{3+}, Cd^{2+}, Ni^{2+}, Zn^{2+}, and Co^{2+} in 1.0 *N* KCl.

the determination of their concentrations may also not be very accurate because of the overlapping of some waves. The ac polarographic technique only permits resolution of ac waves when the half-wave potentials of the constituents differ by 40 mV. Further, the method is strictly applicable only to moderately fast reversible reactions whose rate constants fall between 10^{-2} cm/s and 10^{-1} cm/s. The ac summit peak height does not always vary linearly with the concentration of the depolarizer.

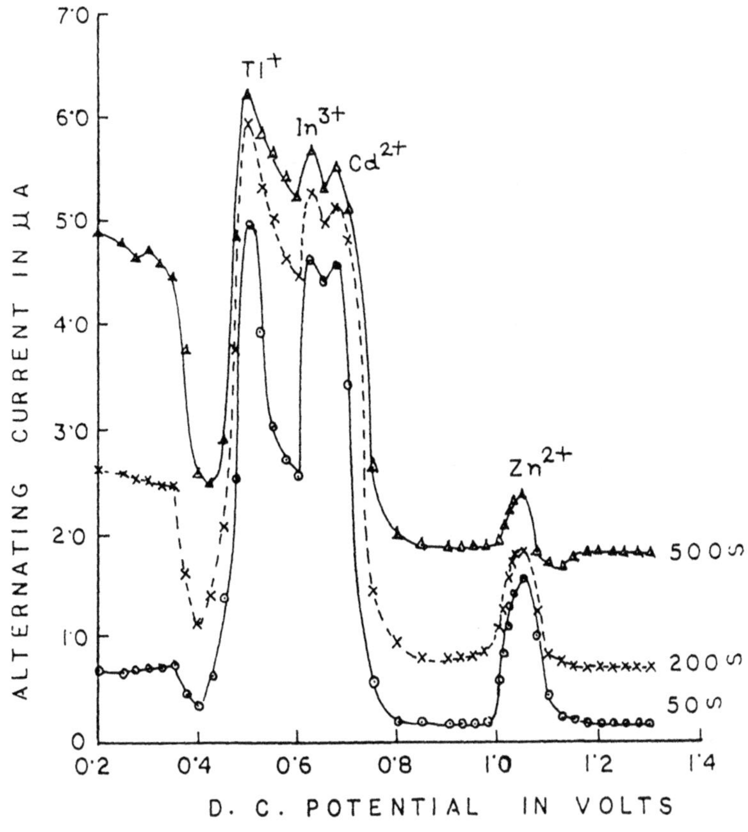

Figure 19. AC polarograms of a mixture containing 1 mM concentrations of Pb^{2+}, Tl^{+}, In^{3+}, Cd^{2+}, Ni^{2+}, Zn^{2+}, and Co^{2+} in 1.0 *N* KCl.

The FR polarograms obtained for the mixture containing the seven components (Fig. 20) has the following distinct features in comparison to dc and ac polarograms:

(i) When FR polarographic studies are carried out with a supporting electrolyte alone (in the absence of depolarizer), the rectified current is found to be zero at all potentials (Fig. 20) unlike the ac polarogram for the supporting electrolyte alone, in which some appreciable initial amount of ac is invariably present.

(ii) Distinct FR summit peaks corresponding to each metal ion are obtained (coincident with the half-wave potential) even

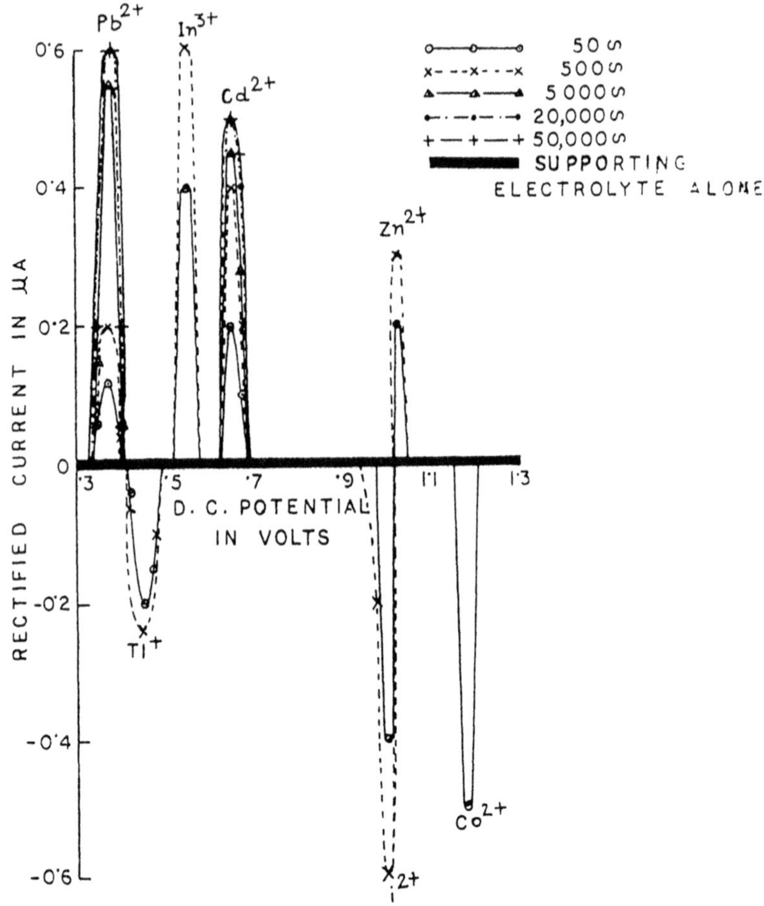

Figure 20. FR polarograms of a mixture containing 1 mM concentrations of Pb^{2+}, Tl^{+}, In^{3+}, Cd^{2+}, Ni^{2+}, Zn^{2+}, and Co^{2+} in 1.0 *N* KCl.

when the half-wave potentials of the components present differ by 20 mV (as can be seen in case of reduction of nickel and zinc ions). This is a great advantage in the qualitative analysis of mixtures containing several constituents.

(iii) For those reactions which are very fast, such as the reduction of lead and cadmium ions, the FR summit peaks even appear up to very high frequencies (50 kHz) whereas for very slow reactions, such as the reduction of Co(II), the FR summit peak is only conspicuous at 50 Hz. Thus, it is obvious that the frequency dependence of FR summit peaks gives an indication of the comparative rates of reduction of the different constituents present in the mixture.

(iv) For some ions, i.e., Pb(II), In(III), Cd(II), and Zn(II), the FR summit peaks are seen in the positive quadrant whereas those corresponding to the reduction of Tl(I), Ni(II), and Co(II) appear in the negative region. This shows that it can be concluded from FR polarograms that those ions whose summit peaks appear in the positive region should have transfer coefficients greater than 0.5, whereas those whose summit peaks fall in the negative quadrant should have transfer coefficients less than 0.5. This also explains how a better resolution of FR summit peaks, even when the half-wave potentials of the constituents differ only by 20 mV, is achieved.

(v) Since the FR summit peak current at any frequency varies linearly with the concentration of the depolarizer, the quantitative estimation of each of the constituents should be possible.

(vi) By applying the formulations for the faradaic rectification under the condition $C_O^0 D_O^{1/2} = C_R^0 D_R^{1/2}$ (which is met at the half-wave potential), it is possible to determine the value of the transfer coefficient (α) and of the apparent rate constant (k_a^0) for the reaction.

(vii) FR polarographic studies are applicable to reversible reactions, quasi-reversible reactions, and irreversible reactions. Thus, FR polarography can be used to study all kinds of reactions and is not limited to either irreversible reactions, as is the case for dc polarography, or moderately fast reactions, as for ac polarography. The method has been applied for the determination of kinetic parameters of some inorganic depolarizers in different supporting electrolytes. The results for each depolarizer will be discussed separately.

(ii) *Single-Electron Charge Transfer Reactions*

(a) *Tl(I)/d.m.e.*

The FR polarogram of 1.0 mM Tl(I) in 1.0 N $NaClO_4$ is shown in Fig. 21. The FR summit peaks at all ac frequencies coincide with

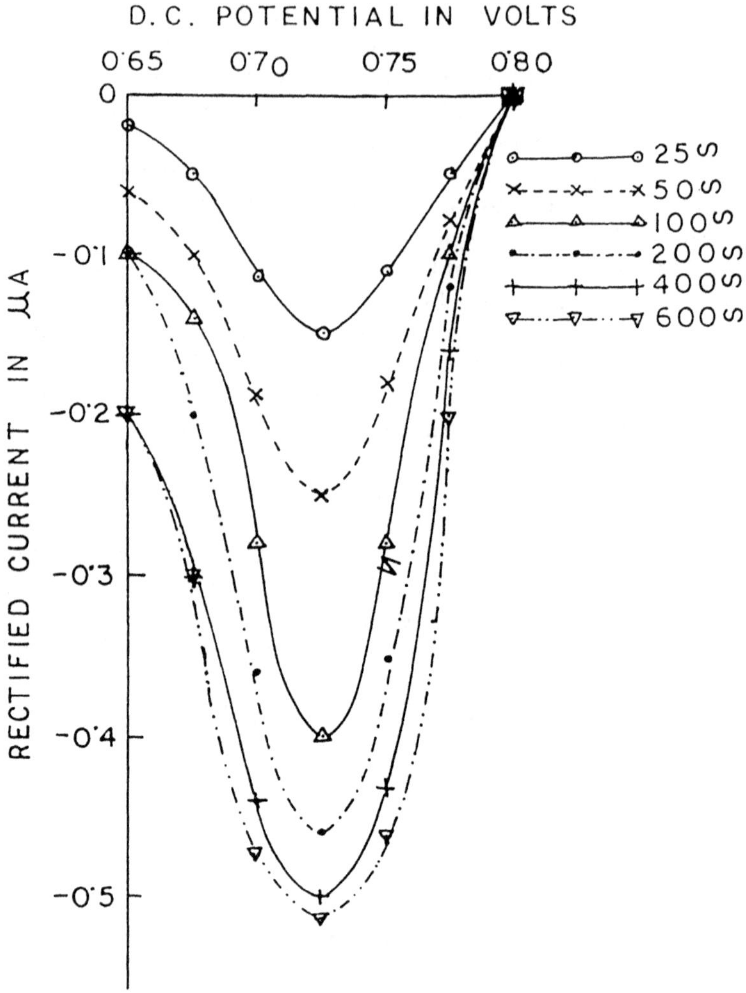

Figure 21. FR polarograms of 1.0 mM Tl(I) in 1.0 *N* $NaClO_4$.

the dc half-wave potential of Tl(I) in 1.0 *N* $NaClO_4$ and the summit peak is conspicuous even up to 600 Hz, indicating that the reaction should be moderately fast. The kinetic parameters are obtained[83,84] from the plot of $\Delta E_\infty / V_A^2$ versus $\omega^{-1/2}$ (Fig. 22). When $C_O^0 D_O^{1/2} = C_R^0 D_R^{1/2}$ (the conditions met at the half-wave potential), the faradaic

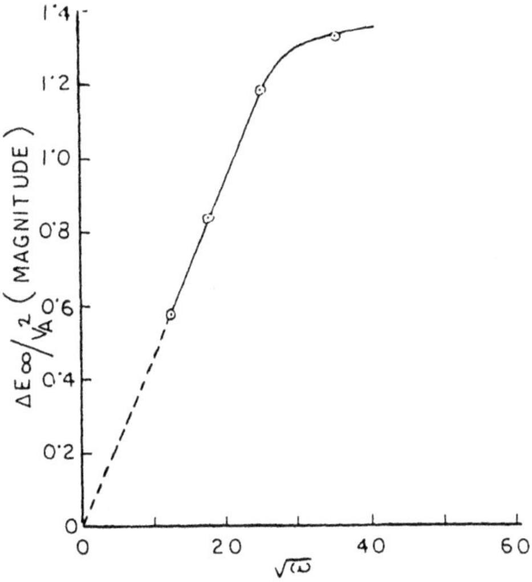

Figure 22. $\Delta E_\infty / V_A^2$ versus $\omega^{-1/2}$ plot for 1.0 mM Tl(I) in 1.0 N $NaClO_4$.

rectification theory[3,4] gives the following expressions:

$$\Delta E_\infty = \frac{nF}{RT} V_A^2 \frac{(2\alpha - 1)}{4} \tag{29}$$

$$\Delta E_\infty = \frac{nF}{RT} V_A^2 \frac{(2\alpha - 1)}{4} \frac{1}{2k_a^0} \left(\frac{\omega D}{2}\right)^{1/2} \tag{30}$$

where all the notations are as usual. Knowing V_A, the ac potential at the electrode/solution interface, and the constant value of $\Delta E_\infty / V_A^2$ obtained in the high-frequency region (Fig. 22), the value of α is obtained from Eq. (29). Substituting the values of α, D, the diffusion coefficient of Tl(I) in 1.0 N $NaClO_4$, and the slope corresponding to the linear relationship in the low-frequency region (Fig. 22) in Eq. (30), the value of k_a^0 is determined.

Similar FR polarograms and $\Delta E_\infty / V_A^2$ versus $\omega^{-1/2}$ plots are obtained for Tl(I) in other supporting electrolytes, and the respective values of α and k_a^0 have been obtained. The kinetic parameters are given in Table 6. As the FR polarograms in all the supporting electrolytes are obtained on the negative side of the abscissa, the transfer coefficient is expected to be less than 0.5, and it lies between

Table 6
Kinetic Parameters of Tl(I) and Cu(I) in Various Supporting Electrolytes[a]

Reaction	Supporting electrolyte	$D_0 \times 10^6$ (meas.) (cm/s)	α (Cathodic meas.)	$k_a^0 \times 10^2$ (cm/s)	$k_t^0 \times 10^2$ (cm/s)
Tl(I)/d.m.e.[b]	1.0 *N* KNO_3	11.57	0.45	3.5	—
	1.0 *N* KCl	11.32	0.40	3.2	1.3
	0.5 *N* K_2SO_4	16.50	0.455	4.0	—
	1.0 *N* $NaClO_4$	7.90	0.43	2.9	—
Cu(I)/d.m.e.[c]	1.0 *N* KBr	4.5	0.55	2.2	—
	1.0 *N* KNO_3	4.1	0.54	1.3	0.38
	1.0 *N* KCl	4.8	0.535	2.0	0.81
	0.5 *N* K_2SO_4	4.0	0.54	2.0	—
	1.0 *N* $NaClO_4$	4.4	0.55	2.0	0.97

[a] Temperature = 27 ± 1°C; concentration of the depolarizer = 1.0 mM.
[b] Refs. 52, 83, and 84.
[c] Refs. 83 and 85.

0.40 and 0.46 (in different supporting electrolytes). The rate constant is of the order of 10^{-2} cm/s and varies from 2.9×10^{-2} to 4.0×10^{-2} cm/s in different supporting electrolytes. It should be noted that these values are of lower order than those reported by Barker[7] using the faradaic rectification method at radio frequencies. Very few references are available in the literature for this reaction. The values of the apparent standard rate constant vary in different supporting electrolytes in the following order:

$$SO_4^{2-} > NO_3^- > Cl^- > ClO_4^-$$

The values of k_a^0 have been corrected for the double layer in a KCl medium (as ϕ_2 data were available only in this medium) by applying the Frumkin theory. They are given in Table 6.

(*b*) *Cu(I)/d.m.e.*

Cu(II) is reduced in KCl and KBr media in two steps, i.e., Cu(II) → Cu(I) in the vicinity of zero potential and Cu(I) → Cu(metal) at −0.225 and −0.15V, respectively. As the ac polarograms and FR polarograms are recorded in different supporting electrolytes beyond −0.10 V, the summit peaks obtained in both cases (Fig. 23) correspond to the reduction of Cu(I) → Cu(metal). The FR summit peaks are conspicuous up to 600 Hz in all supporting electrolytes and are obtained in the positive region at the half-wave potential corresponding to the second electron charge transfer reaction. This shows that the value of α should be greater than 0.5 and the reaction should be moderately fast.[85] The kinetic parameters are obtained in the same way as for the reduction of Tl(I) ions and are given in Table 6. The value of α lies between 0.535 and 0.55 and the rate constant ranges from 1.3×10^{-2} to 2.2×10^{-2} cm/s in different supporting electrolytes. The rate constants obtained are consistent with those obtained by faradaic impedance and ac polarographic methods.[86,87]

The double-layer correction has been applied in perchlorate, nitrate, and chloride media and the true rate constant, k_t^0, varies

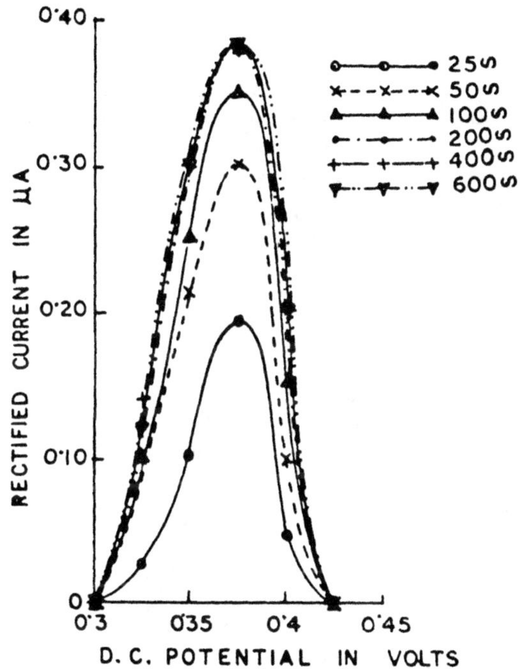

Figure 23. FR polarograms of 1.0 mM Cu(II) in 1.0 N KNO_3.

in the following order:

$$ClO_4^- > Cl^- > NO_3^-$$

It is interesting to note that the rate constant is appreciably lower in the nitrate medium as compared to that in the other supporting electrolytes.

(iii) *Two-Electron Charge Transfer Reactions*

(a) *Pb(II)/d.m.e.*

The FR summit peak has been obtained at the dc half-wave potential in the respective supporting electrolytes (Fig. 24). The

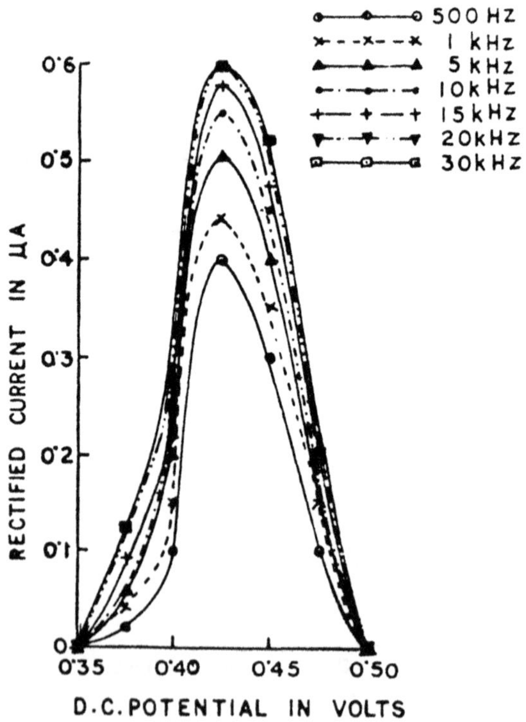

Figure 24. FR polarograms of 1.0 mM Pb(II) in 1.0 *N* KCl.

kinetic parameters have been obtained using the usual two equations [Eqs. (29) and (30)]. The value of α obtained[88] lies between 0.52 and 0.53 and k_a^0 ranges from 0.34×10^{-2} to 0.98×10^{-2} cm/s (Table 7). The values of α are comparable to those reported by Agarwal,[41] but k_a^0 is somewhat lower than given in the literature.[7,41] It should be noted that the value of the rate constant in the KCl medium is fairly high and is not as low as reported by Barker.[7] The high value in the KCl medium is due to excessive adsorption of chloride ion, as has been reported by Sluyters-Rehbach *et al.*[89] The k_a^0 values in KNO_3 and KBr are comparable to those reported in the literature. However, they are not as low as that reported by Barker.[7]

The rate constants in different supporting electrolytes vary in the following order:

$$Cl^- > ClO_4^- > I^- > NO_3^- > Br^-$$

The k_a^0 values in chloride, bromide, and percholate media have been corrected by applying the Frumkin theory for the double-layer

Table 7
Kinetic Parameters of Pb^{2+} and Cd^{2+} in Various Supporting Electrolytes[a]

Reaction	Supporting electrolyte	$D_O \times 10^6$ (meas.) (cm^2/s)	α (Cathodic meas.)	$k_a^0 \times 10^2$ (cm/s)	$k_t^0 \times 10^2$ (cm/s)
Pb^{2+}/d.m.e.[b]	1.0 *N* KNO_3	7.5	0.53	0.39	—
	1.0 *N* KCl	9.0	0.53	0.98	1.08
	1.0 *N* KBr	8.6	0.525	0.34	0.38
	1.0 *N* KI	—	0.52	0.45[c]	—
	1.0 *N* $NaClO_4$	7.2	0.525	0.56	0.50
Cd^{2+}/d.m.e.[d]	1.0 *N* KF	4.5	0.54	0.48	0.16
	1.0 *N* KNO_3	6.4	0.53	0.41	—
	1.0 *N* KCl	7.4	0.53	1.02	1.10
	1.0 *N* KBr	7.7	0.55	0.57	0.70
	1.0 *N* KI	11.3	0.54	0.90	—
	1.0 *N* $NaClO_4$	5.8	0.535	0.27	—

[a] Temperature = 27 ± 1°C; concentration of the depolarizer = 1.0 mM.
[b] Refs. 83 and 88.
[c] The value of k_a^0 is obtained assuming that the value of D in a KI medium is the same as that in a KBr medium. The direct determination of the value of D in a KI medium is not possible as PBI_2 is not appreciably soluble.
[d] Refs. 38 and 83.

correction as ϕ_2 data are available in the literature.[90-92] The corrected rate constant values vary in the following order:

$$Cl^- > ClO_4^- > Br^-$$

The k_t^0 is obtained in the chloride medium corresponding to the complex $PbCl^+$ (other complexes possible are $PbCl_2$, $PbCl_3^-$, and $PbCl_4^{2-}$), as it correlates with the corrected values in the other supporting electrolytes. With the increase of charge on the complex from neutral to negative values, the rate constant will be enhanced.

(*b*) *Cd(II)/d.m.e.*

The reduction of cadmium ions is a fast reaction and has been widely studied. The faradaic rectification method using high-frequency modulated signals has been applied for the determination of kinetic parameters of this reaction by Barker,[7] Van Der Pol *et al.*,[54] and Agarwal using a hanging amalgam drop electrode.[41] The FR polarograms of Cd(II) in 1.0 *N* KF are shown in Fig. 25. From the polarograms, it can be seen that all the FR summit peaks, even up to 50 kHz, coincide with the half-wave potential (which is about −0.77 V). The rate constant and the value of α are obtained from $\Delta E_\infty / V_A^2$ versus $\omega^{-1/2}$ plots using Eqs. (29) and (30). The value of α in different supporting electrolytes lies between 0.53 and 0.55 and its is not as high in potassium iodide as has been reported in the literature.[7,41] k_a^0 lies between 0.27×10^{-2} and 1.02×10^{-2} cm/s (Table 7). In sodium perchlorate, the reaction is slowest whereas it is very fast in potassium chloride. The effect of different supporting electrolytes on the apparent rate constant is in the following order:

$$Cl^- > I^- > Br^- > F^- > NO_3^- > ClO_4^-$$

The values of the rate constants obtained are fairly comparable to those given in the literature, and the present technique appears to be quite reliable for determination of kinetics parameters of fast reactions.

The k_a^0 values in KF, KCl, and KBr have been corrected by applying the Frumkin theory for the double-layer correction as ϕ_2 data were available in the literature for these media. The k_t^0 values obtained in Cl^- and Br^- media correspond to $CdCl^+$ and $CdBr^-$ complexes, respectively, as they correlate[93] with corrected values

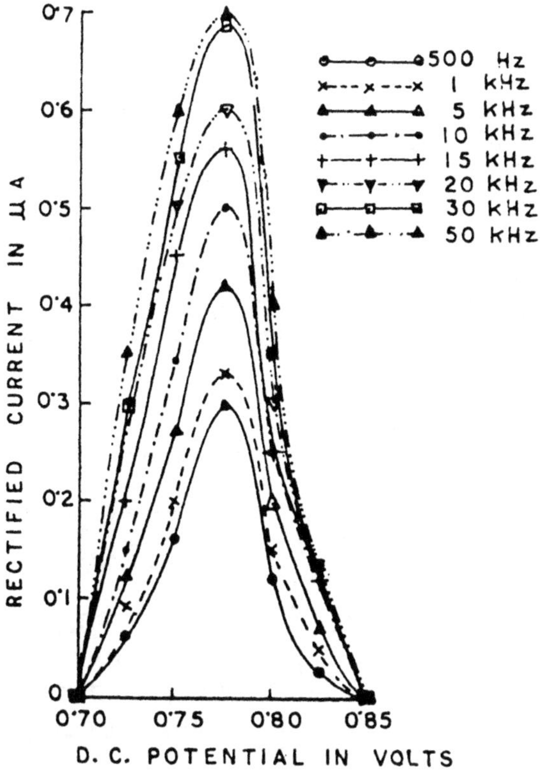

Figure 25. FR polarograms of 1.0 mM Cd(II) in 1.0 *N* KF.

in the other supporting electrolytes. The k_t^0 values in Cl^-, Br^-, and F^- also vary in the above order.

(*c*) *Zn*(*II*)/*d.m.e.*

The reduction of zinc ions at d.m.e. has widely been studied and the reaction has been reported to be quasi-reversible.[94] Van Der Pol and co-workers[54] studied this reaction by the faradaic rectification polarographic technique using high-frequency modulated signals. The kinetic parameters have been evaluated by the present technique using unmodulated audio frequency alternating current and the results have been compared with those given in the literature.[83]

The plot of $\Delta E_{\infty}/V_A^2$ versus $\omega^{-1/2}$ for reduction of zinc ion in 1.0 N KCl is shown in Fig. 26. At high frequencies (400 Hz and above), $\Delta E_{\infty}/V_A^2$ tends to be constant, and the value of α is obtained from Eq. (29). Similarly, the rate constant is determined from the slope of the plot using Eq. (30). The kinetic parameters in different supporting electrolytes are given in Table 8. The value of α lies between 0.52 and 0.535 and k_a^0 is in the range of 2.2×10^{-2} cm/s to 2.8×10^{-2} cm/s. It is interesting to note that the rate constant in

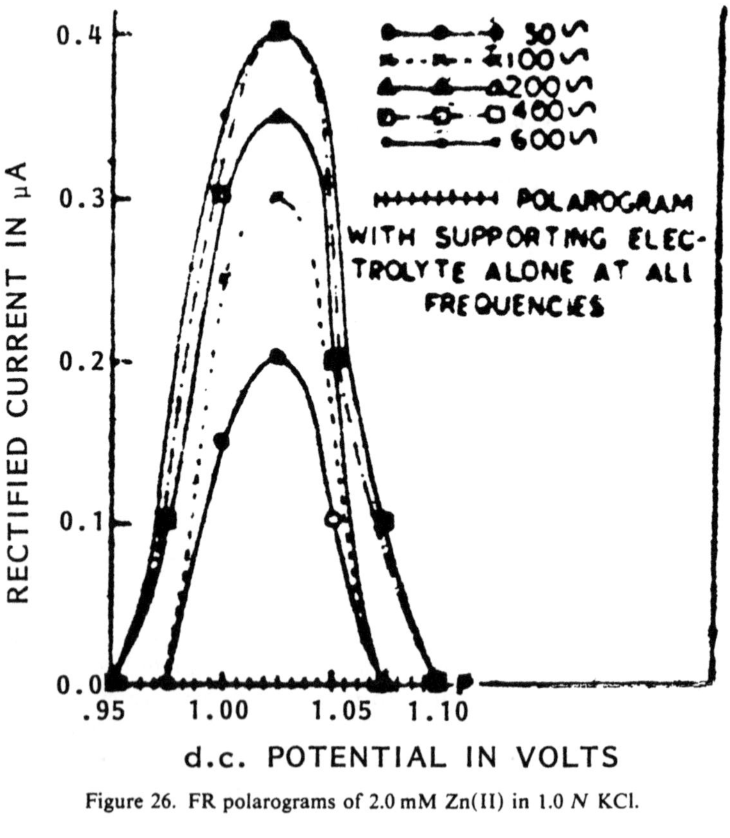

Figure 26. FR polarograms of 2.0 mM Zn(II) in 1.0 N KCl.

Table 8
Kinetic Parameters of Cr(III), In(III), Zn(II), and Co(II) in Various Supporting Electrolytes[a]

Reaction	Supporting electrolyte	$D_O \times 10^6$ (meas.) (cm^2/s)	α (Cathodic meas.)	$k_a^0 \times 10^2$ (cm/s)	$k_i^0 \times 10^2$ (cm/s)
Cr(III)/d.m.e.[b]	1.0 *N* KCl	9.5	0.47	1.4	—
	1.0 *N* KNO_3	7.7	0.47	0.23	—
	1.0 *N* KBr	7.8	0.45	1.6	—
	1.0 *N* K_2SO_4	4.7	0.47	0.54	—
	1.0 *N* $NaClO_4$	6.4	0.45	1.2	—
In(III)/d.m.e.[c]	1.0 *N* KCl	5.6	0.52	8.0	1.07
Zn(II)/d.m.e.[d]	1.0 *N* KCl	8.4	0.52	2.5	—
	1.0 *N* KNO_3	7.8	0.52	2.8	—
	1.0 *N* KBr	6.7	0.525	2.6	—
	0.5 *N* K_2SO_4	4.8	0.535	2.2	—
	1.0 *N* $NaClO_4$	4.6	0.53	2.4	—
Co(II)/d.m.e.[e]	1.0 *N* KF		0.48	0.54	0.048

[a] Temperature = 25°C; concentration of the depolarizer = 1 mM.
[b] Refs. 83 and 99.
[c] Refs. 82 and 83.
[d] Refs. 81 and 83.
[e] Refs. 82 and 83.

a sulfate medium is the lowest whereas in bromide and nitrate media it is a high as reported in the literature.[94] The values of the kinetic parameters in different supporting electrolytes follow the order

$$KNO_3 > KBr > KCl > NaClO_4 > K_2SO_4$$

Van Der Pol *et al.*[54] have reported the same order for the values of the rate constant for the second electron charge transfer step.

(*d*) *Co*(*II*)/*d.m.e.*

The dc polarograms of Co(II) in a noncomplexation state have been obtained[95-98] in 1.0 *N* KCl, 1.0 *N* KF, and 1.0 *M* sodium citrate + 0.1 M NaOH media. The FR polarograms have been obtained in the latter two media (Figs. 27 and 28). It may be noted that, as reported in the literature,[96] no dc step is obtained in a 1.0 M sodium citrate + 0.1 M NaOH medium, except a catalytic wave which is due to the fact that its half-wave potential is close to the potential at which catalytic discharge of hydrogen occurs. In FR polarography, well-defined polarograms are obtained uninterrupted by catalytic discharge of hydrogen. This will be of great advantage in such studies.

The Co(II) reaction in 1.0 N KF is quasi-reversible as is evident from the FR polarograms. The FR summit peaks are quite distinct even up to 600 Hz (Fig. 28). The kinetic parameters for this reaction have been obtained[82] from the $\Delta E_\infty / V_A^2$ versus $\omega^{-1/2}$ plot using Eqs. (29) and (30). The value of α is 0.48 and k_a^0 is 5.4×10^{-3} cm/s

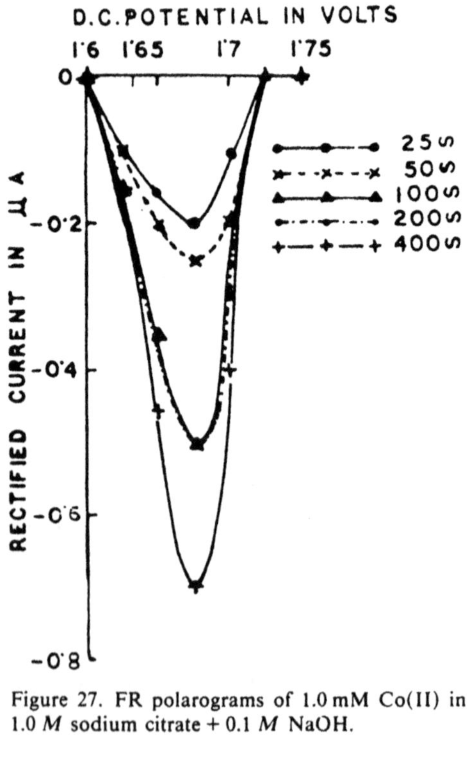

Figure 27. FR polarograms of 1.0 mM Co(II) in 1.0 M sodium citrate + 0.1 M NaOH.

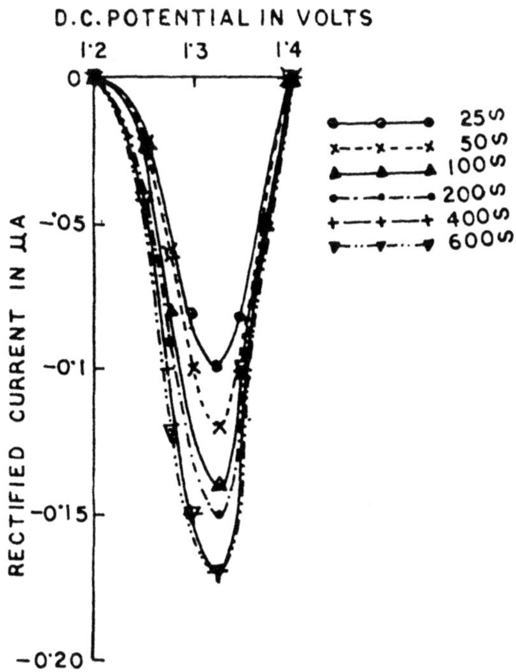

Figure 28. FR polarograms of 1.0 mM Co(II) in 1.0 *N* KF + 0.01% gelatin.

(Table 8). The value of k_a^0 has been corrected for the double layer in KF[20] and the value of k_t^0 obtained is 0.48×10^{-3} cm/s.

(*iv*) *Three-Electron Charge Transfer Reactions*

(*a*) *Cr*(*III*)/*d.m.e.*

Generally, two steps are obtained in the dc polarograms for the reduction of Cr(III) in different supporting electrolytes except in a KNO_3 medium. It is interesting to note that only one FR summit peak is obtained in different supporting electrolytes except in a 1.0 *N* KCl medium. FR polarograms of Cr(III) in 1.0 *N* KCl at different frequencies of the ac are shown in Fig. 29. The FR polarogram shows two FR summit peaks at 0.88 V and 1.45 V.

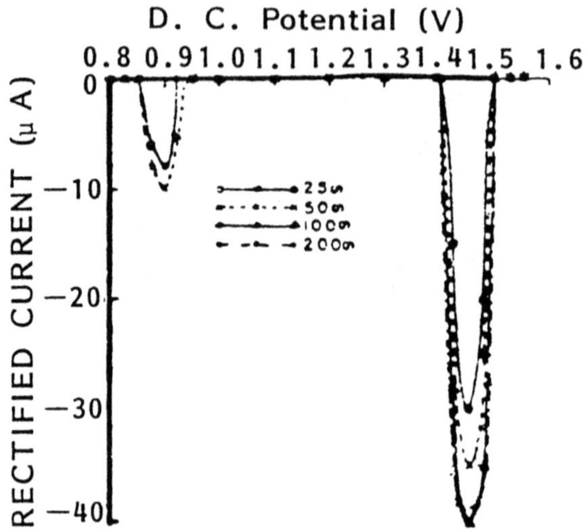

Figure 29. FR polarograms of 1.0 mM of Cr(III) in 1.0 *N* KCl.

corresponding to the two steps of the reaction, i.e., Cr(III) → Cr(II) and Cr(II) → Cr, respectively. As the height of the FR summit peak for the first step of the reaction is very small and appears only up to 50 Hz, this reaction is irreversible, and it is not possible to determine its rate constant. The FR summit peak corresponding to the second step of the reaction is conspicuous even up to 200 Hz, and thus it has been possible to determine the values of α and k_a^0 for this reaction in different supporting electrolytes.[99] The value of α lies between 0.45 and 0.47 and k_a^0 ranges from 0.23×10^{-2} to 1.6×10^{-2} cm/s (Table 8).

The rate constant is moderately fast in bromide, chloride, and perchlorate media and varies in the following order:

$$Br^- > Cl^- > ClO_4^-$$

In nitrate and sulfate media, Cr(III) is not very stable and the reaction Cr(II) → Cr tends to be irreversible. The rate constants obtained in these media are tentative because of excessive adsorption of chromous ions in these media. The rate constants have been corrected for the double layer in 1.0 *N* KCl and 1.0 *N* KNO_3 media.

(*b*) *In*(*III*)/*d.m.e.*

The dc polarogram of In(III) gives only one step and the ac polarogram shows one ac summit peak in 1.0 *N* KCl at 0.65 V, indicating that there is only one step in the reduction in(III) → In. In the FR polarogram in 1.0 *N* KCl (Fig. 30), the FR summit peak is obtained at 0.65 V and the peak is conspicuous even up to 1000 Hz. This shows that the reaction should be moderately fast. The values of the kinetic parameters obtained[82] are $k_a^0 = 8.0 \times 10^{-2}$ cm/s and $\alpha = 0.52$ (Table 8). It is evident that this reaction is not as fast as has been reported by Barker,[7] who found a rate constant of 1.5 cm/s. The reaction is also not irreversible as has invariably been reported

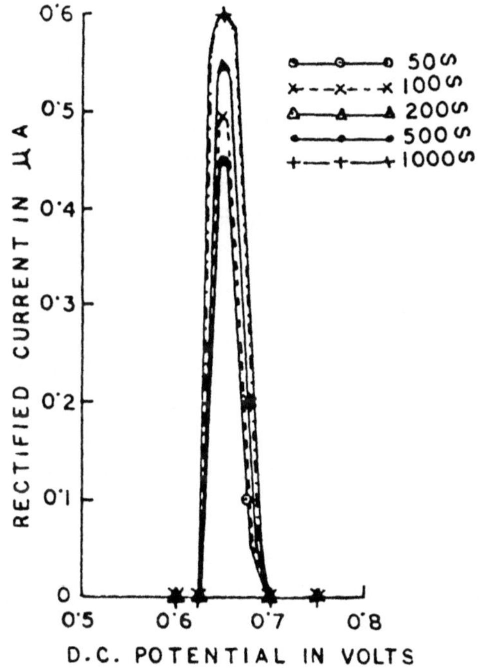

Figure 30. FR polarograms of 1.0 mM of In(III) in 1.0 *N* KCl.

in the literature.[100,101] Jain[102] has recently shown that this reaction is quasi-reversible in halides, which corroborates the above results.

2. Studies Using Organic Compounds as Depolarizers

Polarographic studies of organic compounds are very complicated. Many of the compounds behave as surfactants, most of them exhibit multiple-electron charge transfer, and very few are soluble in water. The measurement of the capacitance of the double layer, the cell resistance, and the impedance at the electrode/solution interface presents many difficulties. To examine the versatility of the FR polarographic technique, a few simple water-soluble compounds have been chosen for the study. The results obtained are somewhat exciting because the FR polarographic studies not only help in the elucidation of the mechanism of the reaction in different stages but also enable the determination of kinetic parameters for each step of reduction.

The experimental cell, the measurement technique, and the method used for purification of the solutions are the same as given in Section IV (1(i)). The organic substances studied are vanillin, isatin, and 5-nitrobenzimidazole. Each of them will be dealt with in detail.

(*i*) *Reduction of Vanillin*

The electrolytic reduction of vanillin was studied by Shima.[103] He reported that it involves a two-electron charge transfer leading to the formation of the final product, vanillyl alcohol. Brdicka[104] and Zuman[105] have independently studied the dc polarography of this compound at varying pH's. Suzuki[106] has reported that vanillin is irreversibly reduced, as it gives no ac polarogram.

The reduction of vanillin has been studied by the FR polarographic technique.[107] The FR summit peaks are conspicuous up to 200–250 Hz at all pH's. With an increase in pH, the FR summit peak potential tends to be more cathodic. The FR wave is invariably obtained on the negative side of the abscissa, indicating that α is less than 0.5. The FR polarograms at pH 10.5 are shown in Fig. 31.

The reduction of vanillin is accompanied by a two-electron charge transfer and the mechanism of the reaction can be explained as follows:

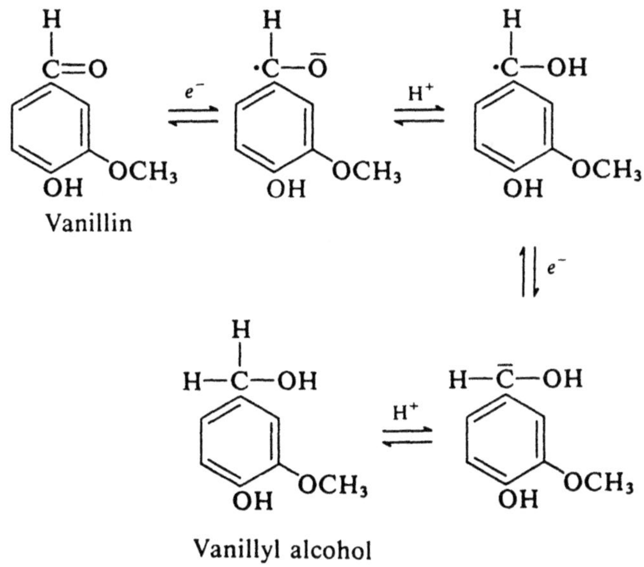

The values of α and the rate constant k_a^0 at varying pH's have been determined[107] and are given in Table 9. The values of α are generally found to decrease with increasing pH and lie between 0.23 and 0.34. The rate constant decreases with an increase in pH except in a neutral medium and it ranges from 3.63×10^{-2} cm/s to 2.42×10^{-2} cm/s.

(*ii*) *Reduction of Isatin*

The polarography of isatin is interesting because the reduction of this compound involves an eight-electron charge transfer and the reduced final product formed is indoline. The dc polarographic study of the compound[108] shows that in fairly acidic media, a two-step reduction occurs. With a decrease in acidity of the medium, a one-step reduction occurs. The FR polarography of isatin at pH 1.1 gives two FR waves, one appearing in the negative quadrant at 0.23 V and the other in the positive quadrant at 0.4 V (Fig. 32). The first wave corresponds to a six-electron charge transfer and the

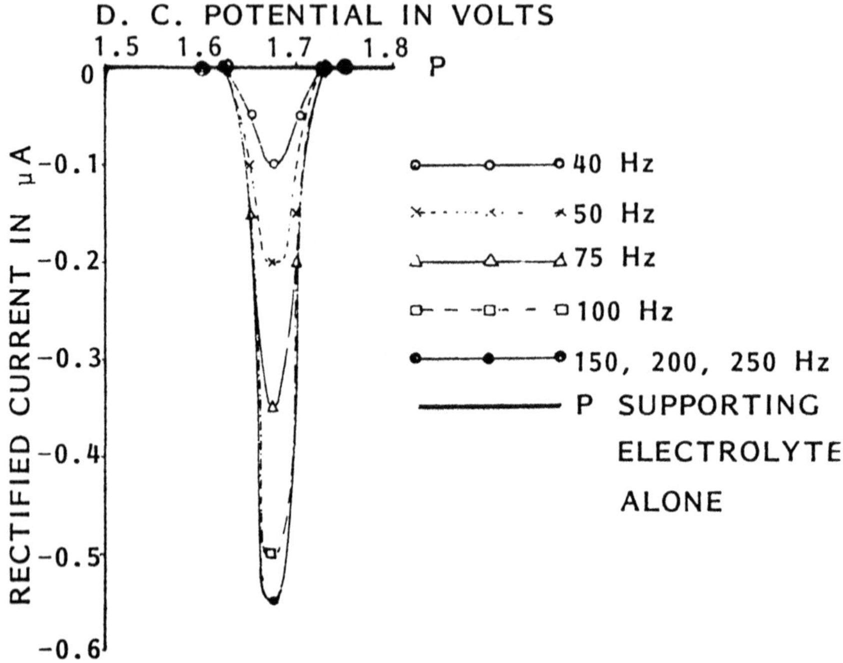

Figure 31. FR polarograms of 1.0 mM vanillin in a buffer of pH 10.5.

second to a two-electron charge transfer. The following mechanism can be given for the reduction in acidic medium[108,109]:

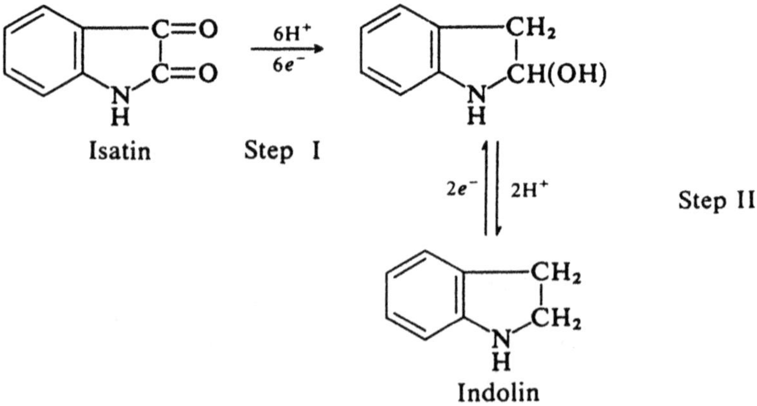

Table 9
Kinetic Parameters of Some Organic Depolarizers[a]

Depolarizer[b]	Diffusion coefficient, $D_2 \times 10^6$ (cm^2/s)	pH of buffer	FR summit Peak potential (V vs. SCE)	α (Cathodic meas.)	k_a^0 (cm/s)
Vanillin	8.2	1.1	−1.09	0.34	3.63×10^{-2}
		4.9	−1.27	0.31	3.45×10^{-2}
		7.0	−1.55	0.25	4.31×10^{-2}
		9.3	−1.60	0.23	3.22×10^{-2}
		10.5	−1.60	0.27	2.42×10^{-2}
5-Nitrobenzimidazole	8.7	7.0	−0.67	0.04	2.6×10^{-2}
Isatin	1.85	1.1	I($6e^-$) −0.23	I0.49	I6.4×10^{-5}
			II($2e^-$) −0.42	II0.75	II6.5×10^{-5}

[a] Temperature = 27 ± 1°C.
[b] Concentration = 1.0 mM.

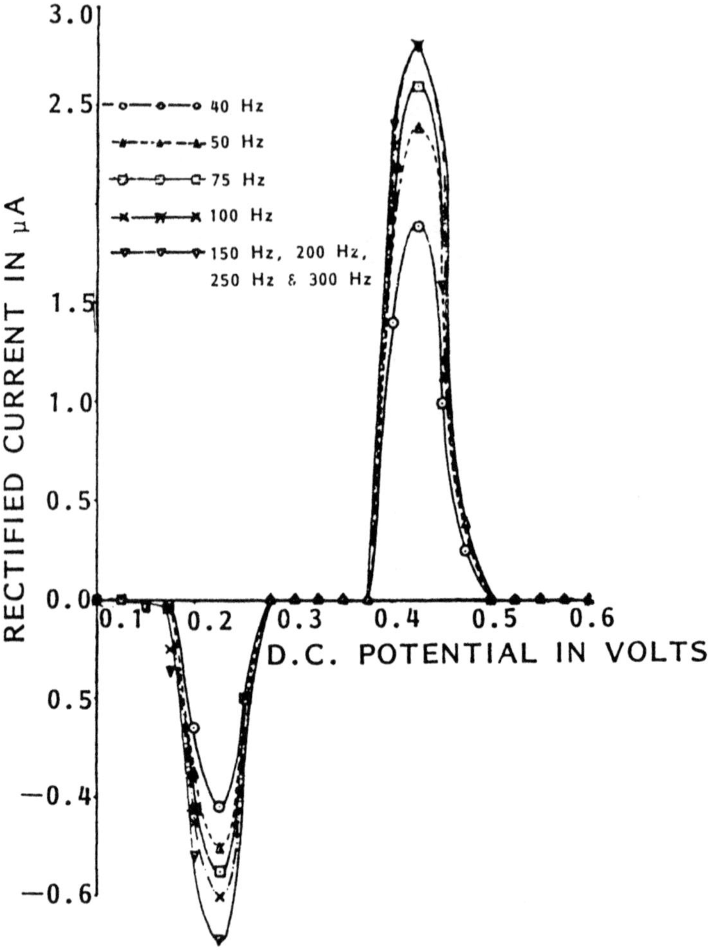

Figure 32. FR polarograms of 1.0 mM of isatin in a buffer of pH 1.1.

Bhargava[109] has reported that no ac polarographic wave is obtained for the first step of the reduction, and for the second step of reduction, the ac wave is conspicuous only at low audio frequencies, indicating that both steps are very slow and irreversible. The diffusion coefficient of isatin in a buffer of pH 1.1 is reported to be 1.85×10^{-6} cm^2/s and the value of n for the second step is 2.

As the first wave appears on the negative side of the abscissa, the value of α is 0.49 whereas for the second wave, appearing on the positive side of the abscissa, the value of α is 0.75. The value of the rate constant[110,111] for the first step of the reaction is 6.4×10^{-5} cm/s whereas for the second step, it is 6.5×10^{-5} cm/s (Table 9).

(*iii*) *Reduction of 5-Nitrobenzimidazole*

The dc polarograms reported for 5-nitrobenzimidazole[112] are of special interest: In a highly acidic medium, the reduction takes place in two steps at −0.14 V and −0.68 V, respectively, whereas at pH 4.0 and above, there is only one step. Bhargava[109] evaluated

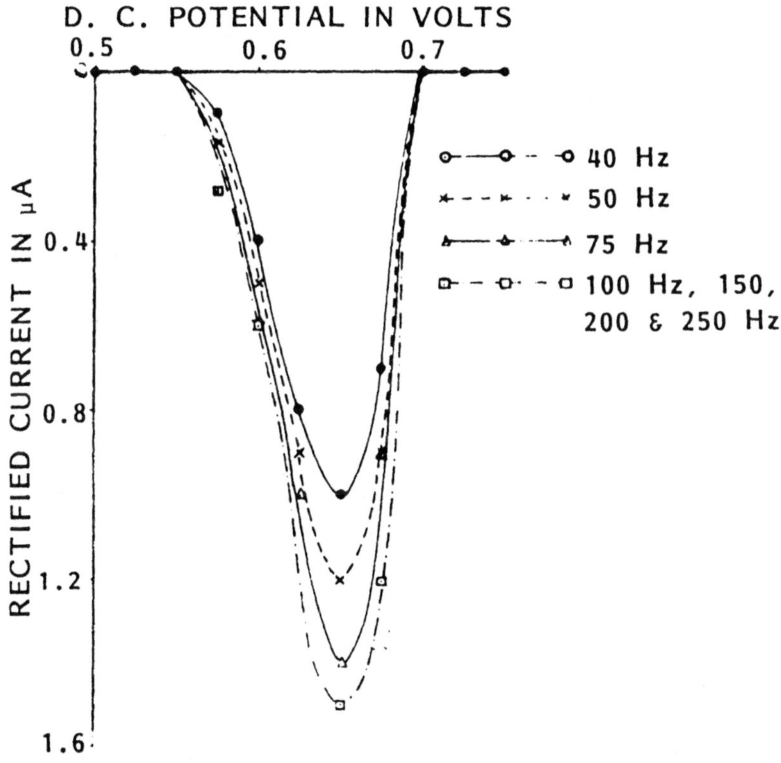

Figure 33. FR polarograms of 1.0 mM 5-nitrobenzimidazole in buffer of pH 7.0.

the value of n from the Ilkovik equation as 6 in a buffer of pH 7.5 and the value of the diffusion coefficient D, as $8.7 \times 10^{-6}\ cm^2/s$. From ac polarographic studies at pH 7.5, Bhargava reported this reaction to be moderately fast.

The FR polarograms of 5-nitrobenzimidazole in a buffer of pH 7.0 are shown in Fig. 33. The FR polarograms are obtained on the negative side of the abscissa, indicating that α should be less than 0.5. The FR summit peak is obtained at the half-wave potential −0.65 V and it is conspicuous up to 100 Hz. As the reaction involves a six-electron charge transfer, it may be assigned the following mechanism:

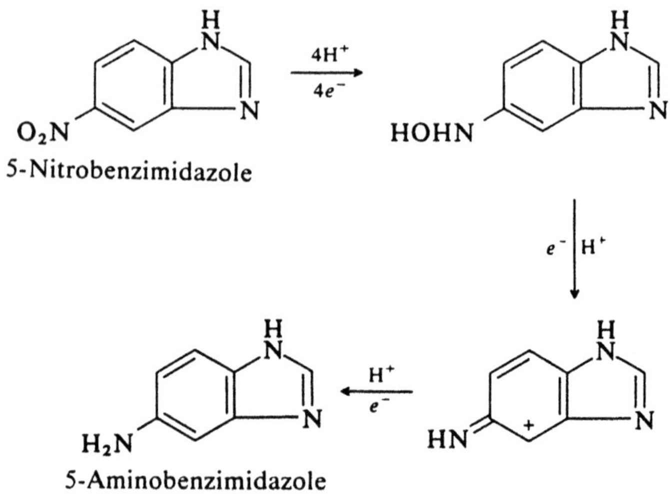

The kinetic parameters (Table 9) obtained for the reaction[110] are $\alpha = 0.04$ and $k_a^0 = 2.6 \times 10^{-2}$ cm/s. The rate constant is of the same order as reported by ac polarography.[109] The low value of the transfer coefficient indicates that the compound is excessively adsorbed at the electrode surface.

V. OTHER APPLICATIONS

Except for the recent developments in measurement of electrode kinetics of multiple-electron charge transfer reactions and in faradaic rectification polarography, very few references are available relating to new applications. Some isolated observations made are of importance, and they should be suitably exploited. Two of them are described below.

1. *Growth formation in epitaxial electrodeposition.* Recently, Sheshadri[113] observed that at small overpotentials caused by faradaic rectification, growth formation occurs in the epitaxial electrodeposition of copper on various copper single-crystal planes.

2. *Study of instantaneous corrosion rates.* Sathyanarayana and Srinivasan[114,115] have extended faradaic rectification studies to the determination of instantaneous corrosion rates of partially reversible metal ion/metal interfaces. They emphasized the practical utility of the method in the fast monitoring of the corrosion of a battery.

VI. CONCLUSIONS

The faradaic rectification effect is a general phenomenon in that it is exhibited in reversible electrode systems, at semiconductor/electrolyte interfaces, and in all kinds of redox couples accompanied by single-electron charge transfer, two-electron charge transfer, and multiple-electron charge transfer. In the past, substantial progress in this field could not be achieved mainly for two reasons: Firstly, the lack of perfection and standardization of experimental techniques for measurements, particularly at metal ion/metal(s), metal ion/d.m.e., redox couple/inert metal(s), and metal ion/amalgram electrode (hanging or dropping) interfaces at audio and radio frequencies, and secondly, for want of a suitable theory applicable to multiple-electron charge transfer reactions.

Earlier studies generally involved the evaluation of kinetic parameters of reactions which are accompanied by single-electron charge transfer.[116] Some reactions involving two-electron charge transfer were also studied, assuming either that both electrons are transferred in a single step or that the slower step in the two-step reaction is in overall control of the rate process. As described in this chapter for the first time, the faradaic rectification theory for multiple-electron charge transfer reactions has been suitably developed and modified, and the kinetic parameters for each step of several multiple-electron charge transfer reactions have been evaluated. Wherever possible, the zero-point method has been applied for the determination of kinetic parameters and the k_a^0 values have been corrected for the double layer. One interesting observation is that in any two-electron charge transfer reaction, if the two rate constants (for the two steps) differ by a factor of 100 or so, it would not be appropriate to assume that the slower reaction is in overall control of the rate process. In any two-electron charge transfer reaction at a metal ion/metal(s) interface, the second-step rate constant is invariably higher than the first-step rate constant if parameters are obtained by eliminating C_R^0, the concentration of the finally reduced species. However, if C_R^0 is taken to be 1.0, k_1^0 is always greater than k_2^0. The kinetic parameters thus obtained by different methods explain the variation in the values reported in the literature. If three or more electrons are involved in the charge transfer, the faradaic rectification theory becomes complicated. The study of two-electron charge transfer reactions in a redox couple reveals that when $C_O^0 > C_R^0$, k_1^0 is greater than k_2^0, and when $C_R^0 > C_O^0$, k_1^0 is less than k_2^0. This explains why the results of earlier workers who carried out studies at varying redox concentration ratios did not tally.

The FR polarographic technique recently developed has the unique distinction of enabling qualitative detection and quantitative estimation (for concentrations as low as 0.1 mM) in mixtures of several components whose half-wave potentials may differ by even less than 20 mV. This method also helps in simultaneous determinations of kinetic parameters of reversible, quasi-reversible, and irreversible reactions. The FR polarographic study of organic compounds will be of immense use in elucidating the mechanism and in evaluating the kinetic parameters of several steps involved in the reactions of these compounds. Most of the organic reactions are generally accompanied by four- to eight-electron charge transfer, and the FR polarographic technique would be very useful in the analysis and study of such complicated reactions. The systematic study of some organic compounds of industrial importance will afford very valuable information that can be used to optimize production processes.

The method can successfully be used in analyses of impurities in metals and alloys, for estimation of minor elements in monomolecular films of oxide layers of Fe–Cr–Ni alloys, for detection of metal impurities in environmental pollution, for studying the depression of high-grade semiconducting materials and for analysis of the corrosion products of contact junction diodes used in microelectronic circuits. Much sophistication is desirable on the instrumental side so as to incorporate an automatic recording device to make an FR polarograph suitable for wider applications and common use.

Recently, Reinmuth[117] has emphasized the importance of a second-order method in the study of the kinetics of more complex reactions. Of the several nonlinear relaxation methods, the potentialities and applications of the faradaic rectification method are many and varied. The study of electrode processes in fused salt electrolytes (which are usually fast) and the rectification in nonaqueous solvents, in supersaturated and supercooled solutions of redox couples, and in fused organic semiconductors seem to be very promising. The applicability of the method in fast monitoring of corrosion, more exhaustive studies at different semiconductor/electrolyte interfaces, and investigations of rectification through polyelectrolytes and single crystals deserve attention. The study of the photovoltaic effect produced by ultraviolet,[118] infrared, and laser radiation on the electrode/solution interface and their influence on nonlinear effects would be very interesting.

Studies on the subject are still in their infancy and a more active pursuit and exploration is needed to investigate the various fields just thrown open. The results presented in this chapter mainly comprise details of our own studies in the field, as practically no references from other workers are available.

ACKNOWLEDGMENT

The author expresses his appreciation and gratitude to all those who have significantly contributed to the development of this new field, bringing it to the forefront as a potential tool for the study of fast electrode kinetics.

APPENDIX A

Case I: C_R^0 Is Kept Constant and C_O^0 Varies

Let the concentration of the reductant be C_R^0 and that of the oxidant be C_O^0, $C_O^{0\prime}$, $C_O^{0\prime\prime}$ for the three different redox ratios in any experiment. the oxidation potential at any concentration can then be written as

$$E_1 = E_0 - \frac{RT}{nF}\ln\frac{C_O^0}{C_R^0}$$

$$E_2 = E_0 - \frac{RT}{nF}\ln\frac{C_O^{0\prime}}{C_R^0}$$

$$E_3 = E_0 - \frac{RT}{nF}\ln\frac{C_O^{0\prime\prime}}{C_R^0}$$

where E_1, E_2, and E_3 are the oxidant potentials corresponding to the three respective concentrations. Again,

$$E_1 - E_2 = \frac{RT}{nF}\ln\frac{C_O^{0\prime}}{C_O^0} \quad \text{and} \quad E_1 - E_3 = \frac{RT}{nF}\ln\frac{C_O^{0\prime\prime}}{C_O^0}$$

The concentration of the intermediate species after the first electron charge transfer at the three redox ratios can be represented as $C_{R_1}^0$, $C_{R_1}^{0\prime}$, and $C_{R_1}^{0\prime\prime}$, respectively. The relationship between them can be given as

$$E_1 - E_2 = \frac{RT}{nF}\ln\frac{C_{R_1}^{0\prime}}{C_{R_1}^0} \quad \text{and} \quad E_1 - E_3 = \frac{RT}{nF}\ln\frac{C_{R_1}^{0\prime\prime}}{C_{R_1}^0}$$

Assuming that $C_{R_1}^{0\prime} = aC_{R_1}^0$ and $C_{R_1}^{0\prime\prime} = bC_{R_1}^0$, Eq. (12) corresponding to the other two redox ratios can be written as

$$\left[\frac{4RT}{F}\frac{\Delta E_\infty'}{V_A^2} - (2\alpha - 1)\right]\frac{\omega^{1/2}a^{1/2}(C_{R_1}^0)^{1/2}D_O^{1/2}}{2^{1/2}}$$

$$= [C_O^{0\prime}(1-\alpha) - a\alpha C_{R_1}^0]\frac{k_1^0}{(C_O^{0\prime})^\alpha}$$

$$+ [aC_{R_1}^0(1-\alpha) - \alpha C_R^0]\frac{k_2^0}{(C_R^0)^{1-\alpha}} \qquad \text{(a)}$$

$$\left[\frac{4RT}{F}\frac{\Delta E''_\infty}{V_A^2} - (2\alpha - 1)\right]\frac{\omega^{1/2}b^{1/2}(C^0_{R_1})^{1/2}D_O^{1/2}}{2^{1/2}}$$
$$= [C_O^{0''}(1-\alpha) - b\alpha C^0_{R_1}]\frac{k_1^0}{(C_O^{0''})^\alpha}$$
$$+ [bC^0_{R_1}(1-\alpha) - \alpha C^0_R]\frac{k_2^0}{(C_R^0)^{1-\alpha}} \quad \text{(b)}$$

where $\Delta E'_\infty$ and $\Delta E''_\infty$ are the rectification potentials for the other two redox concentrations. The second term on the right-hand side in each of the above equations can be approximately taken as equal. On subtracting Eq. (a) from Eq. (12) and rearranging, the expression obtained is

$$\left\{\left[\frac{4RT}{F}\frac{\Delta E_\infty}{V_A^2} - (2\alpha - 1)\right]\right.$$
$$\left. - a^{1/2}\left[\frac{4RT}{F}\frac{\Delta E'_\infty}{V_A^2} - (2\alpha - 1)\right]\right\}\frac{\omega^{1/2}(C^0_{R_1})^{1/2}D_O^{1/2}}{2^{1/2}}$$
$$= \{(C_O^{0'})^\alpha[C_O^0(1-\alpha) - \alpha C^0_{R_1}]$$
$$- (C_O^0)^\alpha[C_O^{0'}(1-\alpha) - a\alpha C^0_{R_1}]\}\frac{k_1^0}{(C_O^0)^\alpha(C_O^{0'})^\alpha} \quad \text{(c)}$$

Similarly, on subtracting Eq. (b) from Eq. (12), one gets

$$\left\{\left[\frac{4RT}{F}\frac{\Delta E_\infty}{V_A^2} - (2\alpha - 1)\right]\right.$$
$$\left. - b^{1/2}\left[\frac{4RT}{F}\frac{\Delta E''_\infty}{V_A^2} - (2\alpha - 1)\right]\right\}\frac{\omega^{1/2}(C^0_{R_1})^{1/2}D_O^{1/2}}{2^{1/2}}$$
$$= [(C_O^{0''})^\alpha\{C_O^0(1-\alpha) - \alpha C^0_{R_1}\}$$
$$- (C_O^0)^\alpha\{C_O^{0''}(1-\alpha) - b\alpha C^0_{R_1}\}]\frac{k_1^0}{(C_O^0)^\alpha(C_O^{0''})^\alpha} \quad \text{(d)}$$

Putting

$$\frac{4RT}{F}\frac{\Delta E_\infty}{V_A^2} - (2\alpha - 1) = d'$$
$$\frac{4RT}{F}\frac{\Delta E'_\infty}{V_A^2} - (2\alpha - 1) = e$$

$$\frac{4RT}{F}\frac{\Delta E''_\infty}{V_A^2} - (2\alpha - 1) = f$$

$$C_O^{0''^\alpha} C_O^{0'^\alpha} = g; \qquad C_O^{0''^\alpha} C_O^{0^\alpha} = h; \qquad C_O^{0'^\alpha} C_O^{0^\alpha} = i$$

On dividing [Eq. (c)] by [Eq. (d)], one gets

$$\frac{(d' - a^{1/2}e)}{(d' - b^{1/2}f)} = \frac{(1-\alpha)(gC_O^0 - hC_O^{0'}) + \alpha C_{R_1}^0(ha - g)}{(1-\alpha)(gC_O^0 - iC_O^{0''}) + \alpha C_{R_1}^0(ib - g)}$$

On cross multiplication and transferring the $C_{R_1}^0$ terms on one side, the expression obtained is

$$\begin{aligned} C_{R_1}^0 = (1-\alpha)[&(d' - b^{1/2}f)(gC_O^0 - hC_O^{0'}) \\ &- (d' - a^{1/2}e)(gC_O^0 - iC_O^{0''})]/ \\ &\alpha[(d' - b^{1/2}f)(g - ha) \\ &- (d' - a^{1/2}e)(g - ib)] \end{aligned} \tag{e}$$

Substituting the experimentally determined value of α, $C_{R_1}^0$ can be obtained as all other terms are known. Now on substituting the value of $C_{R_1}^0$ and all other terms in Eq. (c), the value of k_1^0 can be obtained. For determining the value of k_2^0, the values of k_1^0 and $C_{R_1}^0$ are substituted in Eq. (a).

Case II: C_O^0 Remains Constant and C_R^0 Varies

Considering the case when the experiment at three different concentrations of redox ratios is being carried out, with C_O^0 constant and the concentration of the reductant varying as C_R^0, $C_R^{0'}$, and $C_R^{0''}$. The oxidation potential is given by

$$E_1 = E_0 - \frac{RT}{nF}\ln\frac{C_O^0}{C_R^0}$$

$$E_2 = E_0 - \frac{RT}{nF}\ln\frac{C_O^0}{C_R^{0'}}$$

$$E_3 = E_0 - \frac{RT}{nF}\ln\frac{C_O^0}{C_R^{0''}}$$

or

$$E_2 - E_1 = \frac{RT}{nF}\ln\frac{C_R^{0'}}{C_R^0} \quad \text{and} \quad E_3 - E_1 = \frac{RT}{nF}\ln\frac{C_R^{0''}}{C_R^0}$$

If the concentration of the intermediate species, after the first electron transfer, varies in the three cases as $C^0_{R_1}$, $C^{0\prime}_{R_1}$, and $C^{0\prime\prime}_{R_1}$, respectively, then the relationships

$$E_2 - E_1 = \frac{RT}{nF} \ln \frac{C^{0\prime}_{R_1}}{C^0_{R_1}} \quad \text{and} \quad E_3 - E_1 = \frac{RT}{nF} \ln \frac{C^{0\prime\prime}_{R_1}}{C^0_{R_1}}$$

will hold.

Considering that $C^{0\prime}_{R_1} = aC^0_{R_1}$, $C^{0\prime\prime}_{R_1} = bC^0_{R_1}$ and putting

$$\omega^{1/2} C^{0^{1/2}}_{R_1} D^{1/2}_O / 2^{1/2} = j; \qquad C^0_O(1-\alpha) = k; \qquad \alpha C^0_{R_1} = l$$

$$K^0_1 / C^{0^\alpha}_O = m; \qquad C^0_{R_1}(1-\alpha) = n$$

[p. 183, Eq. (12)] corresponding to the three redox ratios can be written[39] as

$$d'j = (k - l)m + (n - \alpha C^0_R) k^0_2 / C^{0^{(1-\alpha)}}_R \tag{f}$$

$$ej = (k - al)m + (an - \alpha C^{0\prime}_R) k^0_2 / C^{0\prime^{(1-\alpha)}}_R \tag{g}$$

$$fj = (k - bl)m + (bn - \alpha C^{0\prime\prime}_R) k^0_2 / C^{0\prime\prime^{(1-\alpha)}}_R \tag{h}$$

where $\Delta E'_\infty$ and $\Delta E''_\infty$ are the rectification potentials for the respective redox concentrations. The first term on the right-hand side in each of the above three equations can be approximately taken as equal. Subtracting Eq. (g) from Eq. (f) and Eq. (h) from Eq. (f) and rearranging the terms, two equations are obtained. In these equations, the right-hand side terms and the left-hand side terms of the first equation are divided by the respective terms of the second equation (as has been shown in Case I).

Putting $C^{0\prime\prime^{(1-\alpha)}}_R \cdot C^{0\prime^{(1-\alpha)}}_R = o$; $C^{0^{(1-\alpha)}}_R \cdot C^{0\prime\prime^{(1-\alpha)}}_R = p$; $C^{0^{(1-\alpha)}}_R \cdot C^{0\prime^{(1-\alpha)}}_R = q$; the final expression obtained is

$$\frac{(d' - a^{1/2}e)}{(d' - b^{1/2}f)} = \frac{n(o - ap) + \alpha(pC^{0\prime}_R - oC^0_R)}{n(o - bq) + \alpha(qC^{0\prime\prime}_R - oC^0_R)} \tag{i}$$

On cross multiplication and transferring $C^0_{R_1}$ terms on one side,

$$C^0_{R_1} = \frac{\alpha[(d' - b^{1/2}f)(pC^{0\prime}_R - oC^0_R) - (d' - a^{1/2}e)(qC^{0\prime\prime}_R - oC^0_R)]}{(1-\alpha)[(d' - a^{1/2}e)(o - bq) - (d' - b^{1/2}f)(o - ap)]} \tag{j}$$

Substituting the experimentally determined value of α, $C^0_{R_1}$ can be obtained as all other terms are known. The value of k^0_2 is determined by substituting the value of $C^0_{R_1}$ and α in the equation obtained after subtracting Eq. (g) from Eq. (f). Subsequently, by substituting the values of k^0_2, $C^0_{R_1}$, and α in Eq. (f), the value of k^0_1 can be determined.

APPENDIX B

The oxidant potentials at varying concentrations can be written as

$$E_1 = E_0 - \frac{RT}{nF} \ln \frac{C^0_O}{C^0_R}$$

$$E_2 = E_0 - \frac{RT}{nF} \ln \frac{C^{0^{I}}_O}{C^0_R}$$

$$E_3 = E_0 - \frac{RT}{nF} \ln \frac{C^{0^{II}}_O}{C^0_R}$$

$$E_4 = E_0 - \frac{RT}{nF} \ln \frac{C^{0^{III}}_O}{C^0_R}$$

$$E_5 = E_0 - \frac{RT}{nF} \ln \frac{C^{0^{IV}}_O}{C^0_R}$$

Now from the above relationships

$$E_1 - E_2 = \frac{RT}{nF} \ln \frac{C^{0^{I}}_O}{C^0_O}$$

$$E_1 - E_3 = \frac{RT}{nF} \ln \frac{C^{0^{II}}_O}{C^0_O}$$

$$E_1 - E_4 = \frac{RT}{nF} \ln \frac{C^{0^{III}}_O}{C^0_O}$$

$$E_1 - E_5 = \frac{RT}{nF} \ln \frac{C^{0^{IV}}_O}{C^0_O}$$

If the concentrations of the intermediate species vary at the five redox ratios as $C^{0}_{R_{II}}$, $C^{0^{I}}_{R_{II}}$, $C^{0^{II}}_{R_{II}}$, $C^{0^{III}}_{R_{II}}$, and $C^{0^{IV}}_{R_{II}}$ after the first electron charge transfer and $C^{0}_{R_{I}}$, $C^{0^{I}}_{R_{I}}$, $C^{0^{II}}_{R_{I}}$, $C^{0^{III}}_{R_{I}}$, and $C^{0^{IV}}_{R_{I}}$ after the second electron charge transfer, then the relationships

$$E_1 - E_2 = \frac{RT}{nF}\ln\frac{C^{0^{I}}_{R_{II}}}{C^{0}_{R_{II}}} = \frac{RT}{nF}\ln\frac{C^{0^{I}}_{R_{I}}}{C^{0}_{R_{I}}} = \frac{RT}{nF}\ln a$$

$$E_1 - E_3 = \frac{RT}{nF}\ln\frac{C^{0^{II}}_{R_{II}}}{C^{0}_{R_{II}}} = \frac{RT}{nF}\ln\frac{C^{0^{II}}_{R_{I}}}{C^{0}_{R_{I}}} = \frac{RT}{nF}\ln b$$

$$E_1 - E_4 = \frac{RT}{nF}\ln\frac{C^{0^{III}}_{R_{II}}}{C^{0}_{R_{II}}} = \frac{RT}{nF}\ln\frac{C^{0^{III}}_{R_{I}}}{C^{0}_{R_{I}}} = \frac{RT}{nF}\ln c$$

$$E_1 - E_5 = \frac{RT}{nF}\ln\frac{C^{0^{IV}}_{R_{II}}}{C^{0}_{R_{II}}} = \frac{RT}{nF}\ln\frac{C^{0^{IV}}_{R_{I}}}{C^{0}_{R_{I}}} = \frac{RT}{nF}\ln d$$

will hold for the concentration ratios relating to the intermediate species after the first and second steps of electron charge transfer, as explained earlier.

From the above relationships, it would be appropriate to express the concentration of the intermediate species by the following relationship

$$C^{0^{I}}_{R_{II}} = aC^{0}_{R_{II}}; \qquad C^{0^{II}}_{R_{II}} = bC^{0}_{R_{II}};$$

$$C^{0^{III}}_{R_{II}} = cC^{0}_{R_{II}}; \qquad C^{0^{IV}}_{R_{II}} = dC^{0}_{R_{II}}$$

$$C^{0^{I}}_{R_{I}} = aC^{0}_{R_{I}}; \qquad C^{0^{II}}_{R_{I}} = bC^{0}_{R_{I}};$$

$$C^{0^{III}}_{R_{I}} = cC^{0}_{R_{I}}; \qquad C^{0^{IV}}_{R_{I}} = dC^{0}_{R_{I}}$$

Now if ΔE_∞, ΔE^{I}_∞, ΔE^{II}_∞, ΔE^{III}_∞, and ΔE^{IV}_∞ are the rectified potentials for five different redox concentrations, and putting

$$4RT\Delta E^{I}_\infty / FV_A^2 - (2\alpha - 1) = d^{I};$$

$$4RT\Delta E^{II}_\infty / FV_A^2 - (2\alpha - 1) = d^{II};$$

$$4RT\Delta E^{III}_\infty / FV_A^2 - (2\alpha - 1) = d^{III};$$

$$4RT\Delta E^{IV}_\infty / FV_A^2 - (2\alpha - 1) = d^{IV};$$

$$D_O^{1/2}\omega^{1/2}/2^{1/2} = r; \qquad \alpha C^{0}_{R_{II}} = s;$$

$$C^{0}_{R_{II}}(1 - \alpha) = t; \qquad C^{0^{I}}_{O}(1 - \alpha) = u;$$

$$C_O^{0^{\text{II}}}(1-\alpha) = v; \qquad C_O^{0^{\text{III}}}(1-\alpha) = \omega;$$
$$C_O^{0^{\text{IV}}}(1-\alpha) = x; \qquad k_1^0/C_{R_{\text{II}}}^{0(1-\alpha)} = Y_1;$$
$$k_2^0/C_{R_{\text{II}}}^{0\alpha} \cdot C_{R_{\text{I}}}^{0(1-\alpha)} = Y_2; \qquad k_3^0/C_{R_{\text{I}}}^{0\alpha} \cdot C_R^{0(1-\alpha)} = Y_3,$$

then Eq. (14), p. 185, corresponding to each redox ratio can be written as,

$$d'r = (k-s)Y_1/C_O^{0\alpha} + (t-l)Y_2 + (n-\alpha C_R^0)Y_3 \qquad \text{(a)}$$
$$d^{\text{I}}r = (u-as)Y_1/a^{(1-\alpha)} \cdot C_O^{0^{\text{I}\alpha}} + (t-l)Y_2$$
$$+ (an-\alpha C_R^0)Y_3/a^\alpha \qquad \text{(b)}$$
$$d^{\text{II}}r = (v-bs)Y_1/b^{(1-\alpha)} \cdot C_O^{0^{\text{II}\alpha}} + (t-l)Y_2$$
$$+ (bn-\alpha C_R^0)Y_3/b^\alpha \qquad \text{(c)}$$
$$d^{\text{III}}r = (\omega-cs)Y_1/C^{(1-\alpha)} \cdot C_O^{0^{\text{III}\alpha}} + (t-l)Y_2$$
$$+ (cn-\alpha C_R^0)Y_3/C^\alpha \qquad \text{(d)}$$
$$d^{\text{IV}}r = (x-ds)Y_1/d^{(1-\alpha)} \cdot C_O^{0^{\text{IV}\alpha}} + (t-l)Y_2$$
$$+ (dn-\alpha C_R^0)Y_3/d^\alpha \qquad \text{(e)}$$

On subtracting each of Eqs. (b), (c), (d), and (e) from Eq. (a) and assuming for the sake of simplification that the coefficients of the term k_2^0 are almost equal, the expressions obtained are

$$(d'-d^{\text{I}})r = \{(k-s)/C_O^{0\alpha} - (u-as)/a^{(1-\alpha)}C_O^{0^{\text{I}\alpha}}\}Y_1$$
$$+ \{n-\alpha C_R^0 - (an-\alpha C_R^0)/a^\alpha\}Y_3 \qquad \text{(f)}$$
$$(d'-d^{\text{II}})r = \{(k-s)/C_O^{0\alpha} - (v-bs)/b^{(1-\alpha)} \cdot C_O^{0^{\text{II}\alpha}}\}Y_1$$
$$+ \{n-\alpha C_R^0 - (bn-\alpha C_R^0)/b^\alpha\}Y_3 \qquad \text{(g)}$$
$$(d'-d^{\text{III}})r = \{(k-s)/C_O^{0\alpha} - (\omega-cs)/c^{(1-\alpha)} \cdot C_O^{0^{\text{III}\alpha}}\}Y_1$$
$$+ \{n-\alpha C_R^0 - (cn-\alpha C_R^0)/c^\alpha\}Y_3 \qquad \text{(h)}$$
$$(d'-d^{\text{IV}})r = \{(k-s)/C_O^{0\alpha} - (x-ds)/d^{(1-\alpha)} \cdot C_O^{0^{\text{IV}\alpha}}\}Y_1$$
$$+ \{n-\alpha C_R^0 - (dn-\alpha C_R^0)/d^\alpha\}Y_3 \qquad \text{(i)}$$

Putting,

$$\{n-\alpha C_R^0 - (an-\alpha C_R^0)/a^\alpha\} = A;$$
$$\{n-\alpha C_R^0 - (bn-\alpha C_R^0)/b^\alpha\} = B;$$
$$\{n-\alpha C_R^0 - (cn-\alpha C_R^0)/c^\alpha\} = C;$$
$$\{n-\alpha C_R^0 - (dn-\alpha C_R^0)/d^\alpha\} = D$$

Now multiplying [Eq. (f)] by B and [Eq. (g)] by A and subtracting

$$r\{(d' - d^{\mathrm{I}})B - (d' - d^{\mathrm{II}})A\}$$
$$= Y_1[\{(k - s)/C_O^{0\alpha} - (u - as)/a^{(1-\alpha)}C_O^{0\mathrm{I}\alpha}\}B$$
$$- \{(k - s)/C_O^{0\alpha} - (v - bs)/b^{(1-\alpha)}C_O^{0\mathrm{II}\alpha}\}A] \qquad \text{(j)}$$

Similarly multiplying [Eq. (f)] by C and [Eq. (h)] by A and subtracting

$$r\{(d' - d^{\mathrm{I}})C - (d' - d^{\mathrm{III}})A\}$$
$$= Y_1[\{(k - s)/C_O^{0\alpha} - (u - as)/a^{(1-\alpha)} \cdot C_O^{0\mathrm{I}\alpha}\}C$$
$$- \{(k - s)/C_O^{0\alpha} - (\omega - cs)/c^{(1-\alpha)}C_O^{0\mathrm{III}\alpha}\}A] \qquad \text{(k)}$$

Again multiplying [Eq. (f)] by D and [Eq. (i)] by A and subtracting

$$r\{(d' - d^{\mathrm{I}})D - (d' - d^{\mathrm{IV}})A\}$$
$$= Y_1[\{k - s)/C_O^{0\alpha} - (u - as)/a^{(1-\alpha)}C_O^{0\mathrm{I}\alpha}\}D$$
$$- \{(k - s)/C_O^{0\alpha} - (x - ds)/d^{(1-\alpha)}C_O^{0\mathrm{IV}\alpha}\}A] \qquad \text{(l)}$$

Dividing [Eq. (j)] by [Eq. (k)], Y_1, i.e., $k_1^0/C_{R_{\mathrm{II}}}^{0(1-\alpha)}$, can be eliminated and putting

$$(d' - d^{\mathrm{I}}) = z; \qquad \{(k - s)/C_O^{0\alpha}\} = E;$$
$$\{E - (u - as)/a^{(1-\alpha)} \cdot C_O^{0\mathrm{I}\alpha}\} = E',$$

one gets,

$$\{zB - (d' - d^{\mathrm{II}})A\}/\{zC - (d' - d^{\mathrm{III}})A\}$$
$$= \frac{[E'B - \{E - (v - bs)/b^{(1-\alpha)} \cdot C_O^{0\mathrm{II}\alpha}\}A]}{[E'C - \{E - (\omega - cs)/c^{(1-\alpha)}C_O^{0\mathrm{III}\alpha}\}A]} \qquad \text{(m)}$$

Similarly on dividing [Eq. (j)] by [Eq. (l)],

$$\{zB - (d' - d^{\mathrm{II}})A\}/\{zD - (d' - d^{\mathrm{IV}})A\}$$
$$= \frac{[E'B - \{E - (v - bs)/b^{(1-\alpha)} \cdot C_O^{0\mathrm{II}\alpha}\}A]}{[E'D - \{E - (x - ds)/d^{(1-\alpha)} \cdot C_O^{0\mathrm{IV}\alpha}\}A]} \qquad \text{(n)}$$

In [Eqs. (m) and (n)] after cross multiplication and transferring all the terms on one side and equating them to 0, two identical equations are obtained as

$$\begin{aligned}
&C_{R_{\rm I}}^{0^2}(1-\alpha)^3\cdot F + C_{R_{\rm I}}^{0^2}C_{R_{\rm II}}^{0}\alpha(1-\alpha)^2\cdot G \\
&\quad + C_{R_{\rm I}}^{0}C_{R}^{0}\alpha(1-\alpha)^2\cdot H + C_{R_{\rm I}}^{0}C_{R_{\rm II}}^{0}C_{R}^{0}\alpha^2(1-\alpha)I \\
&\quad + C_{R_{\rm II}}^{0}\cdot C_{R}^{0^2}\alpha^3\cdot J + C_{R}^{0^2}\alpha^2(1-\alpha)\cdot K = 0
\end{aligned} \tag{o}$$

$$\begin{aligned}
&C_{R_{\rm I}}^{0^2}(1-\alpha)^3\cdot F' + C_{R_{\rm I}}^{0^2}C_{R_{\rm II}}^{0}\alpha(1-\alpha)^2\cdot G' \\
&\quad + C_{R_{\rm I}}^{0}C_{R}^{0}\alpha(1-\alpha)^2\cdot H' + C_{R_{\rm I}}^{0}C_{R_{\rm II}}^{0}C_{R}^{0}\alpha^2(1-\alpha)I' \\
&\quad + C_{R_{\rm II}}^{0}\cdot C_{R}^{0^2}\cdot\alpha^3 J' + C_{R}^{0^2}\alpha^2(1-\alpha)\cdot K' = 0
\end{aligned} \tag{p}$$

where F, G, H, I, J, K and F', G', H', I', J' and K' are the coefficients of the respective terms in the Eqs. (o) and (p).

The coefficients of $C_{R_{\rm I}}^{0^2}$ and $C_{R}^{0^2}$ in the Eqs. (o) and (p) are equated, i.e.,

$$F = F' \quad \text{and} \quad K = K'$$

On simplifying them, the value of $C_{R_{\rm II}}^{0}$ can be obtained.

On substituting the value of $C_{R_{\rm II}}^{0}$ in Eq. (m), the value of $C_{R_{\rm I}}^{0}$ can be calculated. On substituting the values of $C_{R_{\rm II}}^{0}$ and $C_{R_{\rm I}}^{0}$ in Eq. (j), the value of k_1^0 can be obtained,

$$k_1^0 = \frac{\{B(d'-d^{\rm I})r - (d'-d^{\rm II})Ar\}C_{R_{\rm II}}^{0}}{[E'B - A\{E-(v-bs)/b^{(1-\alpha)}C_O^{{\rm II}^\alpha}\}]} \tag{q}$$

From Eq. (f), the value of k_3^0 can be calculated,

$$k_3^0 = \{(d'-d^{\rm I})r - k_1^0 E' C_{R_{\rm I}}^{0^\alpha}\cdot C_R^{0^{(1-\alpha)}}/C_{R_{\rm II}}^{0^{(1-\alpha)}}\}/A \tag{r}$$

Now from Eq. (a), the value of k_2^0 is determined,

$$\begin{aligned}
k_2^0 = {}&[d'r\{C_O^0(1-\alpha) - \alpha C_{R_{\rm II}}^{0}\}k_1^0/C_O^{0^\alpha}C_{R_{\rm II}}^{0^{(1-\alpha)}} \\
&- \{C_{R_{\rm I}}^{0}(1-\alpha) - \alpha C_R^0\}k_3^0/C_{R_{\rm I}}^{0^\alpha}C_R^{0^{(1-\alpha)}}]\frac{C_{R_{\rm II}}^{0^\alpha}C_{R_{\rm I}}^{0^{(1-\alpha)}}}{C_{R_{\rm II}}^{0}(1-\alpha) - \alpha C_{R_{\rm I}}^{0}}
\end{aligned} \tag{s}$$

where as given above, the values of the different terms used are

$$(d'-d^{\rm I})r = (\Delta E_\infty/V_A^2 - \Delta E_\infty^{\rm I}/V_A^2)4RT\omega^{1/2}D_O^{1/2}/2^{1/2}F$$

$$(d'-d^{\rm II})r = (\Delta E_\infty/V_A^2 - \Delta E_\infty^{\rm II}/V_A^2)4RT\omega^{1/2}D_O^{1/2}/2^{1/2}F$$

$$d'r = \{4RT\Delta E_\infty/V_A^2 F - (2\alpha-1)\}\omega^{1/2}D_O^{1/2}/2^{1/2}$$

$$A = [C^0_{R_I}(1-\alpha) - \alpha C^0_R - \{aC^0_{R_I}(1-\alpha) - \alpha C^0_R\}/a^{\alpha}]$$

$$B = [C^0_{R_I}(1-\alpha) - \alpha C^0_R - \{bC^0_{R_I}(1-\alpha) - \alpha C^0_R\}/b^{\alpha}]$$

$$E = \{C^0_O(1-\alpha) - \alpha C^0_{R_{II}}\}/C_O^{0^{\alpha}}$$

$$E' = [E - \{C^{0'}_O(1-\alpha) - a\alpha C^0_{R_{II}}\}/a^{(1-\alpha)} \cdot C_O^{0'^{\alpha}}]$$

NOTATION

$C^0_O, C^0_R, C^0_{R_I}, C^0_{R_{II}}$	bulk concentration of oxidant, bulk concentration of reductant, concentration of the two intermediate species in the electrode reaction
D_O, D_R	diffusion coefficients of species O and R
α_c, α_a	cathodic transfer coefficient and anodic transfer coefficient
I_{MTR}	total rectification current due to mass transfer
I_0, I^0_a	exchange current density and apparent exchange current density
q_0	$-I_\omega/nFC^0_O D_O^{1/2}\omega^{1/2}$
q_R	$I_\omega/nFC^0_R D_R^{1/2}\omega^{1/2}$
z	nFV_A/RT
n	number of electrons involved in charge transfer
V_A	amplitude of ac voltage at the interface
θ	phase difference between the alternating current and applied alternating voltage
ω	$2\pi f$ (f is the frequency of the alternating current)
F	Faraday's constant
R	gas constant
T	temperature
I_{FR}	total rectification current
I_ω	$-\dfrac{RT}{nF}\dfrac{z}{Z_\omega}\dfrac{1}{(\alpha_c+\alpha_a)}$

Z_ω	faradaic impedance
R_r	reaction resistance
σ	Warburg coefficient
ΔE_∞	shift in mean equilibrium potential (rectified potential)
p	$\dfrac{2^{1/2}\omega^{1/2}(C_O^0)^\alpha(C_R^0)^{(1-\alpha)}D_O^{1/2}D_R^{1/2}}{k^0(C_O^0 D_O^{1/2} + C_R^0 D_R^{1/2})}$
α	transfer coefficient of the forward process
$\Delta E_{\infty_{\mathrm{I}}}, \Delta E_{\infty_{\mathrm{II}}}, \Delta E_{\infty_{\mathrm{III}}}$	rectification potential due to the first-step, second-step, and third-step reaction, respectively
$C_{R_{\mathrm{I}}}^0, C_{R_{\mathrm{II}}}^0$	concentrations of the intermediate species formed intermittently after first and second electron charge transfer
k_1^0, k_2^0, k_3^0	rate constants for the first-step, second-step, and third-step reactions
E_1, E_2, E_3, E_4, E_5	oxidation potentials corresponding to oxidant concentrations C_O^0, $C_O^{0'}$, $C_O^{0''}$, $C_O^{0'''}$, and $C_O^{0^{\mathrm{IV}}}$, respectively
$C_{R_{\mathrm{I}}}^{0\prime}, C_{R_{\mathrm{II}}}^{0\prime}$	concentrations of intermediate species corresponding to $C_O^{0\prime}$, $C_O^{0\prime\prime}$
$\Delta E_\infty, \Delta E'_\infty, \Delta E''_\infty$	rectification potentials corresponding to concentrations C_O^0, $C_O^{0\prime}$, and $C_O^{0\prime\prime}$
$C_R^0, C_R^{0\prime}, C_R^{0\prime\prime}$	concentrations of reductants when C_O^0 is kept constant
ω_1, ω_2	zero-point frequencies corresponding to two different redox concentration ratios

REFERENCES

[1] K. S. G. Doss and H. P. Agarwal, *J. Sci. Ind. Res.* **9B** (1950) 280.
[2] K. S. G. Doss and H. P. Agarwal, *Proc. Indian Acad. Sci.* **34** (1951) 229.
[3] K. S. G. Doss and H. P. Agarwal, *Proc. Indian Acad. Sci.* **34A** (1951) 263.
[4] K. S. G. Doss and H. P. Agarwal, *Proc. Indian Acad. Sci.* **35A** (1952) 45.
[5] K. B. Oldham, *J. Electrochem. Soc.* **107** (1960) 766.
[6] Iu. A. Vdovin, *Dokl. Akad. Nauk SSSR* **120** (1958) 554.
[7] G. C. Barker, in *Transactions of the Symposium on Electrode Process*, Ed. by E. Yeager, Wiley, New York, 1961, p. 325.
[8] G. C. Barker, *Anal. Chim. Acta* **18** (1958) 118.

[9] H. Matsuda and P. Delahay, *J. Am. Chem. Soc.* **82** (1960) 1547.
[10] P. Delahay, M. Senda, and C. H. Weis, *J. Phys. Chem.* **64** (1960) 960.
[11] P. Delahay, M. Senda, and C. H. Weis, *J. Am. Chem. Soc.* **83** (1961) 312.
[12] M. Senda, H. Imai, and P. Delahay, *J. Phys. Chem.* **65** (1961) 1253.
[13] H. Imai and P. Delahay, *J. Phys. Chem.* **66** (1962) 1108.
[14] H. Imai and P. Delahay, *J. Phys. Chem.* **66** (1962) 1683.
[15] H. Breyer and H. H. Bauer, in *Alternating Current Polarography and Tensametry*, Interscience, New York, 1963.
[16] E. Yeager and J. Kuta, in *Physical Chemistry*: An *Advanced Treatise*, Vol. 9, Ed. by H. Eyring, Academic, New York, 1970, Part A, p. 345.
[17] P. Delahay, in *Advances in Electrochemistry and Electrochemical Engineering*, Vol. 1, Ed. by P. Delahay and C. W. Tobias, Interscience, New York, 1961, pp. 233, 277, 291, 307.
[18] A. A. Pilla, *Bull. Soc. France, Electricians*, IV, No. 37 (1963) 24.
[19] H. P. Agarwal, *Agra Univ. Res. J. (Sci.)* **14** (Pt. 1) (1965) 213.
[20] H. P. Agarwal *D. A. V. College Res. J.* **1** (1954) 32.
[21] H. Imai, *Rev. Polarog. (Japan)* **10** (1962) 209.
[22] H. Imai, S. Inonye, and T. Tanaka, *Rev. Polarog. (Japan)* **14** (1967) 147.
[23] John S. C. Chiang, *Diss. Abstr. USA* **29**(6) (1968) 1949.
[24] W. A. Brocke and H. W. Nurenberg, *Z. Instrumentenk* **75(a)** (1967) 291.
[25] G. C. Barker, R. L. Faircloth, and A. W. Gardner, *Nature* **181** (1958) 247.
[26] G. C. Barker, in *Polarography*, Ed. by G. J. Hills, Macmillan, London, 1964, p. 25.
[27] J. Cakenbergha, *Bull. Soc. Chim. Belg.* **60** (1951) 3.
[28] J. Paynter and W. H. Reinmuth, *Anal. Chem.* **34** (1962) 1335.
[29] H. H. Bauer and P. J. Elving, *Anal. Chem.* **30** (1958) 341.
[30] D. E. Smith and W. H. Reinmuth, *Anal. Chem.* **33** (1962) 482.
[31] M. A. V. Devanathan, *Electrochim. Acta* **17** (1972) 1755, 1683.
[32] K. B. Oldham, *Trans. Faraday Soc.* **53** (1957) 229.
[33] S. K. Rangarajan, *J. Electroanal. Chem.* **1** (1960) 396.
[34] D. C. Grahame, *J. Electroanal. Chem.* **99C** (1952) 370.
[35] S. K. Rangarajan, *J. Electroanal. Chem.* **56** (1974) 1.
[36] S. K. Rangarajan, *J. Electroanal. Chem.* **56** (1974) 27.
[37] A. K. N. Reddy, in *Electrochemistry—The Past Thirty and Next Thirty Years*, Ed. by Harry Bloom and Felix Gutman, Plenum Press, New York, 1975, p. 195.
[38] H. P. Agarwal and P. Jain, *Electrochim. Acta* **26** (1981) 621.
[39] P. K. Jain, *Faradaic rectification studies of electrode solution interfaces*, Ph.D. thesis, Bhopal University, 1981.
[40] H. P. Agarwal, *Electroanalytical Chemistry*, Vol. 7, Ed. by A. J. Bard, Marcel Dekker, New York, 1974, p. 161.
[41] H. P. Agarwal, *Electrochim. Acta* **16** (1971) 1395.
[42] H. P. Agarwal, *Proceedings of DAE Symposium on Interactions at Electrode-Electrolyte Interfaces*, 1982, M-1, p. 179.
[43] H. P. Agarwal and S. Qureshi, *Electrochim. Acta* **19** (1974) 349.
[44] J. O'M. Bockris and H. Kita, *J. Electrochem. Soc.* **109** (1968) 2021.
[45] J. O'M. Bockris and M. Enyo, *Trans. Faraday Soc.* **58** (1962) 1287.
[46] J. Hurlen, *Acta Chem. Scand.* **15** (1962) 630.
[47] L. N. Nekrasov and N. P. Berezina, *Dokl. Akad. Nauk SSSR* **142** (1962) 855.
[48] O. R. Brown and H. R. Thirsk, *Electrochim. Acta* **10** (1965) 383.
[49] I. M. Pearson and G. F. Schrader, *Electrochim. Acta* **13** (1968) 2021.
[50] N. A. Hampson and R. J. Latham, *Trans. Faraday Soc.* **66** (1970) 3131.
[51] H. P. Agarwal and P. K. Jain, *Indian. J. Chem.* **16A** (1978) 126.
[52] H. P. Agarwal and P. K. Jain, *Electrochim. Acta*, **30** (1985) 1243.

[53] L. Gaiser and K. E. Heusler, *Electrochim. Acta* **15** (1970) 161.
[54] F. Van Der Pol, M. Rehbach-Sluyters, and J. H. Sluyters, *J. Electroanal. Chem. Interfacial Electrochem.* **58** (1975) 117.
[55] T. Hurlen and P. Karl Fischer, *J. Electroanal. Chem. Interfacial Electrochem.* **61** (1975) 165.
[56] P. P. Sagpet, P. Homsi, V. Plichonet, and J. Badoz-Lambling, *Electrochim. Acta* **20** (1975) 819.
[57] J. P. Sagpet, V. Plichonet, and J. Badoz-Lambling, *Electrochim. Acta* **20** (1975) 825.
[58] S. Armalis and A. Levinskas, *Electrokhimiya* **12**(1) (1976) 1967.
[59] H. P. Agarwal, *J. Electrochem. Soc.* **110** (1963) 237.
[60] H. P. Agarwal and S. Qureshi, *J. Electroanal. Chem.* **75** (1977) 697.
[61] Z. Galus and R. N. Adams, *J. Phys. Chem.* **67** (1963) 866.
[62] V. V. Emelyanenko and A. M. Skundin, *Elektrokhimiya* **11**(a) (1975) 1335.
[63] A. A. Baranov, G. A. Simakan, E. A. Erin, V. N. Kosyakov, G. A. Timofev, and A. G. Rykov, *Radiokhimiya* **21**(1) (1970) 59.
[64] S. V. Gorbachev and E. I. Martynycheva, *Tr. Mosk. Khim.-Tekhnol. Inst.* **67** (1970) 267.
[65] S. V. Gorbachev, E. I. Martynycheva, E. P. Agasyan, A. I. Kamenvev, and L. A. Dunaev, *Westn. Mosk. Univ. Khim.* **17**(3) (1976) 382.
[66] H. P. Agarwal and S. Qureshi, *Electrochim. Acta* **21** (1976) 465.
[67] S. Qureshi, *Study of kinetics of electrode reaction.* Ph.D. thesis, Bhopal University, 1974.
[68] V. A. Tyagai and G. Ya. Kolbasov, *Sov. Electrochem.* **6** (1970) 462.
[69] A. Kozlowska, P. K. Wrone, and Z. Galus, *Bull. Acad. Pol. Sci. Ser. Sci. Chim.* **22**(10) (1974) 917.
[70] S. P. Bukhman and I. O. Krol, *Nauk Kaz. SSR, Ser. Khim.* **25**(3) (1975) 19.
[71] K. J. Vetter and G. Thiemke, *Z. Electrochem.* **64** (1960) 805.
[72] S. Toshima, H. Okaniwa, and M. Nishijima, *Chem. Abstr.* **66** (1966) 110966.
[73] G. Fasco, Poraico and Maria, *Bul. Stint. Tech. Inst. Polit. Timibora,* **20**(2) (1975) 233.
[74] H. P. Agarwal and S. Qureshi, *Indian J. Chem.* **14A** (1976) 565.
[75] A. Frumkin and A. Titenskaja, *Zh. Fiz. Khim.* **31** (1957) 485.
[76] A. Frumkin and N. Polyanooskaja, *Zh. Fiz. Khim.* **32** (1958) 257.
[77] J. E. B. Randles, *Disc. Faraday Soc.* **1** (1947) 11.
[78] F. Van Der Pol., M. Sluyters-Rehbach, and J. H. Sluyters, *J. Electroanal. Chem. Interfacial Electrochem.* **40** (1972) 209.
[79] F. Van Der Pol. M. Sluyters-Rehbach, and J. H. Sluyters, *J. Electroanal Chem. Interfacial Electrochem.* **41** (1973) 512; **45** (1973) 377.
[80] H. P. Agarwal and M. Saxena, *Proceedings of the International Symposium on Industrial Electrochemistry* (*SAEST India*), 1976, 13.
[81] H. P. Agarwal and M. Saxena, *Indian J. Chem.* **16A** (1978) 123.
[82] H. P. Agarwal and M. Saxena, *Indian J. Chem.* **16A** (1978) 754.
[83] M. Saxena, *Studies on electrode processes using a.c. polarography, faradaic rectification and transitional potential decay techniques*, Ph.D. thesis, Bhopal University, 1978.
[84] H. P. Agarwal and M. Saxena, Extended Abstract, 31st Meeting of ISE, Venice, Italy, 1980, AI, 145.
[85] H. P. Agarwal and M. Saxena, Extended Abstract, 29th Meeting of ISE, Budapest, Hungary, 1978.
[86] J. E. B. Randles and K. N. Somerton, *Trans. Faraday Soc.* **48** (1952) 952.
[87] T. Kambara and T. Ishio, *Rev. Polarog.* (*Japan*) **9** (1961) 30.

[88] H. P. Agarwal and M. Saxena, Extended Abstract, 30th Meeting of ISE, Trondheim, Norway, (1979) 292.
[89] M. Sluyters-Rehbach, B. Timmer, and J. H. Sluyters, *J. Electroanal. Chem.* **15** (1967) 151.
[90] D. C. Ghrame and R. Parsons, *J. Am. Chem. Soc.* **83** (1961) 129.
[91] J. Lowrence, R. Parsons, and R. Payne, *J. Electroanal. Chem.* **16** (1968) 193.
[92] R. Payne, *J. Electrochem. Soc.* **113** (1966) 999 and unpublished data.
[93] J. K. Frischmann and A. Timnick, *Anal. Chem.* **39** (1967) 507.
[94] N. Tanka and R. Tamamushi, *Electrochim. Acta* **9** (1964) 963.
[95] I. M. Kolthoff and J. J. Lingane, *Polarography*, Interscience, New York, 1941, p. 480.
[96] J. J. Lingane and H. Kerlinger, *Ind. Eng. Chem.* (*Analedu*) **13**, (1941) 77.
[97] L. Meites and T. Meites, *Polarographic Techniques*, Interscience, New York, 1955, p. 303.
[98] P. W. West, J. F. Dean, and E. J. Breda, *Collect. Czech. Chem. Commun.* **13** (1948) 1.
[99] H. P. Agarwal and M. Saxena, *Trans. SAEST Karaikudi* **12** (1977) 258.
[100] H. Imai, *J. Sci. Hiroshima Univ., Ser A* **22** (1958) 191.
[101] S. Inouye and H. Imai, *Bull. Chem. Soc. Jpn.* **33** (1960) 149.
[102] D. S. Jain, *J. Electrochem. Soc. India* **24** (1975) 189.
[103] G. Shima, *Mem. Coll. Sci. Kyoto Imp. Univ.* **A11** (1928) 419.
[104] R. Brdicka, *Cas. Cis. Lek* **58** (1945) 38.
[105] P. Zuman, *Collect. Czech. Chem. Commun.* **15** (1950) 1138.
[106] M. Suzuki, *Mem. Coll. Agr. Kyoto Univ., Chem. Ser.* **28** (1951) 67.
[107] H. P. Agarwal and M. Saxena, in *Proceedings of the 2nd International Symposium on Industrial Oriented Basic Electrochemistry*, 1980, p. 1.
[108] W. C. Sumpter, Y. L. Williams, P. H. Wilken, and B. L. Willoughby, *J. Org. Chem.* **14** (1949) 713.
[109] M. B. Bhargava, *A.C. polarography of organic compounds*, Ph.D. thesis, Bhopal University, 1972.
[110] H. P. Agarwal and M. Saxena, in *Proceedings II* (*Abstracts*), *J. Heyrovsky Memorial Congress on Polarography*, Prague, Czechoslovakia, 1980.
[111] H. P. Agarwal, in *Proceedings of the International Symposium on Recent Aspects of Electroanalytical Chemistry and Electrochemical Technology*, Panjab University, Chandigarh, Dec. 1982.
[112] I. M. Kolthoff and J. J. Lingane, *Polarography*, Vol. 2, Interscience, New York, 1952.
[113] B. S. Sheshadri, *Surf. Tech.* **4**(3) (1976) 223.
[114] S. Sathyanarayana and R. Srinivasan, *Br. Corros. J.* **12**(4) (1977) 217.
[115] S. Sathyanarayana and R. Srinivasan, *Br. Corros. J.* **12**(4) (1977) 221.
[116] H. P. Agarwal, *Electrochim. Acta* **17** (1972) 285.
[117] W. H. Reinmuth, *J. Electroanal. Chem* **34** (1972) 297.
[118] G. C. Barker, A. W. Gardner, and D. C. Sammon, *J. Electrochem. Soc.* **113**(1) (1966) 1182.
[119] H. P. Agarwal and P. K. Jain, Extended Abstract, 35th Meeting of ISE, Berkeley, California, (1984) 936.
[120] H. P. Agarwal and P. K. Jain, *Bull. Electrochem.* **2**(2) (1986) 185.

4

X Rays as Probes of Electrochemical Interfaces

Héctor D. Abruña

Department of Chemistry, Cornell University, Ithaca, New York 14853

I. INTRODUCTION

The study of the structure of the electrode/electrolyte (or more generally the solid/electrolyte) interface[1,2] represents a problem of both fundamental and practical importance in electrochemistry and many other interfacial disciplines since its properties greatly affect (and often control) reactivity. Its importance relates to a broad range of problems including corrosion, catalysis, fuel cells, the potential and ionic gradients at charged interfaces including colloids and biological membranes, and many others. This problem has, until recently, proved very elusive to experimental study due to the difficulty of applying structure-sensitive techniques to the study of surfaces in contact with a condensed phase. Thus, most of our knowledge of the structure of the electrode/solution interface is based on indirect evidence which relies primarily on theoretical models to explain thermodynamic, spectroscopic, and kinetic data.

In recent years,[3,4] however, there has been renewed interest in the study of the electrode/solution interface due in part to the development of new spectroscopic techniques such as surface-enhanced Raman spectroscopy,[5-7] electrochemically modulated infrared reflectance spectroscopy and related techniques,[8,9] second-harmonic generation,[10-12] and others which give information about the identity and orientation of molecular species in the interfacial region. Other techniques such as ellipsometry,[13] electroreflectance and differential reflectance spectroscopy[14] have been used to follow adsorption, film formation, and surface reaction.

All of these techniques, although powerful, do not reveal the structure and geometric arrangement of atomic species at the interface. Thus, in spite of its importance, our knowledge of the structure of the electrode/solution interface at the atomic level is still very rudimentary.

As mentioned previously, this can be attributed in part to the lack of structure-sensitive techniques that can operate in the presence of a condensed phase. Ultrahigh-vacuum (UHV) surface spectroscopic techniques such as low-energy electron diffraction (LEED), Auger electron spectroscopy (AES), and others have been applied to the study of electrochemical interfaces, and a wealth of information has emerged from these *ex situ* studies on well-defined electrode surfaces.[15-17] However, the fact that these techniques require the use of UHV precludes their use for *in situ* studies of the electrode/solution interface. In addition, transfer of the electrode from the electrolytic medium into UHV introduces the very serious question of whether the nature of the surface examined *ex situ* has the same structure as the surface in contact with the electrolyte and under potential control. Furthermore, any information on the solution side of the interface is, of necessity, lost.

From the foregoing, it is clear that particle spectroscopies (e.g., Auger, LEED) are unsuited for *in situ* studies of the electrode/solution interface. Photons, on the other hand, have very large propagation distances, and photons in the X-ray region are suitable probes of the atomic structure of interfacial species. The main difficulty with these measurements has been the low intensities available in conventional X-ray sources. The advent of synchrotron radiation[18] has dramatically changed the outlook. As a result, a number of experiments based on the use of X rays as probes can now be employed in the study of electrochemical interfaces. These include EXAFS (extended X-ray absorption fine structure), XANES (X-ray absorption near edge structure), X-ray standing waves, and surface diffraction.

In addition, recent instrumental developments have made it possible to perform kinetic measurements on the millisecond time scale.

Clearly, the application of these techniques to the study of electrochemical interfaces will allow a much deeper understanding and correlation of structure/reactivity patterns.

In this chapter, I will try to present an introduction to these various techniques with emphasis on EXAFS and X-ray standing waves and their application to the study of electrochemical interfaces. Each technique will be treated from theoretical and experimental points of view, and selected examples from the literature will be employed to illustrate their application to the study of electrochemical interfaces.

II. X RAYS AND THEIR GENERATION

X rays comprise that portion of the electromagnetic spectrum which lies between ultraviolet and gamma rays. The range of wavelengths is typically from about 0.01 to 100 Å. Because of their very short wavelengths, X rays are powerful probes of atomic structure.

X rays have been traditionally produced by impinging an electron beam (at energies from about 20 to 50 keV) onto a target material such as copper, molybdenum, or tungsten. The sudden deceleration of the electron beam by the target material gives rise to a broad spectrum of emission termed bremsstrahlung. The energy of the emitted X rays is given by

$$\lambda = (hc)/(eV) = 12{,}400/V \tag{1}$$

where λ is in angstroms, and V is the accelerating voltage. The minimum wavelength of emission is obtained when all of the electron energy is converted to an X-ray photon. The intensity and wavelength distribution of this bremsstrahlung are both a function of the accelerating voltage, the current, and the target material. Figure 1 shows typical emission envelopes for a tungsten target at various accelerating voltages.

When the accelerating voltage reaches a specific value (dependent on the nature of the target material), the electrons from the beam are capable of knocking out core-level electrons from the target material, thus giving rise to core vacancies. These are quickly filled by electrons in upper levels and this results in the emission of X-ray photons of characteristic energies which depend on the

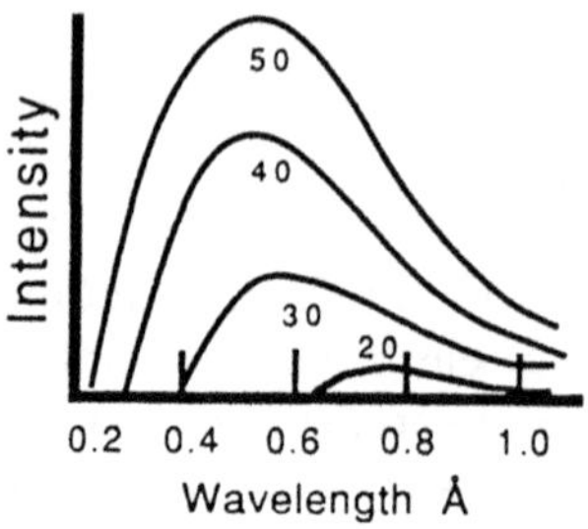

Figure 1. Wavelength distribution of the radiation from an X-ray tube with a tungsten target. The numbers above each curve refer to the accelerating voltages (in keV) employed.

nature of the target. The energies and intensities of characteristic lines depend on the nature of the core hole generated (e.g., *K*- or *L*-shell vacancy) as well as the level from which the electron that fills the vacancy originates. Figure 2 presents a schematic of some of the more important X-ray emission lines. These so-called characteristic lines are much more intense than the bremsstrahlung emission and are superimposed on the latter as very sharp emissions.

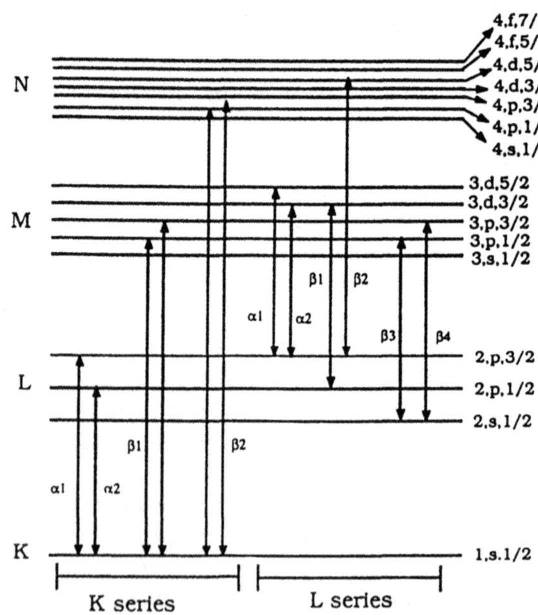

Figure 2. Partial energy level diagram depicting part of the *K*- and *L*-series lines.

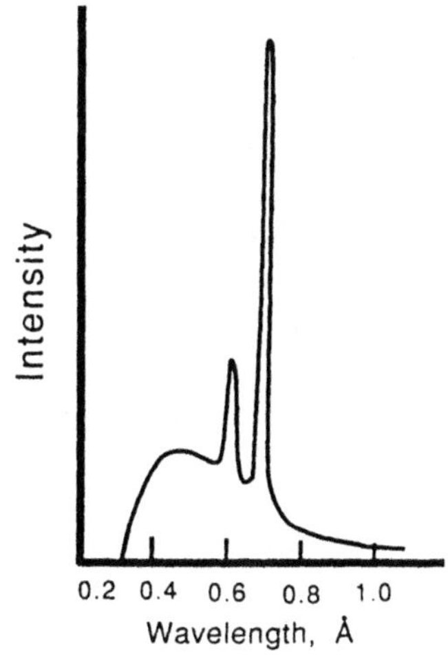

Figure 3. Wavelength distribution of the radiation emitted from a molybdenum target X-ray tube operated at 35 keV.

Figure 3 shows an example of characteristic emissions from a Mo target.

A variant of the typical X-ray tube described above is the rotating anode, which is capable of generating much higher intensities. Although rotating anode sources can and have been used in EXAFS experiments, the intensities are of such magnitude that data acquisition for extended periods of time is required and their application is furthermore limited to bulk samples.

An alternative and the most generally employed source of X rays for EXAFS experiments is that obtained from synchrotron sources based on electron (or positron) storage rings.

III. SYNCHROTRON RADIATION AND ITS ORIGIN

No single development has influenced the field of EXAFS spectroscopy more than the development of synchrotron radiation sources, particularly those based on electron (or positron) storage rings. These provide a continuum of photon energies at intensities that can be from 10^3 to 10^6 higher than those obtained with X-ray tubes, thus dramatically decreasing data acquisition times as well as making other experiments feasible. We will consider very briefly the fundamental aspects of synchrotron radiation.

The most attractive features of synchrotron radiation have been summarized by Winick[19] as:

a. High intensity
b. Broad spectral range
c. High polarization
d. Pulsed time structure
e. Natural collimation
f. Small-source-spot size
g. Stability

We know from Maxwell's equations that whenever a charged particle undergoes acceleration, electromagnetic waves are generated. An electron in a circular orbit experiences an acceleration toward the center of the orbit and as a result emits radiation in an axis perpendicular to the motion.

As pointed out by Tomboulian and Hartman,[20] one can distinguish two general cases and these relate to whether or not an electron is orbiting at relativistic speeds. At nonrelativistic speeds ($v \ll c$), the pattern of the emitted radiation resembles a doughnut (Fig. 4A). However, at relativistic speeds ($v \simeq c$), the radiation pattern is highly peaked (Fig. 4B) and one can think of an orbiting searchlight as a good approximation. This natural collimation effect gives rise to very high fluxes on small targets.

Since the accelerated electrons are constantly emitting radiation, we need to resupply the energy if they are to remain in orbit. This is typically done with high-power RF cavities. The spectral distribution of synchrotron radiation depends on a number of factors, and two that are particularly important are the electron energy E [expressed in GeV (10^9 eV)] and the bending radius R (in meters) of the orbit. A parameter that relates these is the so-called critical wavelength given by

$$\lambda_c \text{ (in angstroms)} = 5.6R/E^3$$

In general, useful fluxes are obtained at wavelengths down to $\lambda_c/4$ although in this region the output decreases precipitously. Figure 5 presents some flux curves for the Cornell High Energy Synchrotron Source (CHESS) operated at various energies. A

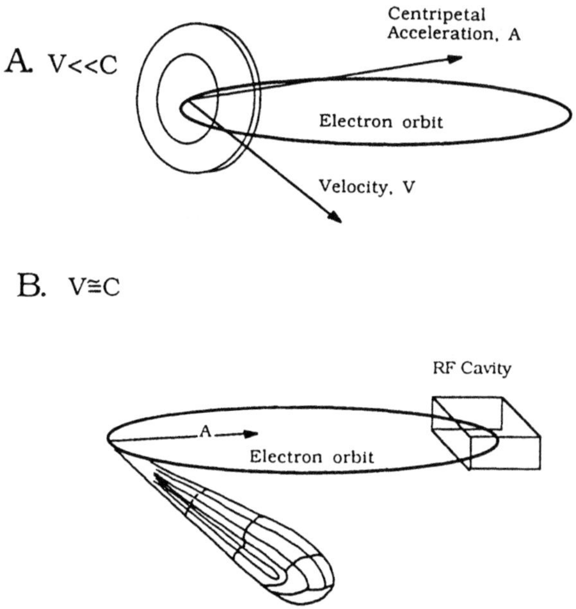

Figure 4. Radiation pattern emitted from orbiting electrons when the velocities are much smaller than (A) or comparable to (B) the speed of light. (Adapted from Ref. 20.)

closely related parameter is the critical energy and this represents the midpoint of the radiated power. That is, half of the radiated power is above and below this energy.

The fact that λ_c is proportional to the bending radius is used in so-called insertion devices such as wiggler and undulator magnets.† Although a description of these is beyond the scope of this chapter, the basic principle behind these is to make the electron beam undergo sharp serpentine motions (thereby having a very short radius of curvature). The net effect is to increase the flux and the critical energy (see topmost curve in Fig. 5).

Another very important property of synchrotron radiation is its very high degree of polarization. The radiation is predominantly polarized with the electric field vector parallel to the acceleration

† For an introductory discussion, see Ref. 21.

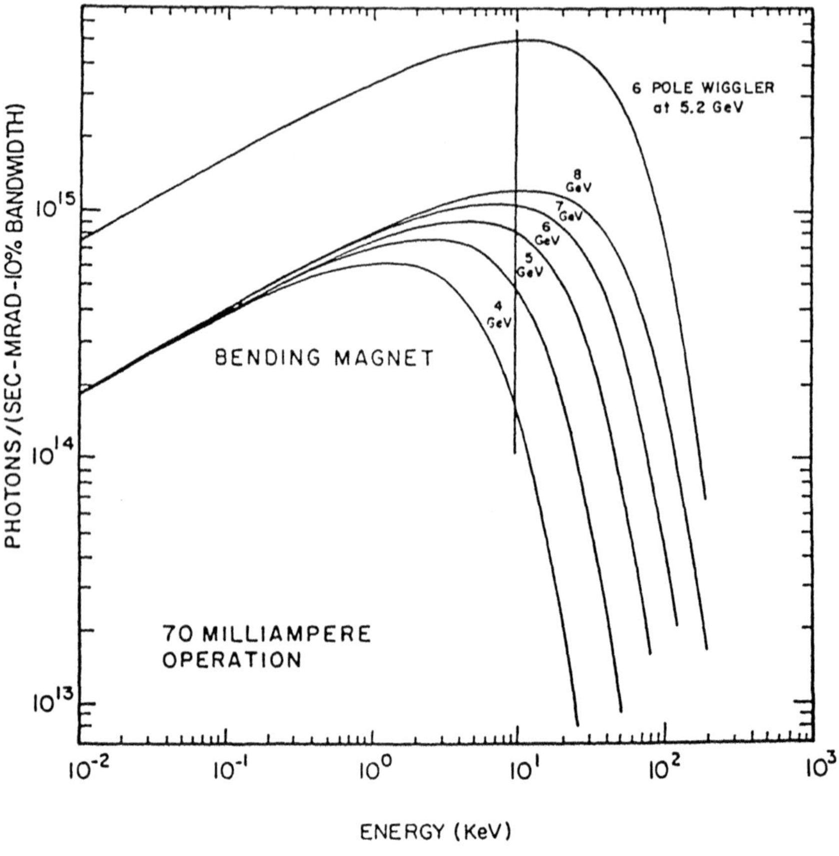

Figure 5. Photon flux as a function of energy for the Cornell High Energy Synchrotron Source (CHESS) operated at various accelerating voltages. The topmost curve is the radiation profile from a 6 pole wiggler magnet. (Figure courtesy of the Laboratory for Nuclear Studies at Cornell University.)

vector. Thus, in the plane of the orbit, the radiation is 100% plane polarized. Elliptical polarization can be obtained by going away from the plane. However, intensities also decrease significantly.

Finally, the pulsed time structure, useful for kinetic studies, arises from the fact that in a storage ring the electrons are orbiting in bunches. The specific energy, the number of bunches, and the circumference of the storage ring dictate the exact time structure.

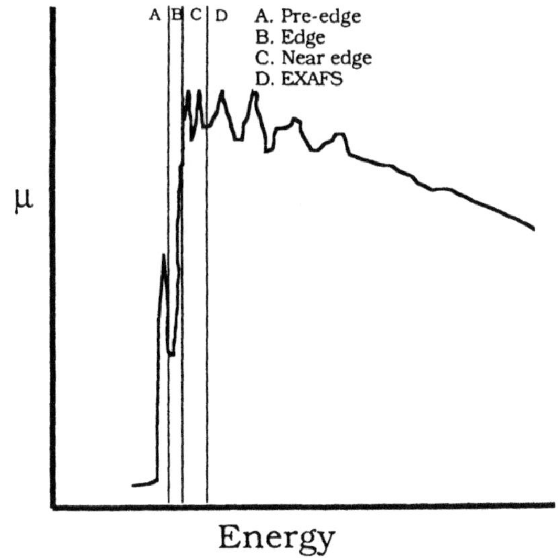

Figure 6. Figure depicting the various regions in an X-ray absorption spectrum.

IV. INTRODUCTION TO EXAFS AND X-RAY ABSORPTION SPECTROSCOPY

Extended X-ray absorption fine structure† (EXAFS) refers to the modulation in the X-ray absorption coefficient beyond an absorption edge. Such modulations can extend up to about 1000 eV beyond the edge and have a magnitude of typically less than 20% of the edge jump.

In order to gain a basic grasp of the EXAFS phenomenon, it is perhaps better to begin by considering the general features observed in an X-ray absorption spectrum. Analogous to obtaining the UV–vis spectrum of a molecule, when we perform an X-ray absorption experiment we measure the absorbance of a sample (typically expressed as an absorption coefficient μ) as we vary the incident photon energy (Fig. 6). In general, as we increase the

† There have been numerous reviews of EXAFS over the last ten years. A selected number of leading references are given as Refs. 22–30.

energy there is a monotonic decrease in the absorption coefficient. However, when the incident X-ray energy is enough to photoionize a core-level electron, there is an abrupt increase in the absorption coefficient and this is termed an absorption edge. There are absorption edges that correspond to the various atomic shells and subshells. For example, a given atom will have one K absorption edge, three L edges, five M edges, etc., with the energies decreasing in the expected order $K > L > M$ (see Fig. 2).

As we continue to scan to higher energies beyond the edge, we can encounter two different situations depending on whether or not the species that we are investigating has near neighbors (typically at 5 Å or closer). If there are no near neighbors, the absorption coefficient will again decrease in a monotonic fashion until its next absorption edge or that of another element present in the sample is encountered.

In the presence of one or more near neighbors, there will be modulations in the absorption coefficient as we scan out to energies about 1000 eV beyond the edge. The modulations present at energies from about 40 eV to 1000 eV beyond the edge are termed EXAFS.

The phenomenon of EXAFS has been known since the 1930s through the work of Kronig[31] who stated that the oscillations are due to the modification of the final state of the photoelectron by near neighbors. Although up to the 1960s there was controversy as to whether short- or long-range order was responsible for the effect, it is now universally accepted that it is the presence of short-range order that gives rise to the EXAFS oscillations.

The absorption coefficient is a measure of the probability that a given X-ray photon will be absorbed and, therefore, depends on the initial and final states of the electron. The initial state is very well defined as it corresponds to the localized core level. The final state is represented by the photoionized electron, which can be visualized as an outgoing photoelectron wave that originates at the center of the absorbing atom and that, for S core levels, has spherical symmetry. In the presence of near neighbors, this photoelectron wave can be backscattered (Fig. 7) so that the final state will be given by the sum of the outgoing and backscattered waves. It is the interference (recall that we are changing the wavelength of the photoelectron wave as we scan the energy) between the outgoing and backscattered waves that gives rise to the EXAFS oscillations.

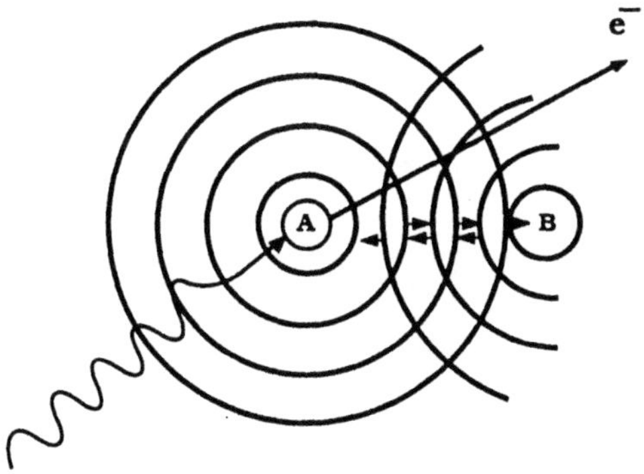

Figure 7. Depiction of origin of EXAFS. An X-ray photon is absorbed by A, resulting in the photoionization of a core-level electron represented as an outgoing (→) photoelectron wave which is backscattered (←) by a near neighbor, B.

Thus, in a very simple approximation we can see that the frequency of the EXAFS oscillations will depend on the distance between the absorber and its near neighbors, whereas the amplitude of the oscillations will depend on the numbers and type of neighbors as well as their distance from the absorber. From an analysis of the EXAFS, one should be able to obtain information on near-neighbor distances, numbers, and types. A further advantage of EXAFS is that it can be applied to all forms of matter—solids, liquids, and gases—and that in the case of solids, single crystals are not required. In addition, one can focus on the environment around a particular element by employing X-ray energies around an absorption edge of the element of interest without regard for the rest of the elements in the sample.

The simple description of EXAFS given above is based on the so-called single-electron, single-scattering formalism.[32-37] Here it is assumed that for sufficiently high energies of the photoelectrons, one can make the plane wave approximation and, in addition, only single backscattering events will be important. This is the reason why the EXAFS is typically considered for energies higher than 40 eV beyond the edge since in this energy region the above approximations hold well.

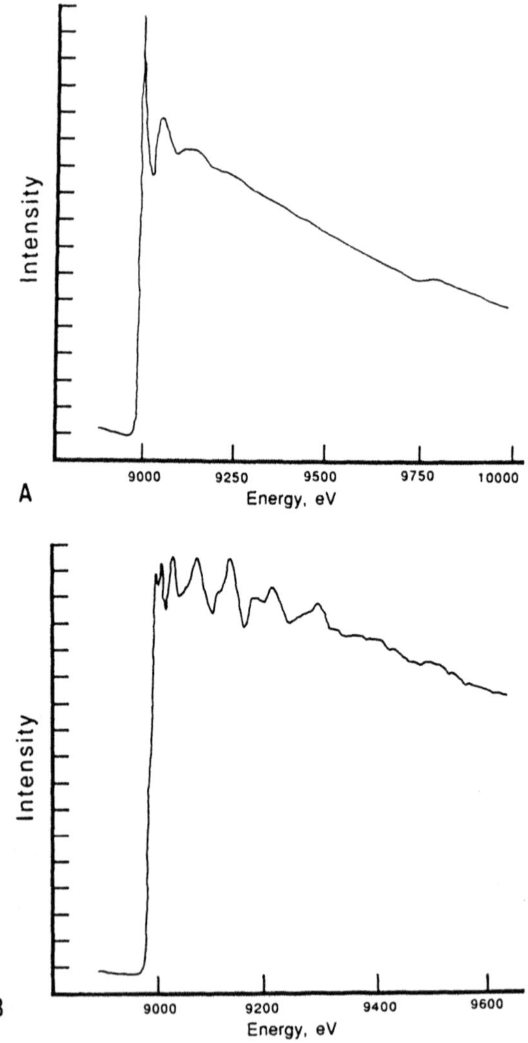

Figure 8. Absorption spectrum for (A) $CuSO_4$ and (B) Cu foil.

In addition to the EXAFS region, Fig. 6 shows that there are also three other regions: the pre-edge, edge, and near-edge regions.

Below or near the edge, there can be absorption peaks due to excitations to bound states. In some cases these can be so intense so as to dominate the edge region. For example, Fig. 8A shows an EXAFS spectrum of $CuSO_4$, and the appearance of a very intense and sharp transition near the edge (often called "white line") is immediately apparent. This peak is associated with a $1s \rightarrow 3d$ localized transition. Because of the nature of the transitions, the pre-edge region is rich in information pertaining to the energetic location of orbitals, site symmetry, and electronic configuration.

The position of the edge contains information concerning the effective charge of the absorbing atom. Thus, its location can be correlated with the oxidation state of the absorber in a way that is analogous to XPS measurements. For example in Figs. 8A (copper sulfate) and 8B (copper foil) the edge position for Cu is shifted to lower energies (by about 2 eV), consistent with the change in oxidation state from +2 to 0. Such shifts can be very diagnostic in the assignment of oxidation states.

Finally, we consider the near-edge region generally called XANES (X-ray absorption near edge structure). [Note: when using UV or soft X rays this region is generally called near-edge X-ray absorption fine structure (NEXAFS).] In this region of the spectrum the photoelectron wave has very small momentum, and as a result, the plane-wave as well as the single-electron single-scattering approximations are no longer valid. Instead, one must consider a spherical photoelectron wave as well as the effects of multiple scattering. Because of this factor, the photoelectron wave can sample much of the environment around the absorber. This region of the spectrum is thus very rich in structure (see Fig. 6); however, the theoretical modeling is very complex. However, increased attention is being given by theoreticians, and in the not so distant future, we will be able to obtain much information from this region.

V. THEORY OF EXAFS

We will consider the theoretical description of EXAFS based on the single-scattering short-range order formalism. The EXAFS can be expressed as the normalized modulation of the absorption coefficient as a function of energy:

$$\chi(E) = [\mu(E) - \mu_0(E)]/\mu_0(E) \tag{2}$$

Here $\mu(E)$ is the total absorption coefficient at energy E and μ_0 is the smooth atomlike absorption coefficient. In order to be able to extract structural information from the EXAFS, we need to go from the energy to the wave vector form using the formulation:

$$k = \sqrt{2m(E - E_0)}/\hbar \tag{3}$$

where E_0 is defined as the threshold energy (typically close but not necessarily coincident with the energy at the absorption edge) (*vide infra*). In wave vector form, the EXAFS can be expressed as:

$$\chi(k) = \sum_j \underbrace{\frac{1}{kr_j^2} N_j S_i(k) F_j(k) \exp^{-2\sigma_j^2 k^2} \exp^{-2r_j/\lambda(k)}}_{\text{a}} \underbrace{\sin[2kr_j + \phi_j(k)]}_{\text{b}} \tag{4}$$

Here I have divided the expression into two terms which correspond to an amplitude factor (a) and an oscillatory component (b). I will now consider each of these terms in some detail.

1. Amplitude Term

The amplitude term

$$\frac{1}{kr_j^2} N_j S_i(k) F_j(k) \exp^{-\sigma_j^2 k^2} \exp^{-2r_j/\lambda(k)} \tag{5}$$

can be subdivided into two main components: a maximum amplitude term and an amplitude reduction factor.

For a given shell, the maximum amplitude is given by the product of the number (N) of the j type of scatterer times its respective backscattering amplitude, $F_j(k)$. This maximum amplitude is then reduced by a series of amplitude reduction factors which are considered below.

(i) Many-Body Effects

The $S_i(k)$ term takes into account amplitude reduction due to many-body effects and includes losses in the photoelectron energy due to electron shake-up (excitation of other electrons in the absorber) or shake-off (ionization of low-binding-energy electrons in the absorber) processes.

(ii) Thermal Vibrations and Static Disorder

Photoionization and therefore EXAFS takes place on a time scale that is much shorter than that of atomic motions so the experiment samples an average configuration of the neighbors around the absorber. Therefore, we need to consider the effects of thermal vibration and static disorder, both of which will have the effect of reducing the EXAFS amplitude. These effects are considered in the so-called Debye-Waller factor which is included as

$$\exp^{-2\sigma_j^2 k^2} \tag{6}$$

This can be separated into static disorder and thermal vibrational components:

$$\sigma_j^2 = \sigma_{\text{vib}}^2 + \sigma_{\text{stat}}^2 \tag{7}$$

It is generally assumed that the disorder can be represented by a symmetric Gaussian-type pair distribution function and that the thermal vibration will be harmonic in nature.

Experimentally, one can only measure a total σ. However, the two contributions can be separated by performing a temperature dependence study of σ or by having an *a priori* knowledge of σ_{vib} from vibrational spectroscopy.

Whereas there is little that one can do to overcome the effects of static disorder, the effects of thermal vibration can be significantly decreased by performing experiments at low temperatures, and, in fact, many solid samples are typically run at liquid nitrogen temperatures just to minimize such effects. An example of the effect of thermal vibration can be ascertained in Fig. 8A, where the EXAFS amplitude decreases precipitously due to the large vibrational amplitude of the Cu—O bond. In general, failure to consider the effects of thermal vibration and static disorder can result in large errors in the determination of coordination numbers (number of neighbors) and interatomic distances.[38-40]

(*iii*) *Inelastic Losses*

Photoelectrons that experience inelastic losses will not have the appropriate wave vector to contribute to the interference process. Such losses are taken into account by an exponential damping factor,

$$\exp^{-2r_j/\lambda(k)} \tag{8}$$

where r is the interatomic distance and $\lambda(k)$ is the electron mean free path. This damping term limits the range of photoelectrons in the energy region of interest, and is in part responsible for the short-range description of the EXAFS phenomenon.

In general, it is the product of all of the above-mentioned factors that will give rise to the observed amplitude.

2. Oscillatory Term

The oscillatory part of the EXAFS takes into account the relative phases between the outgoing and backscattered waves and, as a result, includes the interatomic distance between absorber and scatterer. The phase shifts can be visualized by considering that the outgoing photoelectron wave will experience the absorbing atom's phase shift $\delta_i(k)$ on its outward trajectory. It will then experience the near neighbor's phase shift $\alpha_s(k)$ upon scattering and the absorbing atom's phase shift once again. There is in addition a $2kr$ term which represents twice the interatomic distance between absorber and scatterer. Thus, the oscillatory part of the EXAFS is given by

$$\sin[2kr + 2\delta_i(k) + \alpha_s(k)] \tag{9}$$

Since the accuracy of the determination of interatomic distances depends largely on the appropriate determination of the relative phases, a great deal of attention has been given to this aspect. The problem can be simply stated as follows: when the outgoing photoelectron wave is backscattered by a near neighbor, it is the neighbor's electron cloud and not its nucleus that is largely responsible for the scattering. As a result, the distance obtained will always be necessarily shorter than the true interatomic distance and therefore needs to be corrected. The correction can be achieved by *ab initio* calculation of the phases involved or alternatively it can be determined experimentally through the use of model compounds. A more thorough discussion of phase correction will be given further on.

3. Data Analysis

The basic intent behind any EXAFS data analysis is to be able to extract information related to interatomic distances, numbers, and types of backscattering neighbors. In order to accomplish this, there are a number of steps involved in the data analysis, and these include:

i. Background subtraction and normalization
ii. Conversion to wave vector form
iii. *k* weighing
iv. Fourier transforming and filtering
v. Fitting for phase
vi. Fitting for amplitude

(*i*) *Background Subtraction and Normalization*

The first step in the analysis is the background subtraction. However, since the smooth or atomlike absorption (that is, the absorption for an isolated atom) is in general not available, it is generally assumed that the smooth part of $\mu(E)$ is a good approximation to $\mu_0(E)$.

Background removal routines typically employ polynomial splines of some order (typically second or third order). These are defined over a series of intervals with the constraint that the function and a stipulated number of derivatives be continuous at the intersection between intervals. In addition, the observed EXAFS oscillations need to be normalized to a single-atom value and this is generally done by normalizing the data to the edge jump.

(*ii*) *Conversion to Wave Vector Form*

In order to extract structural information, we need to convert the EXAFS expressed in terms of energy to wave vector form. To do this, however, we need to choose a value for the threshold energy E_0. This is important because of its effect on the phase of the EXAFS oscillations, expecially at low k values. The difficulty in determining E_0 arises from the fact that there is no way of identifying an edge feature with E_0. A procedure proposed by Lee and Beni[41] is to treat E_0 as an adjustable parameter in the data analysis whose value is changed until the observed phase shifts are in good agreement with theoretical values. When good model compounds are available, the use of a fixed value for E_0 works well.[42-44] However, in many cases it is difficult to assess *a priori* whether a given material is a good model compound.

Figure 9A depicts data after background subtraction, normalization, and conversion to wave vector form.

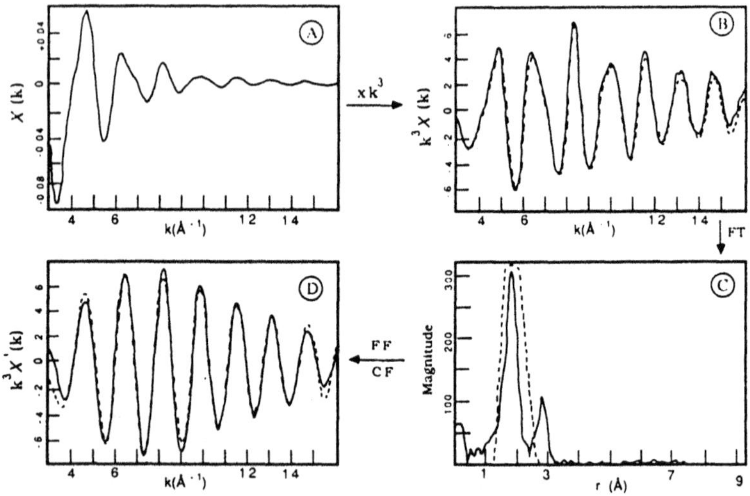

Figure 9. Data reduction and data analysis in EXAFS spectroscopy. (A) EXAFS spectrum $\chi(k)$ versus k after background removal. (B) The solid curve is the weighted EXAFS spectrum $k^3\chi(k)$ versus k (after multiplying $\chi(k)$ by k^3). The dashed curve represents an attempt to fit the data with a two-distance model by the curve-fitting (CF) technique. (C) Fourier transformation (FT) of the weighted EXAFS spectrum in momentum (k) space into the radial distribution function $\rho_3(r')$ versus r' in distance space. The dashed curve is the window function used to filter the major peak in Fourier filtering (FF). (D) Fourier-filtered EXAFS spectrum $k^3\chi'(k)$ versus k (solid curve) of the major peak in (C) after back-transforming into k space. The dashed curve attempts to fit the filtered data with a single-distance model. (From Ref. 25, with permission.)

(*iii*) *k Weighing*

Once the data have been transformed to wave vector form, they are usually multiplied by some power of k, typically, k^2 or k^3. Such a factor cancels the $1/k$ factor in Eq. (4) as well as the $1/k^2$ dependence of the backscattering amplitude at large values of k. Figure 9B depicts multiplication by k^3.

This step is important in that it prevents the large-amplitude oscillations (typically present at low k) from dominating over the smaller ones (typically at high k). This is critical since the determination of interatomic distances depends on the frequency and not the amplitude of the oscillations. Other approaches having the same effect have also been employed.[37,47]

(*iv*) *Fourier Transforming and Filtering*

Examination of the EXAFS formulation in wave vector form reveals that it consists of a sum of sinusoids with phase and amplitude. Sayers *et al.*[32] were the first to recognize the fact that a Fourier transform of the EXAFS from wave vector space (k or direct space) to frequency space (r) yields a function that is qualitatively similar to a radial distribution function and is given by:

$$\phi(r) = \frac{1}{\sqrt{2\pi}} \int_{k_{min}}^{k_{max}} k^n \chi(k) \exp^{(2ikr)} dk \tag{10}$$

Such a function exhibits peaks (Fig. 9C) that correspond to interatomic distances but are shifted to smaller values (recall the distance correction mentioned above). This finding was a major breakthrough in the analysis of EXAFS data since it allowed ready visualization. However, because of the shift to shorter distances and the effects of truncation, such an approach is generally not employed for accurate distance determination. This approach, however, allows for the use of Fourier filtering techniques which make possible the isolation of individual coordination shells (the dashed line in Fig. 9C represents a Fourier filtering window that isolates the first coordination shell). After Fourier filtering, the data is back-transformed to k space (Fig. 9D), where it is fitted for amplitude and phase. The basic principle behind the curve-fitting analysis is to employ a parameterized function that will model the observed EXAFS and then the various parameters are adjusted until the fit is optimized.

(*v*) *Fitting for Phase*

Accurate distance determinations depend critically on the accurate determination of phase shifts. The two general approaches to this problem are theoretical and empirical determination. The two main approaches to the theoretical calculations of phase shifts have been the Hartree-Fock[36,48] (HF) and Hartree-Fock-Slater[47,49] (HFS) methods. The first treatment begins with tabulated atomic wave functions, and the HF equation of the atom plus the external electron is solved by iteration. In the HFS (or local density functional) approach, the atom is replaced by an electron gas of varying density. In general, both of these approaches are too involved for general use. Teo and Lee[50] used the theoretical approach of Lee and Beni[37] to calculate and tabulate theoretical phase shifts for the majority of elements. Use of these theoretical phase shifts requires the use of an adjustable E_0 in the data analysis (*vide supra*).

The second, and more commonly employed, approach is the empirical one based on the use of model compounds and the concept of phase transferability. This approach consists of employing a compound of known structure and which has the same absorber/backscatter combination as that of the material of interest. The EXAFS spectrum of the known compound (typically called model compound) is obtained and the oscillatory part of the EXAFS is fitted to the expression in Eq. (9). Since r is known in this case, the phase shift can be determined. Typically, the phase shift is parameterized as a quadratic expression. Implicit in this treatment is the applicability of phase transferability, meaning that for a given absorber/scatterer combination, the phase shifts can be transferred to any compound with the same absorber/scatterer combination without regard to chemical effects such as ionicity or covalency of the bonds involved. This is based on the idea that at sufficiently high kinetic energies for the photoelectron (e.g., about 50 eV above threshold), the EXAFS scattering processes are largely dominated by core electrons and thus the measured phase shifts are insensitive to chemical effects. Thus, determination of the phase shift for an absorber/scatterer pair in a system of known r allows for the determination of the distance in an unknown having the same atom pair. This was thoroughly demonstrated by Citrin *et al.* in a study of germanium compounds.[51]

With good-quality data and appropriately determined phase shifts, distance determination by EXAFS are typically good to ±0.01 Å and sometimes better in favorable cases.

(*vi*) *Fitting for Amplitude*

Amplitude fitting is employed in order to determine the types and numbers of backscattering atoms around a given absorber. The problem can be divided in two parts, namely, the identification of the types of backscatterers and the determination of their numbers. In the first case, if we have no clue as to the probable nature of the backscatterer, identification is difficult, especially among atoms that have similar atomic number, e.g., N and O. This is because the backscattering amplitudes are not a very strong function of atomic number. For example, Fig. 10 shows the backscattering amplitude for various elements as calculated by Teo and Lee.[50] It

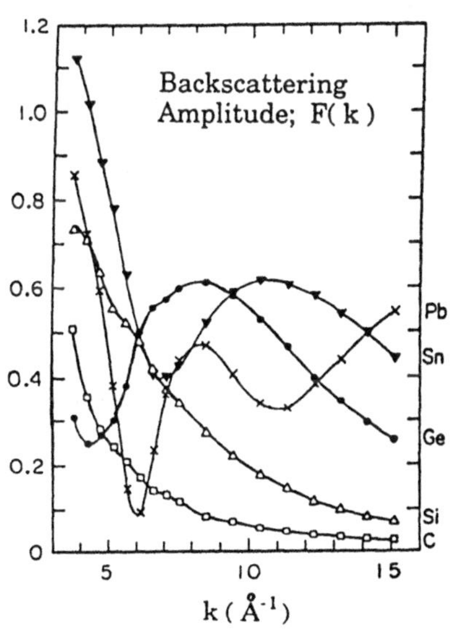

Figure 10. Backscattering amplitude as a function of wave vector for C, Si, Ge, Sn, and Pb. (Adapted from Ref. 50.)

is clear that for Si and C the differences are small and so these two elements would be difficult to differentiate. However, for the case of a heavy-atom backscatterer, there is typically a resonance in the backscattering amplitude (this is analogous to the Ramsauer-Townsend effect) so that differentiation between light and heavy backscatters can be readily made.

It should be mentioned that when a peak from a Fourier transform is filtered and back-transformed to k space, the envelope represents the backscattering amplitude for the near neighbor involved.

If the identity of the backscatterer is known, then the interest is in determining the number of near neighbors. In this case, one needs to compare the amplitude of the EXAFS of the material of interest (unknown) to that for a compound of known coordination number and structure. However, unlike transferability of phase, which is generally regarded as an excellent approximation, the transferability of amplitude is not. This is because there are many factors that affect the amplitude and, except for the case of model compounds of very similar structures, these will not necessarily (and often will not) be the same. As a result, determination of coordination numbers (near neighbors) is usually no better than ±20%.

VI. SURFACE EXAFS AND POLARIZATION STUDIES

EXAFS is fundamentally a bulk technique due to the high penetration of high-energy X rays. In order to make it surface sensitive, one can take one of two general approaches. In the first case, if one knows *a priori* that the specific element of interest is present only at the surface, then a conventional EXAFS measurement will necessarily give surface information. Alternatively, one can employ detection techniques or geometries such that the detected signal arises predominantly from the surface or near-surface regions.[52-56] These include electron detection and operating at angles of incidence that are below the critical angle of the particular material. These aspects will be discussed further in the experimental section. In addition, there have been a number of reviews of this matter, with Citrin's[56] being the most comprehensive.

For studies on single-crystal surfaces, surface EXAFS offers an additional experimental handle, namely, the polarization dependence of the signal. As mentioned previously, synchrotron radiation is highly polarized with the plane of polarization lying in the plane of orbit. Since only those bonds whose interatomic vector lies in the plane of polarization of the beam will contribute to the observed EXAFS, polarization dependence studies can provide a wealth of information concerning adsorption sites and near-neighbor geometries. This is especially significant since it is very difficult (if at all feasible) to obtain this type of information from conventional EXAFS measurements.

For a near-neighbor shell of atoms (N_i) whose interatomic vector with the absorber makes some angle θ_j relative to the plane of polarization, one can relate the effective coordination number (N_i^*) and the true coordination number through[57]

$$N_i^* = 3 \sum_j^{N_i} \cos^2 \theta_j \tag{11}$$

Polarization-dependent surface EXAFS measurements have provided some of the best-defined characterizations of adsorbate structures.

VII. EXPERIMENTAL ASPECTS

1. Synchrotron Sources

There are a number of experimental factors to be considered in a surface EXAFS experiment. First of all, one needs access to a synchrotron source (for the reasons previously mentioned) with significant flux in the hard X-ray region. In the United States, three such facilities exist and these are:

a. Cornell High Energy Synchrotron Source (CHESS)

b. Stanford Synchrotron Radiation Laboratory (SSRL)

c. National Synchrotron Light Source (NSLS) at Brookhaven National Laboratory

[It should be mentioned that another synchrotron source, the Advanced Photon Source (APS) will be built at the Argonne National Laboratory and should be operational in the mid-1990s.]

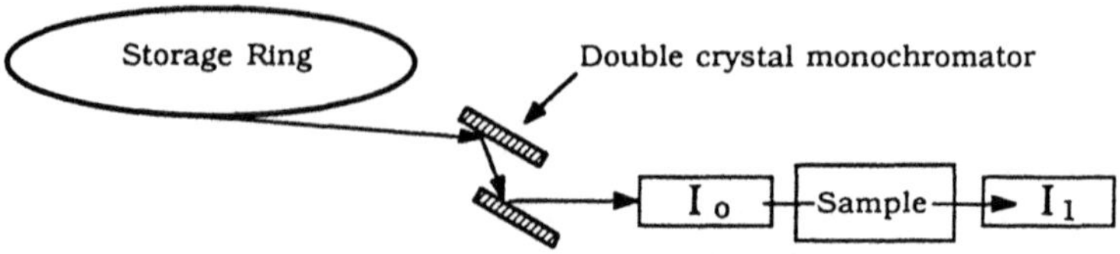

Figure 11. Schematic diagram of a transmission EXAFS experiment. I_0 and I_1 refer to the incident and transmitted intensities, respectively.

In addition, one needs to pay close attention to detection schemes and the design of specialized equipment. Of these, I will focus on detection schemes at this time and will defer the discussion of design of systems for electrochemical measurements to later sections dealing with specific experiments.

2. Detection

Detection schemes are usually dictated by the concentration of the species of interest, the nature of the sample, and the experiment. All of these aspects have been considered in great detail by Lee *et al.*[26] so I will only mention some of the most important aspects. In general, the measurement of any parameter that can be related to the absorption coefficient can be employed in a detection scheme.

(*i*) *Transmission*

For concentrated or bulk samples a transmission experiment is both the simplest and the most effective. In essence, one measures the X-ray intensities incident and transmitted through a thin and uniform film of the material. Careful analysis of signal-to-noise ratio considerations indicates that optimal results are obtained when the sample thickness is of the order of 2.5 absorption lengths. Since in this case a simple Beer's law applies, the data are usually plotted as $\ln(I/I_0)$ versus E. The intensities are measured using ionization chambers in conjunction with high-gain electrometers (see Fig. 11).

(*ii*) *Fluorescence*

For dilute samples, where absorption of the X-ray beam by the element of interest would be very low, a transmission geometry cannot be employed. Instead, fluorescence detection is the method of choice.[58-60] Fluorescence can be employed as a detection mode because the characteristic X-ray fluorescence intensity depends on the number of core holes generated, which, in turn, depends on the absorption coefficient. Fluorescence detection is much more sensitive than transmission because one is measuring the signal over an essentially zero background. Typically, the incident and fluorescent beams impinge and leave the surface at 45° (the X-ray beam and the detector are, of course, at 90°). The detector can be either an ionization chamber or a solid state detector. The former is much simpler to implement whereas the latter gives the best resolution. A filter (to minimize the contributions from elastic and Compton scattering) and soller slits are typically placed in front of the detector. The filter material is chosen so as to have an absorption edge that falls between the excitation energy and the energy of the characteristic X-ray photon from the element of interest. Thus, the filter is generally made from the $Z-1$ or $Z-2$ element, where Z represents the atomic number of the element of interest. For example, for CuK_α detection, a nickel filter is employed. In this way the characteristic fluorescence is only slightly attenuated whereas both the elastic and Compton intensities are greatly reduced. There is the problem, however, that often the K_β emission from the filter material is energetically very close to the K_α emission from the element of interest.

A solid state detector, either Si(Li) (lithium-drifted silicon) or intrinsic germanium, offers the ability to discriminate on the basis of energy. The resolution can be as good as 150 eV, but it degrades somewhat with increasing detector area. The main problem with a solid state detector is that it has a limited count rate (approximately 15 kcps). Since it "sees" a wide range of photon energies from which one chooses the region of interest, it can take significant amounts of time to obtain adequate statistics. In addition, the cost of solid state detectors and associated electronics is much higher than that of ion chambers.

(iii) Electron Yield

Electron yield—Auger, partial, or total—can also be employed as detection means since again it depends on the generation of core holes.[53,54] Because of the very small mean free paths of electrons, electron yield detection is very well suited for surface EXAFS measurements. However, for this very same reason, *in situ* studies of electrochemical interfaces are precluded. Details of electron yield EXAFS have been discussed by a number of authors.[52-56]

(iv) Reflection

When one has a planar surface, one can take advantage of X-ray optics to enhance surface sensitivity.[61,62] The most important is specular or mirror reflection, and this is due to the fact that at X-ray energies the index of refraction of matter is slightly less than one and is given by:

$$n = 1 - \delta - i\beta \tag{12}$$

with

$$\delta = (1/2\pi)(e^2/mc^2)(N_0\rho/A)[Z + \Delta f']\lambda^2$$

$$\beta = \lambda\mu/4\pi$$

where e^2/mc^2 is the classical electron radius, $(N_0\rho)/A$ is the number of atoms per unit volume, N_0 is Avogadro's number, ρ is the density, A is the atomic weight, Z is the atomic number, and λ is the wavelength. The term $[Z + \Delta f']$ is the real part of the scattering factor (including the so-called dispersion term f') and is essentially equal to Z. The imaginary part of the index of refraction, β, is related to absorption, where μ is the linear absorption coefficient. Considering an X-ray beam incident on a smooth surface and Snell's law, one obtains that the critical angle for total reflection is given by:

$$\theta_{\text{crit}} = \sqrt{2\delta} \tag{13}$$

δ is of the order of 10^{-5} and θ_{crit} is typically of the order of a few milliradians (recall that 17.4 millirad equals 1°). Thus, as long as the beam is incident below this critical angle, it is totally reflected and only an evanescent wave penetrates the substrate. This has two very important consequences. First of all, the penetration depth is of the order of 20 Å and thus one can significantly discriminate in favor of a surface-contained material. Compton and elastic scattering are also minimized. In addition, the reflection enhances the local intensity by as much as a factor of 4 as well as the effective "path length." All of these factors combined serve to enhance the surface sensitivity of the technique and, when combined with solid state fluorescence detection, allow for the detection of less than monolayer amounts of material.[60]

(*v*) *Dispersive Arrangements*

Up to this point, the experimental techniques described were based on the use of monochromator crystals and following the signal of interest as the energy of the incident photon was scanned. This conventional mode of operation suffers from the fact that only a very narrow range of wavelengths is used at a given time, so that obtaining a full spectrum requires a significant amount of time, thus precluding real-time kinetic studies of all but the slowest of reactions. The alternative is to employ a dispersive arrangement[63-68] where, by the use of focusing optics, a range of energies can be brought to focus on a spot. The exact energy spread will depend on the specific optical elements employed but a range of 500 to 600 eV represents a realistic value. Coupling this with a photodiode array allows for the simultaneous use of the full range of wavelengths, and thus a spectrum can be obtained in periods as short as milliseconds rather than minutes. This is of great significance because a number of relevant dynamic processes take place on this time scale. The application of this approach to electrochemical studies will be discussed in a later section.

VIII. EXAFS STUDIES OF ELECTROCHEMICAL SYSTEMS

In order to simplify the discussion, the EXAFS studies on electrochemical systems reported to date will be divided into the categories listed below:

1. Oxide films
2. Monolayers
3. Adsorption
4. Spectroelectrochemistry

Further distinction will be made as to whether a given study was conducted *in situ* or *ex situ*, where by *in situ* I will mean that the electrode is both in contact with an electrolyte solution and under potential control.

1. Oxide Films

The study of passive films on electrode surfaces is an area of great fundamental and practical relevance. Despite decades of intensive investigations, there still exists a great deal of controversy as to the exact structural nature of passive films, especially when they are formed in the presence or absence of glass-forming additives such as chromium.

One of the main sources of controversy is that many of the structural studies performed have been on dried films and, as pointed out by O'Grady,[69] this results in the determination of the structure of dehydrated films whose structure can be significantly different from that of hydrated ones.

The use of surface EXAFS in the study of passive films represents a natural application of the technique and, in fact, the studies by Kruger and co-workers[70-73] on the passive film on iron represent the first reported.

In their earliest report, Kruger and co-workers vacuum deposited iron films onto glass slides and subsequently oxidized the films in either nitrite or chromate solution. They obtained the EXAFS spectra for the oxidized films employing a photocathode ionization chamber and compared these with spectra for γ-FeO(OH), γ-Fe_2O_3, and Fe_3O_4. Although these studies were not *in situ*, they did not require evacuation of the samples and therefore represent an intermediate situation between dehydrated film and *in situ* experiments. The spectra for Fe, Fe_3O_4, and the nitrite- and chromate-generated passive films are shown in Figs. 12A and 12B. The near-edge region for the nitrite-generated film showed evidence of an enhancement, similar to that observed for Fe_3O_4, indicative of an increase in the density of available final states with *p* character. Such an enhancement is absent in the chromate-formed films. These results point to a more covalent bonding in the chromate versus the nitrite passivated films.

Upon Fourier transforming of the data, two peaks corresponding to Fe—O and Fe—Fe distances are obtained (Fig. 13). The

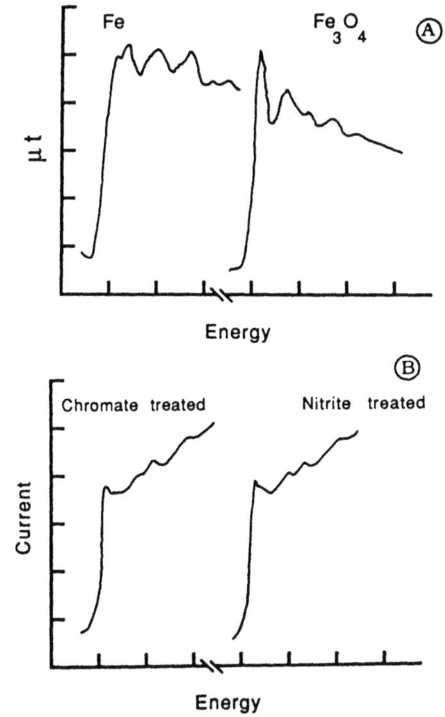

Figure 12. Absorption spectra of (A) Fe and Fe_3O_4 and (B) iron films after treatment in chromate and nitrite solutions. (From Ref. 71, with permission.)

peaks in the Fourier transform of the chromate-generated film are much less well resolved than those for the nitrite films, and this is ascribed to the presence of a glassy structure associated with the chromium. From a comparison of the edge jump for Fe and Cr, the authors estimate that the films have about 12% Cr.

These results point to the importance of hydration effects on the structure of passive films on iron. However, these results were obtained *ex situ* and therefore are subject to some uncertainty.

Most recently, these same authors have employed an *in situ* cell (Fig. 14) for carrying out these experiments. Again they studied nitrite- and chromate-passivated films. The results obtained in this case are quite different from the *ex situ* measurements. In addition,

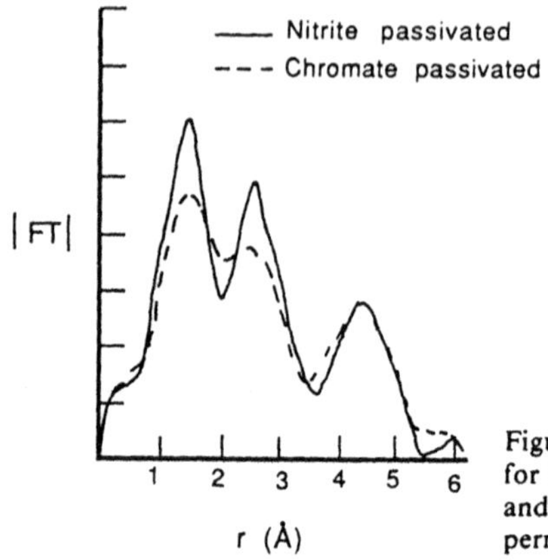

Figure 13. Fourier transform of the EXAFS for iron films after treatment in chromate and nitrite solutions. (From Ref. 71, with permission.)

the spectral features of the *in situ* measurements for both nitrite- and chromate-passivated films are quite similar and these are most easily ascertained from the derivative of the near-edge region (Fig. 15).

The Fourier transforms for both films are again quite similar, but as for the *ex situ* measurements, the chromate-passivated films appear to have a more glassy structure. It should be mentioned that these studies employed a rather limited data range which makes spectral differentiation difficult.

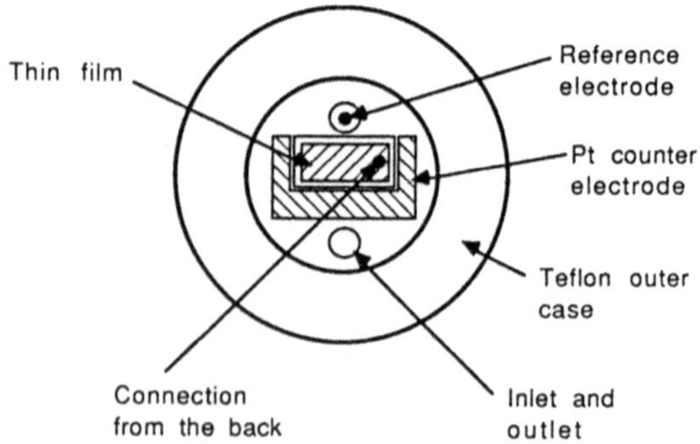

Figure 14. *In situ* cell for performing EXAFS studies on passivated iron films. (From Ref. 72, with permission.)

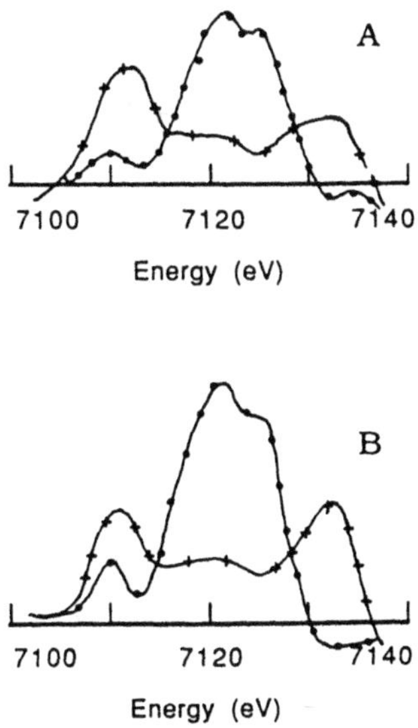

Figure 15. Derivatives of the near-edge region of spectra for nitrite- (A) and chromate- (B) passivated iron films under *in situ* (+) and *ex situ* (●) conditions. (From Ref. 72, with permission.)

Hoffman and Kordesch[74,75] have presented a series of studies on the passive films on iron with particular attention to cell design. They have employed a so-called bag cell that allows for the *in situ* passivation and/or cathodic protection of the iron films. These were deposited onto gold films deposited on Melinex.

In addition, they employed a setup where the working electrode is partially immersed in solution and continuously rotated. In this way, they could expose the electrode with only a very thin film of electrolyte covering the electrode.

Under these conditions, they were able to obtain spectra (Fig. 16) of the film as prepared, of a cathodically protected film, and of a film passivated in borate solution at 1.3 V.

From these studies, they concluded that the passive film had an Fe—O coordination with six near neighbors at a distance of 2 ± 0.1 Å. Although a higher signal-to-noise ratio is required to refine the structure, the approach followed by these authors appears most appropriate since they were able to reduce the deposited films to the metallic state and subsequently oxidize them. It would be most interesting to ascertain how the structure of the passive film varies through sequential reduction/passivation cycles.

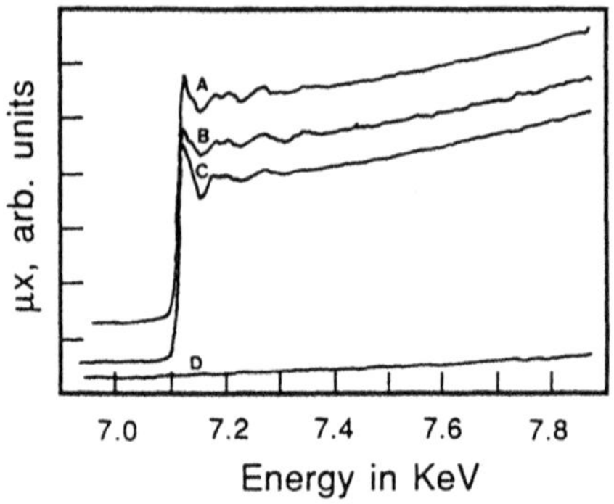

Figure 16. Fluorescence detected X-ray absorption spectra for a 4-nm iron film in emersion cell. Spectra are for: (A) dry film; (B) cathodically protected film; (C) passivated film; (D) background electrolyte. (From Ref. 74, with permission.)

Forty and co-workers[76] have investigated the passive films formed on iron and iron–chromium alloys upon immersion in sodium nitrite solution. They have investigated these films in a wet environment (which they term *in situ*) as well as after dehydration. For a FeCr alloy (13% Cr) they find that the structure of the wet film is analogous to that of γ-FeOOH but with a higher degree of disorder, consistent with the Mössbauer results of O'Grady.[69] Upon dehydration, the structure of the passive films transforms to one that is closer to that of γ-Fe_2O_3 but with reduced long-range order. These authors also looked at the chromium edge and found that the local structure around chromium in the passive films was similar to that of Cr_2O_3. They concluded that the presence of chromium in alloys stabilized the γ-FeOOH-like layers against dehydration, thus forming a glassy-like structure which enhances the stability of the passive film.

Froment and co-workers have employed reflexafs[77] (reflection EXAFS) for studying passive films on iron[78] and nickel.[79,80] The experiment consists of measuring the ratio of the reflected and incident intensities as a function of energy. Although an EXAFS spectrum can be obtained from such a measurement, the process is somewhat involved since the reflectivity is a complex function of angle of incidence, refractive index, and energy. They report some preliminary data for passive films on iron and nickel but do not derive extensive conclusions as their major intent was to demonstrate the applicability of reflexafs. Also concerning reflexafs from a general point of view, Heald and co-workers[81,82] have made a careful comparison of reflexafs versus measurements at grazing incidence with fluorescence detection. They conclude that, in general, the latter offers enhanced sensitivity for studies of monolayers. However, the reflexafs technique can be applied in a dispersive arrangement (that is, with a broad range of energies incident on the sample simultaneously), allowing for faster data acquisition and the possibility of performing kinetic studies on the millisecond time scale (*vide infra*).

The most extensive study of the nickel oxide electrode is that of McBreen *et al.*,[83] who employed an *in situ* cell in a transmission mode (see cell in Fig. 17). The study of nickel oxide is complicated by the numerous species present and their interconversion. McBreen

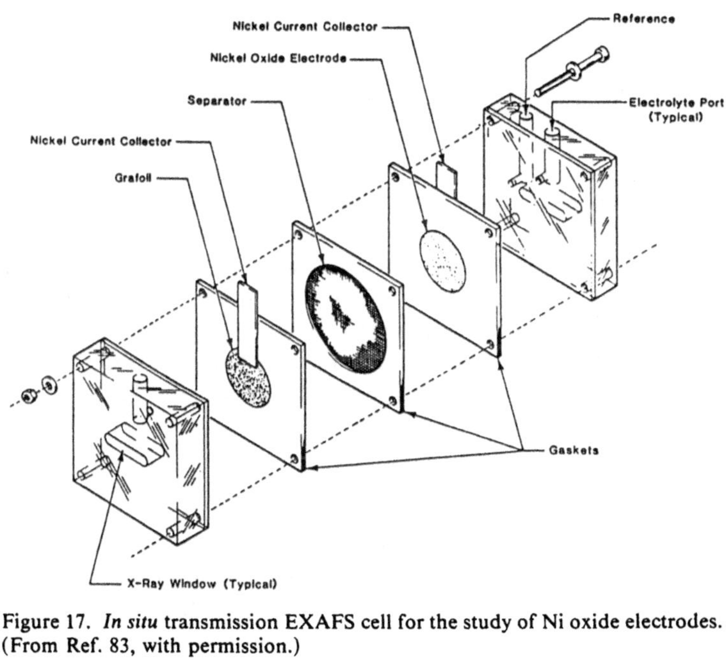

Figure 17. *In situ* transmission EXAFS cell for the study of Ni oxide electrodes. (From Ref. 83, with permission.)

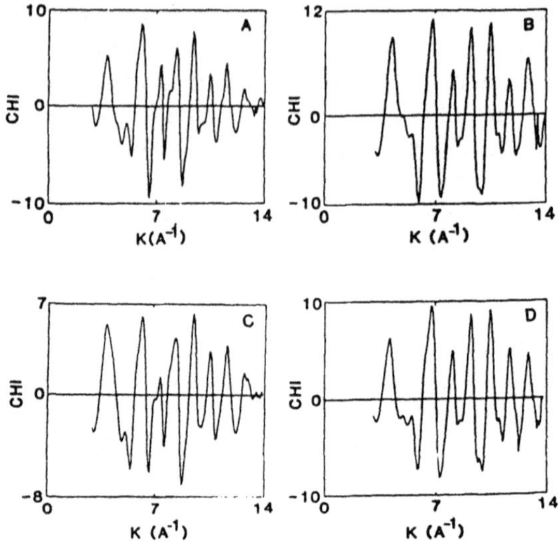

Figure 18. Raw EXAFS spectra for: (A) dry $Ni(OH)_2$ electrode; (B) a $Ni(OH)_2$ electrode charged once; (C) an electrode discharged once; (D) an electrode charged twice. (From Ref. 83, with permission.)

et al. found that the as-prepared β-$Ni(OH)_2$ has the same structure within the x–y plane as that determined by X-ray diffraction experiments but with a significant degree of disorder along the c-axis. Oxidation to the trivalent state results in contraction of the Ni—O and Ni—Ni distances along the x–y plane. Re-reduction of this material yields a structure that is similar to that of the freshly prepared $Ni(OH)_2$. Repeated oxidation-reduction cycles result in an increased disorder which is believed to be responsible for facilitating the electrochemical oxidation to the trivalent state.

Figure 18 shows some of the spectra reported. These are fully consistent with the above statements.

2. Monolayers

The study of electrochemically deposited monolayers poses the strictest experimental constraints since the signals will be necessarily very low. On the other hand, these studies can provide much detail on interfacial structure at electrode surfaces as well as on the effects of solvent and supporting electrolyte ions. The use of underpotential deposition[84] allows for the precise control of the coverage of electrodeposited layers up to a monolayer. This represents a unique family of systems with which to probe electrochemical interfacial structure *in situ*.

We and others have been involved in the study of such systems including Cu/Au(111),[85,86] Ag/Au(111),[87] Pb/Ag(111),[88] and Cu/Pt(111).[89] The first three systems involved the use of epitaxially deposited metal films on mica as electrodes.[90–92] Such deposition gives rise to electrodes with well-defined single-crystalline structures. In the last case a bulk platinum single crystal was employed. Because of the single-crystalline nature of the electrodes, polarization dependence studies could be used to ascertain surface structure.

The best-characterized system to date is the underpotentially deposited copper on gold. In this case we were able to obtain EXAFS spectra of a deposited monolayer with the polarization of the X-ray beam either perpendicular (Fig. 19A) or parallel (Fig. 19B) to the plane of the electrode. A number of salient features can be pointed out. First of all, the copper atoms appear to be located at threefold hollow sites (i.e., three gold near neighbors) on the gold (111) surface with copper near neighbors. The Au/Cu and Cu/Cu distances obtained are 2.58 and 2.91 ± 0.03 Å, respectively. This last number is very similar to the Au/Au distance in the (111) direction, suggesting a commensurate structure. Most surprising, however, was the presence of oxygen as a scatterer at a distance of 2.08 ± 0.02 Å. From analysis and fitting of the data we obtain that the surface copper atoms are bonded to an oxygen from either water or sulfate anion from the electrolyte. That there might be water or sulfate in contact with the copper layer is not surprising; however, such interactions generally have very large Debye–Waller factors so that typically no EXAFS oscillations (or heavily damped oscillations) are observed (see the spectrum for $[Cu(H_2O)_6]^{2+}$ in Fig. 8A). The fact that the presence of oxygen (from water or electrolyte) at a very well-defined distance is observed is indicative of a significant interaction and underscores the importance of *in situ* studies.

A pictorial representation of this system is shown in Fig. 20 where the source of oxygen is presented as water. However, it

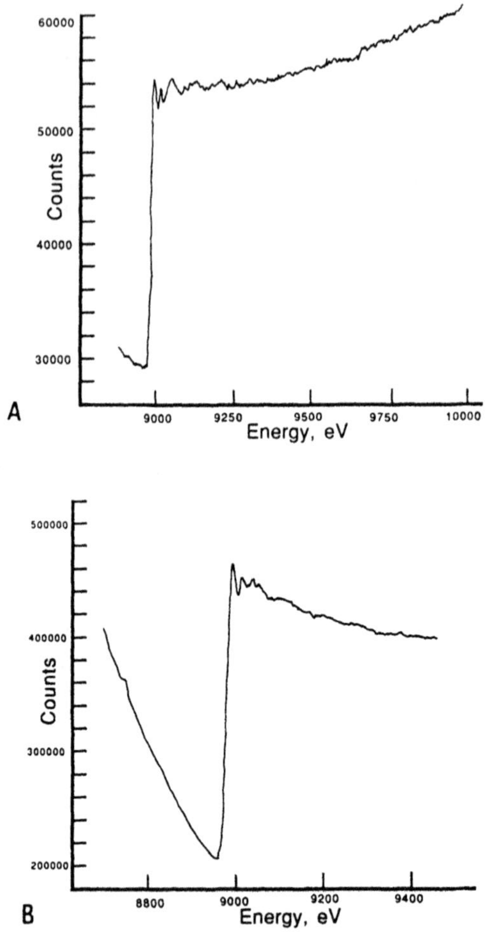

Figure 19. *In situ* X-ray absorption spectrum for a copper upd monolayer on a gold (111) electrode with the polarization of the X-ray beam being perpendicular (A) or parallel (B) to the electrode surface.

should be mentioned that from the EXAFS experiment we cannot rule out sulfate anions as the source of oxygen. In fact, experiments by Kolb and co-workers[93] indicate that sulfate may be present since at the potential for monolayer deposition the electrode is positive of the potential of zero charge, E_{PZC}, so that sulfate would be

Figure 20. Schematic representation of the structure of a copper upd monolayer on a gold (111) electrode surface. The copper atoms sit at three-fold hollow sites on the gold surface and water molecules are bonded to the copper atoms.

present to counterbalance the surface charge. We can, in addition, follow the edge features to ascertain changes in oxidation state. Figure 21A shows the edge region for the deposited monolayer while Fig. 21B shows the spectrum after stripping of the copper monolayer. The appearance of the characteristic "white" line (resonance near the edge) as well as the edge shift to higher energies are fully consistent with the oxidation state assignments.

Studies of Ag on Au(111)[87] yield very similar results in terms of the structure of the deposited monolayer (i.e., the silver atoms are bonded to three surface gold atoms and are located at three-fold hollow sites forming a commensurate layer) with again strong interaction by oxygen from water or electrolyte (perchlorate).

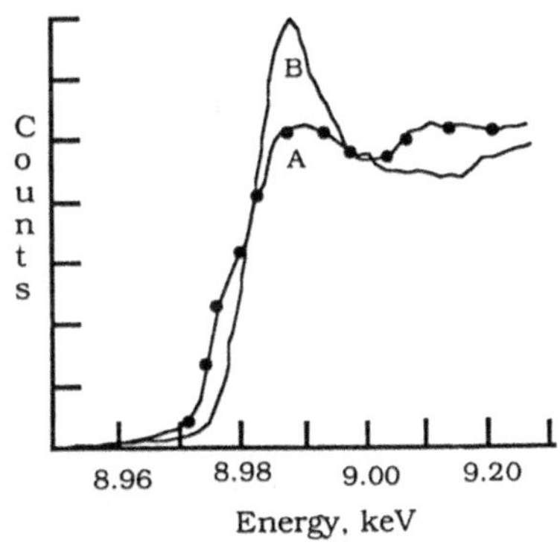

Figure 21. Near-edge spectra for (A) copper upd monolayer on a gold (111) electrode surface and (B) after electrochemically stripping the copper monolayer.

Melroy and co-workers[88] recently reported on the EXAFS spectrum of Pb underpotentially deposited on silver (111). In this case, no Pb/Ag scattering was observed and this was ascribed to the large Debye–Waller factor for the lead as well as to the presence of an incommensurate layer. However, data analysis as well as comparison of the edge region of spectra for the underpotentially deposited lead, lead foil, lead acetate, and lead oxide indicated the presence of oxygen from either water or acetate (from electrolyte) as a backscatterer.

They were also able to perform a potential dependence study at −0.53 and −1.0 V. They found that the Pb—O distance increases from 2.33 ± 0.02 to 2.38 ± 0.02 Å upon changing the potential from −0.53 to −1.0 V versus Ag/AgCl (Fig. 22). This is consistent with the negatively charged electrode repelling a negatively charged or strongly dipolar adsorbate.

Most recently, we have been able to obtain the *in situ* surface EXAFS spectrum of a half-monolayer of underpotentially deposited copper on a bulk Pt(111) single crystal pretreated with iodine. The spectrum shown in Fig. 23 is a bit noisy (due to limited number of scans) but at least five well-defined oscillations can be observed. Preliminary data analysis indicates that the copper adatoms sit on threefold hollow sites with copper neighbors at 2.80 ± 0.03 Å. This distance is very close to the Pt—Pt distance in the (111) direction and indicates the presence of a commensurate

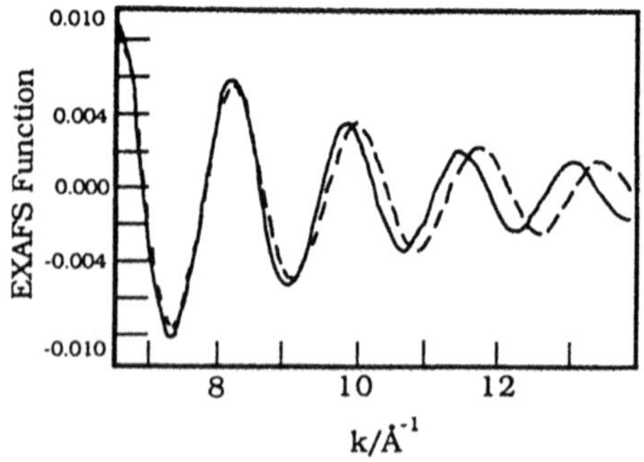

Figure 22. Fourier-filtered data for a lead upd monolayer on a silver (111) electrode at two applied potentials. Solid curve, −1.0 V; dashed curve, −0.53 V. (From Ref. 88, with permission.)

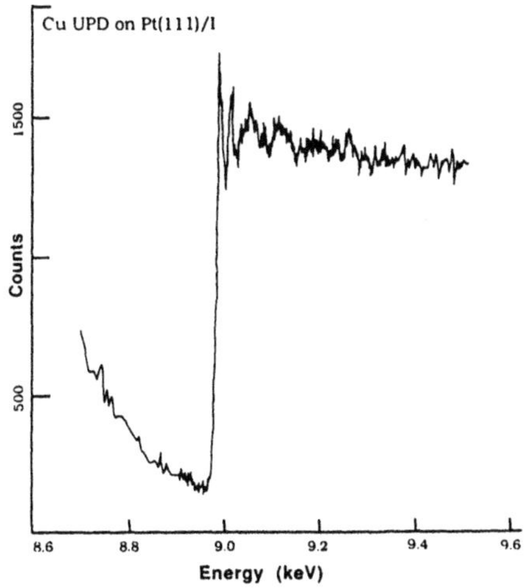

Figure 23. *In situ* X-ray absorption spectrum for half a monolayer of copper underpotentially deposited on a bulk Pt (111) electrode pretreated with iodine.

layer. The fact that such well-defined two-dimensional structures are present at half a monolayer coverage is a strong indication that the electrodeposition occurs by initial clustering rather than by random decoration of the surface with subsequent coalescence.

3. Adsorption

We have studied the adsorption of iodine on Pt(111) electrodes[94] from solutions containing iodide with the intent of following changes in the coverage as a function of applied potential (that is, constructing an *in situ* adsorption isotherm). We were further encouraged to carry out these experiments by the recent report by Hubbard and co-workers[95] on the determination of iodine adsorption by following Auger electron intensities.

We find that after appropriate normalization, our data (Fig. 24A) agree quite well with the results presented by Hubbard for potentials negative of +0.40 V (Fig. 24B). A particularly gratifying result was the change in coverage—from $\frac{1}{3}$ to $\frac{4}{9}$, in going from −0.3 to +0.2 V. An even more interesting and tantalizing finding was that at potentials positive of +0.40 V we observe a significant

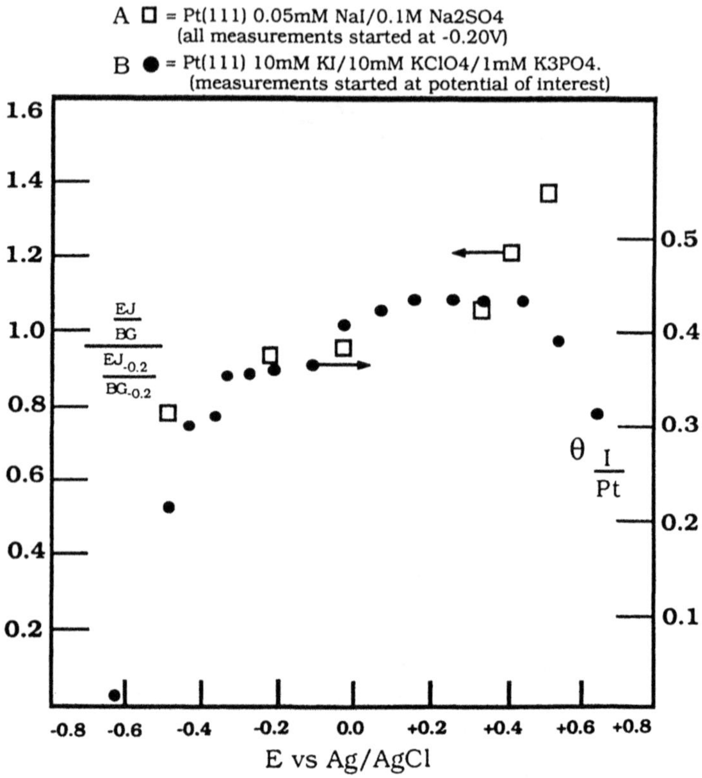

Figure 24. Iodine on Pt adsorption isotherm obtained *in situ* by following the edge-jump intensity at the iodine *K*-edge as a function of potential (A) and comparison with data obtained by Hubbard *et al.* via Auger intensities (B). (Adapted from Ref. 95.)

enhancement of the iodine fluorescence signal, whereas Hubbard reports no significant changes up to about +0.70 V where faradaic oxidation of iodine to iodate takes place. We believe that the local increase in the iodine/iodide concentration is produced by a faradaic charge flow, followed by association of the oxidation product(s) with the adsorbed iodine layer at the given potential. It appears that the associated layer does not survive the transfer to vacuum employed in the *ex situ* study and again underscores the importance of *in situ* measurements.

4. Spectroelectrochemistry

EXAFS and XANES techniques have been applied in the more traditional type of spectroelectrochemical experiments where a thin-layer cell configuration is employed. Drawing from extensive experience with the related UV-vis measurements, Heineman in collaboration with Elder[96,97] were the first to report on an *in situ* EXAFS spectroelectrochemistry experiment. Their first cell design employed gold minigrid electrodes similar to those typically employed in traditional UV-vis experiments. They studied the ferro cyanide/ferricyanide couple in each oxidation state by monitoring the absorption about the iron K-edge using fluorescence detection. A typical spectrum and related Fourier transform are presented in Fig. 25. From analysis of their data, they were able to determine

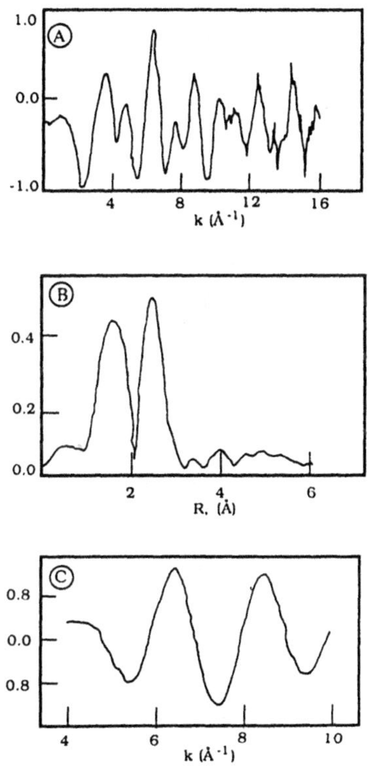

Figure 25. EXAFS data for $K_3[Fe(CN)_6]$: (A) k^2-weighted EXAFS; (B) Fourier transform of (A) showing Fe—C and Fe—N peaks; (C) Fourier-filtered back-transformation of the Fe—C peak. (From Ref. 97, with permission.)

that for Fe(II) there are 7.4 carbon atoms at 1.97 ± 0.01 Å, whereas for Fe(III) the numbers are 6.8 and 1.94 ± 0.01 Å, respectively. Since coordination number determination is usually no better than 20%, the numbers they find are in good agreement with the known value of 6. More interesting is the fact that the observed contraction of the Fe—C bond upon oxidation is contrary to results based on crystallographic studies. This points to the importance of *in situ* measurements since by means of the applied potential one can prevent changes in the oxidation state of the species being studied.

In addition, the determination of metal–ligand bond distances in solution and their oxidation state dependence is critical to the application of electron transfer theories since such changes can contribute significantly to the energy of activation through the so-called inner-sphere reorganizational energy term.

These authors have also developed a cell[98] (Fig. 26) that employs reticulated vitreous carbon as a working electrode and they find that such a design allows for much faster electrolysis. Using such a cell, they have studied the $[Ru(NH_3)_6]^{3+/2+}$ couple

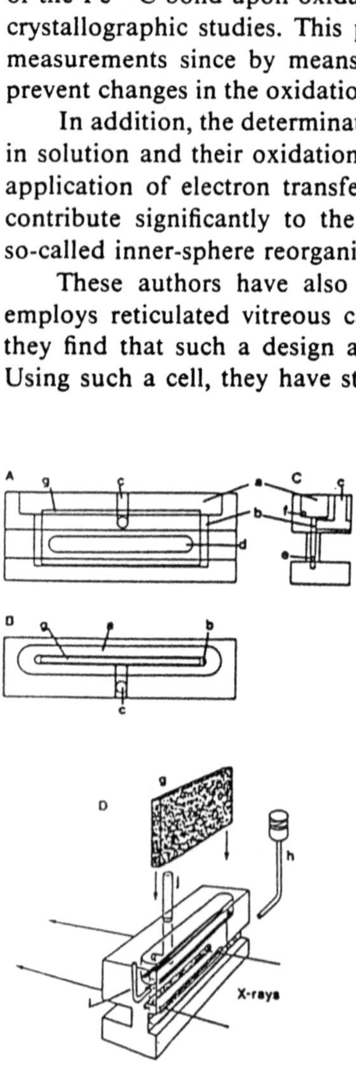

Figure 26. EXAFS spectroelectrochemical cell: (A) front view, (B) top view, (C) side view, (D) assembly; (a) auxiliary electrode compartment, (b) working electrode well, (c) reference electrode compartment, (d) X-ray window, (e) inlet port, (f) auxiliary electrode lead, (g) RVC working electrode, (h) Pt syringe needle inlet and electrical contact, (i) Pt wire auxiliary electrode, (j) Ag/AgCl(3*M* NaCl) reference electrode. (From Ref. 98, with permission.)

and a cobalt(III/II) sepulchrate as well as the Fe—C distance in cytochrome *c*.

Most recently, they have developed[99] a cell configuration for the study of modified electrodes that employs, as a working electrode, colloidal graphite deposited onto kapton tape (typically employed as a window material). Such an arrangement minimizes attenuation due to the electrolyte solution.

Antonio *et al.*[100] have performed an *in situ* EXAFS spectroelectrochemical study of heteropolytungstate anions.

We have also performed some *in situ* EXAFS measurements on chemically modified electrodes.[101] Specifically, we have studied films of $[Ru(v\text{-}bpy)_3]^{2+}$ (v-bpy is 4-vinyl-4′-methyl-2,2′-bipyridine) electropolymerized onto a platinum electrode and in contact with an acetonitrile/0.1 *M* tetrabutylammonium perchlorate (TBAP) solution and under potential control. The experimental setup consists of a thin-layer configuration employing a thin (6 μm) Teflon window and grazing incidence so as to take advantage of the total external reflection effects mentioned previously. We have focused on determining the lower limit of detection as well as trying to ascertain any differences in the metal-ligand bond distances for the electropolymerized films as a function of coverage when compared to the parent complex. Figures 27A and 27B show spectra for electrodes modified with one and five monolayers of the complex, respectively, whereas Fig. 27C shows the spectrum for bulk $[Ru(bpy)_3]^{2+}$. In Fig. 27A one can ascertain that only the most prominent features of the spectrum of the parent compound (Fig. 27C) are present. (It should be mentioned that a monolayer of $[Ru(v\text{-}bpy)_3]^{2+}$ represents about 5.4×10^{13} molecules/cm^2, which is about 5% of a metal monolayer. This is mentioned since it is the metal centers that give rise to the characteristic fluorescence employed in the detection.) However, at a coverage of five monolayers of complex (Fig. 27B) the spectrum is essentially indistinguishable from that of the bulk material. This indicates that the structure of electroactive polymer films can be obtained at relatively low coverages and this should have important implications in trying to identify the structures of reactive intermediates in electrocatalytic reactions at chemically modified electrodes.

Furthermore, one can monitor changes in oxidation state by the shift in the edge position. For example, Fig. 28 shows that upon

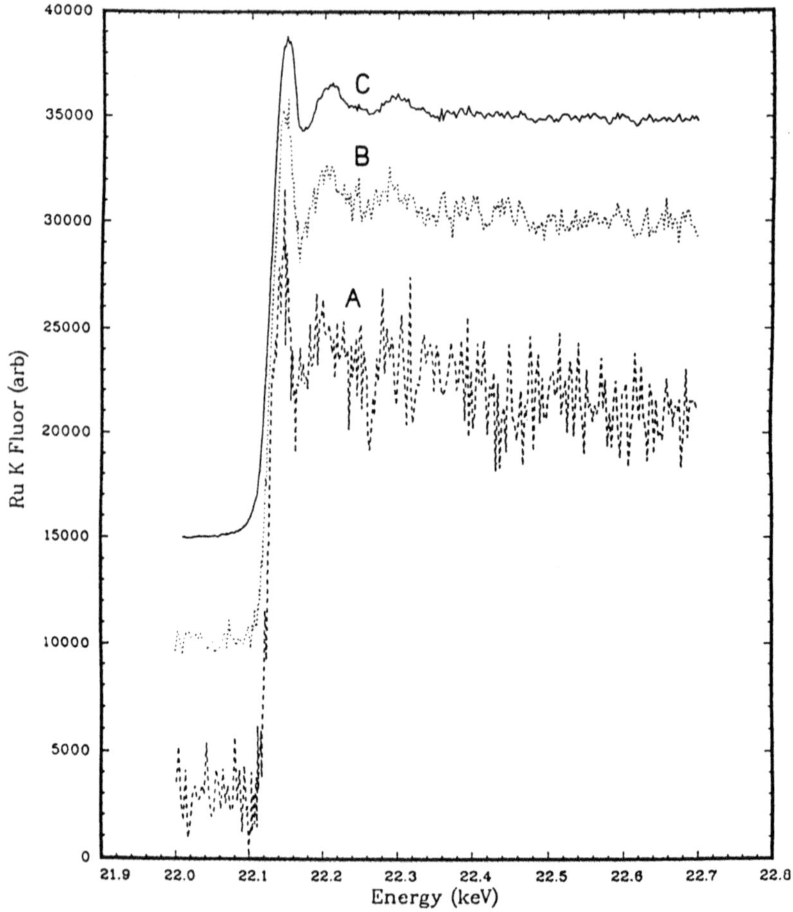

Figure 27. *In situ* X-ray absorption spectra around the ruthenium *K*-edge for an electrode modified with (A) one and (B) five monolayers of $[Ru(v\text{-}bpy)_3]^{2+}$. (C) Spectrum of bulk $[Ru(bpy)_3]^{2+}$.

oxidation of the polymer film (at +1.50 V) from Ru(II) to Ru(III), the edge position shifts to higher energy by about 1.5 eV. Thus, one can determine the oxidation state of the metal inside a polymer film on an electrode surface.

Tourillon and co-workers[102-110] have also reported on a number of spectroelectrochemical studies, especially of electrodeposition

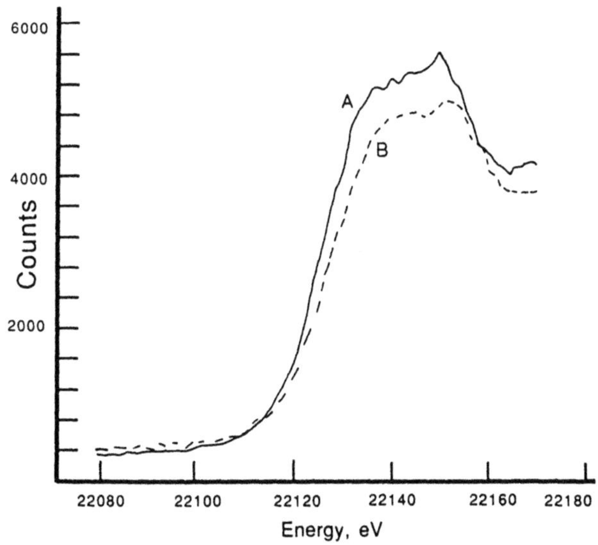

Figure 28. *In situ* X-ray absorption of the near-edge region for an electrode modified with a polymeric film of $[Ru(v\text{-}bpy)_3]^{2+}$ and its potential dependence. Applied potentials were +0.7 V (A) and +1.5 V (B).

of metals, particularly copper, on electrodes modified with poly(3-methylthiophene). What sets their experiments apart is the use of a dispersive approach (Fig. 29). In such a setup, focusing optics (employing a bent crystal) are employed so as to have a range of energies (as wide as 500 eV) come to a tight focal spot at the sample. The beam then impinges a photodiode array with 1024 pixels so that all energies are monitored at once. The net result is to significantly decrease data acquisition times so that spectra can be obtained in times as short as a few milliseconds. Thus, this opens up tremendous possibilities in terms of kinetic and dynamic studies. One of the more impressive results using this approach is shown in Fig. 30, which shows spectra obtained around the copper edge for a poly(3-methylthiophene) film (on a platinum electrode) doped with Cu^{2+} ions. The potential of the electrode is stepped so as to reduce the Cu^{2+} ions to Cu^{1+} and subsequently to Cu^0. The spectra shown in the figure were taken at 7-s intervals and the transitions from Cu^{2+} to Cu^{1+} and then to metallic copper are clearly evident.

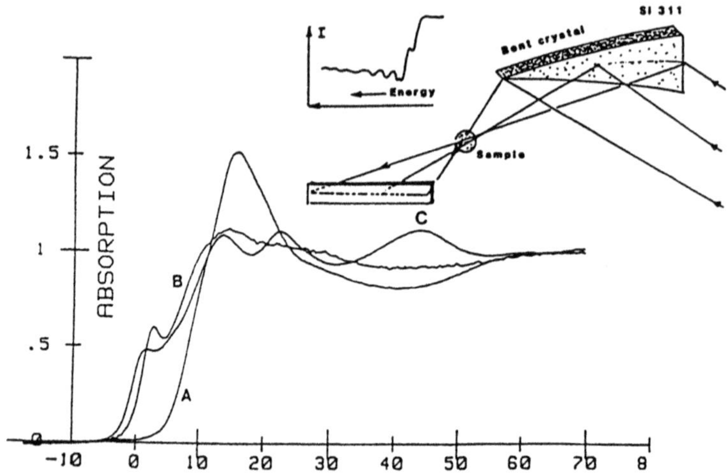

Figure 29. Schematic of a dispersive EXAFS setup as well as spectra for: (A) Cu^{2+}; (B) Cu^{1+}; (C) copper foil. (From Ref. 105, with permission.)

These authors have also employed this technique for the study of other metallic inclusions into poly(3-methylthiophene) films, including Ir, Au, and Pt. Thus, this type of arrangement could open up new exciting possibilities in terms of kinetic studies.

IX. X-RAY STANDING WAVES

1. Introduction

The X-ray standing wave (XSW) technique represents an extremely sensitive tool for determining the position of impurity atoms within a crystal or adsorbed onto crystal surfaces.[111,112] This technique is based on the X-ray standing wave field that arises as a result of the interference between the coherently related incident and Bragg diffracted beams from a perfect crystal. In the vicinity of a Bragg reflection (Fig. 31), an incident plane wave (with wave vector k_0) and a reflected wave (with wave vector k_H) interfere to generate a standing wave with a periodicity equivalent to that of the (h, k, l) diffracting planes. The standing wave develops not only in the diffracting crystal, but also extends well beyond its surface.

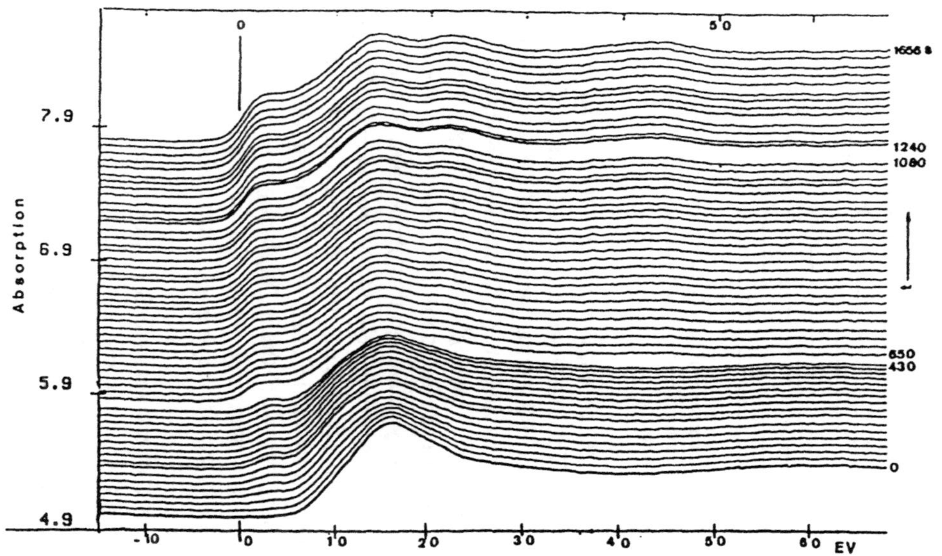

Figure 30. *In situ* measurements of the time evolution of the Cu K-edge when a platinum electrode coated with a polymeric film of poly (methylthiophene) is cathodically polarized in an aqueous solution containing 50 mM $CuCl_2$. (From Ref. 105, with permission.)

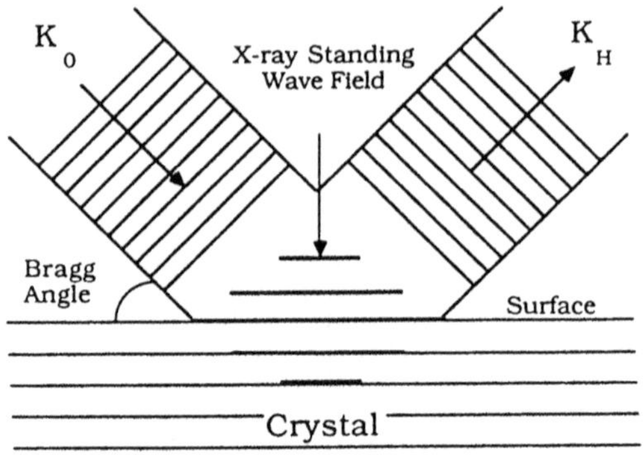

Figure 31. Depiction of the X-ray standing wave field formed by the interference between incident and Bragg reflected beams.

Estimates of this coherence length range to values as large as 1000 Å from the interface.[113] The nodal and antinodal planes of the standing wave are parallel to the diffracting planes and the nodal wavelength corresponds to the d-spacing of the diffracting planes. As the angle of incidence is advanced through the strong Bragg reflection, the relative phase between the incident and reflected plane waves (at a fixed point) changes by π. Due to this phase change, the antinodal planes of the standing wave field move in the $-H$ direction by one-half of a d-spacing, from a position halfway between the (h, k, l) diffracting planes (low-angle side of the Bragg reflection) to a position that coincides with them (high-angle side of the Bragg reflection). Thus, the standing wave can be made to sample an adsorbate or overlayer at varying positions above the substrate interface.

For an atomic overlayer which is positioned parallel to the diffracting planes, the nodal and antinodal planes of the standing wave will pass through the atom plane as the angle of incidence is advanced. Using an incident beam energy at or beyond the absorption edge of the atoms in the overlayer, the fluorescence emission yield will be modulated in a characteristic fashion (Fig. 32) as the substrate is rocked in angle. The phase and amplitude of this modulation (or so-called coherent position and coherent fraction) are a measure of the mean position $\langle Z \rangle$ and width $\sqrt{\langle Z \rangle^2}$ of the distribution of atoms in the overlayer. The Z-scale is mod d and

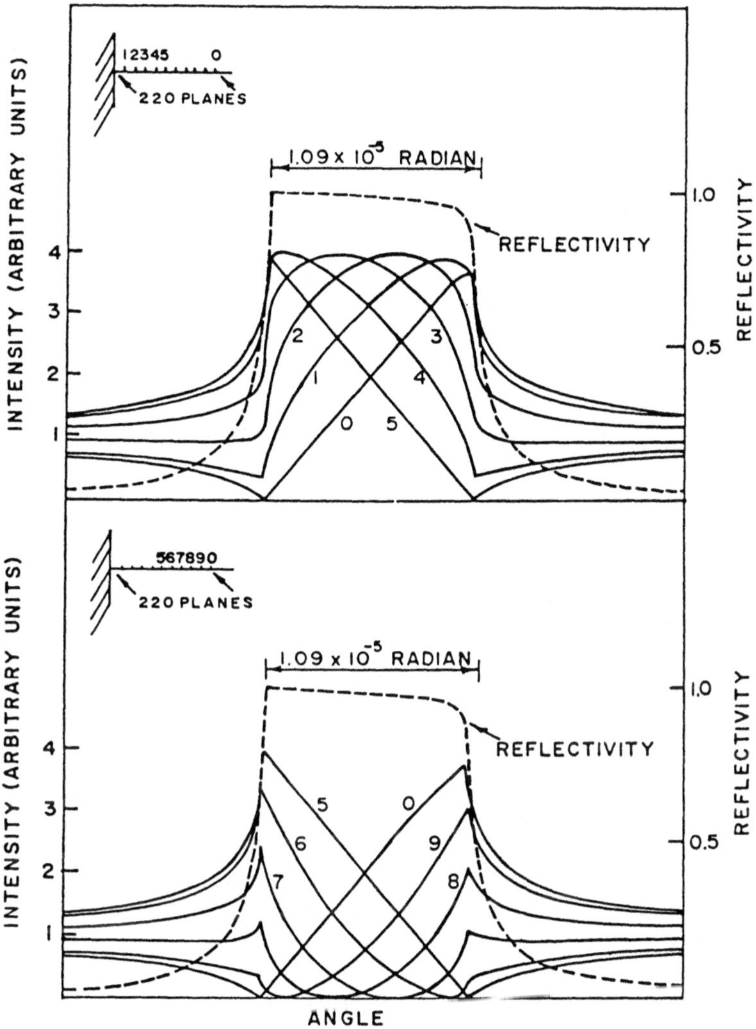

Figure 32. X-ray field intensities at extended Ge (220) lattice positions (0–9) for a perfectly collimated incident X-ray beam. An atomic adlayer whose center falls on one of these positions would have its characteristic fluorescence intensity modulated in the same fashion. The dashed curve represents the Bragg reflectivity profile. (From M. J. Bedzyk, Ph. D thesis, SUNY Albany, 1982.)

points in the direction normal to the diffraction planes. Standing wave measurements of $\langle Z \rangle$ and $\sqrt{\langle Z \rangle^2}$ can be accurate to within 1% and 2% of the d-spacing, respectively.[114] Golovchenko and co-workers[115] have applied this technique to the study of surface adsorbates, monitoring the angular dependence of the fluorescence yield of bromine chemisorbed onto a Si(111) crystal. It should be mentioned that these experiments were performed while the crystal was covered by a thin film of methanol, pointing to the feasibility of performing experiments at the solid–liquid interface. Bedzyk and Materlik[114] used the angular dependence of the fluorescence yield of bromine to relate its position to the Ge(111) and Ge(333) diffracting planes. In addition, they demonstrated the feasibility of using higher-order reflection for determining the thermal vibration amplitude of the bromine adsorbate.

One of the problems associated with the implementation of the standing wave technique is the fact that it requires the use of perfect or nearly perfect crystals. This presents a problem especially for relatively soft materials such as copper, gold, silver, and platinum, which are not only very difficult to grow in such high quality, but are also very difficult to maintain in that state. Thus, most experiments have been performed on silicon or germanium single crystals.

An alternative to the use of perfect crystals is the use of synthetic layered microstructures (LSMs).[116] These devices are prepared by the sequential deposition of alternate layers of materials, typically of high- and low-electron-density elements such as W/C, Mo/C, W/Si, and Pt/C, onto a smooth substrate material such as silicon, germanium, or float glass. Recent results have hinted at the possibility of growing these devices on cleaved mica. The number of layer pairs is such that the d-spacing of the synthetic multilayer dictates the diffracting properties of the interface. Although originally developed as X-ray mirrors for regions of the spectrum where natural crystals are not effective, these devices have found widespread use in X-ray standing wave experiments, and, in fact, numerous applications have appeared.[117]

The advantages that accrue from the use of LSMs are manifold. Firstly, these devices are physically robust and can be handled without undue provision. They can be tailored to contain materials of interest and, furthermore, one can vary the d-spacing of the diffracting planes, thus varying the probing distance of the technique. Their use also simplifies experimental design since the angular reflection widths for LSMs are of the order of milliradians as opposed to microradians for crystals.

2. Experimental Aspects

One of the main difficulties with X-ray standing wave measurements is that they are experimentally very demanding. Although the experimental setup is not particularly complex, alignment of the sample relative to the beam is critical. A typical setup is shown in Fig. 33 and consists of an incident beam monitor I_0, a sample stage, a reflected beam monitor I_R, and a detector at 90° relative to the X-ray beam. Of particular importance in this experiment is the angular resolution of the sample stage since a typical reflection width will be of the order of tens of microradians for a single crystal and a few milliradians for LSMs. In both cases, however, high angular resolution is required if we are to have a well-resolved reflectivity profile. In addition, when measuring fluorescence from an adsorbate layer, care must be taken to accurately subtract background radiation.

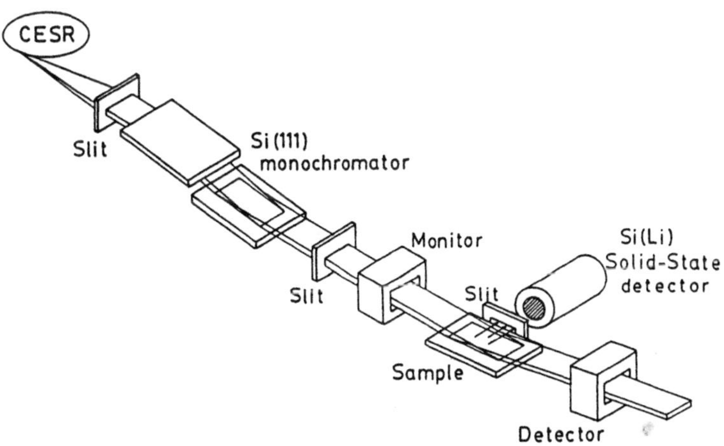

Figure 33. Experimental setup for X-ray standing wave measurements on an LSM.

3. X-Ray Standing Wave Studies at Electrochemical Interfaces

Due to the experimental difficulties involved, there have been only three reports of XSW measurements at electrochemical interfaces. Materlik and co-workers have studied the underpotential deposition of thallium on single-crystal copper electrodes under both *ex situ*[119] and *in situ*[120] conditions. In addition, they report results from studies in the absence and presence of small amounts of oxygen.

In the *ex situ* studies, the thallium layer was electrodeposited and the electrode was subsequently removed from solution and placed inside a helium-filled box where the XSW experiments were carried out.

For the *in situ* studies, an electrochemical cell was designed to hold the nearly perfect copper crystal in contact with a thin layer (20 to 50 μm) of electrolyte. Figures 34 and 35 show the cells employed in the *ex situ* and *in situ* experiments, respectively. In addition, Fig. 34 shows the voltammetric traces obtained for the deposition of Tl in the presence and absence of oxygen. In the

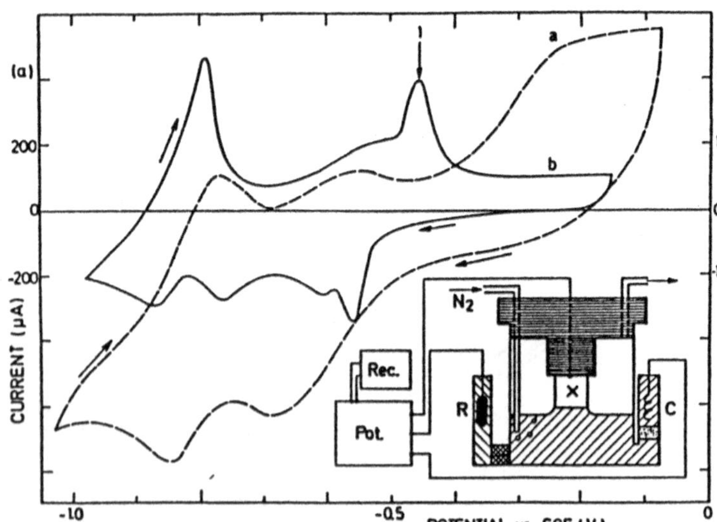

Figure 34. Voltammograms for Tl deposition onto a copper single crystal in the presence (a) and absence (b) of traces of oxygen. Inset: electrochemical cell. (From Ref. 120, with permission.)

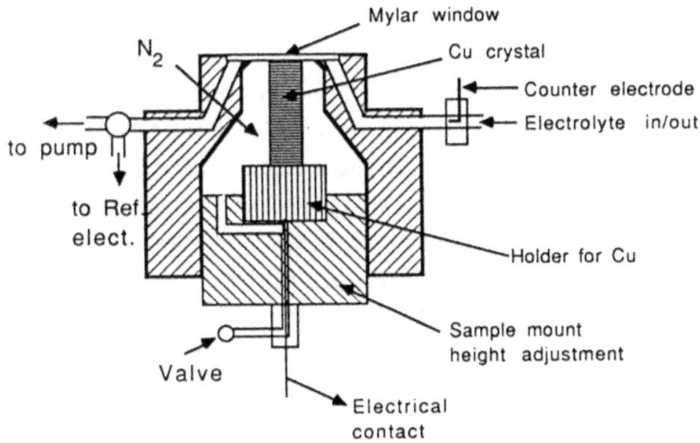

Figure 35. Electrochemical cell for *in situ* X-ray standing wave measurements. (From Ref. 120, with permission.)

experiments, both the reflectivity and the Tl fluorescence intensity were monitored simultaneously. Figure 36 shows the results for the *ex situ* study.

From an analysis of their data, Materlik and co-workers were able to determine that for the *ex situ* case and in the absence of oxygen, the thallium atoms are located at twofold sites at a mean distance of 2.67 ± 0.02 Å. For the *in situ* case and again in the

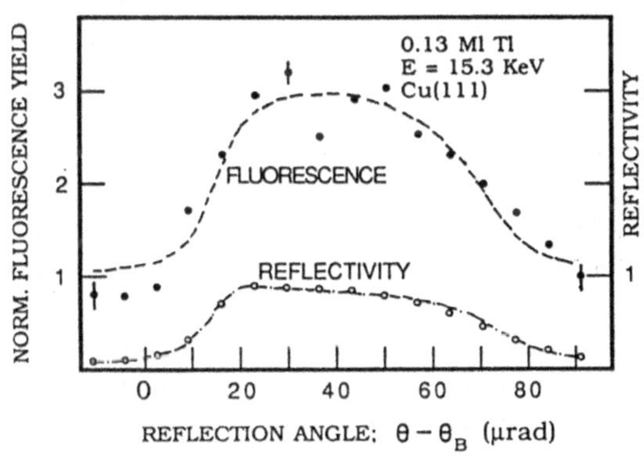

Figure 36. Angular dependence of the Cu(111) reflectivity and the normalized Tl_L fluorescence yield. Points represent experimental data; curves are least-squares fits. (From Ref. 119, with permission.)

absence of oxygen, the data are consistent with the thallium atoms being at 2.58 ± 0.02 Å, but at threefold sites.

In the presence of oxygen, there is a significant contraction of the mean distance of the thallium to 2.27 ± 0.04 Å. This is ascribed to a surface reconstruction of the copper induced by the adsorbed oxygen which results in an inward shift of the copper surface atoms by about 0.3 Å. This is consistent with low-energy ion-scattering studies. In general, these studies are of great significance since they demonstrate the applicability of the X-ray standing wave technique to the *in situ* study of electrochemical interfaces, even employing single crystals.

We have performed some experiments on the use of LSMs in the investigation of electrochemical interfaces.[121] The system that we have studied involves the adsorption of iodide onto a platinum/carbon LSM followed by the electrodeposition of a layer of copper. The LSM sample consisted of 15 platinum/carbon layer pairs with each layer having 26 and 30 Å of platinum and carbon, respectively, with platinum as the outermost layer. We used 9.2-keV radiation from the Cornell High Energy Synchrotron Source (CHESS) to excite *L*-level and *K*-level fluorescence from the iodide and copper, respectively. Initially, the LSM was contacted with a 35 mM aqueous solution of sodium iodide for 15 min. It was then studied by the X-ray standing wave technique. The characteristic iodine *L* fluorescence could be detected and its angular dependence was indicative of the fact that the layer was on top of the platinum surface layer. A well-developed reflectivity curve (collected simultaneously) was also obtained. Following this, the LSM was placed in an electrochemical cell and half a monolayer of copper was electrodeposited. The LSM (now with half a monolayer of copper and a monolayer of iodide) was again analyzed by the X-ray standing wave technique. Since the incident X-ray energy (9.2 keV) was capable of exciting fluorescence from both the copper and the iodide, the fluorescence intensities of both elements (as well as the reflectivity) were obtained simultaneously. The results presented in Fig. 37 show the reflectivity curve and the modulation of the iodide and copper fluorescence intensities. The most important feature is the noticeable phase difference between the iodide and copper modulation, i.e., the location of the iodide and copper fluorescence maxima, with the copper maximum being to the right of the iodide

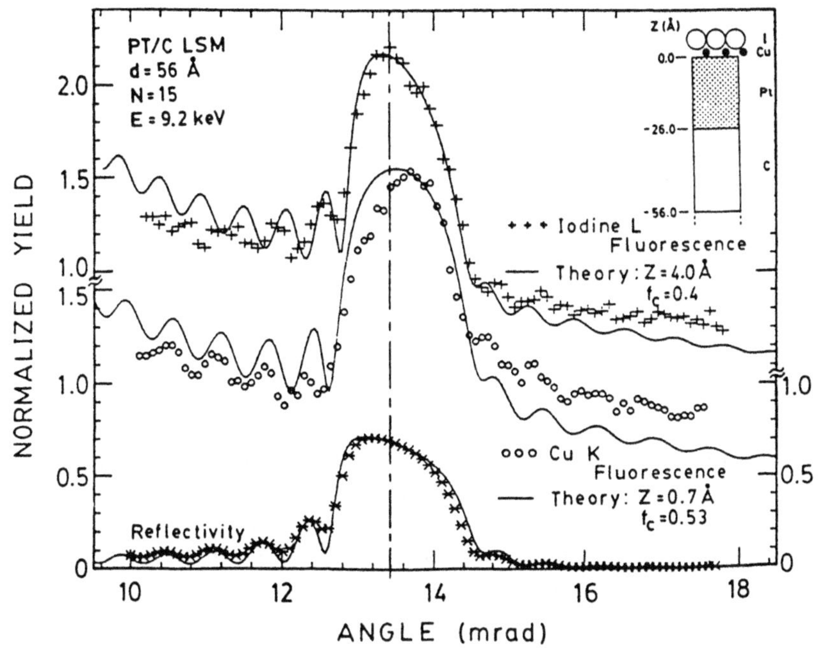

Figure 37. Experimental results and least-squares fits of data (solid lines) for a Pt/C LSM covered with an electrodeposited layer of copper and an adsorbed layer of iodine. Topmost curve: I_L fluorescence; middle curve: copper K_α fluorescence; bottom curve: reflectivity.

maximum. Since the antinodes move inward as the angle increases, the order in which these maxima occur can be unambiguously interpreted as meaning that the copper layer is closer than the iodide layer to the surface of the platinum. Since the iodide had been previously deposited on the platinum, this represents unequivocal evidence of the displacement of the iodide layer by the electrodeposited copper. Similar findings based on Auger intensities and LEED patterns have been previously reported by Hubbard and co-workers.[122,123] In addition, from an analysis of the copper fluorescence intensity, we were able to determine that the electrodeposited layer had a significant degree of coherence (53%) with the underlying substrate (Fig. 37). The experimental difficulties associated with the X-ray standing wave technique and the need to prepare single crystals of high perfection have limited its wide-

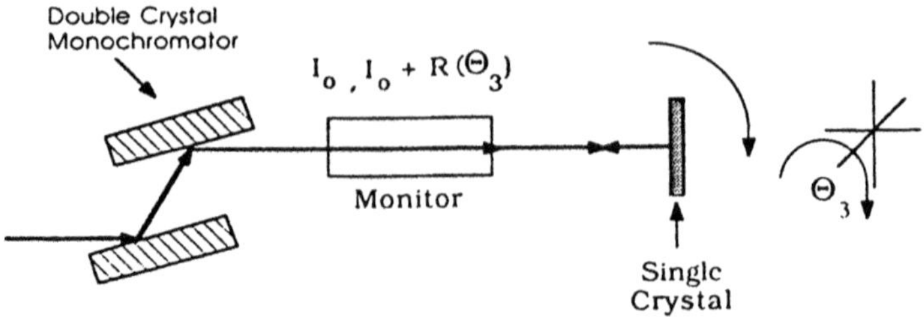

Figure 38. Experimental setup for back-reflection X-ray standing wave measurements.

spread applicability. Recent studies by Woodruff *et al.*[124] and us[125] have revealed that in a backscattering geometry (Fig. 38) the reflection widths of single crystals such as platinum are of the order of millradians (rather than microradians). This means that single crystals with a broader mosaic spread can be employed for X-ray standing wave measurements. Thus, with the use of this geometry for single crystals or the use of LSMs, I am certain that many more studies employing this technique will be carried out.

X. X-RAY DIFFRACTION

In addition to surface EXAFS and X-ray standing waves, X-ray diffraction can be employed in the study of electrochemical interfaces. Although an extensive treatment of X-ray diffraction techniques is beyond the scope of this chapter, some brief statements are appropriate.

A number of surface diffraction techniques can be employed in the structural study of electrochemical interfaces, depending on the details of the system under study. For bulk materials or thick films (such that the X-ray beam only samples that layer) conventional diffraction experiments can be performed and, in fact, a number of *in situ* X-ray diffraction studies of this type have been reported.[126-129] In the case of thin films or monolayers, two different techniques can be employed and these are the reflection–diffraction technique introduced by Marra and Eisenberger[130-132] and the technique based on surface truncation rods.[133] In the first case, the incident X-ray beam impinges on the sample at an angle below the critical angle so that the penetration depth of the X ray is very shallow (approximately 20–50 Å), allowing for the study of very thin films on a substrate. This technique has been employed by Fleischmann and co-workers in a variety of investigations including studies of iodine adsorbed on graphite,[134,135] lead underpotentially deposited on silver,[135,136] $Ni(OH)_2$,[135] and hydrogen and CO adsorption on platinum.[137] They introduced the variation of performing potential modulation experiments (akin to the procedures employed in FT-IR studies of electrochemical interfaces) to obtain difference diffractograms which were then interpreted in terms of differences in structure at the two potentials.

For the case of surface truncation rods, the technique is based on the detection of diffraction peaks between Bragg peaks. Although this requires careful alignment and some *a priori* knowledge of the structure, monolayer sensitivity can be achieved. In fact, Samant *et al.*[138] have recently performed an *in situ* surface diffraction study of lead underpotentially deposited on silver employing this technique along with grazing incidence diffraction. It is clear that this technique will also find widespread use in the near future.

XI. CONCLUSIONS AND FUTURE DIRECTIONS

The use of X rays is providing a rare glimpse of the *in situ* structure of electrochemical interfaces, and as these experiments become more widespread, a wide range of phenomena will be explored. I am certain that these studies will provide the basis for a better understanding and control of electrochemical reactivity.

ACKNOWLEDGMENTS

Our work was generously supported by the Materials Science Center at Cornell University, the National Science Foundation, the Office of Naval Research, the Army Research Office, and the Dow Chemical Company. Special thanks to Dr. Michael J. Bedzyk (CHESS) and to Dr. James H. White as well as to Michael Albarelli, Mark Bommarito, Dr. Martin McMillan, and David Acevedo. The work on the copper and silver underpotentially deposited on gold was in collaboration with Dr. Owen Melroy and Dr. Joseph Gordon (IBM, San Jose, California) and Professor Lesser Blum (University of Puerto Rico).

REFERENCES

[1] M. J. Sparnaay, "The Electrical Double Layer," in *The International Encyclopedia of Physical Chemistry and Chemical Physics*, Vol. 14, Pergamon Press, Glasgow, 1972.

[2] J. O'M. Bockris, B. E. Conway, and E. Yeager, *Comprehensive Treatise of Electrochemistry*, Vol. 1, Plenum Press, New York, 1980.

[3] T. E. Furtak, K. L. Kliewar, and D. W. Lynch, Eds., *Proceedings of the International Conference on Non-Traditional Approaches to the Study of the Solid Electrolyte Interface, Surf. Sci.* **101** (1980).

[4] See *J. Electroanal. Chem.* **150** (1983).

[5] M. Fleischmann, P. J. Hendra, and A. J. McQuillan, *Chem. Phys. Lett.* **26** (1974) 173; *J. Electroanal. Chem.* **65** (1975) 933.

[6] D. J. Jeanmarie and R. P. Van Duyne, *J. Electroanal. Chem.* **84** (1977) 1.

[7] R. P. Van Duyne, in *Chemical and Biological Applications of Lasers*, Vol. 4, Ed. by C. B. Moore, Academic, New York, 1979.

[8] S. Pons, *J. Electroanal. Chem.* **150** (1983) 495.

[9] A. Bewick, *J. Electroanal. Chem.* **150** (1983) 481.

[10] C. K. Chen, T. F. Heinz, D. Ricard, and Y. R. Shen, *Phys. Rev. Lett.* **46** (1981) 1010.

[11] R. M. Corn and M. Philpott, *J. Chem. Phys.* **81** (1984) 4138.

[12] G. L. Richmond, *Surf. Sci.* **147** (1984) 115.

[13] J. D. E. McIntyre, in *Advances in Electrochemistry and Electrochemical Engineering*, Vol. 9, Ed. by R. H. Muller, Wiley-Interscience, New York, 1973.

[14] D. M. Kolb and H. Gerischer, *Electrochim. Acta* **18** (1973) 987.

[15] A. T. Hubbard, *Accounts. Chem. Res.* **13** (1980) 177.

[16] E. B. Yeager, *J. Electroanal. Chem.* **128** (1981) 1600.

[17] P. N. Ross, *Surf. Sci.* **102** (1981) 463.

[18] H. Winick and S. Doniach, Eds., *Synchrotron Radiation Research*, Plenum Press, New York, 1980.

[19] H. Winick, in *Synchrotron Radiation Research*, Ed. by H. Winick and S. Doniach, Plenum Press, New York, 1980, p. 11.

[20] D. H. Tomboulian and P. Hartman, *Phys. Rev.* **102** (1956) 1423.

[21] H. Winick, G. Brown, K. Halbach, and J. Harris, *Physics Today* (May 1981) 50.

[22] E. A. Stern, *Sci. Am.* **234**(4) (1976) 96.

[23] P. Eisenberger and B. M. Kincaid, *Science* **200** (1978) 1441.

[24] S. P. Cramer and K. O. Hodgson, *Prog. Inorg. Chem.* **25** (1979) 1.

[25] B. K. Teo, *Accounts. Chem. Res.* **13** (1980) 412.

[26] P. A. Lee, P. H. Citrin, P. Eisenberger, and B. M. Kincaid, *Rev. Mod. Phys.* **53** (1981) 769.

[27] B. K. Teo and D. C. Joy, Eds., *EXAFS Spectroscopy; Techniques and Applications*, Plenum Press, New York, 1981.

[28] A. Bianconi, L. Inoccia, and S. Stippich, Eds., *EXAFS and Near Edge Structure*, Springer-Verlag, Berlin, 1983.

[29] K. O. Hodgson, B. Hedman, and J. E. Penner-Hahn, Eds., *EXAFS and Near Edge Structure III*, Springer-Verlag, Berlin, 1984.

[30] B. K. Teo, *EXAFS: Basic Principles and Data Analysis*, Springer-Verlag, Berlin, 1986.
[31] R. de L. Kronig, *Z. Phys.* **70** (1931) 317; **75** (1932) 191, 468.
[32] D. E. Sayers, E. A. Stern, and F. W. Lytle, *Phys. Rev. Lett.* **27** (1971) 1204.
[33] E. A. Stern, *Phys. Rev. B* **10** (1974) 3027.
[34] E. A. Stern, D. E. Sayers, and F. W. Lytle, *Phys. Rev. B* **11** (1975) 4836.
[35] C. A. Ashby and S. Doniach, *Phys. Rev. B* **11** (1975) 1279.
[36] P. A. Lee and J. B. Pendry, *Phys. Rev. B* **11** (1975) 2795.
[37] P. A. Lee and G. Beni, *Phys. Rev. B* **15** (1977) 2862.
[38] P. Eisenberger and G. S. Brown, *Solid State Commun.* **29** (1979) 481.
[39] T. M. Hayes, J. W. Allen, J. Tauc, B. G. Giessen, and J. J. Hauser, *Phys. Rev. Lett.* **40** (1978) 1282.
[40] T. M. Hayes, J. B. Boyce, and J. L. Beeby, *J. Phys. C* **11** (1978) 2931.
[41] P. A. Lee and G. Beni, *Phys. Rev. B* **15** (1977) 2682.
[42] T. M. Hayes, P. N. Sen, and S. H. Hunter, *J. Phys. C* **9** (1976) 4357.
[43] S. P. Cramer, J. H. Dawson, K. O. Hodgson, and L. P. Hager, *J. Am. Chem. Soc.* **100** (1978) 7282.
[44] S. P. Cramer, W. O. Gillum, K. O. Hodgson, L. E. Mortenson, E. I. Stiefel, J. R. Chisnell, J. W. Brill, and V. K. Shah, *J. Am. Chem. Soc.* **100** (1978) 3814.
[45] S. P. Cramer, K. O. Hodgson, E. I. Stiefel, and W. E. Newton, *J. Am. Chem. Soc.* **100** (1978) 2748.
[46] T. Tullius, P. Frank, and K. O. Hodgson, *Proc. Natl. Acad. Sci. USA* **75** (1978) 4069.
[47] C. A. Ashby and S. Doniach, *Phys. Rev. B* **11** (1975) 1279.
[48] R. F. Pettifer and P. W. McMillan, *Phil. Mag.* **35** (1977) 871.
[49] P. Lagarde, *Phys. Rev. B* **13** (1976) 741.
[50] B. K. Teo and P. A. Lee, *J. Am. Chem. Soc.* **101** (1979) 2815.
[51] P. H. Citrin, P. Eisenberger, and B. M. Kincaid, *Phys. Rev. Lett.* **36** (1976) 1346.
[52] E. A. Stern, *J. Vac. Sci. Technol.* **14** (1977) 461.
[53] U. Landman and D. L. Adams, *J. Vac. Sci. Technol.* **14** (1977) 466.
[54] J. Stohr, in *Emission and Scattering Techniques Studies of Inorganic Molecules, Solids and Surfaces*, Ed. by P. Day, Reidel, Dordrecht, 1981.
[55] J. Hasse, *Appl. Phys. A* **38** (1985) 181.
[56] P. H. Citrin, *J. Phys.* (*Paris*), *Colloq.* **47** (C8), (1986) 437.
[57] P. A. Lee, *Phys. Rev. B* **13** (1976) 5261.
[58] J. Jaklevic, J. A. Kirby, M. P. Klein, A. S. Robertson, A. S. Brown, and P. Eisenberger, *Solid State Commun.* **23** (1977) 679.
[59] J. B. Hastings, P. Eisenberger, B. Lengler, and M. L. Perlman, *Phys. Rev. Lett.* **43** (1979) 1807.
[60] S. M. Heald, E. Keller, and E. A. Stern, *Phys. Lett.* **103A** (1984) 155.
[61] R. W. James, *The Optical Principles of the Diffraction of X-rays*, Oxbow Press, Woodbridge, Connecticut, 1982.
[62] See also: D. H. Bilderback, *SPIE Proc.* **315** (1982) 90.
[63] E. Dartyge, A. Fontaine, A. Jucha, and D. Sayers, in *EXAFS and Near Edge Structure III*, Ed. by K. O. Hodgson, B. Hedman, and J. E. Penner-Hahn, Springer-Verlag, Berlin, 1984, p. 472.
[64] A. M. Flank, A. Fontaine, A. Jucha, M. Lemmonier, and C. Williams, in *EXAFS and Near Edge Structure*, Ed. by A. Bianconi, L. Inoccia, and S. Stippich, Springer-Verlag, Berlin, 1983, p. 405.
[65] D. E. Sayers, D. Bazin, H. Dexpert, A. Jucha, E. Dartyge, A. Fontaine, and P. Lagarde, *EXAFS and Near Edge Structure*, Ed. by A. Bianconi, L. Froccia, and S. Stippich, Springer-Verlag, Berlin, 1983, p. 209.

[66] H. Oyanagi, T. Matsushita, U. Kaminaga, and H. Hashimoto, *J. Phys. (Paris), Colloq* **47** (C8), (1986) 139.
[67] S. Saigo, H. Oyanagi, T. Matsushita, H. Hashimoto, N. Yoshida, M. Fujimoto, T. Nagamura, *J. Phys. (Paris), Colloq.* **47** (C8) (1986) 555.
[68] J. Mimault, R. Cortes, E. Dartyge, A. Fontaine, A. Jucha, D. Sayers, in *EXAFS and Near Edge Structure III* Ed. by K. O. Hodgson, B. Hedman, and J. E. Penner-Hahn, Springer-Verlag, Berlin, 1984, p. 47.
[69] W. E. O'Grady, *J. Electrochem. Soc.* **127** (1980) 555.
[70] G. G. Long, J. Kruger, D. R. Black, and M. Kuriyama, *J. Electrochem. Soc.* **130** (1983) 240.
[71] G. G. Long, J. Kruger, D. R. Black, and M. Kuriyama, *J. Electroanal. Chem.* **150** (1983) 603.
[72] G. G. Long, J. Kruger, and M. Kuriyama, in *Passivity of Metals and Semiconductors*, Ed. by M. Froment, Elsevier, Amsterdam, 1983, p. 139.
[73] J. Kruger, G. G. Long, M. Kuriyama, and A. I. Goldman, in *Passivity of Metals and Semiconductors*, Ed. by M. Froment, Elsevier, Amsterdam, 1983, p. 163.
[74] M. E. Kordesch and R. W. Hoffman, *Nucl. Instrum. Methods Phys. Res.* **222** (1984) 347.
[75] R. W. Hoffman, in *Passivity of Metals and Semiconductors*, Ed. by M. Froment, Elsevier, Amsterdam, 1983, p. 147.
[76] A. J. Forty, M. Kerkar, J. Robinson, and M. Ward, *J. Phys. (Paris), Colloq.* **47** (C8), (1986) 1077.
[77] G. Martens and P. Rabe, *Phys. Status Solidi A* **58** (1980) 415.
[78] L. Bosio, R. Cortes, A. Defrain, M. Froment, and A. M. Lebrun, in *Passivity of Metals and Semiconductors*, Ed. by M. Froment, Elsevier, Amsterdam, 1983, p. 131.
[79] L. Bosio, R. Cortes, A. Defrain, and M. Froment, *J. Electroanal. Chem.* **180** (1984) 265.
[80] L. Bosio, R. Cortes, and M. Froment, in *EXAFS and Near Edge Structure III*, Ed. by K. O. Hodgson, B. Hedman, and J. E. Penner-Hahn, Springer-Verlag, Berlin, 1984, p. 484.
[81] S. M. Heald, J. M. Tranquada, and H. Chen, *J. Phys. (Paris), Colloq.* **47** (C8), (1986) 825.
[82] E. A. Stern, E. Keller, O. Petitpierre, L. E. Bouldin, S. M. Heald, and J. Tranquada, in *EXAFS and Near Edge Structure III*, Ed. by K. O. Hodgson, B. Hedman, and J. E. Penner-Hahn, Springer-Verlag, Berlin, 1984, p. 261.
[83] J. McBreen, W. E. O'Grady, K. I. Pandya, R. W. Hoffman, and D. E. Sayers, *Langmuir* **3** (1986) 428.
[84] D. M. Kolb, in *Advances in Electrochemistry and Electrochemical Engineering*, Vol. 11, Ed. by H. Gerischer and C. Tobias, Pergamon Press, New York, 1978, p. 125.
[85] L. Blum, H. D. Abruña, J. H. White, M. J. Albarelli, J. G. Gordon, G. Borges, M. Samant, and O. R. Melroy, *J. Chem. Phys.* **85** (1986) 6732.
[86] O. R. Melroy, M. G. Samant, G. C. Borges, J. G. Gordon, L. Blum, J. H. White, M. J. Albarelli, M. McMillan, and H. D. Abruña, *J. Phys. Chem.* **92** (1988) 4432.
[87] J. H. White, M. J. Albarelli, H. D. Abruña, L. Blum, O. R. Melroy, M. Samant, G. Borges, and J. G. Gordon, submitted.
[88] M. G. Samant, G. L. Borges, J. G. Gordon, O. R. Melroy, and L. Blum, *J. Am. Chem. Soc.* **109** (1987) 5970.
[89] J. H. White, M. J. Albarelli, M. McMillan, G. M. Bommarito, D. Acevedo, and H. D. Abruña, submitted.
[90] D. W. Pashley, *Phil. Mag.* **4** (1959) 316.

[91] E. Grunbaum, *Vacuum* **24** (1973) 153.
[92] K. Reichelt and H. O. Lutz, *J. Cryst. Growth* **10** (1971) 103.
[93] M. Zei, G. Qiao, G. Lehmpfhul, and D. M. Kolb, *Ber. Bunsenges. Phys. Chem.* **91** (1987) 349.
[94] J. H. White, M. J. Albarelli, M. McMillan, M. G. Bommarito, D. Acevedo, and H. D. Abruña, unpublished results.
[95] F. Lu, G. N. Salaita, H. Baltruschat, and A. T. Hubbard, *J. Electroanal. Chem.* **222** (1987) 305.
[96] D. A. Smith, M. J. Heeg, W. R. Heineman, and R. C. Elder, *J. Am. Chem. Soc.* **106** (1984) 3053.
[97] D. A. Smith, R. C. Elder, and W. R. Heineman, *Anal. Chem.* **57** (1985) 2361.
[98] H. D. Dewald, J. W. Watkins, R. C. Elder, and W. R. Heineman, *Anal. Chem.* **58** (1986) 2968.
[99] W. R. Heineman, private communication.
[100] M. R. Antonio, J. S. Wainwright, and O. J. Murphy, private communication.
[101] M. J. Albarelli, J. H. White, M. McMillan, M. G. Bommarito, and H. D. Abruña, *J. Electroanal. Chem.* **248** (1988) 77.
[102] G. Tourillon, E. Dartyge, H. Dexpert, A. Fontaine, A. Jucha, P. Lagarde, and D. E. Sayers, *J. Electroanal. Chem.* **178** (1984) 357.
[103] G. Tourillon, E. Dartyge, H. Dexpert, A. Fontaine, A. Jucha, P. Lagarde, and D. E. Sayers, *Surf. Sci.* **156** (1985) 536.
[104] H. Dexpert, P. Lagarde, and G. Tourillon, in *EXAFS and Near Edge Structure III*, Ed. by K. O. Hodgson, B. Hedman, and J. E. Penner-Hahn, Springer-Verlag, Berlin, 1984, p. 400.
[105] G. Tourillon, E. Dartyge, A. Fontaine, and A. Jucha, *Phys. Rev. Lett.* **57** (1986) 603.
[106] E. Dartyge, A. Fontaine, G. Tourillon, and A. Jucha, *J. Phys. (Paris), Colloq.* **47** (C8) (1986) 607.
[107] E. Dartyge, C. Depautex, J. M. Dubuisson, A. Fontaine, A. Jucha, P. Leboucher, and G. Tourillon, *Nucl. Instrum. Methods Phys. Res.* **A246** (1986) 452.
[108] E. Dartyge, A. Fontaine, G. Tourillon, R. Cortes, and A. Jucha, *Phys. Lett.* **113A** (1986) 384.
[109] A. Fontaine, E. Dartyge, A. Jucha, and G. Tourillon, *Nucl. Instrum. Methods Phys. Res.* **A253** (1987) 519.
[110] G. Tourillon, H. Dexpert, and P. Lagarde, *J. Electrochem. Soc.* **134** (1987) 327.
[111] B. W. Batterman, and H. Cole, *Rev. Mod. Phys.* **36** (1964) 681.
[112] B. W. Batterman, *Phys. Rev.* **133** (1964) A759.
[113] T. W. Barbee and J. H. Underwood, *Optics Commun.* **48**, (1983) 161.
[114] M. J. Bedzyk and G. Materlik, *Phys. Rev. B* **31**, (1985) 4110.
[115] P. L. Cowan, J. A. Golovchenko, and M. F. Robbins, *Phys. Rev. Lett.* **44** (1980) 1680.
[116] T. W. Barbee, in *Low Energy X-ray Diagnostics*—1981, Ed. by D. T. Atwood and B. L. Henke, AIP Press, New York, 1981.
[117] T. W. Barbee and W. K. Warburton, *Mater. Lett.* **3** (1984) 17.
[118] A. Iida, T. Matsushita, and T. Ishikawa, *Jpn. J. Appl. Phys.* **24** (1985) L675.
[119] G. Materlik, J. Zegenhagen, and W. Uelhoff, *Phys. Rev. B* **32** (1985) 5502.
[120] G. Materlik, M. Schmah, J. Zegenhagen, and W. Uelhoff, *Ber. Bunsenges. Phys. Chem.* **91** (1987) 292.
[121] M. J. Bedzyk, D. Bilderback, J. H. White, H. D. Abruña, and M. G. Bommarito, *J. Phys. Chem.* **90** (1986) 4926.
[122] J. L. Stickney, S. D. Rosasco, and A. T. Hubbard, *J. Electrochem. Soc.* **131** (1984) 260.

[123] J. L. Stickney, S. D. Rosasco, B. C. Schardt, and A. T. Hubbard, *J. Phys. Chem.* **88** (1984) 251.
[124] D. P. Woodruff, D. L. Seymour, C. F. McConville, C. E. Riley, M. D. Crapper, N. P. Prince, and R. G. Jones, *Phys. Rev. Lett.* **58** (1987) 1460.
[125] M. J. Bedzyk and H. D. Abruña, unpublished results.
[126] S. U. Falk, *J. Electrochem. Soc.* **107** (1960) 661.
[127] A. J. Salkind, C. J. Venuto, and S. U. Falk, *J. Electrochem. Soc.* **111** (1964) 493.
[128] K. Machida and M. Enyo, *Chem. Lett.* (1986) 1437.
[129] G. Nazri and R. H. Müller, *J. Electrochem. Soc.* **132** (1985) 1385.
[130] W. C. Marra, P. Eisenberger, and A. Y. Cho, *J. Appl. Phys.* **50** (1979) 6927.
[131] P. Eisenberger and W. C. Marra, *Phys. Rev. Lett.* **46** (1981) 1081.
[132] W. C. Marra, P. H. Fuoss, and P. Eisenberger, *Phys. Rev. Lett.* **49** (1982) 1169.
[133] I. K. Robinson, *Phys. Rev. B* **33** (1986) 3830.
[134] M. Fleischmann, P. J. Hendra, and J. Robinson, *Nature* **288** (1980) 152.
[135] M. Fleischmann, A. Oliver, and J. Robinson, *Electrochim. Acta* **31** (1986) 899.
[136] M. Fleischmann, P. Graves, I. Hill, A. Oliver, and J. Robinson, *J. Electroanal. Chem.* **150** (1983) 33.
[137] M. Fleischmann and B. W. Mao, *J. Electroanal. Chem.* **229** (1987) 125.
[138] M. G. Samant, M. F. Toney, G. L. Borges, L. Blum, and O. R. Melroy, *J. Am. Chem. Soc. J. Phys. Chem.* **92** (1988) 220.

5

Electrochemical and Photoelectrochemical Reduction of Carbon Dioxide

Isao Taniguchi

Department of Applied Chemistry, Faculty of Engineering, Kumamoto University, Kurokami, Kumamoto 860, Japan

I. INTRODUCTION

The reduction of carbon dioxide has been a subject of active interest for more than a century.[1] Especially in recent years, electrochemical and photoelectrochemical reduction of carbon dioxide has been extensively studied.[2-4] This is because this reaction has several attractive features. In view of the increasing possibility of unavailability of oil and other fossil fuels in the near future,[5,6] alternative fuels have to be produced from abundant resources such as carbon dioxide and water. Carbon dioxide reduction is also an important branch of C_1 chemistry. In addition, the effect of recent excessive production of carbon dioxide on the future climate of the Earth is being seriously discussed,[7] and carbon dioxide reduction to organic raw materials or fuels would help to reduce this type of atmospheric pollution as well. Carbon dioxide reduction can be used as a suitable reaction for energy storage, as is required, for instance, in the conversion of solar to storable chemical energy.[8,9] Moreover, formic acid, which is one of the reduction products of carbon dioxide, has been proposed as a convenient means of hydrogen storage.[10]

From a fundamental viewpoint, carbon dioxide reduction is a model reaction which can help us to understand better the mechanism of natural photosynthesis.[11] Development of artificial photosynthetic systems, by mimicking functions of green plants, is one of our ultimate goals. Interest has also been devoted to carbon dioxide reduction as a model of the geochemical carbon cycle[12] and of prebiological photosynthesis.[13,14] Also, since the reduction of carbon dioxide does not take place easily, the development of effective catalysts is required; such research would lead to an insight into the activation of stable molecules and would yield information about reaction pathways for many-electron transfer that provide a saving of energy and high efficiency. Thus, reduction of carbon dioxide has been related to fascinating aspects in various fields of chemistry.

Here, some recent studies of the electrochemical and photoelectrochemical reduction of carbon dioxide as well as some other related subjects will be reviewed and discussed. Attention is focused especially on the work done in the last ten years, to avoid duplication of previous review articles.[2,4]

II. ELECTROCHEMICAL REDUCTION OF CARBON DIOXIDE

1. Reduction of Carbon Dioxide at Metal Electrodes

Electrochemical reduction of carbon dioxide usually requires a large overvoltage and competes with hydrogen evolution, resulting in a low power efficiency. Carbon dioxide, however, is actually reduced electrochemically at highly negative potentials (ca. −2.0 V versus SCE or more negative). Thus, the reaction has been carried out at metal electrodes having a high hydrogen overvoltage, such as Hg and Pb. The principal product of CO_2 reduction in aqueous solutions was reported to be formic acid (or formate ion), and high current efficiencies, up to near 100% were achieved at an amalgamated Zn electrode under a high CO_2 pressure.[15]

Udapa *et al.*[16] showed that CO_2 was reduced to formic acid at a mercury electrode in a 0.05 *M* phosphate buffer (pH 6.8) solution. A current efficiency of 81.5% was obtained at a current density of 20 mA/cm^2 and a cell voltage of 3.5 V. On the other hand, Bewick and Greener[17] reported that malate and glycolate were produced at Hg and Pb electrodes, respectively, using aqueous quartenary ammonium salt electrolytes; unfortunately, the results have never been reproduced.

Instead of mercury, Ito *et al.*[18] examined systematically some *sp* metals, such as Zn Pb, Sn, In, and Cd (5*N* purity), as the cathode materials for CO_2 reduction (Fig. 1). Using an In electrode at 3.9 mA/cm^2, the highest current efficiency (92%) for formic acid production was obtained in an aqueous Li_2CO_3 solution; the potentials at which CO_2 was reduced at an In electrode were ca. 400 mV less negative than those at an Hg electrode.

A little later, Russell *et al.*[19] tried to obtain methanol from carbon dioxide by electrolysis. Reduction of carbon dioxide to formate ion took place in a neutral electrolyte at a mercury electrode. On the other hand, formic acid was reduced to methanol either in a perchloric acid solution at a lead electrode or in a buffered formic acid solution at a tin electrode. The largest faradaic efficiency for methanol formation from formic acid was ca. 12%, with poor reproducibility, after passing 1900 C in the perchloric acid solution at Pb in a very narrow potential region (−0.9 to −1.0 V versus SCE). In the buffered formic acid solution (0.25 *M* HCOOH + 0.1 *M*

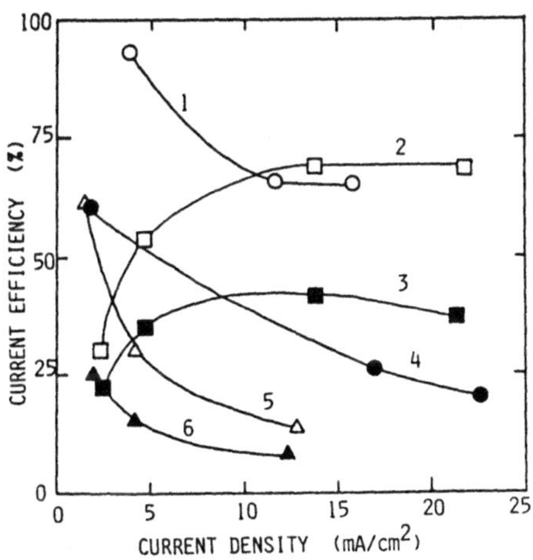

Figure 1. Current efficiencies for reduction of CO_2 to formic acid in a 0.1 *M* Li_2CO_3 solution at 25 ± 1°C at various electrodes.[18] (1) In; (2) Sn (previously anodized); (3) Sn; (4) Zn; (5) Pb; (6) Cd.

$NaHCO_3$, pH 3.8), good faradaic efficiencies up to 100% were obtained, but again only in a narrow potential region (−0.68 to −0.72 V versus SCE) with a current density of less than 4 μA/cm^2. No formaldehyde was produced. Furthermore, the reduction of formaldehyde to methanol took place at an Hg electrode at a current density of ca. 10 mA/cm^2 with faradaic efficiencies of more than 90% in a basic solution (0.1 *M* HCHO + 0.1 *M* Na_2CO_3, pH ~11). No dimerization product such as ethylene glycol was observed. Thus, the authors concluded that the direct reduction of CO_2 to methanol by electrolysis was difficult because of the differences in optimum conditions for electrolysis of CO_2, formic acid, and formaldehyde. Hori *et al.*[20] reported that the current efficiencies for formic acid formation in a CO_2-saturated 0.5 *M* $NaHCO_3$ solution by constant-current electrolysis at 16 mA/cm^2 decreased in the order In > Sn > Zn > Pb > Cu ≫ Au, and the partial currents for CO_2 reduction were linearly dependent on the hydrogen overvoltage of the metals[21] (Fig. 2). Later, tin and indium electrodes were again examined by Kapusta and Hackerman,[22] and carbon dioxide reduction to formic acid was confirmed to proceed with a high current efficiency (ca. 90%), although the overall power efficiency was very low due to the high overpotential of the reaction. Reduction of formic acid was observed only on a tin electrode at low current densities.

More recently, Hori *et al.*[23] reported that reduction of CO_2 by galvanostatic electrolysis (5 mA/cm^2) at Au and Ag electrodes gave

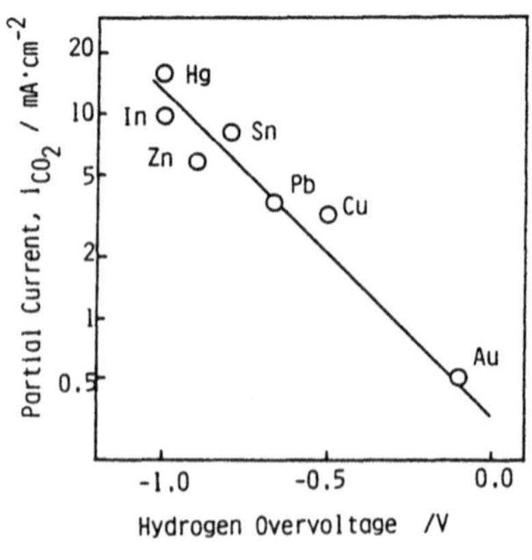

Figure 2. Partial currents for CO_2 reduction at various electrodes during constant-current electrolysis at 16 mA/cm^2 in a 0.5 *M* $NaHCO_3$ solution as a function of hydrogen overvoltage of the metal used.[20] Values[21] of hydrogen overvoltage of the metals were those obtained in acidic solutions at 1 mA/cm^2.

CO as a main product in a 0.5 *M* $KHCO_3$ solution (Table 1). At a Cu electrode, CH_4 was produced in appreciable amounts (Table 1); production of CH_4 was favored below 20°C, and at 0°C the faradaic efficiency reached ca. 65%, while at 40°C the amount of C_2H_4 formed increased to ca. 20% in faradaic efficiency with increasing hydrogen evolution. At Ru electrodes,[24] CO_2 was reduced to methane, methanol, and CO in a CO_2-saturated 0.2 *M* Na_2SO_4 solution at ca. 60°C and ca. −0.55 V (versus SCE) with current densities less than 0.5 mA/cm^2. Also, Mo electrodes were used[25] for CO_2 reduction at room temperature: Methanol was reported to be a major product of electrolysis at −0.7 to −0.8 V for 0.2 *M* Na_2SO_4 (pH 4.2) and at −0.57 to −0.67 V (versus SCE) for 0.05 *M* H_2SO_4 (pH 1.5), although current densities obtained were rather small (ca. 0.1 mA/cm^2) and prolonged electrolysis showed a remarkable decrease in current efficiency for methanol formation. Interestingly, when Mo electrodes were pretreated by cycling in a CO_2-saturated 0.2 *M* Na_2SO_4 solution between −1.2 and 0.2 V (versus SCE), faradaic efficiencies of greater than 100% for methanol formation were observed, indicating that some chemical reaction of CO_2 at the electrode surface is involved in the methanol formation. Taniguchi *et al.*[26] also confirmed the formation of CO at Au and Ag electrodes, hydrocarbons at a Cu electrode, and methanol at an Mo electrode, but the current efficiencies of these products, especially the hydrocarbons and methanol, were much smaller than those previously reported. Current efficiencies of the products depended strongly on the surface conditions such as roughness of the electrode and the presence of deposited impurities and, therefore, very small amounts of metallic impurities should be removed from the electrolyte; for example, when ca. 100 ppm of Bi^{3+} was present in the electrolyte, formic acid became a main product (ca. 60% faradaic efficiency) at a Cu electrode in an aqueous CO_2-saturated 0.5 *M* $KHCO_3$ solution at a constant current of 5 mA/cm^2. These results indicate that metal electrodes themselves have interesting catalytic activities which effect the distribution of the reduction products of CO_2; this may be caused by the difference in adsorption behavior of CO_2, H_2, and the intermediates involved in CO_2 reduction pathways.

In aqueous solutions, solubility of the reactant, i.e., CO_2, is rather low (ca. 38 mmol/liter in water at 25°C and 1 atm of CO_2),

Table 1
Results of Cathodic Reduction of CO_2 at Various Electrodes[a,b]

Electrode	Electrode potential (V versus SHE)	Faradaic efficiency (lower limit/upper limit) (%)				
		$HCOO^-$	CO	CH_4	H_2	Total
Cd[c]	−1.66 ± 0.02	65.3/67.2	6.2/11.1	0.2	14.9/22.2	93/100
Sn[c]	−1.40 ± 0.04	65.5/79.5	2.4/4.1	0.1/0.2	13.4/40.8	94/110
Pb[c]	−1.62 ± 0.03	72.5/88.8	0.3/0.6	0.1/0.2	3.8/30.9	94/100
In[c]	−1.51 ± 0.05	92.7/97.6	0.9/2.2	0.0	1.6/4.5	93/102
Zn[c]	−1.56 ± 0.08	17.6/85.0	3.3/63.3	0.0	2.2/17.6	90/98
Cu[d]	−1.39 ± 0.02	15.4/16.5	1.5/3.1	37.1/40.0	32.8/33.0	87/92
Ag[d]	−1.45 ± 0.02	1.6/4.6	61.4/89.9	0.0	10.4/35.3	99/106
Au[d]	−1.14 ± 0.01	0.4/1.0	81.2/93.0	0.0	6.7/23.2	100/105
Ni[d]	−1.39	0.3	0.0	1.2	96.3	98
Fe[d]	−1.42	2.1	1.4	0.0	97.5	101

[a] Ref. 23.
[b] Concentration of $KHCO_3$: 1.0 mol/dm^3 for Cu electrode and 0.5 mol/dm^3 for other electrodes.
[c] Current density: 5.5 mA/cm^2.
[d] Current density: 5.0 mA/cm^2.

and thus the expected maximum current density does not exceed 10 mA/cm^2. When values of 10^{-5} cm^2 s^{-1} for the diffusion coefficient of CO_2 and 10^{-3} cm for the thickness of the diffusion layer are assumed, the maximum current density for CO_2 reduction to formic acid can be estimated to be ca. 7 mA/cm^2. One effective way to increase the concentration of CO_2 in the solution is to use a high CO_2 pressure. Ito *et al.*[27] have examined CO_2 reduction under high CO_2 pressures (0–20 kg/cm^2). In aqueous Li_2CO_3 solutions under CO_2 pressures up to 5 kg/cm^2, the higher the CO_2 pressure, the greater the current density and also the current efficiency for formic acid production. When CO_2 was reduced in aqueous solutions of tetraalkylammonium salts under a CO_2 pressure of 10 kg/cm^2 at In, Sn, Pb, and Pb–Hg electrodes, the main product was formic acid, but small amounts of propionic acid and *n*-butyric acid and a trace amount of oxalic acid were detected as well (Table 2).

In organic aprotic solvents, advantages for CO_2 reduction would be expected; hydrogen evolution which competes with CO_2 reduction can be suppressed, and the solubility of CO_2 is much higher[28-30] in organic solvents than in aqueous solutions, although the latter point has not been stressed. Studies on CO_2 reduction in nonaqueous solvents have been carried out both from electroanalytical[31-33] and electrosynthetic[34-39] viewpoints, but such studies are still limited.

Von Kaiser and Heitz[34] reported the formation of oxalic acid from CO_2 in propylene carbonate (PC) and acetonitrile using a Cr–Ni–Mo (18:10:2)-steel electrode. Addition of small amounts of water gave rise to additional products, such as glycolic, glyoxylic, tertaric, malic, and succinic acids. Use of a platinum electrode favored CO formation. *N,N*-Dimethylformamide (DMF) was also used by Tyssee *et al.*[35] and Gambino and Silvestri.[36] Gressin *et al.*[37] reported that oxalate and carbon monoxide have been observed together with formate in nonaqueous solvents, and addition of small amounts of water favored not only the formation of formate ion, but also glycolate production by further reduction of oxalate. Fischer *et al.*[38] examined oxalic acid production in various organic solvents. Optimum current efficiencies for oxalic acid production (ca. 90%) were obtained by preparative-scale electrolysis using an undivided cell with a sacrificial Zn anode and a stainless steel

Table 2
Products Obtained by Electrochemical Reduction of Carbon Dioxide under High Pressure of 10 kg/cm² Gage[a]

Cathode	Electrolyte	Cathode potential (V)	Quantity of electricity (C)	Current efficiency[b] (%)	Concentration of carboxylic acid ($\times 10^{-3}$ mol/dm³)					
					Formic acid	Oxalic acid	Malonic acid	Propionic acid	n-Butyric acid	Unknown
Pb	0.10 *M* $TBABr_{aq}$	−1.6	1000	34.9	90			1.8		
		−1.6	1964	31.8	162			9.7		
		−1.8	3627	27.4	257	t[c]		3.9		
		−2.2	2000	45.1	232			2.9		
	0.10 *M* $TEAP_{aq}$	−1.7	2000	68.6	355	4.4				
		−1.8	2000	70.3	364	t				
		−1.8	4213	47.8	522	t	t			$t_1t_2t_3$
		−1.9	2800	47.8	347				t	
	0.18 *M* $TEAP_{aq}$	−1.9	2000	76.0	394	t				
		−2.3	2583	68.6	451	t				
Pb-Hg	0.10 *M* $TEAP_{aq}$	−1.7	2001	60.6	314	t				
		−2.0	1650	86.0	368	t				
Sn	0.10 *M* $TBABr_{aq}$	−1.6	200	34.7	18	t				
		−1.8	200	49.5	26	t				
		−2.0	2000	42.3	269			1.8		
		−2.0	4000	38.4	398				1.7	
	0.18 *M* $TEAP_{aq}$	−1.6	2501	48.8	316	t				
		−1.8	4566	42.4	502		t			
	0.10 *M* $TEAP_{aq}$	−1.6	2000	50.5	262	t				
		−2.0	2000	30.0	155	t		t		
In	0.10 *M* $TEAP_{aq}$	−1.5	2000	49.0	254	t				
		−2.0	200	65.0	34	t				
	0.10 *M* $TBABr_{aq}$	−1.7	2000	34.0	176	t		1.5		
		−2.0	200	65.2	34	t			t	t

[a] Ref. 27.
[b] Current efficiency of HCOOH formation.

Table 3
Electrochemical Reduction Products of CO_2 at Various Cathodes in 0.10 *M* TEAP/DMSO[a]

Cathode	Potential (V)	Current efficiency[b] (%)	Concentrations ($\times 10^{-3}$ *M*)[c]						
			Oxalic acid	Tartaric acid	Malonic acid	Glycolic acid	Formic acid	Propionic acid	*n*-Butyric acid
In	−1.5	1.8	1.6		1.5	1.7	2.8		2.2
	−1.7	1.6	0.8		0.8	0.8	3.2		
Zn	−2.0	4.8	0.7	0.7	1.2	7.9	11.5	6.0	
Sn	−2.0	8.2	2.8		3.8	2.1	17.2	1.0	2.5
Pb	−2.0	52.3	110.0	4.2	0.8	4.2	20.1	4.9	2.3

[a] Ref. 40.
[b] Total current efficiency of $H_2C_2O_4$ and HCOOH formation.
[c] When catholytes used are 20 cm^3 and quantity of electricity passed is 965 C.

cathode in acetonitrile with tetrabutylammonium perchlorate. Later, the performance of various cathode materials, catholyte formulations, and cell designs were investigated[39] for the formation of oxalic acid from CO_2 in DMF under both atmospheric and elevated pressures of CO_2.

Ito *et al.*[40] examined the electrochemical reduction of CO_2 in dimethylsulfoxide (DMSO) with tetraalkylammonium salts at Pb, In, Zn, and Sn under high CO_2 pressures. At a Pb electrode, the main product was oxalic acid with additional products such as tartaric, malonic, glycolic, propionic, and *n*-butyric acids, while at In, Zn, and Sn electrodes, the yields of these products were very low (Table 3), and carbon monoxide was verified to be the main product; even at a Pt electrode, CO was mainly produced in nonaqueous solvents such as acetonitrile and DMF.[41] Also, the products in propylene carbonate[42] were oxalic acid at Pb, CO at Sn and In, and substantial amounts of oxalic acid, glyoxylic acid, and CO at Zn, indicating again that the reduction products of CO_2 depend on the electrode materials used.

2. Mechanisms of Electrochemical Reduction of Carbon Dioxide

Mechanisms of carbon dioxide reduction in both aqueous and nonaqueous solutions have been studied mainly at metal electrodes.

Van Rysselberghe *et al.*[43] reported the fundamentally very important result that carbon dioxide (CO_2) molecules, not bicarbonate (HCO_3^-) or carbonate (CO_3^{2-}) ions, were the reacting species at a mercury electrode. Recently, Hori and Suzuki[44] have studied the electrolytic reduction of bicarbonate ion at Hg in aqueous $NaHCO_3$ and Na_2CO_3 mixtures; formate ion was formed, and the authors concluded that the dissociation of bicarbonate took place to give the electroactive species, CO_2 molecules. The maximum limiting current, controlled by the dissociation of HCO_3^- to CO_2 ($k_d = 6.8 \times 10^{-4}\ s^{-1}$), was estimated to be ca. 4 mA/cm^2.

The works of Eyring and co-workers[45] were the first to discuss CO_2 reduction on mercury in detail on the basis of the analysis of polarization curves obtained in aqueous solutions. The polarization curve showed two regions with different Tafel slopes: region I (low overvoltage with a Tafel slope of ca. 90 mV/decade) and region II (high overvoltage with a Tafel slope of more than 200 mV/decade).

The reaction orders with respect to CO_2 concentration were almost zero and one in regions I and II, respectively. From these results, the following mechanism was proposed:

$$CO_2 + e^- \rightarrow CO^{\cdot-}_{2ads} \tag{1}$$

$$CO^{\cdot-}_{2ads} + H_2O \rightarrow HCO^{\cdot}_{2ads} + OH^- \tag{2}$$

$$HCO^{\cdot}_{2ads} + e^- \rightarrow HCO_2^- \tag{3}$$

The overall reaction was

$$CO_2 + H_2O + 2e^- \rightarrow HCO_2^- + OH^- \tag{4}$$

where Eq. (3) was the rate-determining step (r.d.s.) in region I, while in region II the first step was the r.d.s.

Ito *et al.*[18] supported the above reaction pathways for various cathode materials, such as In, Sn, Cd, and Pb, from the similarity in Tafel slopes. Hori and Suzuki[46] verified the above mechanism in various aqueous solutions on Hg. Russell *et al.*[19] also agreed with the above mechanism. Adsorbed $CO_2^{\cdot-}$ anion radical was found as an intermediate at a Pb electrode using modulated specular reflectance spectroscopy.[47] This intermediate underwent rapid chemical reaction in an aqueous solution; the rate constant for protonation was found to be 5.5 $M^{-1}\,s^{-1}$, and the coverage of the intermediate was estimated to be very low (0.02).

Kapusta and Hackerman[22] also reported that the reaction pathways on Sn and In electrodes were similar to those postulated on an Hg electrode. In the case of the Sn electrode, two Tafel regions were observed. One was at potentials less negative than −1.45 V versus SCE with a slope of 115 mV/decade, and the other was at more negative potentials than −1.45 V with a slope of 320 ± 20 mV. At the In electrode, a single Tafel line with a slope of 140 mV was obtained. The reaction order was one with respect to CO_2 concentration over the entire potential range tested, and the partial current for CO_2 reduction in the region of low overvoltage was independent of the pH of the solution (pH 1–6.5). These results were similar to those observed by the Zakharyan *et al.*[48] The coverage of the electrode by adsorbed species was confirmed to be less than 5% by both capacitance and potential decay measurements.[22] Thus, the observed Tafel slope, ca. 120 mV/decade, and the independence of pH observed for the partial current for CO_2

reduction were interpreted in terms of Eqs. (1)-(3), with the discharge of CO_2 [Eq. (1)] as the r.d.s., by assuming Langmuir adsorption isotherms for the adsorbed species. Since the product of CO_2 reduction was formate ion, using the value of the reversible potential for $CO_2 + H_2O + 2e^- \rightarrow HCO_2^- + OH^-$, -0.76 V vs. SCE at pH 7, the exchange current densities (A/cm^2) were estimated to be 5×10^{-11} (at Hg), 1×10^{-9} (at Sn), and 1×10^{-8} (at In).

Results of photoemission experiments also support the reaction mechanism on both In and Sn as being the same as on Hg,[49] and indirect chemical reduction of CO_2 with adsorbed hydrogen atoms formed by electrochemical reduction of H^+ was suggested to be involved. An electrochemically modulated IR spectroscopic investigation at a Pt electrode also showed that the reduction of CO_2 involved a reaction with adsorbed hydrogen.[50] The laser photoelectronic emission technique has recently been used to elucidate the reduction mechanism of CO_2 on mercury.[51] $CO_{2\,ads}^{\bar{\cdot}}$ was the only intermediate of the two-electron reduction of CO_2 on mercury at pH values greater than 3, and the electrode coverage by the intermediate at potentials more negative than -1.8 V (versus SCE) was reported to be negligible (ca. 10^{-5}). Also, the free energies of the electrode reactions CO_2 (gas) + $e^- \rightarrow CO_2^{\bar{\cdot}}$ (aq) and CO_2 (gas) + $e^- \rightarrow CO_{2\,ads}^{\bar{\cdot}}$ were calculated to be 2.05 ± 0.12 and 1.60 eV, respectively. Later, Vassiliev *et al.*[52] also examined the mechanism and kinetics of electroreduction of CO_2 using various metals with high and moderate hydrogen overvoltage. Electrochemical and photoemission measurements showed basically the same mechanism as that described above; the effect of the electrode material was explained in terms of the adsorption of CO_2 molecules initially and of the anion radicals, $CO_2^{\bar{\cdot}}$, produced as well as the effect of electrode potential on adsorption behavior of these species. From these results, the optimal electrodes for electroreduction of CO_2 were suggested to be those with moderate hydrogen overvoltages; on these electrodes, CO_2 reduction took place near the zero-charge potential, where the maximum adsorption of CO_2 on the electrode can be expected.

As described above, the mechanism for CO_2 reduction in aqueous solutions proposed by Eyring and co-workers[45] has been widely accepted. However, there remains a rather large difference between theoretical and observed values for the Tafel slope,[45] and recent measurements on the intermediate and its coverage on the electrode are not completely consistent with this mechanism. Additional precise experiments and discussion would be required.

In nonaqueous solvents, reduction pathways of CO_2 should be different from those in aqueous solutions since different products are obtained. In dry propylene carbonate, the optical absorption spectrum observed during reduction of CO_2 at a Pb electrode showed two strong bands[47]: The band at 285 nm in PC was attributed to $CO_2^{\bar{\cdot}}$, and the band at longer wavelengths to a second intermediate, but not oxalate. Thus, the following mechanism, involving a step in which $CO_2^{\bar{\cdot}}$ is attacked by a CO_2 molecule [Eq. (6)], was suggested:

$$CO_2 + e^- \rightarrow CO_2^{\bar{\cdot}}\,(\text{slow}) \tag{5}$$

$$CO_2 + CO_2^{\bar{\cdot}} \rightarrow (CO_2)_2^{\bar{\cdot}} \tag{6}$$

$$(CO_2)_2^{\bar{\cdot}} + e^- \rightarrow (CO_2^-)_2 \tag{7}$$

The first step [Eq. (5)] was postulated to be rate determining because of the Tafel slope of 107 mV/decade and the first-order dependence of the reduction current on the CO_2 concentration. The second-order rate constant of Eq. (6) was estimated to be $7.5 \times 10^3\ M^{-1}\,s^{-1}$.

Lamy *et al.*[32] measured the standard potential and kinetic parameters for electrochemical reduction of CO_2 (1–8 mM) in *N,N*-dimethylformamide, from which residual water was deactivated using an active alumina suspension, by cyclic voltammetry with a high scan rate (4400 V/s). They reported a standard redox potential for $CO_2/CO_2^{\bar{\cdot}}$ of −2.21 V versus SCE, a standard heterogeneous rate constant for electron transfer of 6×10^{-3} cm s^{-1}, a transfer coefficient of 0.4, and a dimerization rate constant for $CO_2^{\bar{\cdot}}$ of $10^7\ M^{-1}\,s^{-1}$. In this case, dimerization of $CO_2^{\bar{\cdot}}$ was considered, and the values obtained indicate that the r.d.s. in the overall CO_2 reduction was the charge transfer to CO_2. More recently, Amatore and Saveant[53] have discussed the mechanism of CO_2 reduction in media of low proton availability, such as DMF, on the basis of product distribution with the aid of a theory[54] relating the product distribution to the intrinsic (rates, diffusion coefficients) and operational (concentrations, current densities, thickness of diffusion layer) parameters of the system. Three competing reaction pathways were suggested: (1) Oxalate formation through self-coupling of $CO_2^{\cdot-}$ (k = ca. $10^7\ M^{-1}\ s^{-1}$), (2) carbon monoxide formation via an oxygen–carbon coupling of $CO_2^{\cdot-}$ with CO_2 ($k = 3.2 \times 10^3\ M^{-1}\ s^{-1}$), and (3) formate formation through protonation of $CO_2^{\cdot-}$ by residual or added water ($k = 7.7 \times 10^2\ M^{-1}\ s^{-1}$), followed by homogeneous electron transfer from $CO_2^{\cdot-}$:

$$CO_2 + e^- \rightleftarrows CO_2^{\cdot-} \tag{8}$$

Oxalate formation:

$$2CO_2^{\cdot-} \rightarrow (CO_2^{\cdot-})_2 \tag{9}$$

Carbon monoxide formation:

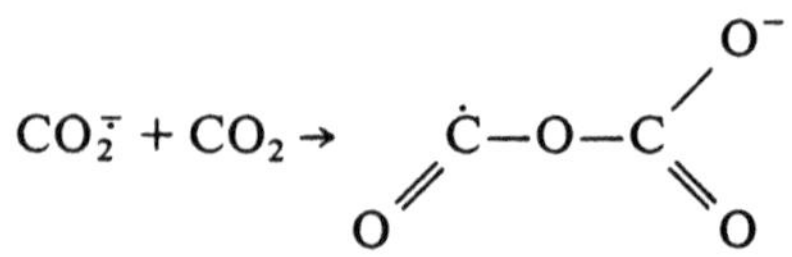

(10)

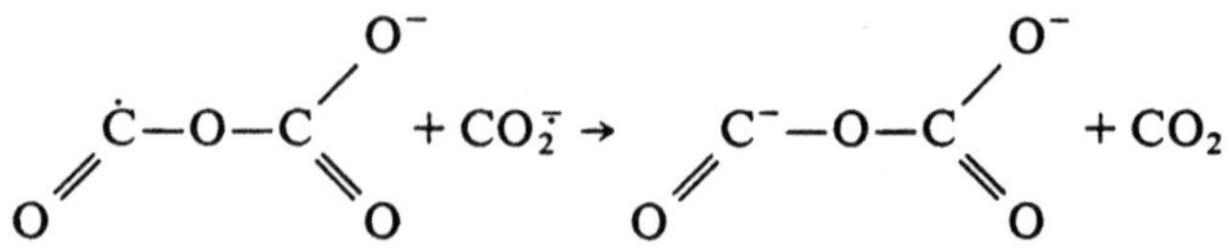

(11)

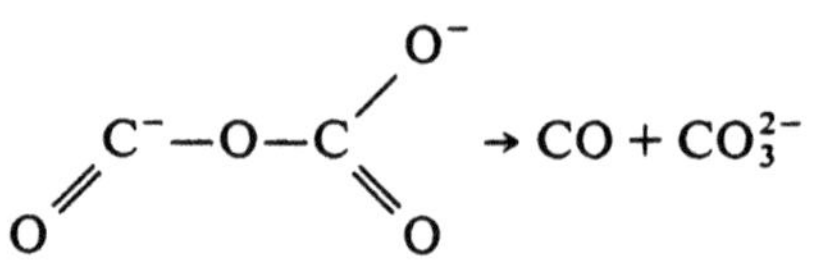

(12)

Formate formation:

$$CO_2^{\cdot-} + H_2O \rightarrow HCOO\cdot + OH^- \tag{13}$$

$$HCOO\cdot + CO_2^{\cdot-} \rightarrow HCOO^- + CO_2 \tag{14}$$

From the analysis of the experimental results, the authors concluded that the reaction pathway suggested by Aylmer-Kelly *et al.*[47] for oxalate formation via carbon–carbon coupling of $CO_2^{\cdot-}$ with a CO_2 molecule was unlikely, but that the dimerization given by Eq. (9) was more probable. This conclusion seems to be in good agreement with ESR data[55] and with the fact that the anion radical produced by oxidation of oxalate ion was unstable in acetonitrile[56]; if oxalate were formed by coupling of $CO_2^{\cdot-}$ with a CO_2 molecule followed by further electron transfer, the oxalate anion radical would be more stable. Moreover, the authors pointed out that the reason why oxalate formation was higher on lead than on mercury (where CO was the main product) was not due to a specific chemical effect of the metals on the course of CO_2 reduction, but to a difference in the current density used for electrolysis. More recent studies of CO_2 reduction in several nonaqueous solvents including DMF and acetonitrile using various metal electrodes[41,42] showed that Pb seems to have a certain catalytic activity for oxalate formation that reinforces the effect of current density on the distribution of reduction products of CO_2. Vassiliev *et al.* have reported[57] that in aprotic solvents the r.d.s. was the electron transfer to $(CO_2)_2^{\dot{-}}$ at Hg, Pb, Sn, In, and Pt electrodes in the first Tafel region, while in the second Tafel region the first electron transfer to an adsorbed CO_2 molecule was the r.d.s. The effect of potential on adsorption of CO_2 and anion radicals and on repulsion of negatively charged radicals was also suggested. Furthermore, on Pt and Rh, tightly chemisorbed species were proposed.

As described above, the mechanism of CO_2 reduction is still unclear. The establishment of unequivocal reaction pathways on the basis of a better understanding of each elementary step would make it possible to develop an effective means for reduction of CO_2. Investigations employing recently developed spectroscopic techniques, such as FTIR spectroscopy,[58] electrochemically modulated IR spectroscopy (EMIRS),[47] *in situ* IR reflection spectroscopy,[59] photoemission spectroscopy,[22,51,52,57] and Raman spectroscopy[60] and SERS[61] for adsorbed species, together with further studies by techniques used so far, such as flash photolysis,[62,63] ESR,[64] and electrochemical measurements,[65,66] would help to clarify the reaction intermediates and the reaction mechanism. Recently, EMIRS and FTIR reflection-absorption spectroscopy (FTIRRAS) have been used extensively to study the adsorption and oxidation behavior of CO,[67,68] formic acid,[69] and methanol[70,71] at various metal electrodes. Although the studies done to date are not directly intended to clarify the mechanism of CO_2 reduction, knowledge about these species at the electrode surface would be helpful to understand the fundamental steps of CO_2 reduction at the molecular level; these studies showed that dissociative adsorption of the species which appear in the reduction pathways of CO_2 on the electrode surface plays an important role in the mechanism of CO_2 reduction, and thus in the reduction product distribution.

For the reaction of formic acid in aqueous solutions, the adsorption of formic acid on the electrode was suggested to be the rate-determining step at Sn and Pb electrodes.[19] The point of zero charge (pzc) of the electrode was suggested to have a significant importance, and cadmium was recommended as the cathode material for formic acid reduction because of its low value of the pzc. Kapusta and Hackerman[22] examined formic acid reduction to methanol at Sn and In electrodes. The highest current efficiency (ca. 95%) for reduction to methanol was obtained at a tin electrode at a low current density (ca. 5 $\mu A/cm^2$), corresponding to a potential of −0.95 V versus SCE in a 0.5 *M* HCOOH + 0.5 *M* HCOONa solution. However, the formation of an organometallic complex on the electrode surface accelerated hydrogen evolution and the current efficiency of formic acid reduction decreased with time during electrolysis. From photoelectrochemical measurements, the authors concluded that the rate-determining step in HCOOH reduction at both Sn and In electrodes was the first electron transfer to the HCOOH molecule, and an HCO• radical was involved as an adsorbed intermediate.

For reduction of formaldehyde on mercury, the Tafel slope decreased with an increase in either the formaldehyde concentration (at constant pH) or the pH of the solution (at constant HCHO concentration).[19] The experimental results were in basic agreement with the previously proposed mechanism,[72,73] where formaldehyde is present predominantly in an electroinactive hydrated form, methylene glycol, which undergoes base-catalyzed dehydration to give electroactive formaldehyde. From the Tafel slopes obtained (66–36 mV at pH 6.8–13.0), the authors concluded that it was not the first electron transfer to the electroactive formaldehyde molecule but the final step to methanol that was rate determining (the so-called CECE mechanism), although polyoxymethylene glycol present as an impurity in formaldehyde made the mechanistic study complicated and ambiguous. Electrolysis of formaldehyde at a tin electrode[74] was also examined, but a tin complex formed on the electrode surface made hydrogen evolution more favorable than HCHO reduction. At a bright Pd cathode,[75] electrochemical reduction of bicarbonate, HCO_3^-, was suggested to take place.

3. Pathways for Carbon Dioxide Reduction

If carbon dioxide is reduced directly to give products of interest, the reduction potentials for the half-cell reactions in an aqueous solution of pH 7 are as follows:

$$CO_2(g) + 2H^+ + 2e^- \rightarrow HCOOH(aq) \qquad E^0 = -0.61 \tag{15}$$

$$CO_2(g) + 4H^+ + 4e^- \rightarrow HCHO(aq) + H_2O \qquad E^0 = -0.48 \tag{16}$$

$$CO_2(g) + 6H^+ + 6e^- \rightarrow CH_3OH(aq) + H_2O \qquad E^0 = -0.38 \tag{17}$$

$$CO_2(g) + 8H^+ + 8e^- \rightarrow CH_4(g) + 2H_2O \qquad E^0 = -0.24 \tag{18}$$

$$CO_2(g) + 2H^+ + 2e^- \rightarrow CO(g) + H_2O \qquad E^0 = -0.52 \tag{19}$$

$$2CO_2(g) + 2H^+ + 2e^- \rightarrow H_2C_2O_4(aq) \qquad E^0 = -0.90 \tag{20}$$

where the standard redox potentials[2] are given in volts versus NHE at pH 7.0, and (g) and (aq) denote the gaseous state and aqueous solution, respectively.

In aprotic solvents, the reduction of carbon dioxide to form carbon monoxide and carbonate ion[2] is also possible:

$$2CO_2(g) + 2e^- \rightarrow CO(g) + CO_3^{2-} \qquad E^0 = -1.07 \tag{21}$$

These reactions can be easily combined, if necessary, with an anodic reaction such as oxygen evolution to estimate the thermodynamic standard free energy, ΔG:

$$2H_2O \rightarrow O_2 + 4H^+ + 4e^- \qquad E^0 = 0.81 \text{ (at pH 7)} \tag{22}$$

The values of E^0 for Eqs. (15)–(20) indicate that if multielectron reductions of CO_2 take place, for example, by using suitable catalysts, the potentials required are much less negative than that for single-electron transfer, $CO_2/CO_2^{\cdot-}$, and are also less negative than the potentials actually required for CO_2 reduction (ca. −2.0 V or more negative). Thus, thermodynamically, but not kinetically, CO_2 reduction is comparable in difficulty to hydrogen evolution. For this reason, attempts to find catalysts for multielectron reactions and to use semiconductor electrodes as multielectron donors have been subjects of active interest. Further understanding of the elementary step in CO_2 reduction would, hopefully, lead toward sophisticated methods for CO_2 reduction.

A calculation of the temperature dependence of the free energy for the reactions in Eqs. (15)–(18), and hence the electrochemical potential, showed that with an increase in temperature, formic acid formation became more unfavorable.[4] In the case of formaldehyde, methanol, and methane formation, the calculation indicated a positive shift in the reduction potential, but of very small magnitude: ca. 30 mV for a temperature change from 300 to 500 K, and ca. 20 mV from 500 to 1200 K.[4]

4. Reduction of Carbon Dioxide at Semiconductor Electrodes in the Dark

Semiconductor electrodes seem to be attractive and promising materials for carbon dioxide reduction to highly reduced products such as methanol and methane, in contrast to many metal electrodes at which formic acid or CO is the major reduction product. This potential utility of semiconductor materials is due to their band structure (especially the conduction band level, where multielectron transfer may be achieved)[76] and chemical properties (e.g., CO_2 is well known to adsorb onto metal oxides and/or noble metal-doped metal oxides to become more active states[77-81]). Recently, several reports dealing with CO_2 reduction at *n*-type semiconductors in the dark have appeared, as described below.

Augustynski and co-workers[82] showed by cyclic voltammetry that at an *n*-TiO_2 electrode in a 0.5 *M* KCl solution (pH 6), CO_2 reduction took place at potentials less negative than those for hydrogen evolution. At both *n*-TiO_2 and ruthenium (1 at. %)-doped TiO_2 electrodes, reduction of carbon dioxide to methanol was achieved by long-term electrolysis at −0.9 V (versus SCE), although no information about the faradaic yield of methanol was given. Inoue *et al.*[83] also reported that CO_2 was reduced at a TiO_2 electrode to a mixture of formic acid, formaldehyde, methanol, and methane in acidic solutions.

Recently, however, Tinnemans *et al.*[84] have questioned the results of Augustynski and co-workers[82] and claimed that the larger current obtained in a CO_2-saturated solution compared to a N_2-saturated solution was not due to CO_2 reduction, but rather to hydrogen evolution, with the potentials at which hydrogen evolution occurred shifted toward less negative potentials because of the change in pH in the vicinity of the electrode (Fig. 3). Also, they reported that long-term electrolysis at −1.0 V versus SCE of a CO_2-saturated 0.1 *M* acetate buffer solution at a TiO_2/RuO_2 (0.5 wt %) cathode gave a mixture of formaldehyde and methanol but with a current efficiency of at most ca. 1%. Thus, the authors suggested that CO_2 reduction took place by reaction with adsorbed hydrogen generated by the photoassisted decomposition of water. Reactions of formic acid and formaldehyde with adsorbed hydrogen at polycrystalline semiconductor materials were also suggested.[85]

In commenting on the observations of Tinnemans *et al.*,[84] Augustynski remarked[86] that the importance of the marked affinity of the hydrated TiO_2 for CO_2 was apparent from anodic peaks observed on the voltammograms obtained on the reverse sweep after scanning up to a sufficient negative potential at which CO_2

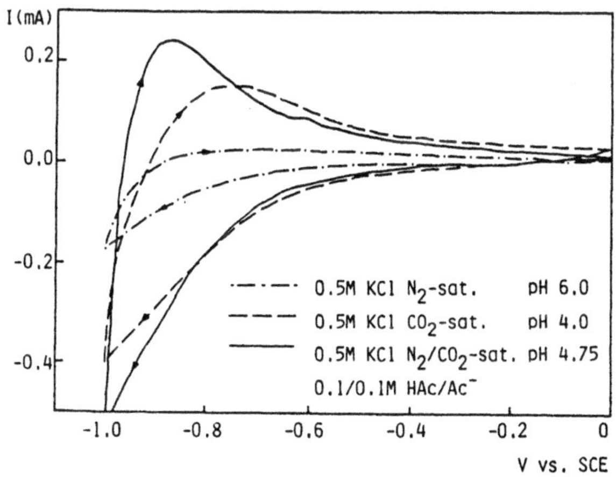

Figure 3. Cyclic voltammograms at a TiO_2/RuO_2 (0.5 wt %) electrode in various solutions.[84] Scan rate: 0.01 V/s.

reduction occurred. In addition, the following differences between the results of his group and those of Tinnemans *et al.* were pointed out: (1) the potential range used by Tinnemans *et al.*[84] for cyclic voltammetry was too narrow to find the reduced species of CO_2 at the electrode surface, (2) the acetate and formate buffer solutions used by Tinnemans *et al.* introduced confusion in interpreting the i–E curves because the species contained within these buffer solutions underwent cathodic reduction at potentials less negative than the hydrogen evolution reaction, and (3) the nature of the oxide electrodes used by these two groups would be different from each other because of the different preparation procedures employed.

In this connection, cyclic voltammetric measurements on the electrochemical reduction of CO_2 at n-TiO_2 and platinized TiO_2 film electrodes were reported a little later by Augustynski and co-workers.[87] The existence of two electrochemically detectable species resulting from CO_2 reduction was suggested by anodic peaks on the cyclic voltammograms (Fig. 4). Unfortunately, however, no

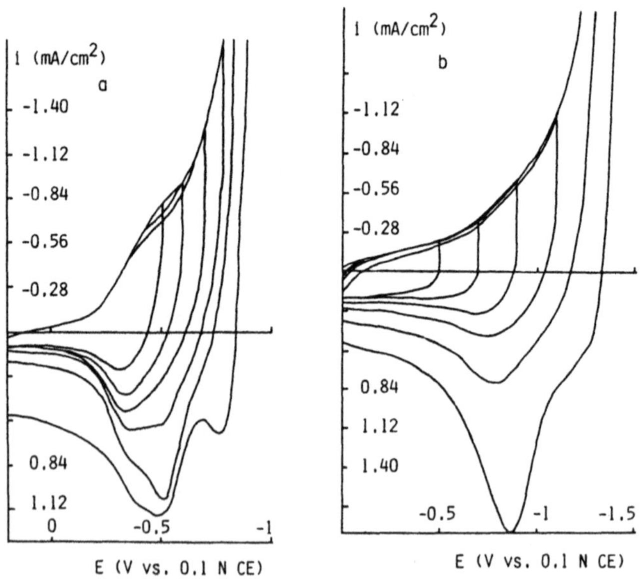

Figure 4. Cyclic voltammograms at Pt-treated TiO_2 (a) and TiO_2 (b) electrodes in a CO_2-saturated 0.5 M KCl solution at 40°C.[87] Scan rate: 0.05 V/s.

quantitative analysis of the products of CO_2 reduction was given in this report and the above conflict is still not completely resolved. More recently, Miles *et al.*[88] have concluded that in aqueous solutions containing formic acid, the predominant reaction at TiO_2 and other metal electrodes such as Ti, In, Ag, and Pt is the reduction of H_3O^+ rather than HCOOH or HCO_3^-; electrochemical conversion of CO_2 into methanol via formic acid is unlikely. To understand better the mechanism of CO_2 reduction at semiconductor electrodes and to develop effective semiconductor cathodes for CO_2 reduction, further experiments including quantitative analysis of products are required.

Canfield and Frese[89] used *n*-GaAs for CO_2 reduction to methanol. The faradaic efficiency for methanol formation was examined as a function of crystal face, electrolyte, and current density. The highest efficiency (almost 100%) for methanol formation was obtained using the As(111) face of *n*-GaAs in a 0.2 *M* reagent-grade Na_2SO_4 solution at a current density of 0.16 to 0.2 mA/cm^2 (−1.2 to −1.4 V versus SCE). On the other hand, when a solution of ultrapure (99.999%) Na_2SO_4 prepared with 1.6×10^7 Ω-cm water was used, no methanol was obtained. At the Ga(111) face of *n*-GaAs, methanol was produced in both electrolytes described above at lower faradaic efficiencies (14–80%). Also, the (100) and (110) faces of *n*-GaAs gave low yields of methanol (0–14%). Methanol synthesis was reported to be limited by a surface chemical step involving adsorbed hydrogen and an unidentified intermediate, such as CO_{ad}, —COH, —CH—OH, —CH_2—OH, and —O—CH_3, with a chemical rate constant of 6.1×10^{-5} A/cm^2 for the rate-determining step. Frese and Canfield[90] also observed the effectiveness of a surface pretreatment of *n*-GaAs (111) with Ru(III) for CO_2 reduction to methanol. Later, Sears and Morrison[91] indicated that GaAs dissolved (by corrosion) in carbonic acid solutions reacted with CO_2 to form hydroxides of gallium and arsenic plus methanol (or possibly formaldehyde).

Since indium is one of the most effective metals for electrochemical reduction of CO_2, *n*-TiO_2 on which indium had been electrodeposited was examined.[92] Enhancement of the faradaic efficiency of CO_2 reduction by one order of magnitude or more compared to that at undoped *n*-TiO_2 was observed, but the product detected was mainly hydrogen with a small amount (<5%

in faradaic efficiency) of formic acid in aqueous tetraethylammonium perchlorate solutions.

These results suggest that further characterization of the electrode surface is required to obtain more reproducible results, but, on the other hand, surface modification would be a very promising approach to the reduction of CO_2 with high selectivity and efficiency.

Although further experiments are required to establish the utility of semiconductor electrodes for CO_2 reduction directly to give highly reduced compounds, the results of gas phase reactions using semiconductor materials support the potential reduction of CO_2 to highly reduced products. For example, Hemminger *et al.*[93] reported that methane was directly obtained from gaseous water and CO_2 adsorbed on strontium titanate (111) crystals that were in contact with platinum foils by illumination with light of energy greater than the band gap of the semiconductor or by heating to 420 K in the dark; surface Ti^{3+} ions were proposed to act as a catalyst.

In nonaqueous solvents, little has been published to date dealing with CO_2 reduction at *n*-type semiconductors. Tinnemans *et al.*[84] suggested oxalic acid formation from CO_2 in DMF and DMSO at *n*-TiO_2/RuO_2 (0.5 wt %) by cyclic voltammetry. At an *n*-TiO_2 electrode, CO was obtained as the main product,[92] with a faradaic efficiency of ca. 80%, by electrolysis at −2.5 V versus Ag/Ag^+ in CO_2-saturated acetonitrile with 0.1 *M* Et_4NClO_4, while with the introduction of Pb onto an *n*-TiO_2 electrode, oxalate was also obtained. However, deposition of small amounts of metals such as In, Pt, Rh, Pd, and Ru onto an *n*-TiO_2 electrode did not significantly affect the reduction product of CO_2; again CO was formed predominantly.

Thus, although the potential required for polarization would be much larger at *n*-type semiconductors than at illuminated *p*-type semiconductors and despite the fact that not all *n*-type semiconductors can be used because of corrosion (or reduction) of semiconductor materials themselves, the use of *n*-type semiconductors to examine CO_2 reduction seems to be indicated because the cathodic current is much larger (the electron is the major carrier for *n*-type semiconductors), approaching that of metal electrodes, compared to the photocurrent obtained at illuminated *p*-type semiconductors, so that long-term electrolysis to give detectable amounts of products requires less time. Knowledge about the effective surface structure for CO_2 reduction at *n*-type semiconductors thus obtained would also be applicable for *p*-type semiconductor electrodes.

III. PHOTOELECTROCHEMICAL REDUCTION OF CARBON DIOXIDE

1. Reduction of Carbon Dioxide at Illuminated *p*-Type Semiconductor Electrodes

At illuminated *p*-type semiconductors, light energy brings about a shift of the applied cathode potential at which CO_2 reduction takes place toward a less negative potential by the photovoltaic effect.[94] Thus, light energy can be used to reduce the apparent overpotential of CO_2 reduction.

Halmann reported in 1978 the first example of the reduction of carbon dioxide at a *p*-GaP electrode in an aqueous solution (0.05 *M* phosphate buffer, pH 6.8).[95] At −1.0 V versus SCE, the initial photocurrent under CO_2 was 6 mA/cm^2, decreasing to 1 mA/cm^2 after 24 h, while the dark current was 0.1 mA/cm^2. In contrast to the electrochemical reduction of CO_2 on metal electrodes, formic acid, which is a main product at metal electrodes, was further reduced to formaldehyde and methanol at an illuminated *p*-GaP. Analysis of the solution after photoassisted electrolysis for 18 and 90 h showed that the products were 1.2×10^{-2} and 5×10^{-2} *M* formic acid, 3.2×10^{-4} and 2.8×10^{-4} *M* formaldehyde, and 1.1×10^{-4} and 8.1×10^{-4} *M* methanol, respectively. The maximum optical conversion efficiency calculated from Eq. (23) for production of formaldehyde and methanol (assuming 100% current efficiency) was 5.6 and 3.6%, respectively, where the bias voltage against a carbon anode was −0.8 to −0.9 V and 365-nm monochromatic light was used. In a later publication,[4] these values were given as ca. 1% or less, where actual current efficiencies were taken into account [Eq. (24)].

Also, using *n*-TiO_2 as an anode and *p*-GaP as a cathode in 0.1 *M* lithium carbonate solution, under illumination on both electrodes, methanol was produced (3×10^{-3} mol) at a current efficiency of 60% by electrolysis for 16 h at a constant current of 0.5 mA (2.1 mA/cm^2 on the *p*-GaP cathode). During electrolysis, negative bias to the cathode gradually increased from −0.86 to −1.4 V (versus SCE) to maintain this current. Inoue *et al.* reported a little later[83] that formaldehyde and methanol were formed on a *p*-GaP electrode at −1.5 V versus SCE under illumination with light of wavelength shorter than 500 nm in a 0.5 *M* H_2SO_4 solution.

To calculate the optical to chemical energy conversion efficiency, Halmann[95] used the following equation:

$$\text{Optical conversion efficiency (\%)} = 100 I_c[(\Delta H/n) - V_b]/W \tag{23}$$

where I_c is the current density (mA/cm^2), W (mW/cm^2) the incident light intensity, ΔH (eV) the heat of combustion (= 2.962, 2.639, 5.315, and 7.259 for hydrogen, formic acid, formaldehyde, and methanol, respectively), n the number of electrons involved in the reduction of one molecule of reactant to one molecule of product (= 2, 2, 4, and 6 for producing hydrogen, formic acid, formaldehyde, and methanol, respectively), and V_b (V) the bias voltage between the photocathode and the counter electrode.

In a later publication,[96] the standard free energy of formation of the products, ΔG in V, was used instead of ΔH in Eq. (23) so that comparisons could be made with the commonly reported efficiencies of solid state solar cells. For the reduction of carbon dioxide to organic compounds, the optical conversion efficiency of the system is the sum of the efficiencies for each product. Thus, it can be given as

$$\text{Optical conversion efficiency (\%)} = 100 I_c F_i[(\Delta G/n) - V_b]/W \tag{24}$$

where the values of $\Delta G/n$ for reduction of CO_2 to formic acid (liquid), formaldehyde (gas), and methanol (liquid) are 1.48, 1.35, and 1.21 V, respectively.

Another expression which has been used relates the extent of conversion of the total input energy (both electrical and optical) to chemical energy:

$$\text{Power conversion efficiency (\%)} = 100 I_c F_i(\Delta G/n)/(I_c V_b + W) \tag{25}$$

where $I_c V_b$ represents the electrical energy input. Note that use of Eq. (25) results in much larger values than are calculated using Eq. (23) or (24), and no negative values appear.

Canfield and Frese[89] showed that at the As(111) plane of *p*-GaAs and the P(111) plane of *p*-InP, CO_2 was reduced to methanol in aqueous Na_2SO_4 solutions at −1.2 to −1.4 V versus SCE. The highest faradaic efficiency (ca. 80%) was obtained at a photocurrent density of 60 $\mu A/cm^2$ (−1.2 to −1.4 V versus SCE) using a *p*-InP cathode.

Aurian-Blajeni *et al.*[97] examined CO_2 reduction at illuminated (600 mW/cm^2 was used) *p*-GaAs and *p*-GaP under high CO_2 pressures, using a specially designed cell (Fig. 5). The products were mainly formic acid with small amounts of formaldehyde and methanol. The best faradaic efficiency of 80% was obtained under 8.5 atm pressure of CO_2 using *p*-GaP at −1.0 V versus Ag/AgCl in a 0.5 *M* Na_2CO_3 solution. The main difficulty reported was the instability of the electrodes, especially in the case of *p*-GaAs.

Since *p*-GaAs has a suitable band gap (ca. 1.4 eV) for the solar spectrum, it is an attractive material for solar energy utilization. Zafrir *et al.*[96] examined CO_2 reduction at *p*-GaAs in the presence of the vanadium redox couple V(II)/V(III) to overcome the problem of corrosion of the electrode, because *p*-Si and *p*-InP were successfully stabilized by introduction of the redox couple.[98] Since a V(II)/V(III) chloride solution has a violet to blue color, some of the incident light was absorbed by the solution. The highest photocurrent at *p*-GaAS was obtained in 4 *M* HCl solutions having vanadium ion concentrations of less than 0.1 *M*. The highest optical energy conversion efficiency of 0.21%, calculated from Eq. (24), was obtained in a 4 *M* HCl solution containing 0.07 *M* V(II) at 80°C under irradiation with a light flux of 75 mW/cm^2. The products of CO_2 reduction observed were formic acid, formaldehyde, and methanol. In this case, homogeneous catalytic reduction of CO_2 with the vanadous ions and also reduction with adsorbed hydrogen atoms formed by photoelectrolysis of water were suggested to be conceivable possibilities, in addition to direct CO_2 reduction on *p*-GaAs.

An interesting result which questions the necessity of metal ions for catalysis of CO_2 reduction was reported;[99] at a polyaniline-coated *p*-Si electrode, CO_2 was effectively reduced to formic acid

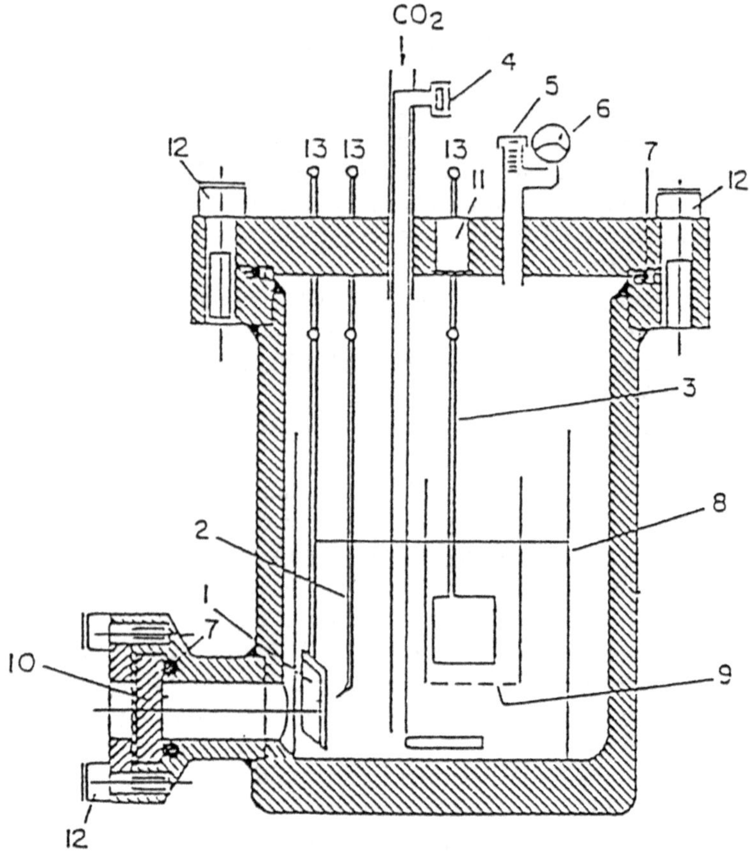

Figure 5. Design of a cell for photoassisted electrolysis of CO_2 under elevated pressures.[97] (1) Photoelectrode; (2) reference electrode; (3) counter electrode; (4) sampling port with septum; (5) pressure regulator; (6) pressure gauge; (7) O-rings; (8) reaction cell; (9) separator; (10) quartz window; (11) insulated connection; (12) bolts; (13) connections to potentiostat.

and formaldehyde in an aqueous CO_2-saturated $LiClO_4$ solution, although the origin of the catalytic activity was unclear.

Taniguchi *et al.*[100] have reported that in the reduction of CO_2 at *p*-GaP in Li_2CO_3 electrolytes, the current efficiency was enhanced by dissolving 15-crown-5 ether in the electrolyte. The proposed reaction pathway involved the initiation of CO_2 reduction by cathodically deposited Li metal on the GaP surface to give $CO_2^{\bar{\cdot}}$:

$$Li + CO_2 \rightarrow Li^+ + COO^{\bar{\cdot}} \quad (26)$$

where the crown ether facilitated the deposition of Li on the *p*-GaP electrode, and adsorbed crown ether retarded the hydrogen evolution reaction. The reduction of CO_2 molecules with lithium metal was confirmed in propylene carbonate to give $CO_2^{\bar{\cdot}}$, which was detected by its absorption spectrum ($\lambda_{max} = 265$ nm). In this case, the optical energy conversion efficiency calculated from Eq. (24) was very small (ca. 0.001%) or negative, and even the power conversion efficiencies [Eq. (25)] were ca. 0.01%.

Recently, results of careful experiments were reported by Ito *et al.*[101] They claimed that formic acid, formaldehyde, and methanol, which had been previously reported as photoelectrochemical reduction products of carbon dioxide, were observed also by photolysis of cell materials, such as electrolytes, including 15-crown-5 ether, and epoxy resin, which has often been used as the molding material of semiconductor electrodes in aqueous solutions. Previously reported reduction products were obtained also under nitrogen with (Table 4) and without (Table 5) a *p*-GaP photocathode under illumination. These precise experiments under improved conditions, where no photolytic products were observed, gave the result that the main reduction product of carbon dioxide at a *p*-GaP photocathode in aqueous electrolytes was formic acid. Thus, many kinds of products reported in previous papers[83,97,100] were suggested to be due to photolysis of cell materials.

The results of Ito *et al.*[101] indicate that careful experiments including enough blank experiments are necessary in studies of photoelectrochemical reduction products of carbon dioxide because, unfortunately, the products observed to date are in very low concentrations. Purification of the carbon dioxide gas itself should also be considered, expecially in experiments in which a continuous flow of CO_2 gas is used. Accumulation of organics which are present as impurities in CO_2 gas is often observed. Purification methods for CO_2 gas used are given in some papers,[95-97,102] but establishment of a common recommended method would be helpful. Also, it may be advisable to reexamine earlier work on CO_2 reduction to exclude meaningless results. In future experiments, the use of labeled $^{13}CO_2$ is to be recommended.

Table 4

Photoelectrolytic Products under N_2 and CO_2 Atmospheres in the Cell with Various Electrolytes and a *p*-GaP Photocathode Molded by Epoxy Resin[a,b]

Atmosphere	Electrolyte	n (μmol)					
		HCOOH	$CH_3OH + HCHO$[c]	$(COOH)_2$	$HOCH_2COOH$	CH_3CHO	C_2H_5OH
N_2	Li_2CO_3	1.5	0.04	0.1	2.1	0.02	0.02
	TEAP	1.3	0.03	t[d]	1.3	0.36	0.06
CO_2	TEABr	2.3	0.12	0.28	4.8	0.9	t
	TBABr	1.9	t	0.18	1.9	0.17	t
	15-Crown-5	6.5	t	0.25	1.5	0.29	t

[a] Ref. 101.
[b] Photoelectrolysis was performed up to 10 C at −1.2 V versus Ag/AgCl electrode by illuminating with a 300-W xenon lamp alone.
[c] Methanol and formaldehyde were represented as $CH_3OH + HCHO$ in Ref. 101 because they were not able to be separated by the steam chromatographic technique; the amount of product was calculated as an amount of methanol.
[d] t: Trace amount of product.

Table 5
Photolytic Products of Various Electrolytes in Quartz Test Tube Illuminated by a 300-W Xenon Lamp Alone or with the Filter Toshiba UV-37, under N_2 Atmosphere[a]

Filter of Xe lamp	Time (h)	Electrolyte	n (μmol) HCOOH	$CH_3OH + HCHO$	$(COOH)_2$	$HOCH_2COOH$	CH_3CHO	C_2H_5OH
Not used	25	Li_2CO_3	n[b]	n	n	n	n	n
	28	TEAP	n	n	n	n	0.24	n
	24	TEABr	n	n	n	n	0.28	n
	26	TBABr	n	n	n	n	0.21	0.14
	29	15-Crown-5	16.7	0.55	0.22	n	0.07	1.01
Used	24	TEAP	n	n	n	n	n	n
	26	TEABr	n	n	n	n	n	n
	24	15-Crown-5	13.1	n	0.22	n	n	n

[a] Ref. 101.
[b] n: No product.

In photoelectrochemical reduction of carbon dioxide, organic solvents and their mixtures with water have also been used. The use of organic solvents has the advantages[103] that (1) competitive hydrogen formation can be suppressed and (2) the increased solubility of CO_2 in nonaqueous solutions[28-30] has similar effects to the use of higher CO_2 pressures.

Guruswamy and Bockris[104] reported that oxalic acid was qualitatively detected in a CO_2-saturated DMF solution after photoelectrolysis using an illuminated *p*-GaP electrode. Taniguchi *et al.*[103,105] have recently shown that the photocurrent–potential curves at a *p*-CdTe electrode under monochromatic light ($\lambda = 600$ nm), in a DMF–0.1 *M* tetrabutylammonium perchlorate (TBAP) solution containing 5% water, shifted markedly (ca. 0.7 V) toward less negative potentials when the bubbling gas was changed from Ar to CO_2 (Fig. 6). Controlled-potential electrolysis (−1.2 to −2.4 V versus SCE) under a CO_2 atmosphere gave CO in high selectivity. The illuminated *p*-CdTe electrode showed much better performance

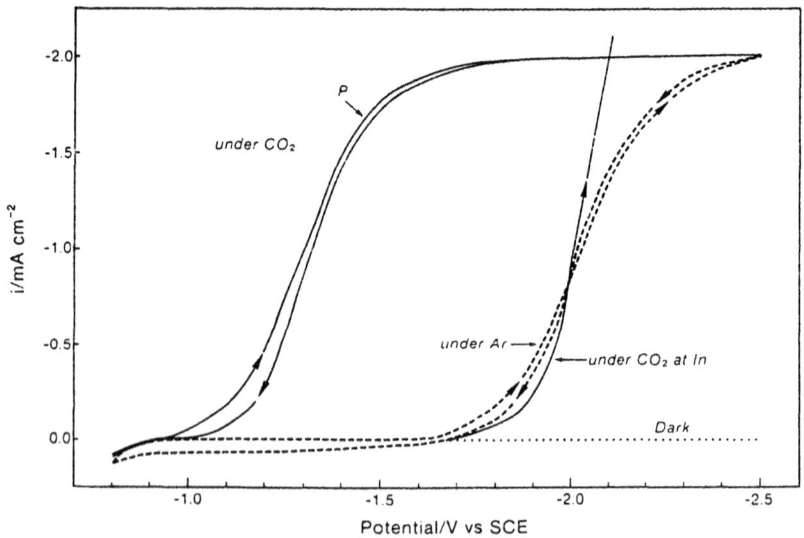

Figure 6. Current–potential curves at a *p*-CdTe electrode in a DMF–0.1 *M* TBAP solution containing 5% water under irradiation with monochromatic light of 600 nm, compared with the current–potential curve in a CO_2 atmosphere at an In electrode.[105] Electrode area: 0.2 cm^2 (*p*-CdTe) and 1 cm^2 (In). Potential scan rate: 0.1 V/s.

than an In electrode, which is the best metal electrode known to date for CO_2 reduction in nonaqueous media. In a later publication, various *p*-type semiconductors were also examined[103] in a DMF–0.1 *M* TBAP solution with 5% water under a CO_2 atmosphere. The photocurrent (quantum efficiency)-potential curves, the quantum efficiency as a function of wavelength, and the results of photoassisted controlled-potential electrolysis at various *p*-type semiconductor electrodes are shown in Figs. 7 and 8 and in Table 6. Among the *p*-type semiconductors tested, *p*-CdTe gave the best results. The *p*-Si electrode can be used over a range of wavelengths of the solar spectrum with high quantum efficiencies, and CO_2 was reduced to CO with high current efficiency at this electrode, but higher negative potentials were required for CO_2 reduction than at *p*-CdTe. At both *p*-InP and *p*-GaP electrodes, the current–potential curves showed a more positive onset potential than at *p*-CdTe, and good solar energy utilization was expected from their quantum efficiency–wavelength relationships. However, at *p*-InP and *p*-GaP electrodes,

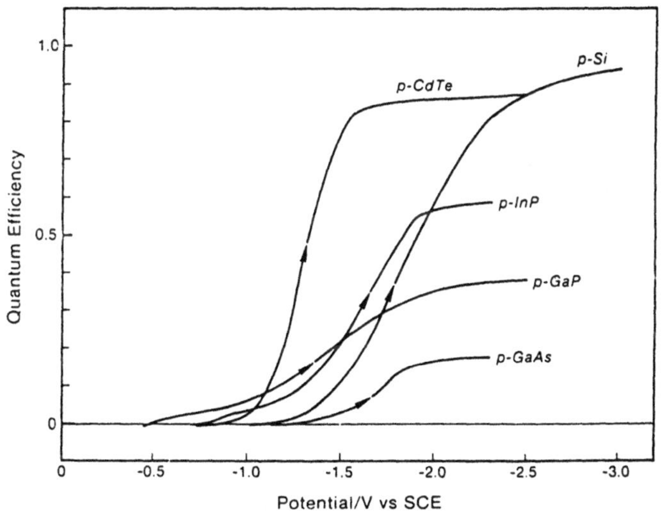

Figure 7. Quantum efficiency versus potential at various *p*-type semiconductors in a DMF-0.1 *M* TBAP solution containing 5% water under a CO_2 atmosphere. Monochromatic light of 600 nm was used for *p*-Si, *p*-InP, *p*-GaAs, and *p*-CdTe, while light of 400 nm was used for *p*-GaP.[103] Scan rate: 0.1 V/s

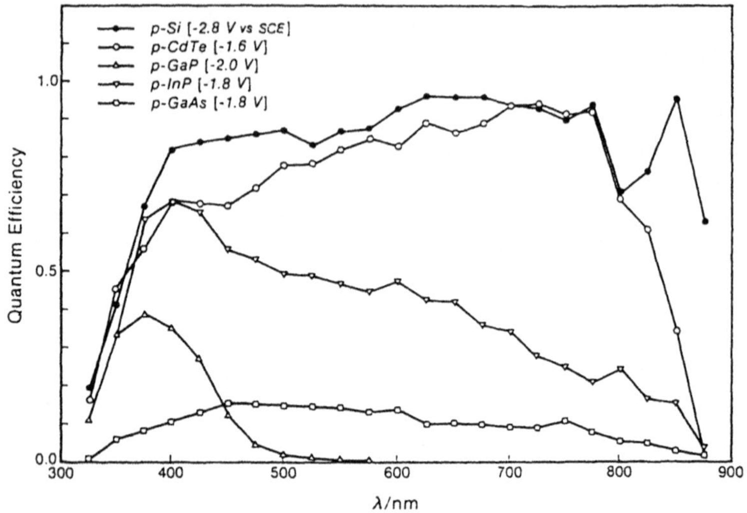

Figure 8. Quantum efficiency as a function of wavelength (λ) for various semiconductors in a DMF-0.1 *M* TBAP solution containing 5% water under CO_2 atmosphere.[103]

the faradaic current efficiency for CO_2 reduction was low and hydrogen evolution occurred as well. At *p*-CdTe, the current efficiency and product selectivity of the reduction of CO_2 to CO were not affected by the water concentration in DMF up to 25%, when 0.1 *M* TBAP was used as the supporting electrolyte. The current efficiency was ca. 90% or more when the CO dissolved in the solutions was taken into account. Various organic solvents can be used for CO_2 reduction to CO at illuminated *p*-CdTe (Table 7). No remarkable difference in either the current efficiency or the product selectivity of CO_2 reduction to CO was observed when various tetraalkylammonium salts were used. For CO_2 reduction to CO in DMF solutions, the reaction order with respect to CO_2 was one and the analysis of the photocurrent-potential curves was consistent with the simple expression for the cathodic photocurrent at a semiconductor electrode when the rate-determining step is charge transfer to CO_2 in the presence of surface states.[76,106,107] On the basis of these results, the following reaction pathways were suggested:

$$CO_2 + e^- \rightarrow CO_2^{\cdot -} \tag{27}$$

Table 6
Photoassisted Controlled-Potential Electrolysis (CPE) in DMF Solutions Containing 5% Water and TBAP under a CO_2 Atmosphere[a,b]

Run	Cathode materials	Cathode potential (V versus SCE)	Electricity passed (C)	CO formed[c]		H_2 formed[c]	
				μmol	Current efficiency (%)	μmol	Current efficiency (%)
1	*p*-Si(100)	−2.0	4.6	16.3	68.3	0.6	2.5
2	*p*-Si(100)	−2.0	11.1	45.9	79.6	1.6	2.7
3	*p*-InP(100)	−1.6	3.6	6.2	33.2	4.5	24.1
4	*p*-InP(100)	−1.6	6.9	13.9	38.9	9.3	26.0
5	*p*-GaP(100)	−1.6	3.4	5.3	30.1	6.6	37.5
6	*p*-GaP(100)	−1.6	6.2	9.1	28.3	9.6	29.9
7	*p*-GaP(100)	−2.0	5.6	8.9	30.7	9.0	31.0
8	*p*-CdTe(100)	−1.6	8.7	35.2	78.1	0.15	0.3
9	*p*-CdTe(100)	−1.6	16.1	62.5	75.0	0.25	0.3

[a] Ref. 103.
[b] For *p*-Si, *p*-InP, and *p*-CdTe, monochromatic light of 600 nm was used, while light of 400 nm was used for *p*-GaP.
[c] Based on the amount in the gas phase.

Table 7
Photoassisted Controlled-Potential Electrolysis (CPE) at *p*-CdTe at −1.6 V versus SCE in Various Solvents under Irradiation with Monochromatic Light of 600 nm[a]

Run	Solvent[b]	Electricity passed (C)	CO formed[c]		H_2 formed[c]	
			μmol	Current efficiency (%)	μmol	Current efficiency (%)
1	DMF–5% H_2O	7.9	33.0	80.6		<0.3
2		20.1	85.2	81.8		<0.3
3	DMSO–5% H_2O	5.7	25.5	86.3		<0.1
4		18.3	74.1	78.1		<0.1
5	MeCN–5% H_2O	6.8	19.8	56.2	7.6	21.6
6		21.7	65.3	58.2	20.4	18.1
7	PC–5% H_2O	7.3	32.5	86.4		<0.2
8		18.6	76.8	79.9		<0.2

[a] Ref. 103.
[b] 0.1 *M* Bu_4NBF_4 was used as a supporting electrolyte. DMF = *N,N*-dimethylformamide, DMSO = dimethylsulfoxide, MeCN = acetonitrile, PC = propylene carbonate.
[c] Based on the amount in the gas phase.

$$CO_2^{\overline{\cdot}} + CO_2 + e^- \rightarrow CO + CO_3^{2-} \quad (28)$$

and/or

$$CO_2^{\overline{\cdot}} + 2H^+ + e^- \rightarrow CO + H_2O \quad (29)$$

To examine the mechanism of CO_2 reduction, it would be useful to detect intermediates. In this connection, FTIR techniques were applied to study an illuminated *p*-CdTe electrode during the photoassisted reduction of CO_2 in acetonitrile.[58] This was the first report on IR spectroscopy applied to an illuminated electrode, and adsorbed $CO_2^{\overline{\cdot}}$ was detected (Fig. 9). The parallelism between the coverage of the intermediate and the photocurrent-potential curve was concluded to be consistent with the mechanism suggested by Amatore and Saveant[53] for CO_2 reduction to CO [Eqs. (8) and (10)-(12)], when we take into account that $CO_2^{\overline{\cdot}}$ was adsorbed on the electrode.

More recently, Ikeda *et al.*[108] have examined CO_2 reduction in aqueous and nonaqueous solvents using metal-deposited *p*-GaP and *p*-InP electrodes under illumination. Metal coatings on these semiconductor electrodes gave much improved faradaic efficiencies for CO_2 reduction. In an aqueous solution, the products obtained were formic acid and CO with hydrogen evolution at Pb-, Zn-, and In-coated electrodes, while in a nonaqueous PC solution, CO was obtained with faradaic efficiencies of ca. 90% at In-, Zn-, and Au-coated *p*-GaP and *p*-InP, and a Pb coating on a *p*-GaP electrode gave oxalate as the main product with a faradaic efficiency of ca. 50% at -1.2 V versus Ag/AgCl.

Bradley *et al.*[109] have combined a *p*-Si photocathode and homogeneous catalysts (tetraazamacrocyclic metal complexes, which had been shown to be effective catalysts for CO_2 reduction at an Hg electrode[110]) to reduce the applied cathode potential. The catalysts showed[111] reversible cyclic voltammetric responses in acetonitrile at illuminated *p*-Si electrodes at potentials significantly more positive (ca. 0.4 V) than those required at a Pt electrode, where the *p*-Si used had surface states in high density and Fermi level pinning[112] occurred. Electrolysis of a CO_2-saturated solution (acetonitrile-H_2O-$LiClO_4$; 1:1:0.1 *M*) in the presence of 180 m*M*

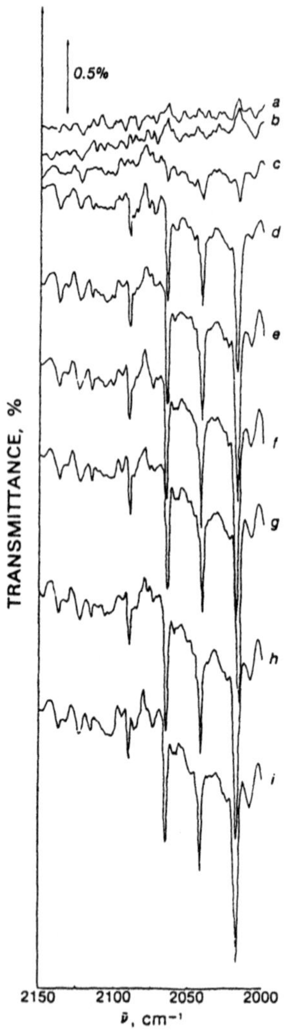

Figure 9. FTIR spectra at an illuminated *p*-CdTe electrode in CO_2-saturated acetonitrile with 0.1 *M* Bu_4NBF.[58] Spectra recorded at (a) −0.9, (b) −1.1, (c) −1.3, (d) −1.5, (e) −1.7, (f) −1.9, (g) −2.1 (h) −2.3, and (i) −2.5 V versus Ag/AgCl.

of the nickel complex of 5,5,7,12,12,14-hexamethyl-1,4,8,11-tetraazacyclotetradecane, $[(Me_6[14]aneN_4)Ni^{2+}]$, as the electron-transfer catalyst at −1.0 V versus SCE gave the best results; carbon monoxide and hydrogen were obtained in a 2:1 molar ratio with current efficiencies of 95 ± 5%. A 750-W tungsten halogen lamp was used as a light source. In a dry acetonitrile-TBAP-$[(Me_6[14]aneN_4)Ni^{2+}]$ solution, CO_2 reduction was observed at *p*-Si at −1.3 V versus SCE, and CO and CO_3^{2-} were the products. Without an electron transfer catalyst, potentials more negative than −1.9 V versus SCE were needed for CO_2 reduction at *p*-Si. The initial step of the reaction pathway was suggested to be an electron transfer from the reduced metal complex to CO_2. The $CO_2^{\cdot-}$ thus formed would then be protonated (in the case of aqueous acetonitrile solutions) and subsequently a second reduction would occur, resulting in the overall reaction represented by Eq. (19), while in dry acetonitrile Eqs. (10)–(12) were considered. Photoassisted electrochemical reduction of CO_2 on *p*-GaAs(111) at −0.95 V (versus SHE) was carried out[113] in aqueous 0.1 *M* $KClO_4$ solution (pH 4.5). Ni^{2+}-cyclam (cyclam = 1,4,8,11-tetraazacyclotetradecane) catalyzed the reduction of CO_2 to CO with a CO/H_2 ratio of ca. 2:1. Similar results were also obtained at a *p*-GaP electrode. Carbera and Abruña[114] have shown that at *p*-Si and polycrystalline thin-film *p*-WSe_2 semiconductor electrodes on which $[Re(CO)_3$(v-bpy)Cl] (v-bpy = 4-vinyl-4′-methyl-2,2′-bipyridine) was incorporated by electropolymerization, photoelectrocatalytic reduction of CO_2 took place in acetonitrile with 0.1 *M* TBAP. CO was the predominant product (>95% current yield). The onset potential for CO_2 reduction was ca. −0.8 V (*p*-Si) and −0.65 V (*p*-WSe_2) versus a sodium chloride saturated calomel electrode (SSCE), while at a Pt electrode coated similarly with the catalyst, CO_2 reduction began to occur at −1.43 V. The turnover numbers of the catalyst at the photocathodes were >450.

2. Photoassisted Reduction of Carbon Dioxide with Suspensions of Semiconductor Powders

In experiments investigating conversion of solar energy to chemical energy, systems of semiconductor suspensions have also been used. These systems have several advantages and disadvantages:

1. No external energy other than light energy is introduced, and light energy can be stored when uphill reactions take place in the solutions.

2. Since each semiconductor particle can be considered as a microphotocell, fast reaction rates can be expected because of the extremely large surface area of the semiconductor on which the reactions take place.

3. Irradiation of the system is much simpler than in the case of a photoelectrochemical cell using photoelectrodes.

4. No supporting electrolyte is required in the solutions.

5. Expensive semiconductor wafers are not required; powders of the semiconductor suffice. This is convenient because various materials which are not available as electrodes can be used.

6. Furthermore, unique reactions would be expected to occur because both oxidation and reduction sites exist close to each other (on the same particle). On the other hand, reverse reactions of the desired ones easily occur, resulting in low energy conversion efficiency.

7. Another disadvantage of using semiconductor powders is the difficulty in obtaining kinetic and thermodynamic data.

Inoue *et al.*[83] showed that CO_2 was reduced to organic compounds such as formic acid, formaldehyde, methanol, and methane in the presence of photosensitive semiconductor powders suspended in water as catalysts. The observed quantum yield of each product was ca. 10^{-4} to 10^{-3}. Powders of WO_3, TiO_2, ZnO, CdS, GaP, and SiC (99.5–99.9999% purity) were used. In each case, 1–2 g of semiconductor powder (200–400 mesh) was suspended in 100 ml of purified water in a glass cell, into which CO_2 gas was bubbled at a rate of 3 liter/min, and the solution was stirred with a magnetic bar. After 7 h of irradiation using SiC powder as a photocatalyst, CO_2 was reduced to give 1 mM formaldehyde and 5.35 mM methanol. Interestingly, the yields of methyl alcohol from photocatalytic reduction of CO_2 increased as the energy levels of conduction band of the semiconductor catalysts became more negative with respect to the redox potential of H_2CO_3/CH_3OH, while in the presence of WO_3, of which conduction band level is more positive than the redox potential of H_2CO_3/CH_3OH, no methanol was produced.

Halmann and Aurian-Blajeni[115] also examined CO_2 reduction by irradiation either with sunlight or a high-pressure Hg lamp of aqueous suspensions of various oxide semiconductors (i.e., TiO_2, Fe_2O_3, WO_3, ZnO, and nontronite, an ion-containing clay mineral).

In the case of TiO_2 (anatase), a light energy to chemical energy conversion of ca. 0.2% was reported. Moreover, in contrast to the above observation of Inoue *et al.*[83], the reduction of CO_2 was reported to occur quite well with WO_3 as a photocatalyst. In a solution of pH 2.3 with 3.9 g/liter of WO_3, with bubbling of CO_2 at a rate of 264 ml/min, irradiation with a 70-W high-pressure mercury lamp gave methanol and formaldehyde at rates of 1.17 and 0.01 μmol/h, respectively. Aurian-Blajeni *et al.*[116] compared the gas–solid process of CO_2 reduction at illuminated semiconductor surfaces with the liquid–solid reaction by illuminating an aqueous suspension of semiconductor powders through which CO_2 was bubbled. The latter reaction gave a much higher efficiency than the former, and aqueous suspensions of $SrTiO_3$ and WO_3 gave a high activity for methanol formation (5–7 μmol/h). Moreover, the authors measured the band gaps of the semiconductors tested by diffuse reflectance spectroscopy and showed a poor relationship between the conduction band levels of the semiconductors and the activity of the semiconductors for the reduction of CO_2.

The effect of a rare earth dopant, such as Eu_2O_3, Sm_2O_3, Nd_2O_3, and CeO_2, for the large-band-gap semiconductors $BaTiO_3$ and $LiNbO_3$ on the photoassisted reduction of CO_2 was examined using aqueous suspension systems.[117] An enhancement in the yield of reduction products, formic acid and formaldehyde, was reported, although optical energy conversion efficiencies were 0.01% or less. Strontium titanate powders treated with various transition metal oxides were also examined.[118] The predominant reduction product was formic acid with smaller amounts of formaldehyde and methanol formed in aqueous solutions. Among the various additives, irridium oxide was the most effective for formic acid formation (the optical to chemical energy conversion efficiency was ca. 0.02%), while for methanol formation, doping with ruthenium oxide was suitable (the highest optical energy conversion efficiency was 0.03%). The lowest doping level, 0.57 mol %, gave the best results. The efficiency of CO_2 reduction increased linearly with the quantity of light absorbed in $SrTiO_3$ doped with various additives. The photoreduction of CO_2 was also examined[119] in aqueous suspensions of TiO_2 powders which had been doped with noble metal oxides and transition elements. RuO_2-doped TiO_2 showed an increase in the rate of methanol production, but the optical to chemical energy conversion efficiency, 100 (heat of combustion of products/incident light flux), was ca. 0.04% at best, and the efficiency declined with prolonged illumination.

The photoassisted reduction of aqueous carbon dioxide in the presence of inorganic minerals has been examined as a model of prebiological photosynthesis,[120] a potential precursor to the photosynthetic fixation of CO_2 by plants.

The semiconductor suspension systems have serious problems; reduction products of CO_2, such as formic acid and methanol, are reoxidized at illuminated semiconductor powders. Also, the recombination of photogenerated holes and electrons lowers the efficiency of the reactions of holes and/or electrons with species in solution.[121,122] Henglein and Gutierrez[123] used semiconductor colloidal particles instead of powders, because the recombination of charge carriers generated in illuminated colloidal particles is relatively slow and thus molecules adsorbed at the colloidal particles are expected to react efficiently. In the presence of small colloidal particles of ZnS, formic acid was formed upon illumination of a 50 m*M* aqueous CO_2 solution containing 5 m*M* sulfite; the yield of formic acid was estimated to be 0.2 molecule per photon adsorbed. Henglein *et al.*[124] also reported that in the presence of an alcohol, especially 2-propanol, as a positive hole scavenger, formic acid was produced in ZnS solutions (2×10^{-4} *M*) containing 50 m*M* CO_2 under irradiation with a Xe lamp (Fig. 10). The quantum yield for the formation of HCOOH reached 0.4 molecule/photon, which corresponds to a quantum efficiency of 80%, when the 2-propanol concentration was 1 *M*. In the absence of alcohol, the formate produced was effectively oxidized to CO_2 at illuminated ZnS. The reaction scheme is illustrated in Fig. 11. The experimental results showed that 2-propanol reacted with a positive hole via a one-hole mechanism to give acetone (80%) and pinacol (20%), whereas two-electron reduction of CO_2 to formate took place in two ways: (1) The $CO_2^{\cdot-}$ radical formed initially was adsorbed too strogly to react with other radicals, and thus the uptake of the second electron occurred, and (2) the electrons produced in the colloidal particles combined with Zn^{2+} to form Zn atoms at the interface (Zn was detected as a product of illumination), and then the Zn atom acted as a two-electron transfer agent toward CO_2. By irradiation with an Hg arc lamp of CO_2-saturated water (25 ml) containing Pt-doped

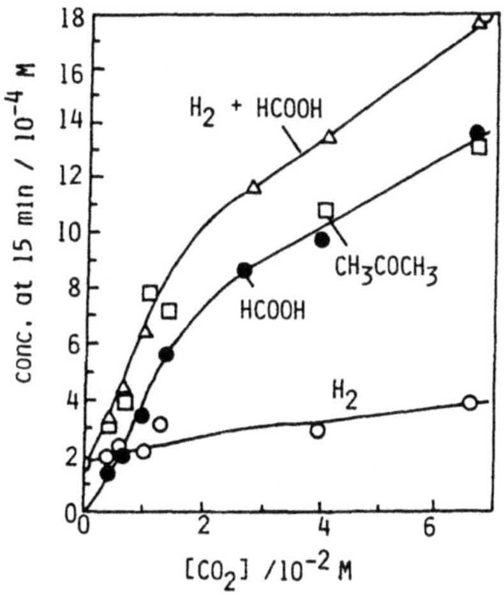

Figure 10. Concentration of various products obtained by illumination of ZnS colloidal particles in aqueous solution in the presence of 1 *M* 2-propanol as a function of CO_2 concentration.[124]

TiO_2 powder (0.1 g), CO_2 was reduced to give methane (1 μmol) and CO (0.25 μmol) after 24 h, while use of colloidal TiO_2 (Pt doped) increased by ca. 80 times the amounts of products with respect to a unit weight of TiO_2.[125]

For CO_2 reduction in powder suspension systems, the use of nonaqueous solvents has not yet been reported.

IV. CATALYSTS FOR CARBON DIOXIDE REDUCTION

Since noncatalyzed carbon dioxide reduction shows a large overpotential and potentials far more negative than −2.0 V versus SCE

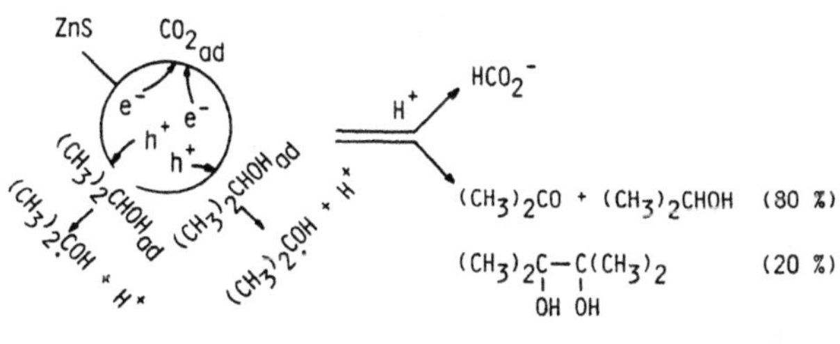

Figure 11. Illustration of the reaction at a ZnS colloidal particle in the presence of CO_2 and 2-propanol.[124] Two-electron and one-hole mechanism for CO_2 reduction and 2-propanol oxidation, respectively, are shown in the figure.

are usually required in preparative-scale electrolysis, a great deal of effort has been devoted to finding effective catalysts for this reaction. As was described in the previous section, from a thermodynamic point of view, CO_2 reduction can take place at much less negative potentials [Eqs. (15)-(20)].

1. Metal Complexes of *N*-Macrocycles

The first catalysts reported for the electroreduction of CO_2 were metallophthalocyanines (M-Pc).[126] In aqueous solutions of tetraalkylammonium salts, current-potential curves at a cobalt phthalocyanine (Co-Pc)-coated graphite electrode showed a reduction current peak whose height was proportional to the CO_2 concentration and to the square root of the potential sweep rate at a given CO_2 concentration. On electrolysis, oxalic acid and glycolic acid were detected, but formic acid was not. Mn and Pd phthalocyanines were inactive, while Cu and Fe phthalocyanines were slightly active. At the potentials used for CO_2 reduction, M-Pc catalysts would be in their dinegative state, and the occupied d_{z^2} orbital of the metal ion in the metallophthalocyanine was suggested to play an important role in the catalytic activity.

Hiratsuka *et al.*[102] used water-soluble tetrasulfonated Co and Ni phthalocyanines (M-TSP) as homogeneous catalysts for CO_2 reduction to formic acid at an amalgamated platinum electrode. The current–potential and capacitance–potential curves showed that the reduction potential of CO_2 was reduced by ca. 0.2 to 0.4 V at 1 mA/cm^2 in Clark–Lubs buffer solutions in the presence of catalysts compared to catalyst-free solutions. The authors suggested that a two-step mechanism for CO_2 reduction in which a CO_2–M-TSP complex was formed at ca. −0.8 V versus SCE, the first reduction wave of M-TSP, and then the reduction of CO_2–M-TSP took place at ca. −1.2 V versus SCE, the second reduction wave. Recently, metal phthalocyanines deposited on carbon electrodes have been used[127] for electroreduction of CO_2 in aqueous solutions. The catalytic activity of the catalysts depended on the central metal ions and the relative order $Co^{2+} > Ni^{2+} \gg Fe^{2+} = Cu^{2+} > Cr^{3+}, Sn^{2+}$ was obtained. On electrolysis at a potential between −1.2 and −1.4 V (versus SCE), formic acid was the product with a current efficiency of ca. 60% in solutions of pH greater than 5, while at lower pH values, methanol was also produced but its current efficiency was less than 5%. Also, carbon electrodes modified by adsorption of cobalt phthalocyanine, Co-Pc, have been reported[128] to be effective for CO_2 reduction in an aqueous solution of pH 5 to give CO (55-60%) with H_2 (35-30%) at a potential ca. 0.3 V more negative than the thermodynamic CO_2/CO redox potential. Oxalate and formate were also detected in the solution, as was previously reported,[102,126] but in only trace amounts, and the major product was gaseous CO. The turnover numbers of the catalyst exceeded 100 s^{-1}, and the dinegative state of Co-Pc was again suggested to be the active form.

Cobalt porphyrin derivatives were also reported[129] to be active for electrochemical reduction of CO_2 to formic acid at an amalgamated Pt electrode. More recently, Becker *et al.* have reported[130] that Ag^{2+} and Pd^{2+} metalloporphyrins acted as homogeneous catalysts for CO_2 reduction in dry CH_2Cl_2; oxalic acid and H_2 (its source was not clear) were produced, but no CO was detected.

Tetraazamacrocyclic complexes[131] of cobalt and nickel were found[110] to be effective in facilitating the reduction of CO_2 at -1.3 to -1.6 V versus SCE (Table 8). An acetonitrile-water mixture and water were used as solvents, while in dry dimethylsulfoxide no catalytic reduction of CO_2 took place. Using an Hg electrode, both CO and H_2 were produced, where total current efficiencies were greater than 90%. The turnover numbers of the catalysts were 2-9 h^{-1}. The catalytic activity lasted for more than 24 h and the turnover numbers of the catalysts exceeded 100. A protic source was required to produce both CO and H_2, and the authors suggested that both products may arise from a common intermediate, which is most likely a metal hydride. The applied potential for CO_2 reduction was further reduced by using illuminated *p*-Si in the presence of the above catalysts.[111]

Tinnemans *et al.*[132] have examined the photo(electro)chemical and electrochemical reduction of CO_2 using some tetraazamacrocyclic Co(II) and Ni(II) complexes as catalysts. CO and H_2 were the products. Pearce and Pletcher[133] have investigated the mechanism of the reduction of CO_2 in acetonitrile-water mixtures by using square planar complexes of nickel and cobalt with macrocyclic ligands in solution as catalysts. CO was the reduction product with no significant amounts of either formic or oxalic acids

Table 8
Results of Electrolysis with Various Macrocycles at Hg[a]

Compound	Electrode potential[b] (V versus SCE)	Average current efficiency[c] (%)	Products ratio[d]	Turnovers per h at 23°C[e]	Solvent system
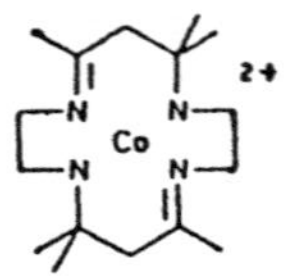 1	−1.6	93	CO/H_2, 1:1	7.8	0.1 *M* KNO_3 in H_2O/CH_3CN 2:1 (v/v) or H_2O only
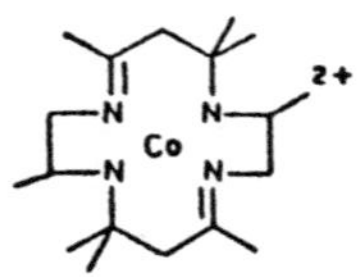 2	−1.5	90	CO/H_2, 1:1	9	0.1 *M* KNO_3 in H_2O/CH_3CN 2:1 (v/v) or H_2O only
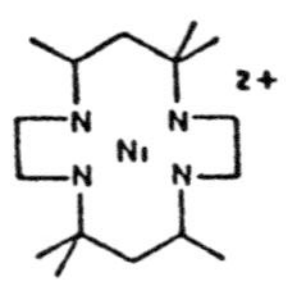 3	−1.6	98	CO/H_2, 2:1	6	0.1 *M* $LiClO_4$ in H_2O/CH_3CN 2:1 (v/v)

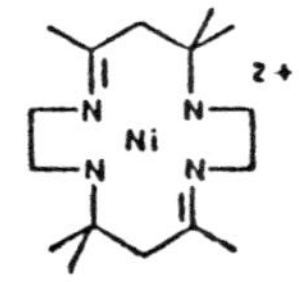 4	−1.5	*f*			
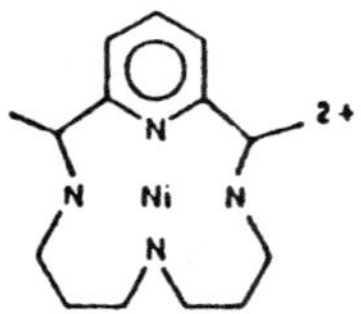 5	−1.3	44	CO	2.1	0.1 *M* KNO_3 in H_2O/CH_3CN 2:1 (v/v)

[a] From Ref. 110.

[b] All Controlled-potential electrolysis (CPE) experiments were carried out at the cathodic $E^{1/2}$ or 0.1 V more negative than the $E^{1/2}_{p,c}$ for the $M^{2+/1+}$ couple in the solvent system used.

[c] Averaged over numerous runs by using the following catalyst concentrations: compounds 1, 2, 3, and 4, 1.2 mM; compound 5, 2.5 mM.

[d] From gas chromatographic data.

[e] Turnovers per hour per mole of catalyst for runs in which the catalyst concentration was 1.4–2.4 mM. A turnover is defined as 1 equiv of electrons passed through the electrolysis cell per mole of catalyst. Since the reduction products require two electrons for their formation, these numbers correspond to twice the moles of product formed per mole of complex per hour.

[f] Although catalysis has been observed with this compound in a number of solvent systems, reliable current efficiencies and rates have not been obtained.

Table 9
Electrocatalytic Reduction of CO_2 by $[Ni^{II}(cyclam)]^{2+}$ in Water[a,b]

Run	Electrocatalyst	E (V versus N.H.E.)	Total volume[c] of CO produced (ml)	Turnover frequency[d] (h^{-1}); overall turnover of Ni	Average current efficiency[e] (%)	Volume of H_2 produced/ml; H_2:CO in gas produced
1	—	−1.05	<0.05	—	—	0.36; >10
2	$NiCl_2 \cdot 6H_2O$	−1.05	<0.05	—	—	1.6; >30
3	Ni(cyclam)Cl_2	−0.90	0.4	0.3; 1.2	36	<0.01; $<2 \times 10^{-2}$
4	Ni(cyclam)Cl_2	−0.95	3.6	2.9; 10.8	82	<0.01; $<3 \times 10^{-3}$
5	Ni(cyclam)Cl_2	−1.00	23.7	18; 77.5	99	<0.01; $<5 \times 10^{-4}$
6	Ni(cyclam)Cl_2	−1.05	35.6	32; 116	96	<0.01; $< 3 \times 10^{-4}$

[a] Ref. 135.
[b] CO_2 (99.995% purity) saturated solutions (75 ml H_2O at 25°C; pH ca. 4.1) containing the electrocatalyst (1.7×10^{-4} *M*) and KNO_3 (0.1 *M*) were placed in a gas-tight electrolysis cell; the working electrode (18 cm^2) was mercury (99.99999% purity). The total volume occupied by the gases in the electrolysis cell was 86 ml. The gases were analyzed by gas chromatography.
[c] After 4 h of electrolysis.
[d] Turnover numbers are calculated from moles of CO produced per mole of electrocatalyst.
[e] Current efficiency $\rho : \rho = (2n_{CO} \times 96\,500/C)$, where n_{CO} = moles of CO produced, C = coulombs passed during the run.

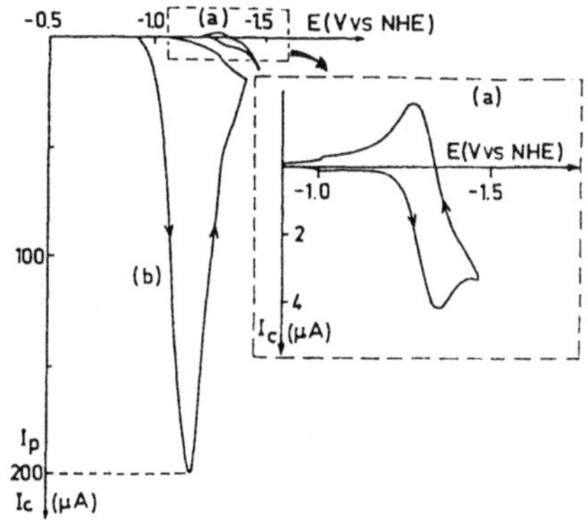

Figure 12. Current-potential curves for Ni(II)-cyclam (1 m*M*) in an aqueous 0.1 *M* $KClO_4$ solution (pH 4.5) under N_2 (a) or CO_2 (b) at a hanging mercury drop electrode.[135] Scan rate: 0.1 V/s.

formed, even in the presence of strong proton donors. Thus, the mechanism of CO_2 reduction in the presence of transition metals is quite different from that at metal electrodes; the monocation of the transition metal reacted with CO_2 to give a species which was further reduced at the electrode, where the protonation step was constrained to occur at an oxygen rather than a carbon site of the intermediate. Bailey *et al.*[134] used a modified Pt electrode with a nickel tetraazaannulene complex electropolymerized film, $(Ni[Me_4Bro_2[14]tetraeneN_4)_n$, as a catalyst. Electrolysis at −1.85 V versus SSCE in CO_2-saturated acetonitrile-0.1 *M* tetraethylammonium perchlorate (TEAP) solutions containing 2 vol % methanol gave formate ion as a product without CO or oxalate.

An extremely selective electrocatalyst for CO_2 reduction to CO in water has recently been found by Beley *et al.*[135] A rather simple Ni complex of 1,4,8,11-tetraazacyclotetradecane, [Ni(II)-cyclam],

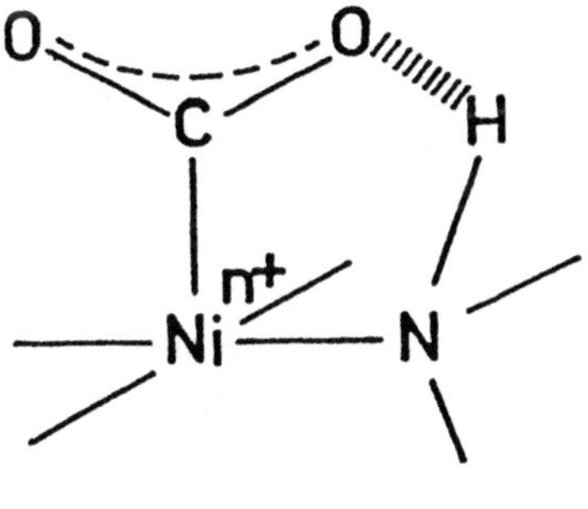

Figure 13. Postulated structure of CO_2-Ni-cyclam complex.[135]

cycles effectively more than 10^3 times with no significant deactivation to give CO selectively, even in water, by electrolysis at −1.05 V (versus NHE) at an Hg electrode (Table 9 and Fig. 12). The importance of the molecular species observed on the electrode surface was shown. Also, the size of the ligand and the presence of a secondary amine group have been suggested to be the origin of the special properties of the electrocatalyst; the former imparts high kinetic and thermodynamic stability to the Ni(II) complex, and the latter could favor CO_2 fixation by hydrogen bonding (N—H···O) in addition to the carbon-to-Ni(I) binding (Fig. 13). Selective reduction of CO_2 to CO in the presence of Ni(II)-cyclam was again observed at a Pb electrode.[136] Also, both Ni(II)-isocyclam (isocyclam = 1,4,7,11-tetraazacyclotetradecane) and cyclams with central metals other than Ni, such as Co, Zn, Cu, and Au, showed much less catalytic activity; use of Rh(III)-cyclam gave formic acid rather than CO as the main product of CO_2 reduction.[136] Unique catalytic properties of metal cyclams, such as Co- and Ni-cyclam, have also been demonstrated in electrocatalytic reduction of nitrogen oxyanions.[137]

2. Iron–Sulfur Clusters

Tetranuclear iron–sulfur clusters of the type $[Fe_4S_4(SR)_4]^{2-}$, where R = $CH_2C_6H_5$ and C_6H_5, were found[138] to catalyze the reduction of CO_2 in DMF solutions. Controlled-potential electrolyses were carried out in a CO_2-saturated 0.1 *M* tetrabutylammonium tetrafluoroborate (TBAT)–DMF solution at a mercury pool cathode. In the absence of a catalyst, CO_2 was substantially reduced only at potentials more negative than −2.4 V versus SCE, while in the presence of a cluster, the reduction took place at around −1.7 V; thus, potential shift of ca. 0.7 V was achieved. The products were analyzed by means of gas chromatography and isotachophoresis. Without a catalyst, oxalate was the main product, and addition of small amounts of water to the DMF solution favored formate production, whereas in the presence of the catalyst, formate was produced predominantly even in a dry DMF solution. This result was interpreted in terms of indirect reduction of CO_2, proceeding by electron transfer from the reduced cluster to CO_2 in the bulk solution and then protonation (tetraalkylammonium ion was considered as a hydrogen source, in part) rather than self-coupling. The catalysts were shown to undergo a two-step reduction to give the trianion and tetraanion, and the latter was suggested, from current–potential curves, to be active for CO_2 reduction. In addition, $[Fe_4S_4(SR)_4]^{2-}$ (R = $PhCH_2$ or Bu) and $[M_2Fe_6S_8(SEt)_9]^{3-}$ (M = Mo or W) have recently been reported[139] as catalysts for CO_2 reduction in DMF. The cubane structure of $[Fe_4S_4(SCH_2Ph)_4]^{2-}$ collapsed rapidly during electrolysis at −2.0 V versus SCE under CO_2. Addition of an excess amount of $PhCH_2SH$ prevented degradation of the cluster, and phenyl acetate was formed (Eq. 30) as a reduction product at faradaic efficiencies of 5–15%:

$$PhCH_2S^- + CO_2 + 2e^- \rightarrow PhCH_2COO^- + S^{2-} \qquad (30)$$

3. Re, Rh, and Ru Complexes

Recently, Hawecker *et al.*[140] have shown $Re(bpy)(CO)_3Cl$ (bpy = 2,2′-bipyridine) to be an efficient homogeneous catalyst for the electrochemical reduction of CO_2 to CO. The complex was originally found to be a catalyst for CO_2 photoreduction by the same group.[141] In a CO_2-saturated DMF–water (9:1) solution (60 ml) containing 2 mg of the catalyst and 0.1 *M* tetraethylammonium chloride (Et_4NCl), electrolysis at −1.25 V (versus NHE) on a glassy carbon electrode gave CO at 98% current efficiency (Table 10). About 300 catalytic cycles without loss of activity were observed in 14 h. A similar Re complex, $Re(vbpy)(CO)_3Cl$ (vbpy = 4-vinyl-4′-methyl-2,2′-bipyridine), also showed[142] electrocatalytic activity for CO_2 reduction to give CO in acetonitrile at a Pt electrode coated with a polymeric film containing the complex; this catalyst was later immobilized on semiconductor photocathodes.[114] No carbonate was produced at this electrode. The turnover numbers obtained were much larger than those observed with Re(bpy)-$(CO)_3Cl$ in solution.[140] More recently, a Pt electrode coated with a polypyrrolic film containing $Re(bpy)(CO)_3Cl$ was used[143] as a catalytic electrode; equal amounts of CO and CO_3^{2-} were produced, as was the case for $Re(bpy)(CO)_3Cl$ in solution.[140] The reason for the difference in the products obtained at $Re(vbpy)(CO)_3Cl$- and

Table 10
Electroreduction of CO_2 Catalyzed by $Re(bipy)(CO)_3Cl$ in $DMF–H_2O$ Solutions[a,b]

Expt.	Medium composition	Electrolysis time (h)	Volume of CO produced (ml)	Coulombs consumed	Average current efficiency (%)[c]
1[d]	$DMF-H_2O(10\%)-NBu_4ClO_4$	3	6.8	55.6	
		7	15.8	129.5	98
		14	31.6	259.9	
2	$DMF-NEt_4Cl$	1	0.96	8.3	
		2	1.7	14.9	
		3	2.4	20.9	92
		4	2.9	25.9	
		5	3.1	28.1	
3	$DMF-H_2O(5\%)-NEt_4Cl$	1	1.6	13.8	
		2	3.0	25.9	
		3	4.6	40.2	93
		4	6.0	51.0	
4	$DMF-H_2O(10\%)-NEt_4Cl$	1	2.4	20.5	
		2	4.7	40.6	
		3.5	8.2	70.9	94
		5	11.7	100.1	
5	$DMF-H_2O(10\%)-NBu_4ClO_4$	1.5	3.6	31.5	
		3	7.2	61.6	94
		5	12.0	102.7	
6[e]	$DMF-H_2O(20\%)-NEt_4Cl$	1	1.6	14.3	
		2	3.2	28.3	91
		3.5	5.5	48.6	

[a] Ref. 140.
[b] CO_2 (99.8% purity) saturated solution (60 ml) containing 25 mg of $Re(bipy)(CO)_3Cl$ (9.0×10^{-4} *M*) and 0.1 *M* supporting electrolyte was placed in a gas-tight electrolysis cell (three-necked, round-bottomed flask equipped with an oil valve); the working electrode in all the experiments was glassy carbon (ca. 10 cm^2). The connection to the working electrode was made with a Pt wire inserted through the lateral part of the flask. The total volume occupied by the gases in the electrolysis cell was 130 ml. All the solutions were electrolyzed at −1.25 V versus N.H.E. and at ca. 25°C. The gases were analyzed by gas chromatography.
[c] Averaged over the total duration of the experiment. Since the reduction product requires two electrons for its formation, 2 equivalents of electrons passed through the electrolysis cell afford 1 mol of CO.
[d] Only 7.5×10^{-5} *M* $Re(bipy)(CO)_3Cl$ was used in this run.
[e] In this experiment, traces of H_2 were detected (about 10 μl after 3.5 h). After this time, a slight, as yet uncharacterized, precipitate appeared.

$Re(bpy)(CO)_3Cl$-modified electrodes has not yet been explained. However, from the cyclic voltammograms of *fac*-$Re(bpy)(CO)_3Cl$ (Fig. 14) and from the intermediate complexes formed by electrolysis in acetonitrile in the presence and absence of CO_2, two different electrocatalytic pathways (Fig. 15) were suggested[144]: initial one-electron reduction of the catalyst at ca. −1.5 V versus SCE followed by the reduction of CO_2 to give CO and CO_3^{2-}, and initial two-electron reduction of the catalyst at ca. −1.8 V to give CO with no CO_3^{2-}. The electrochemistry of $[Re(CO)_3(dmbpy)Cl]$ (dmbpy = 4,4′-dimethyl-2,2′-bipyridine) was investigated[145] to obtain mechanistic information on CO_2 reduction, and the catalytic reac-

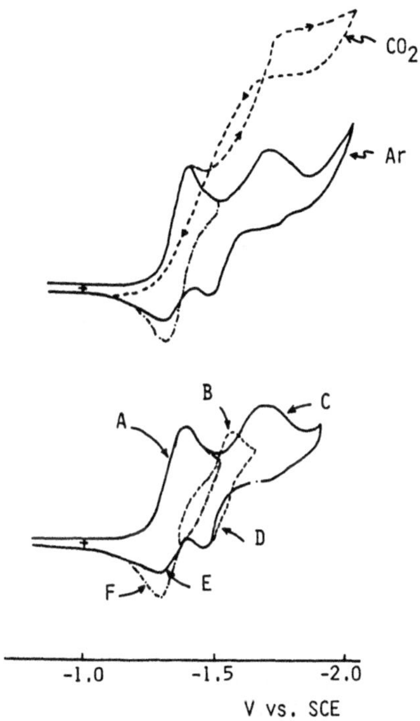

Figure 14. Cyclic voltammograms of *fac*-$Re(bpy)(CO)_3Cl$ in acetonitrile-0.1 *M* Bu_4NPF_6 at a Pt electrode.[144] Scan rate: 0.2 V/s. The lower voltammograms show the switching potential characteristics: A and F, reversible one-electron wave; B and D, redox couple due to a dimer of the complex; C, the second metal-based wave. The upper curves show the effect of CO_2 on the voltammogram. See also Figure 15.

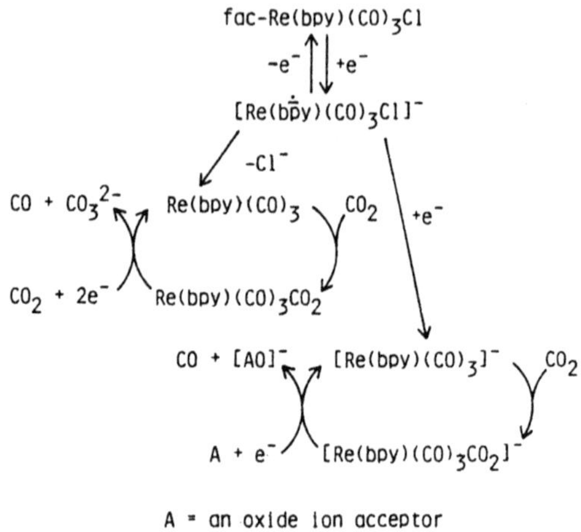

Figure 15. Postulated reaction pathways for CO_2 reduction in the presence of *fac*-Re(bpy)$(CO)_3Cl$.[144]

tion of such Re complexes was suggested to involve a monocoordinated bipyridine intermediate.

Also, it would be worthwhile to investigate catalysts developed for CO_2 reduction in other fields with regard to their possible application to electrochemical and photoelectrochemical reduction of CO_2, and vice versa; in fact, catalysts developed for a particular system have been applied successfully in various related systems as described above.

Rhodium and ruthenium complexes have also been studied as effective catalysts. Rh(diphos)$_2$Cl [diphos = 1,2-bis(diphenylphosphino)ethane] catalyzed the electroreduction of CO_2 in acetonitrile solution.[146] Formate was produced at current efficiencies of ca. 20–40% in dry acetonitrile at ca. −1.5 V (versus Ag wire). It was suggested that acetonitrile itself was the source of the hydrogen atom and that formation of the hydride HRh(diphos)$_2$ as an active intermediate was involved. Rh(bpy)$_3$Cl$_3$, which had been used as a catalyst for the two-electron reduction of NAD^+ (nicotinamide adenine dinucleotide) to NADH by Wienkamp and Steckhan,[147] has also acted as a catalyst for CO_2 reduction in aqueous solutions (0.1 *M* TEAP) at −1.1 V versus SCE using Hg, Pb, In, graphite, and *n*-TiO_2 electrodes.[148] Formate was the main

Table 11
Reduction of CO_2 at Various Electrodes in H_2O–0.1 *M* Et_4NClO_4[a]

Run	Cathode	*E* (V versus SCE)	Rh complex[b] (mM)	*Q* (*C*)	HCOOH (μmol)	Current efficiency (%)	Volume of gas produced (ml)
1	Hg	−1.1	1.0	100	401	77.4	20.0 (H_2)
2	Hg	−1.1	0	180	7	1.9	90.6 (H_2)
3	Pb	−1.1	1.0	100	326	62.9	40.0 (H_2)
4	Pb	−1.8	1.0	100	94	18.1	86.2 (H_2)
5	In	−1.1	1.0	300	870	56.0	
6	In	−1.1	0	150	70	9.0	
7	In	−1.8	1.0	100	101	19.4	72.0 (H_2), 2.5 (CO)
8	GC	−1.1	1.0	300	848	54.5	
9	GC	−1.1	0	7	—	—	
10	*n*-TiO_2	−1.1	1.0	300	905	58.3	
11	*n*-TiO_2	−1.1	0	600	10	0.4	103 (H_2)
12	*n*-TiO_2	−1.8	1.0	300	288	18.5	

[a] Ref. 148.
[b] Rh complex 1 mM = 10 μmol in 10 ml electrolyte; Rh complex = $Rh(bpy)_3Cl_3$. Possible reaction pathways are: $Rh(bpy)_3^{3+} + 2e^- \rightarrow Rh(bpy)_2^+ + bpy$; $Rh(bpy)_2^+ + H_3O^+ \rightarrow [(H^-)Rh^{3+}(bpy)_2(H_2O)]^{2+}$; $[(H^-)Rh^{3+}(bpy)_2(H_2O)]^{2+} + CO_2 \rightarrow [Rh^{3+}(bpy)_2(HCOO^-)(H_2O)]^{2+} \rightarrow Rh(bpy)_2^{3+} + H_2O + HCOO^-$.

product at current efficiencies of ca. 50–80% (Table 11). Bolinger *et al.*[149] have shown that $[Rh(bpy)_2(O_3SCF_3)_2]^+$ and [Ru(trpy)-(dppene)Cl]$^+$ [trpy = 2,2′,2″-terpyridine; dppene = *cis*-1,2-bis(diphenylphosphino)ethylene] work as catalysts in acetonitrile to produce formate and CO, respectively. The following reactions were suggested for the Rh complex:

$$Bu_4N^+ + CO_2 + 2e^- \rightarrow HCO_2^- + Bu_3N + CH_2{=}CHEt \quad (31)$$

and

$$2Bu_4N^+ + 2e^- \rightarrow H_2 + 2Bu_3N + 2CH_2{=}CHEt \quad (32)$$

while for the Ru complex:

$$Bu_4N^+ + 2CO_2 + 2e^- \rightarrow Bu_3N + CH_2{=}CHEt + CO + HCO_3^- \quad (33)$$

indicating that the reduction product depended on the nature of the catalyst.

Tanaka *et al.*[150] reported that in the presence of $[Ru(bpy)_2CO_2](PF_6)_2$, controlled-potential electrolysis of a CO_2-saturated H_2O/DMF (9/1 v/v) solution at −1.5 V versus SCE using an Hg pool electrode gave CO and H_2 in acidic conditions, and formic acid and CO as well as H_2 in alkaline conditions.

From the results described above, CO_2 seems to be reduced to CO in the presence of Re complexes in most cases, while RH complexes give formic acid and Ru complexes give both formic acid and CO, depending on the conditions.

4. Other Catalysts

The electrochemical reduction of aqueous bicarbonate to formic acid,

$$HCO_3^- + 2H^+ + 2e^- \rightarrow HCO_2^- + H_2O \quad (34)$$

using Pd-impregnated polymer-modified electrodes,[151] proceeds at potentials within 80 mV of the thermodynamic one for HCO_3^-/HCO_2^-, −0.76 V versus SCE; the supported Pd catalyst itself was previously used[152] for reduction of HCO_3^- with H_2 to form HCO_2^-, and later a bright Pd cathode was also used.[75]

Ogura *et al.*[153] reduced CO_2 to methanol using the so-called Everitt's salt ($K_2Fe^{2+}[Fe^{2+}(CN)_6]$)-modified electrode by a somewhat complicated but interesting route in the presence of a metal complex, such as Fe(II), Co(II), and Ni(II) complexes of 1-nitroso-2-naphthyl-3,6-disulfonic acid, and additional methanol:

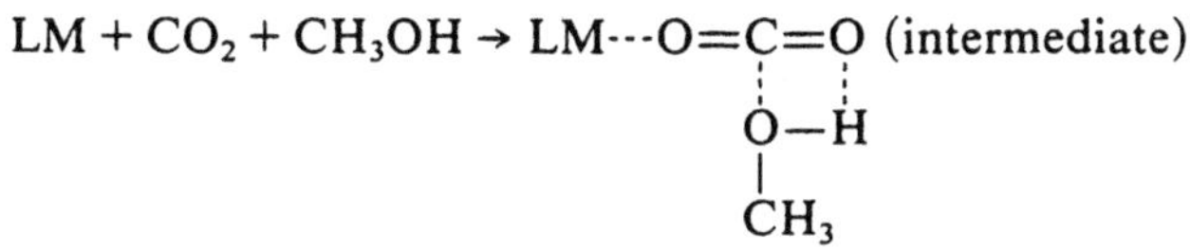

(35)

$$\text{Intermediate} + 6H^+ + 6e^- \rightarrow LM + 2CH_3OH + H_2O \qquad (36)$$

where LM is the metal complex and the electrons are supplied by the Everitt's salt (ES); ES is oxidized to become the so-called Prussian blue (PB, $KFe^{3+}[Fe^{2+}(CN)_6]$), which is again reduced to ES electrochemically. Thus, the overall reaction is given as:

$$CO_2 + 6ES + 6H^+ \rightarrow CH_3OH + 6PB + 6K^+ + H_2O \qquad (37)$$

$$\text{(PB} \xrightarrow{6e^-} \text{ES)}$$

Electrocatalytic reduction of carbon dioxide to C_1–C_3 hydrocarbons with less than 0.2% electrochemical yield was reported[154] at pH 7 in the presence of pyrocatechol, $TiCl_3$, and Na_2MoO_4 at −1.55 V versus SCE.

As described above, many reports published to date indicate that metal complexes are promising catalysts for CO_2 fixation. The catalytic activity is considered basically to be due to a CO_2–catalyst complex formation. Thus, the complexes have to provide a binding site for CO_2, and this can be realized for some catalysts by losing a ligand on reduction of the catalyst at the electrode. Also, the CO_2 molecule is not linear but is rather a bent structure[155,156] in the activated state of the CO_2–catalyst complexes. Theoretical calculations of CO_2–catalyst bonding[157] and general ideas about activation of CO_2 by metal complexes have been summarized in several recent articles.[158,159]

In addition, catalysts for CO_2 reduction based on nonmetallic compounds have also been reported. Taniguchi *et al.*[100] reported that crown ethers such as 15-crown-5 in lithium carbonate solutions enhanced CO_2 reduction, although in this case, the crown ether was proposed to facilitate the electrodeposition of Li metal on the electrode, which catalyzed the reduction of CO_2. A polyaniline-coated *p*-Si photocathode was found[99] to be active for CO_2 reduction in a $LiClO_4$ solution. It would also be useful to incorporate catalysts in a polymer matrix in high density, and modification of electrodes with electropolymerized films in which catalysts are incorporated has been widely employed. Recently, ammonium ion was suggested[160] to mediate the photoassisted reduction of CO_2 to CO at a *p*-CdTe electrode in DMF with small amounts of water. The photocurrent-potential curve shifted toward less negative potentials when the supporting electrolyte was changed from TBAP to NH_4ClO_4, accompanied by a change in Tafel slopes. Under a CO_2 atmosphere, CO was formed by photoassisted electrolysis at a *p*-CdTe electrode, but NH_3 was the product under an Ar atmosphere.

The electron transfer between an electrochemically produced perylene dianion and a CO_2 molecule was also suggested by cyclic voltammetry in a DMF solution.[161] Later, perylene was used[162] in the photochemical fixation of CO_2, as a nonmetal electron carrier to CO_2.

Furthermore, a biological catalyst [formate dehydrogenase (FDH)] combined with an illuminated *p*-InP photocathode was

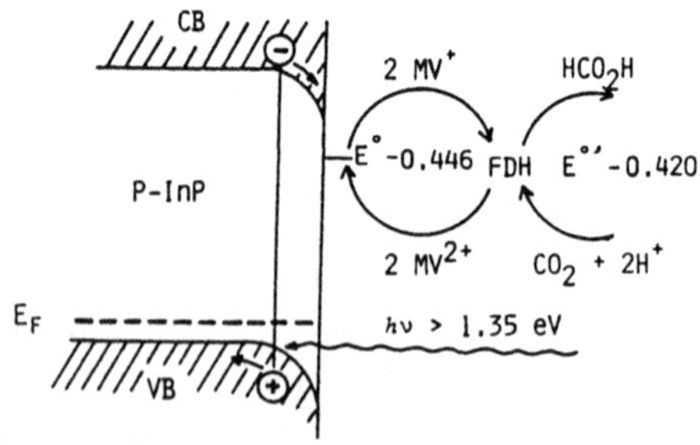

Figure 16. Scheme for the photoelectrochemical reduction of CO_2 at *p*-InP with formate dehydrogenase (FDH) as the catalyst and methyl viologen (MV^{2+}) as the electron transfer mediator.[163]

effective in the reduction of CO_2 to give formic acid.[163] Photogenerated electrons in *p*-InP reduced the enzyme through methyl viologen as a mediator, and then reduction of CO_2 took place (Fig. 16), when photoassisted electrolysis was carried out at 0.05 V (versus NHE). The turnover numbers of the enzyme exceeded 2×10^4, but loss of enzyme activity due to denaturation of the protein occurred.

V. MISCELLANEOUS STUDIES

To establish an effective system for CO_2 reduction, various approaches have to be considered. In this section, miscellaneous studies of CO_2 fixation, other than those involving the usual electrochemical and photoelectrochemical reduction of CO_2, are briefly reviewed.

1. Photochemical Reduction of Carbon Dioxide

Photochemical fixation of carbon dioxide is a function of green plants and some bacteria in nature in the form of photosynthesis. All living organisms on the Earth are indebted directly or indirectly to photosynthesis. Thus, many attempts have been made to simulate the photosynthetic system and make artificial systems, although to date very little success has been achieved.

Tazuke and Kitamura[162] reported the first example of an artificial photosynthetic system based on electron transport sensitization, although the product was not a hydrocarbon, but rather formic acid. Their system is shown schematically in Fig. 17. In this system, the photochemically generated singlet excited state of an aromatic hydrocarbon, such as pyren (Py) or perylene (Pe), was

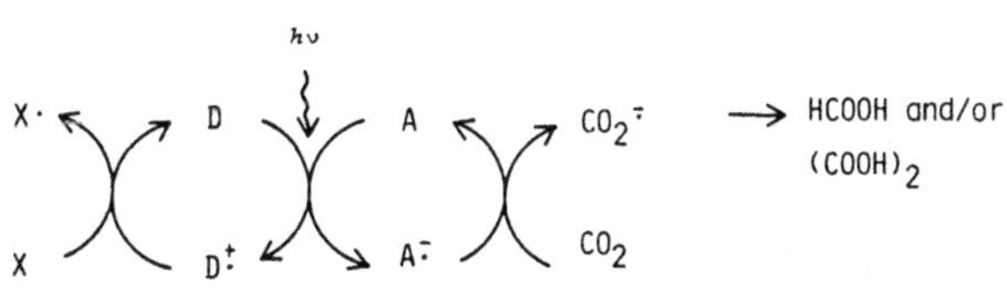

Figure 17. Schematic representation of an artificial photosynthetic system.[162]

the electron donor (D), and 1,4-dicyanobenzene or 9,10-dicyanoanthracene was used as an electron-acceptor (A). The electrons accepted are successively transferred to CO_2 molecules. The species denoted as X in Fig. 17 was not clearly identified but possible candidates considered were OH^-, HCO_3^-, and/or HCOOH. Using this system in an aqueous acetonitrile solution, ca. 1 mmol of formic acid was formed under irradiation with a 300-W high-pressure Hg lamp, with a poor quantum yield ($<10^{-4}$). The main overall reaction was given as

$$CO_2 + 2H_2O \xrightarrow{h\nu} HCOOH + H_2O_2 \tag{38}$$

Later, an improved system for CO_2 photofixation was reported by the same authors.[164] The new system consisted of 6.5×10^{-5} *M* tris(2,2′-bipyridine)ruthenium(II), $Ru(bpy)_3$, as the photosensitive electron donor, methyl viologen (MV^{2+}, 20 m*M*) as the electron acceptor, and triethanolamine (TEOA, 0.6 *M*) as a sacrificial electron donor in a CO_2-saturated aqueous solution (Fig. 18). Under irradiation with a 300-W high-pressure Hg lamp with a $CuSO_4$ chemical filter ($\lambda > 320$ nm), formic acid, which was detected by isotachophoresis, was produced in quantum yields of ca. 0.01%. Recently, however, Kase *et al.*[165] have repeated this experiment using a $^{13}CO_2$ tracer and have claimed that the formic acid obtained was produced not by CO_2 reduction but rather by oxidative cleavage of TEOA.

Lehn and Ziessel[166] have also developed systems for the photochemical reduction of CO_2. These systems are similar to those represented by Fig. 18. Visible-light irradiation of CO_2-saturated aqueous acetonitrile solutions containing $Ru(bpy)_3^{2+}$ as a photosensitizer, cobalt(II) chloride as an electron acceptor, and triethylamine as a sacrificial electron donor gave carbon monoxide and

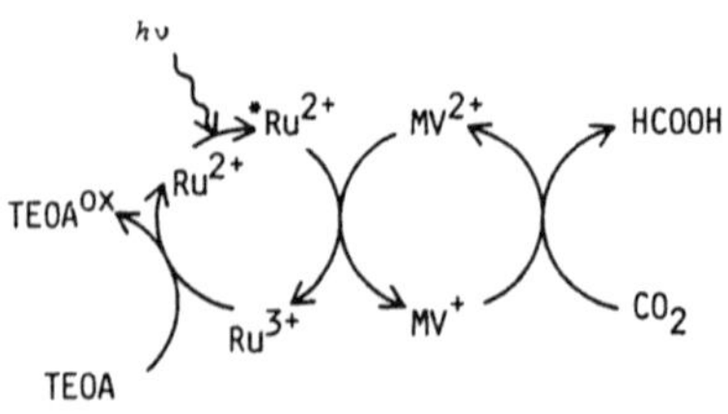

Figure 18. Scheme of an example of an improved photosynthetic system.[164] For other combinations of photosensitizers and electron acceptors, see text.

hydrogen simultaneously. Addition of free bipyridine to the solution decreased CO generation but increased H_2 evolution. When triethylamine was replaced by other NR_3 compounds, the quantity of gas ($CO + H_2$) produced and the CO/H_2 ratio increased in the order $R = Me < Et < Pr$. When triethanolamine was used instead of triethylamine, CO was selectively produced from CO_2 in high yield; after irradiation with a 1-kW Xe lamp with a 400-nm cutoff filter for 22 h, 2.93 ml of CO and 0.12 ml of H_2 were obtained (the quantum yields were not reported). In this system, cobalt ion was an efficient and specific electron mediator for CO_2, and the marked effects of the tertiary amine and of bipyridine were explained in terms of the differences in their coordination to the cobalt ion, which would influence the reaction process. Since $Co(bpy)_3^+$ is known[167] to react with bicarbonate to give insoluble $[Co(bpy)(CO)_2]_2$, which decomposes to liberate CO and H_2 by acidification to $pH < 1$:

$$[Co(bpy)(CO)_2]_2 + 6H^+ \rightarrow 4CO + 2H_2 + 2Co^{2+} + 2bpyH^+ \quad (39)$$

and Co(I) can be photochemically generated using a ruthenium(II) polypyridine complex as a sensitizer, the product (CO) of the photochemical reduction of CO_2 seems to be due to the mediator used. Using the $Rh(bpy)_3^{3+}$ complex, a similar photochemical system for the generation of H_2 by reduction of water can be made,[168] and by introduction of CO_2 into this system, products other than CO may be obtained. Later, $Ru(bpy)_3Cl_2$–TEOA–CO_2 in DMF was shown[169] to represent a catalytic system for the photoreduction of CO_2 to formic acid. No additional electron acceptor, such as Co(II) ion, was used. DMF was a better solvent than acetonitrile. $^{13}CO_2$ was used to verify that the reduction of CO_2 took place, and the following steps were suggested for the reduction process: photogeneration of $Ru(bpy)_3^+$, ligand photolabilization, hydride formation, insertion of CO_2, and release of the formate.

Hawecker *et al.*[141] used $Re(bpy)(CO)_3X$ (X = Cl, Br) complexes as photosensitizers and succeeded in improving markedly the efficiency of CO formation using a system similar to that described above, where DMF was used as a solvent and 2,9-dimethyl-1,10-phenanthroline was added, as a ligand for the cobalt ion, to a solution containing $Ru(bpy)_3^{2+}$, Co^{2+}, and triethanolamine.

This system produced 8 ml of CO and 19 ml of H_2 after irradiation with a 1-kW Xe lamp for 15 h. Using $^{13}CO_2$, the CO produced was verified to come from CO_2. Furthermore, Re(L) $(CO)_3X$ (L = 2,2'-bipyridine or 1,10-phenantroline; X = Cl, Br) complexes were found to act as both photosensitizers and catalysts for CO_2 reduction to CO. Metal carbonyls were considered to be effective because of their well-known activity as catalysts for the water–gas shift reaction.[170] Addition of Cl^- or Br^- stabilized the corresponding $Re(bpy)(CO)_3X$ (X = Cl, Br) complexes and resulted in more efficient production of CO (see runs 2 and 4, 3 and 5, and 8, 9, and 10 in Table 12). The Re complexes were also useful as homogeneous catalysts in the electrochemical reduction of CO_2 at a glassy carbon electrode in aqueous DMF.[140] More recently, an improvement of this system for the photochemical reduction of CO_2 by visible light was examined using $Ru(bpy)_3^{2+}$ and $Re(CO)_3(bpy)Cl$ as cocatalysts,[171] but, unfortunately, the stability of the system was poor for long-term usage.

The effects of transition metals on the photochemical reduction of CO_2 to formaldehyde (0.1%), formaldehyde to methanol (6–8%), and methanol to methane (ca. 10^{-5}%) were examined[172] in aqueous solutions, but the yields were very low as shown in parentheses for each reaction.

As a model of photosynthesis in green plants, platinized chlorophyll *a* dihydrate polycrystals were used.[173] Illumination of Pt-chlorophyll in the presence of CO_2 and water gave formic acid by the reaction

$$2CO_2 + 2H_2O \rightarrow 2HCOOH + O_2 \tag{40}$$

where the products were determined mass spectrometrically. Unfortunately, the photoactivity of the chlorophyll decreased as O_2 was produced due to poisoning.

From the viewpoint of a model of prebiotic chemical evolution and of the primitive atmosphere of the Earth,[174,175] photosynthetic reactions of CO_2 were also examined, and formaldehyde with various nitrogen-containing products was obtained.

For other reports dealing with photochemical fixation of CO_2, Halmann's review[4] is helpful (see also the references cited therein).

This is the present state in the development of chemical systems for artificial photosynthesis. For solar energy conversion and

Table 12

Generation of CO by Photoreduction of CO_2 via Visible Light Irradiation of Solutions Containing $Re(L)(CO)_3X$ and CO_2 in $(HOCH_2CH_2)_3N$–DMF[a,b]

Expt.	Complex	Additive[c]	Irradiation time (h)	Vol. of CO produced (ml)	Turnover number[d]
1	$Re(bipy)(CO)_3Cl$	0	1	6.5	11
2	$Re(bipy)(CO)_3Cl$	0	2	9.7	16
3	$Re(bipy)(CO)_3Cl$	0	4	16.8	27
4	$Re(bipy)(CO)_3Cl$	NEt_4Cl	2	14.5	23
5	$Re(bipy)(CO)_3Cl$	NEt_4Cl	4	30.0	48
6	$Re(bipy)(CO)_3Cl$	NEt_4ClO_4	2	6.4	10
7	$Re(bipy)(CO)_3Cl$	NEt_4Cl[g]	2	14.0	22
8	$Re(bipy)(CO)_3Br$	0	2	7.6	14
9	$Re(bipy)(CO)_3Br$	0	4	11.4	20
10	$Re(bipy)(CO)_3Br$	NBu_4Br	2	12.0	21
11	$Re(bipy)(CO)_3Br$	NBu_4Br	4	16.0	28
12	$Re(Br\text{-}phen)(CO)_3Br$	0	2	2.7	5
13	$Re(Br\text{-}phen)(CO)_3Br$	NBu_4Br	2	3.7	9
14[e]	$Re(bipy)(CO)_3Br$	0	3.5	0.08	—
15[f]	$Re(bipy)(CO)_3Br$	0	6	0.04	—

[a] Ref. 141.

[b] $Re(bipy)(CO)_3$ Cl, 8.7×10^{-4} *M*; $Re(bipy)(CO)_3Br$, 7.9×10^{-4} *M*; $Re(Br\text{-}phen)(CO)_3Br$, 6.6×10^{-4} *M*; Br-phen = 5-bromo-1,10-phenanthroline. 30 ml of solution containing $Re(L)(CO)_3X$ and 160 ml CO_2 (99.8% purity) dissolved in dimethylformamide-$(HOCH_2CH_2)_3N$(5:1) were irradiated with a 250-W halogen lamp (slide projector) fitted with a 400-nm cutoff filter (Schott GG 420).

[c] NEt_4Cl, 2×10^{-2} *M*; NBu_4Br, 10^{-2} *M*.

[d] Obtained by dividing the number of moles of CO produced by the number of moles of $ReL(CO)_3X$.

[e] Experiment carried out without CO_2; formal pH of the solution adjusted to 9.5; 1.1 ml of H_2 generated.

[f] Same conditions as in experiment 14 but adjusted to 'pH' 8.5; 1.3 ml of H_2 generated.

[g] And 25 equiv. of bipy.

storage as well as improvements in efficiency and in product selectivity, further efforts are needed.

2. Reduction of Carbon Monoxide

Since carbon monoxide has recently been found as a main product of CO_2 reduction in many systems, it would be important to convert CO into further reduced products such as methanol; this is feasible because CO is much more reactive than CO_2 and is thus one of the starting materials of C_1 chemistry.[176]

Uribe *et al.*[177] examined the reduction of CO in liquid NH_3–0.1 *M* KI at −50°C, using various working electrodes such as Pt, Ni, C, and Hg. The reaction of CO with electrogenerated solvated electrons produced dimeric species, which precipitated as $K_2C_2O_2$. Electrochemical reduction of CO in an aqueous solution at porous gas-diffusion and wet-proof electrodes of Co, Ni, and Fe was carried out,[178] and C_1 to C_3 hydrocarbons and ethylene were reported to be the products.

CO conversion has been investigated for methanol synthesis. The Fischer–Tropsch reaction[179] proceeds over catalysts at a synthesis gas pressure of near 300 atm and a temperature of near 200°C. Recently, Ogura and Yamasaki[180] have reported that CO was electrochemically reduced selectively to methanol at room temperature at atmospheric pressure, using Everritt's salt (ES) in the presence of a pentacyano iron(II) complex and methanol, as in the case of CO_2 reduction.[153] The catalytic reaction of CO with H_2 is a well-known thermal gas phase reaction.[181–186] Photoelectrochemical reduction of CO has recently been carried out[187] at *p*-Si in aqueous solution to give formaldehyde with a current efficiency of less than 5%. An Fe-porphyrin-deposited *p*-Si electrode sometimes showed a positive effect for CO reduction, while CO reduction was suppressed in the presence of Li^+ ions. Yoneyama *et al.*[188] have shown that at *p*-GaP photocathodes coated with heat-treated Fe(II)-TPP (TPP = tetraphenylporphyrin), CO was reduced to methanol with a current efficiency of ca. 10% in 0.5 *M* H_2SO_4 at a constant current density of 0.5 mA/cm^2. Using a Cu electrode, CO was reduced to methanol[189] at −1.4 V and 25°C in an aqueous 0.1 *M* Et_4NClO_4 solution with a current efficiency of ca. 20% when less than 20 C were passed. Also, in the presence of

1 mM Ni-cyclam, a catalyst[135] for CO_2 reduction to CO, methanol was obtained[189] by electrolysis of an aqueous CO_2-saturated 0.1 *M* Et_4NClO_4 solution at −1.5 V versus SCE using a Cu electrode, although the current efficiency was rather low (ca. 5%); the main reduction product of CO_2 was CO.

3. Thermal Reactions for Carbon Dioxide Reduction

Hydrogenation of CO and CO_2 with H_2 to give organic fuels in gas phase thermal reactions is also an attractive subject, and many catalysts have been developed and used.[181–186,190–194] Basic concepts developed in this field would be applicable to electrochemical and photoelectrochemical CO_2 fixation. For example, as has already been tried,[82,84–87] some noble metals or their oxides can be doped into semiconductor electrodes. For hydrogenation reactions of CO_2 and CO, transition metal catalysts have commonly been used. Hydrogenation of CO_2 to methanol was reported to occur effectively over rhenium catalysts such as Re–ZrO_2 and Re–Nb_2O_5, by gas phase thermal reactions under moderate conditions (10 atm and 160–220°C) with a selectivity for methanol formation of 50–70%.[193] Recently, effective Rh-based catalysts for acetaldehyde formation from CO and H_2 have been reported.[194] Hydrogenation of CO_2 in aqueous solution with a Rh hydride complex was also reported.[195] Also, knowledge[196–201] about the adsorption behavior of CO_2 on catalytic metals (such as Rh and Pt) on various metal oxide supports would be useful to design effective catalysts.

4. Carbon Dioxide Fixation Using Reactions with Other Compounds

Electrochemical carboxylations of organic molecules such as olefins,[202] aromatic hydrocarbons,[203] and alkyl halides[204–206] in the presence of CO_2 have been examined, as one of the subjects of organic electrochemistry.[207]

Attempts have been made to fix CO_2 using organometallic complexes in photochemical reactions. A reversible binding of CO_2 was achieved with the Cu(I) phenylacetylide-phosphine complex,[208] which acted as a reversible CO_2 carrier at ambient temperatures and atmospheric pressure, by CO_2 insertion into the

Cu—C bond of the complex. Recently, activation of CO_2 as an η^1-C metalocarboxylate[209] and photoinduced or thermal insertion of CO_2 into a metal-hydride bond[210] have been reported. Furthermore, some metal (Zn^{2+}, Ni^{2+}, and Cd^{2+}) complexes of tetraazacycloalkanes have been found to take up CO_2 easily in basic alcoholic solutions, and their structures have also been examined.[211] More recently facile insertion of CO_2 into $Rh_2(\mu\text{-}OH)_2$ to yield a carbonate complex of a rather complicated structure has been reported.[212]

VI. SUMMARY AND FUTURE PERSPECTIVES

In this article, recent developments, up to late 1986, on carbon dioxide reduction have been reviewed. These can be summarized as follows:

1. Reduction of carbon dioxide takes place at various metal electrodes. The main products are formic acid in aqueous solutions and oxalate, CO, and formic acid in nonaqueous solutions. An indium electrode is the most potential saving for CO_2 reduction. Due to the difference in optimum conditions between those for CO_2 reduction to formic acid and those for formic acid reduction to further reduced products, direct reduction of CO_2 in aqueous solutions without a catalyst to highly reduced products seems to be difficult at metal electrodes. However, catalytic effects of metal electrodes themselves have recently become more clear; for example, on Cu, methane was detected, while on Ag and Au, CO was produced effectively in aqueous solutions. Furthermore, at a Mo electrode, methanol was obtained. The power efficiency is, however, still low at any electrode.

2. The reaction mechanism of CO_2 reduction is still a subject of discussion, although, in general, the mechanisms proposed by Eyring and co-workers[45] and Amatore and Saveant[53] have proved acceptable for aqueous and nonaqueous solutions, respectively. *In situ* spectroscopic measurement techniques, by which intermediates and their adsorption behavior can be estimated, will become more and more important in better understanding each elementary step of the reaction pathway.

3. If multielectron transfer takes place, the potentials required thermodynamically for CO_2 reduction are much less negative than

the potential for single-electron transfer to CO_2, $CO_2/CO_2^{\overline{\cdot}}$. Some effective catalysts for this purpose have been extensively examined.

4. Semiconductor electrodes are promising for CO_2 reduction to highly reduced products, such as methanol and methane, because the band structures of semiconductors give rise to a pool of electrons (in the conduction band), which may facilitate multielectron transfer. In spite of complications resulting from various conflicting reports, a great deal of progress in the investigation of CO_2 reduction on semiconductor electrodes has been made. Unfortunately, however, the usable currents reported are usually very low, even at *n*-type semiconductors.

5. The photoelectrochemical reduction of CO_2 at illuminated *p*-type semiconductor electrodes is also effective for CO_2 reduction to highly reduced products. The combination of photocathodes with catalysts for CO_2 reduction leads to a marked decrease in the apparent overpotential. At present, however, light to chemical energy conversion efficiencies are still very low, and negative in some cases.

6. The photoassisted reduction of CO_2 with suspended semiconductor powders gives, at present, very low energy efficiencies (at most, ca. 0.01% or less). The use of colloidal semiconductor particles is more efficient in some cases.

7. Various kinds and types of catalysts for CO_2 reduction have been developed. Most of them are metal-based complexes. Metal complexes of N-macrocyclic compounds are promising. Among them, Ni-cyclam is an effective catalyst for selective CO production from CO_2, even in water. Re, Ru, and Rh complexes have been effectively used in recent years. Also, iron–sulfur clusters are interesting as catalysts. Nonmetal catalysts also seem to be possible. The use of polymer-modified electrodes is one of the most interesting aspects because polymers may have catalytic functions and also catalysts of interest can be incorporated in high density into the polymer matrix. Fortunately, immobilization of the catalysts on electrodes sometimes stabilizes their activities, i.e., turnover numbers increase remarkably. (The review by Murray,[213] for example, may be consulted to understand the general features of polymer-modified electrodes.) Although only a few studies focusing on catalysts for electrochemical and/or photoelectrochemical reduction of CO have been reported, the catalytic reduction of CO would

also lead us to develop multifunctional catalysts with which the direct reduction of CO_2 to highly reduced products can be achieved.

8. Considerable progress has been made on CO_2 fixation in photochemical reduction. The use of Re complexes as photosensitizers gave the best results; the reduction product was CO or HCOOH. The catalysts developed in this field are applicable to both the electrochemical and photoelectrochemical reduction of CO_2. Basic concepts developed in the gas phase reduction of CO_2 with H_2 can also be used. Furthermore, electrochemical carboxylation of organic molecules such as olefins, aromatic hydrocarbons, and alkyl halides in the presence of CO_2 is also an attractive research subject. Photoinduced and thermal insertion of CO_2 using organometallic complexes has also been extensively examined in recent years.

9. The following points may be made regarding future potential advances in CO_2 fixation:

(i) Precise analysis of CO_2 reduction products would be helpful to understand better CO_2 reduction pathways. Products both in the gas phase and in solution should also be taken into consideration. A gas-tight cell is useful.

(ii) To eliminate confusion, much attention should be paid to blank experiments and also to CO_2 purification. Photodecomposition of electrolytes as well as cell materials, such as epoxy resins, should be carefully monitored.

(iii) The use of organic solvents is worthwhile because of the high solubility of CO_2 in these solvents, which has an effect similar to that of high CO_2 pressure in aqueous solutions. In fact, much recent successful work has been done by using organic solvents and their mixture with water.

10. As mentioned in the text, it is clear that CO_2 reduction to valuable fuels and/or raw chemicals is still at an early stage and, unfortunately, far from having practical application, especially in the case of solar to storable chemical energy conversion by photoelectrochemical means. Many difficulties remain to be overcome. However, it is also true that there has been much progress in CO_2 reduction in recent years.

It is difficult to comment about the possibility that CO_2 reduction by various means can be a practical process of use in the future, because of the many difficulties to be overcome, especially the low energy conversion efficiency. In fact, work in this area has not yet reached the stage for such a discussion. The maximum efficiency of solar energy conversion at present seems to be, at most, 1% or less, which is far less than that of green plants in nature, and no theoretical limit of energy conversion efficiency for CO_2 reduction using an ideal system has been established yet. For the time being, the electrochemical reduction of CO_2 by using solar cells seems to be more efficient from the particular viewpoint of solar energy utilization. However, this is another subject. At present, more fundamental studies are required to design sophisticated systems for artificial photosynthesis in the future. Finally, it should also be noted that when CO_2 reduction proceeds successfully, it will be sure to have huge benefits. Undoubtedly, a number of new breakthroughs will be required to establish a practical system for artificial photosynthesis and for fuel production. However, the efforts devoted will also have fruitful influences in various fields of chemistry.

NOTE ADDED IN PROOF

CO_2 reduction is currently one of the most attractive subjects of investigation and new publications have been continuously coming out since the first writing of this manuscript. Some of them are included in References 214–222. These recent publications show that CO_2 and CO now show much promise for conversion to highly reduced products such as CH_4, C_2H_4, and methanol. It is likely that a new era of CO_2 reduction is about to begin.

ACKNOWLEDGMENTS

I wish to express my thanks to Professor J. O'M. Bockris for his invitation to write this article. I gratefully acknowledge that this work is based on stimulating discussions with Dr. Benedict Aurian-Blajeni and Professor J. O'M. Bockris, who were my co-workers at Texas A & M University several years ago. I also would like to thank Professor Kaname Ito of the Nagoya Institute of Technology and many other Japanese colleagues as well as Professor Jean-Pierre

Sauvage of CNRS (Strasbourg) for their encouragement and useful discussions about CO_2 fixation over the years. Partial financial support by the Yazaki Memorial Foundation for some work on CO_2 fixation presented in the text is also acknowledged.

REFERENCES

[1] M. E. Royer, *C. R. Acad. Sci.* **70** (1870) 731.
[2] J.-P. Randin, in *Encyclopedia of Electrochemistry of the Elements*, Vol. VII, Ed. by A. J. Bard, Marcel Dekker, New York, 1976, Chap. 1.
[3] J. R. Bolton and D. O. Hall, *Annu. Rev. Energ.* **4** (1979) 353.
[4] M. Halmann, in *Energy Resources through Photochemistry and Catalysis*, Ed. by M. Gratzel, Academic, New York, 1983, Chap. 15.
[5] D. Root and E. Attanasi, *Am. Assoc. Petrol. Geol. Bull.* (1978).
[6] J. O'M. Bockris, *Energy, The Solar-Hydrogen Alternative*, Australia & New Zealand Book Co., Sydney, 1975.
[7] J. Hansen, D. Johnson, A. Lacis, S. Lebedeff, P. Lee, D. Lind, and G. Russel, *Science* **213** (1981) 957.
[8] S. Inoue and N. Yamazaki,Eds., *Organic and Bioorganic Chemistry of Carbon Dioxide*, Kodansha Ltd., Tokyo and Wiley, New York, 1982.
[9] J. R. Bolton, A. F. Haught, and R. T. Ross, in *Photochemical Conversion and Storage of Solar Energy*, Ed. by J. S. Connolly, Academic, New York, 1981, p. 297.
[10] R. Williams, R. S. Crandall, and A. Bloom, *Appl. Phys. Lett.* **33** (1978) 381.
[11] M. Calvin, in *Photochemical Conversion and Storage of Solar Energy*, Ed. by J. S. Connolly, Academic, New York, 1981, p. 1.
[12] G. J. F. Chittenden and A. W. Schwartz, *Biosystems* **14** (1981) 15.
[13] C. Folsome and A. Brittain, *Nature* **291** (1981) 482.
[14] R. D. Brown, in *Origin of Life*, Ed. by Y. Wolman, Reidel, Dordrecht, 1981, p. 1.
[15] F. Fischer and O. Prziza, *Ber. Dtsch. Chem. Ges.* **47** (1914) 256.
[16] K. S. Udapa, G. S. Subramanian, and H. V. K. Udapa, *Electrochim. Acta* **16** (1971) 1593.
[17] A. Bewick and G. P. Greener, *Tetrahedron Lett.* (1969) 4623; (1970) 391.
[18] K. Ito, T. Murata, and S. Ikeda, *Bull. Nagoya Inst. Tech.* **27** (1975) 209.
[19] P. G. Russell, N. Kovac, S. Srinivasan, and M. Steinberg, *J. Electrochem. Soc.* **124** (1977) 1329.
[20] Y. Hori, N. Kamide, and S. Suzuki, *J. Faculty Eng. Chiba Univ.* **32** (1981) 37.
[21] K. J. Vetter, *Electrochemical Kinetics*, Academic, New York, 1967.
[22] S. Kapusta and N. Hackerman, *J. Electrochem. Soc.* **130** (1983) 607.
[23] Y. Hori, K. Kikuchi, and S. Suzuki, *Chem. Lett.* (1985) 1695; Y. Hori, K. Kikuchi, A. Murata, and S. Suzuki, *Chem. Lett.* (1986) 897.
[24] K. W. Frese, Jr. and S. Leach, *J. Electrochem. Soc.* **132** (1985) 259.
[25] D. P. Summers, S. Leach, and K. W. Frese, Jr., *J. Electroanal. Chem.* **205** (1986) 219.
[26] I. Taniguchi, N. Nakashima, K. Ogata, and Y. Shiraishi, unpublished results.
[27] K. Ito, S. Ikeda, and M. Okabe, *Denki Kagaku* **48** (1980) 247; K. Ito, S. Ikeda, T. Iida, and H. Niwa, *Denki Kagaku* **49** (1981) 106.
[28] H. Stephen and T. Stephen, Eds., *Solubilities of Inorganic and Organic Compounds*, Vol. 1, Macmillan, New York, 1963, Part 2, p. 1063.

[29] W. F. Linke, *Solubility of Inorganic and Metal Organic Compounds*, 4th Ed., Van Nostrand, New York, 1958, p. 480.
[30] H. L. Clever and R. Battino, in *Solutions and Solubilities*, Ed. by M. R. J. Cack, Wiley, New York, 1975, p. 386.
[31] J. L. Roberts, Jr. and D. T. Sawyer, *J. Electroanal. Chem.* **9** (1965) 1.
[32] E. Lamy, L. Nadjo, and J.-M. Saveant, *J. Electroanal. Chem.* **78** (1977) 403.
[33] B. R. Eggins and J. McNeill, *J. Electroanal. Chem.* **148** (1983) 17.
[34] U. Von Kaiser and E. Heitz, *Ber. Bunsenges. Phys. Chem.* **77** (1973) 818.
[35] D. A. Tyssee, J. H. Wagenknecht, M. M. Baizer, and J. L. Chruma, *Tetrahedron Lett.* (1972) 4809.
[36] S. Gambino and G. Silvestri, *Tetrahedron Lett.* (1973) 3025.
[37] J. C. Gressin, D. Michelet, L. Nadjo, and J.-M. Saveant, *Nouv. J. Chim.* **3** (1979) 545.
[38] J. Fischer, Th. Lehmann, and E. Heintz, *J. Appl. Electrochem.* **11** (1981) 743.
[39] F. Goodridge and G. Presland, *J. Appl. Electrochem.* **14** (1984) 791.
[40] K. Ito, S. Ikeda, T. Iida, and A. Nomura, *Denki Kaguku* **50** (1982) 463.
[41] I. Taniguchi, N. Nakashima, and K. Ogata, unpublished results.
[42] K. Ito, S. Ikeda, N. Yamauchi, T. Iida, and T. Takagi, *Bull. Chem. Soc. Jpn* **58** (1985) 3027.
[43] P. Van Rysselberghe and G. J. Alkire, *J. Am. Chem. Soc.* **66** (1944) 1801; T. E. Teeter and P. Van Rysselberghe, *J. Chem. Phys.* **22** (1954) 759.
[44] Y. Hori and S. Suzuki, *J. Electrochem. Soc.* **130** (1983) 2387.
[45] W. Paik, T. N. Andersen, and H. Eyring, *Electrochim. Acta* **14** (1969) 1217; J. Ryu, T. N. Andersen, and H. Eyring, *J. Phys. Chem.* **76** (1972) 3278.
[46] Y. Hori and S. Suzuki, *Bull. Chem. Soc. Jpn.* **55** (1982) 660.
[47] A. W. B. Aylmer-Kelly, A. Bewick, P. R. Cantrill, and A. M. Tuxford, *Discuss. Faraday Soc.* **56** (1973) 96.
[48] A. V. Zakharyan, N. V. Osetrova, and Yu. B. Vasilev, *Sov. Electrochem.* **13** (1978) 1568; A. V. Zakharyan, Z. A. Rotenberg, N. V. Osetrova, and Yu. B. Vasilev, *Sov. Electrochem.* **14** (1978) 1317.
[49] D. J. Schiffrin, *Discuss. Faraday Soc.* **56** (1973) 75.
[50] B. Beden, A. Bewick, M. Razaq, and J. Weber, *J. Electroanal. Chem.* **139** (1982) 203.
[51] S. D. Babenko, V. A. Benderskii, A. G. Krivenko, and V. A. Kurmaz, *J. Electroanal. Chem.* **159** (1983) 163.
[52] Yu. B. Vassiliev, V. S. Bagotzky, N. V. Osetrova, O. A. Khazova, and N. A. Mayorova, *J. Electroanal. Chem.* **189** (1985) 271.
[53] C. Amatore and J.-M. Saveant, *J. Am. Chem. Soc.* **103** (1981) 5021.
[54] C. Amatore and J.-M. Saveant, *J. Electroanal. Chem.* **125** (1981) 23.
[55] D. W. Overral and D. H. Whiffen, *Mol. Phys.* **4** (1961) 113.
[56] M. M. Chang, T. Saji, and A. J. Bard, *J. Am. Chem. Soc.* **99** (1977) 5399.
[57] Yu. B. Vassiliev, V. S. Bagotzky, O. A. Khazova, and N. A. Mayorova, *J. Electroanal. Chem.* **189** (1985) 295, 311.
[58] B. Aurian-Blajeni, M. A. Habib, I. Taniguchi, and J. O'M. Bockris, *J. Electroanal. Chem.* **154** (1983) 399.
[59] K. Kunimatsu, *J. Electroanal. Chem.* **140** (1982) 205; *J. Phys. Chem.* **88** (1984) 2195.
[60] H. Finsterholzl, *Ber. Bunsenges. Phys. Chem.* **86** (1982) 797; R. Kruse and E. U. Frank, *Ber. Bunsenges. Phys. Chem.* **86** (1982) 1036.
[61] M. R. Mahoney, M. W. Howard, and R. P. Cooney, *Chem. Phys. Lett.* **71** (1980) 59.
[62] J. P. Keene, Y. Raef, and A. J. Swallow, in *Pulse Radiolysis*, Ed. by M. Evert, J. P. Keene, and A. J. Swallow, Academic, London, 1965, p. 100.

[63] R. Maskiewicz and B. H. J. Bielski, *Biochim. Biophys. Acta* **638** (1981) 153.
[64] B. Aurian-Blajeni, M. Halmann, and J. Manassen, *J. Photochem. Photobiol.* **35** (1982) 157.
[65] Z. Sobkowski, A. Wieckowski, P. Zelenay, and A. Czerwinski, *J. Electroanal. Chem.* **100** (1979) 781.
[66] A. M. Baruzzi, E. P. M. Leiva, and P. Giordano, *J. Electroanal. Chem.* **158** (1983) 103.
[67] K. Kunimatsu, W. G. Golden, H. Seki, and M. R. Philpott, *Langmuir* **1** (1985) 245; K. Kunimatsu, H. Seki, W. G. Golden, J. G. Golden II, and M. R. Philpott, *Langmuir* **2** (1986) 464.
[68] M. A. Chesters, S. F. Parker, and R. Raval, *Surf. Sci.* **165** (1986) 179.
[69] B. Beden, A. Bewick, and C. Lamy, *J. Electroanal. Chem.* **148** (1983) 147.
[70] B. Beden, C. Lamy, A. Bewick, and K. Kunimatsu, *J. Electroanal. Chem.* **121** (1983) 343.
[71] F. Adami, M.-C. Pham, P.-C. Lacaze, and J.-E. Dubois, *J. Electroanal. Chem.* **210** (1986) 295.
[72] D. Barnes and P. Zuman, *J. Electroanal. Chem.* **46** (1973) 323.
[73] S. Clarke and J. A. Harrison, *J. Electroanal. Chem.* **36** (1972) 109.
[74] S. Kapusta and N. Hackerman, *J. Electroanal. Chem.* **138** (1982) 295.
[75] M. Spichiger-Ulmann and J. Augustynski, *J. Chem. Soc., Faraday Trans. 1* **81** (1985) 713.
[76] S. R. Morrison, *Electrochemistry at Semiconductor and Oxidized Metal Electrodes*, Plenum Press, New York, 1980.
[77] G. Ghiotti and E. Garrone, J. Chem. Soc., *Faraday Trans. 1*, **76** (1980) 2102.
[78] S. J. Tauster, S. C. Fung, R. T. K. Baker, and J. A. Horsley, *Science* **211** (1981) 1121.
[79] K. Tanaka and J. M. White, *J. Phys. Chem.* **86** (1982) 3977.
[80] F. Solymosi, A. Erdohelyi, and S. Bansagi, *J. Chem. Soc., Faraday Trans. 1* **77** (1981) 2645.
[81] M. A. Henderson and S. D. Worley, *J. Phys. Chem.* **89** (1985) 1417.
[82] A. Monnier, J. Augustynski, and C. Stalder, *J. Electroanal. Chem.* **112** (1980) 383.
[83] T. Inoue, A. Fujishima, S. Konishi, and K. Honda, *Nature* **277** (1979) 633.
[84] A. H. A. Tinnemans, T. P. M. Koster, O. H. M. W. Thewissen, C. W. Dekreuk, and A. Mackor, *J. Electroanal. Chem.* **145** (1983) 449.
[85] A. H. A. Tinnemans, T. P. M. Koster, O. H. M. W. Thewissen, and A. Mackor, *Nouv. J. Chim.* **6** (1982) 373; *Sol. Energ. R & D Eur. Community, Ser D* **2** (1983) 86.
[86] J. Augustynski, *J. Electroanal. Chem.* **145** (1983) 457.
[87] M. Koudelka, A. Monnier, and J. Augustynski, *J. Electrochem. Soc.* **131** (1984) 745.
[88] M. H. Miles, A. N. Fletcher, G. E. McManis, and L. O. Spreer, *J. Electroanal. Chem.* **190** (1985) 157.
[89] D. Canfield and K. W. Frese, Jr., *J. Electrochem. Soc.* **130** (1983) 1772; K. W. Frese, Jr. and D. Canfield, *J. Electrochem. Soc.* **131** (1984) 2518.
[90] K. W. Frese, Jr. and D. Canfield, Extended Abstracts, Electrochem. Soc. Meeting, San Francisco, 1983, No. 693; *Chem. Eng. News*, Nov. 29 (1983).
[91] W. M. Sears and S. R. Morrison, *J. Phys. Chem.* **89** (1985) 3295.
[92] I. Taniguchi, H. Murakami, and T. Hayashida, unpublished results.
[93] J. C. Hemminger, R. Carr, and G. A. Somorjai, *Chem. Phys. Lett.* **57** (1978) 100.
[94] A. J. Bard, *Science* **207** (1980) 139; *J. Photochem.* **10** (1979) 59.
[95] M. Halmann, *Nature* **275** (1978) 115.
[96] M. Zafrir, M. Ulman, Y. Zuckerman, and M. Halmann, *J. Electroanal. Chem.* **159** (1983) 373.

[97] B. Aurian-Blajeni, M. Halmann, and J. Manassen, *Sol. Energ. Mater.* **8** (1983) 425.
[98] A. Heller, B. Miller, H. J. Lewerenz, and K. J. Bachmann, *J. Am. Chem. Soc.* **102** (1980) 6555; A. Heller, J. H. Lewerenz, and B. Miller, *J. Am. Chem. Soc.* **103** (1981) 200.
[99] B. Aurian-Blajeni, I. Taniguchi, and J. O'M. Bockris, *J. Electroanal. Chem.* **149** (1983) 291.
[100] Y. Taniguchi, H. Yoneyama, and H. Tamura, *Bull. Chem. Soc. Jpn.* **55** (1982) 2034.
[101] K. Ito, S. Ikeda, M. Yoshida, S. Ohta, and T. Iida, *Bull. Chem. Soc. Jpn.* **57** (1984) 583.
[102] K. Hiratsuka, K. Takahashi, H. Sasaki, and S. Toshima, *Chem. Lett.* (1977) 1137.
[103] I. Taniguchi, B. Aurian-Blajeni, and J. O'M. Bockris, *Electrochim. Acta* **29** (1984) 923.
[104] V. Guruswamy and J. O'M. Bockris, *Energ. Res.* **3** (1979) 397.
[105] I. Taniguchi, B. Ajrian-Blajeni, and J. O'M. Bockris, *J. Electroanal. Chem.* **157** (1983) 179.
[106] M. Green, *J. Chem. Phys.* **31** (1959) 200.
[107] V. A. Myamlin and Yu. V. Pleskov, *Electrochemistry of Semiconductors*, Plenum Press, New York, 1967.
[108] S. Ikeda, M. Yoshida, and K. Ito, *Bull. Chem. Soc. Jpn.* **58** (1985) 1353; S. Ikeda and K. Ito, Abstracts of the Symposium on Electrochemistry and Catalytic Process for Carbon Dioxide and Nitrogen Fixation, held at the Institute for Molecular Science, Okazaki, 1986, p. 9 (in Japanese).
[109] M. G. Bradley, T. Tysak, D. J. Graves, and N. A. Vlachopoulos, *J. Chem. Soc., Chem. Commun.* (1983) 349.
[110] B. Fisher and R. Eisenberg, *J. Am. Chem. Soc.* **102** (1980) 7361.
[111] M. G. Bradley and T. Tysak, *J. Electroanal. Chem.* **135** (1982) 153.
[112] A. J. Bard, A. B. Bocarsly, F. R. F. Fan, E. G. Walton, and M. S. Wrighton, *J. Am. Chem. Soc.* **102** (1980) 3671; A. B. Bocarsly, D. C. Bookbinder, R. N. Dominey, N. S. Lewis, and M. S. Writon, *J. Am. Chem. Soc.* **102** (1980) 3683.
[113] M. Beley, J.-P. Collin, J.-P. Sauvage, J.-P. Petit, and P. Chartier, *J. Electroanal. Chem.* **206** (1986) 333.
[114] C. R. Carbera and H. D. Abruña, *J. Electroanal. Chem.* **209** (1986) 101.
[115] M. Halmann and B. Aurian-Blajeni, *Proceedings of the 2nd European Commun. Photovolt. Sol. Energ. Conference*, 1979, p. 682.
[116] B. Aurian-Blajeni, M. Halmann, and J. Manassen, *Sol. Energ.* **23** (1980) 165.
[117] M. Ulmann, B. Aurian-Blajeni, and M. Halmann, *Isr. J. Chem.* **22** (1982) 177.
[118] M. Ulmann, A. H. A. Tinnemans, M. Mackor, B. Aurian-Blajeni, and M. Halmann, *Int. J. Sol. Energ.* **1** (1982) 213.
[119] M. Halmann, V. Katzir, E. Borgarello, and J. Kiwi, *Sol. Energ. Mater.* **10** (1984) 85.
[120] M. Halmann, B. Aurian-Blajeni, and S. Bloch, in *Origin of Life*, Ed. by Y. Wolman, Reidel, Dordrecht, 1981, p. 143.
[121] M. Miyake, H. Yoneyama, and H. Tamura, *J. Catal.* **58** (1979) 22.
[122] M. A. Enriquez and J. P. Fraissard, *J. Chim. Phys.* **78** (1981) 457.
[123] A. Henglein and M. Gutierrez, *Ber. Bunsenges. Phys. Chem.* **87** (1983) 852.
[124] A. Henglein, M. Gutierrez, and Ch.-H. Fisher, *Ber. Bunsenges. Phys. Chem.* **88** (1984) 170.
[125] I. Taniguchi and T. Hayashida, unpublished results.
[126] S. Meshitsuka, M. Ichikawa, and K. Tamaru, *J. Chem. Soc., Chem. Commun.* (1974) 158.
[127] S. Kapusta and N. Hackerman, *J. Electrochem. Soc.* **131** (1984) 1511.
[128] C. M. Lieber and N. S. Lewis, *J. Am. Chem. Soc.* **106** (1984) 5033.

[129] K. Takahashi, K. Hiratsuka, H. Sasaki, and S. Toshima, *Chem. Lett.* (1979) 305.
[130] J. Y. Becker, B. Vainas, R. Eger, and L. Kaufman, *J. Chem. Soc., Chem. Commun.* (1985) 1471.
[131] D. H. Bush, *Acc. Chem. Res.* **11** (1978) 392.
[132] A. H. A. Tinnemans, T. P. M. Koster, D. H. M. W. Thewissen, and A. Mackor, *Recl. Trav. Chim. Pays-Bas* **103** (1984) 288.
[133] D. J. Pearce and D. Pletcher, *J. Electroanal. Chem.* **197** (1986) 317.
[134] C. L. Bailey, R. D. Bereman, D. P. Rillema, and R. Nowak, *Inorg. Chim. Acta* **116** (1986) L45.
[135] M. Beley, J.-P. Collin, R. Ruppert, and J.-P. Sauvage, *J. Chem. Soc., Chem. Commun.* (1984) 1315; M. Beley, J.-P. Collin, R. Ruppert, and J.-P. Sauvage, *J. Am. Chem. Soc.* **108** (1986) 7461.
[136] I. Taniguchi, Abstracts of the Symposium on Electrochemistry and Catalytic Process for Carbon Dioxide and Nitrogen Fixations, held at Institute for Molecular Science, Okazaki, 1986, p. 23 (in Japanese).
[137] I. Taniguchi, N. Nakashima, and K. Yasukouchi, *J. Chem. Soc., Chem. Commun.* (1986) 1814; I. Taniguchi, N. Nakashima, K. Matsushita, and K. Yasukouchi, *J. Electroanal. Chem.* **224** (1987) 199.
[138] M. Tezuka, T. Yajima, A. Tsuchiya, Y. Matsumoto, Y. Uchida, and M. Hidai, *J. Am. Chem. Soc.* **104** (1982) 6834.
[139] M. Nakazawa, Y. Mizobe, Y. Matsumoto, Y. Uchida, M. Tezuka, and M. Hidai, *Bull. Chem. Soc. Jpn.* **59** (1986) 809.
[140] J. Hawecker, J.-M. Lehn, and R. Ziessel, *J. Chem. Soc., Chem. Commun.* (1984) 328.
[141] J. Hawecker, J.-M. Lehn, and R. Ziessel, *J. Chem. Soc., Chem. Commun.* (1983) 536.
[142] T. R. O'Toole, L. D. Margerum, T. D. Westmoreland, W. J. Vining, R. W. Murray, and T. J. Meyer, *J. Chem. Soc., Chem. Commun.* (1985) 1416.
[143] S. Cosnier, A. Deronzier, and J.-C. Moutet, *J. Electroanal. Chem.* **207** (1986) 315.
[144] B. P. Sullivan, C. M. Bolinger, D. Conrad, W. J. Vining, and T. J. Meyer, *J. Chem. Soc., Chem. Commun.* (1985) 1414.
[145] A. I. Breikss and H. D. Abruna, *J. Electroanal. Chem.* **201** (1986) 347.
[146] S. Slater and J. H. Wagenknecht, *J. Am. Chem. Soc.* **106** (1984) 5367.
[147] R. Wienkamp and E. Steckhan, *Angew. Chem.* **94** (1982) 786; *Angew. Chem., Int. Ed. Engl.* **21** (1982) 782.
[148] I. Taniguchi, unpublished results.
[149] C. M. Bolinger, B. P. Sullivan, D. Conrad, J. A. Gilbert, N. Story, and T. J. Meyer, *J. Chem. Soc., Chem. Commun.* (1985) 796.
[150] K. Tanaka, M. Morimoto, and T. Tanaka, *Chem. Lett.* (1983) 901; H. Ishida, K. Tanaka, and T. Tanaka, *Chem. Lett.* (1985) 405.
[151] C. J. Stalder, S. Chao, and M. S. Wrighton, *J. Am. Chem. Soc.* **106** (1984) 3673.
[152] C. J. Stalder, S. Chao, D. P. Summers, and M. S. Wrighton, *J. Am. Chem. Soc.* **105** (1983) 6318; **106** (1984) 2723.
[153] K. Ogura and M. Takagi, *J. Electroanal. Chem.* **201** (1986) 359; **206** (1986) 209; K. Ogura and I. Yoshida, *J. Mol. Catal.* **34** (1986) 67.
[154] G. N. Petrova and O. N. Efimov, *Elektrokhimiya* **19** (1983) 978; *Chem. Abstr.* (1983) 157762 S.
[155] S. Gambarotta, F. Arena, C. Floriani, and P. F. Zanazzi, *J. Am. Chem. Soc.* **104** (1982) 5082.
[156] M. G. Mason and J. A. Ibers, *J. Am. Chem. Soc.* **104** (1982) 5153.
[157] S. Sakaki and A. Dedieu, *J. Organometal. Chem.* **341** (1986) C63.

[158] R. Eisenberg and D. E. Hendriksen, *Adv. Catal.* **28** (1979) 79.
[159] D. J. Darensbourg and A. Kudaroski, *Adv. Organometal.* **22** (1983) 129.
[160] I. Taniguchi, B. Aurian-Blajeni, and J. O'M. Bockris, *J. Electroanal. Chem.* **161** (1984) 385.
[161] H. Lund and J. Simonet, *J. Electroanal. Chem.* **65** (1975) 205.
[162] S. Tazuke and N. Kitamura, *Nature* **275** (1978) 301.
[163] B. A. Parkinson and P. F. Weaver, *Nature* **309** (1984) 148.
[164] N. Kitamura and S. Tazuke, *Chem. Lett.* (1983) 1109.
[165] H. Kase, T. Iida, K. Yamane, and T. Mitamura, *Denki Kagaku* **54** (1986) 437.
[166] J.-M. Lehn and R. Ziessel, *Proc. Natl. Acad. Sci. USA* **79** (1982) 701.
[167] F. R. Keene, C. Creutz, and N. Sutin, *Coord. Chem. Rev.* **64** (1985) 247; C. Creutz and N. Sutin, *Coord. Chem. Rev.* **64** (1985) 321.
[168] M. Kirch, J.-M. Lehn, and J.-P. Sauvage, *Helv. Chim. Acta* **62** (1979) 1345.
[169] J. Hawecker, J.-M. Lehn, and R. Ziessel, *J. Chem. Soc. Chem. Commun.* (1985) 56.
[170] D. J. Darensbourg, A. Rokicki, and M. Y. Darensbourg, *J. Am. Chem. Soc.* **103** (1981) 3224.
[171] H. Hukkanen and T. T. Pakkanen, *Inorg. Chim. Acta* **114** (1986) L43.
[172] B. Akermark, U. Eklund-Westlin, P. Beckstrom, and R. Lof, *Acta Chem. Scand. B* **34** (1980) 27.
[173] D. R. Furge, G. D. Fong, and F. K. Fong, *J. Am. Chem. Soc.* **101** (1979) 3694.
[174] C. E. Folsome, A. Brittain, A. Smith, and S. Chang, *Nature* **294** (1981) 64.
[175] J. P. Pinto, G. R. Gladstone, and Y. L. Yung, *Science* **210** (1980) 183.
[176] R. A. Sheldon, *Chemicals from Synthesis Gas*, Reidel, Dordrecht, 1983.
[177] F. A. Uribe, P. R. Sharp, and A. J. Bard, *J. Electroanal. Chem.* **152** (1983) 173.
[178] G. A. Kolyagin, V. G. Danilov, V. L. Kornienko, I. A. Kedrinskii, and S. P. Gubin, *Elektrokhimiya* **19** (1983) 1004; *CA* 9573 U (1983).
[179] H. H. Sorch, H. Golumbic, and R. B. Anderson, *The Fischer-Tropsch and Related Syntheses*, Wiley, New York, 1951.
[180] K. Ogura and S. Yamasaki, private communication.
[181] T. Iizuka, Y. Tanaka, and K. Tanabe, *J. Catal.* **76** (1982) 1.
[182] S. Polizzotti and J. A. Schwarz, *J. Catal.* **77** (1982) 1.
[183] H. Miura, M. L. McLanghlin, and R. D. Gonzalez, *J. Catal.* **79** (1983) 227.
[184] R. Kirch, M. Kotter, and L. Rickert, *Ber. Bunsenges. Phys. Chem.* **88** (1984) 1054.
[185] M. A. Vannice and C. Sudhakar, *J. Phys. Chem.* **88** (1984) 2429.
[186] K. Kunimori, S. Matsui, and T. Uchijima, *Chem. Lett.* (1985) 359.
[187] S. Yamamura, H. Kojima, and W. Kawai, *J. Electroanal. Chem.* **186** (1985) 309.
[188] H. Yoneyama, K. Wakamoto, N. Hatanaka, and H. Tamura, *Chem. Lett.* (1985) 539.
[189] I. Taniguchi and Y. Shiraishi, unpublished results.
[190] B. A. Sexton and G. A. Somorjai, *J. Catal.* **46** (1977) 167.
[191] G. D. Weatherbee and C. H. Bartholomew, *J. Catal.* **77** (1982) 460.
[192] H. E. Ferkul, D. J. Stauton, J. D. McCowan, and M. C. Baird, *J. Chem. Soc., Chem. Commun.* (1982) 955.
[193] T. Iizuka, M. Kojima, and K. Tanabe, *J. Chem. Soc., Chem. Commun.* (1983) 638.
[194] H. Orita, S. Naito, and K. Tamaru, *J. Chem. Soc., Chem. Commun.* (1984) 150.
[195] T. Yoshida, D. L. Thorn, T. Okano, J. A. Ibers, and S. Otsuka, *J. Am. Chem. Soc.* **101** (1979) 4212.
[196] F. Solymosi, A. Erdohelyi, and T. Bansagi, *J. Chem. Soc., Faraday Trans.* 1 **77** (1981) 2465.
[197] K. Tanaka and J. M. White, *J. Phys. Chem.* **86** (1982) 3977.
[198] A. Amariglio, A. Elbianche, and H. Amariglio, *J. Catal.* **98** (1986) 355.

[199] F. Solymosi and J. Kiss, *Surf. Sci.* **149** (1985) 17.

[200] H. A. C. M. Hendrickx, A. P. J. M. Jongenelis, and B. E. Nieumenhuys, *Surf. Sci.* **162** (1985) 269.

[201] A. Czerwinski, J. Sobokowski, and R. Marassi, *Anal. Lett.* **18** (A14) (1985) 1717; J. Sobkowski and A. Czerwinski, *J. Phys. Chem.* **89** (1985) 365.

[202] D. A. Tyssee and M. M. Baizer, *J. Org. Chem.* **39** (1974) 2819, 2823.

[203] S. Wawzonek and D. Wearring, *J. Am. Chem. Soc.* **81** (1959) 1067.

[204] M. M. Baizer and J. L. Chruma, *J. Org. Chem.* **37** (1972) 1951.

[205] J. W. Wagenknecht, *J. Electroanal. Chem.* **52** (1974) 489.

[206] S. Wawzonek and J. M. Shradel, *J. Electrochem. Soc.* **126** (1979) 401.

[207] M. M. Baizer and H. Lund, Eds., *Organic Electrochemistry*, Marcel Dekker, New York, 1983, Chaps. 6, 20, and 25.

[208] T. Tsuda, Y. Chujo, and T. Saegusa, *J. Chem. Soc., Chem. Commun.* (1975) 963.

[209] T. Forschner, K. Menard, and A. Cutler, *J. Chem. Soc., Chem. Commun.* (1984) 121.

[210] B. P. Sullivan and T. J. Meyer, *J. Chem. Soc., Chem. Commun.* (1984) 1244.

[211] M. Kato and T. Ito, *Inorg. Chem.* **24** (1985) 504, 509; H. Ito and T. Ito, *Bull. Chem. Soc. Jpn.* **58** (1985) 1755.

[212] E. G. Lundquist, K. Folting, J. C. Huffman, and K. G. Caulton, *Inorg. Chem.* **26** (1987) 205.

[213] R. W. Murray, in *Electroanalytical Chemistry*, Vol. 13, Ed. by A. J. Bard, Marcel Dekker, New York, 1984, p. 191.

[214] R. L. Cook, R. C. MacDuff, and A. F. Sammelles, *J. Electrochem. Soc.* **134** (1987) 1873; *ibid.*, **134** (1987) 2375.

[215] S. Ikeda, T. Takagi, and K. Ito, *Bull. Chem. Soc. Jpn.* **60** (1987) 2517.

[216] Y. Hori, A. Murata, R. Takahashi, and S. Suzuki, *J. Am. Chem. Soc.* **109** (1987) 5022.

[217] K. Ogura and H. Uchida, *J. Electroanal. Chem.* **220** (1987) 333.

[218] H. Tanabe and K. Ohno, *Electrochim. Acta* **32** (1987) 1121.

[219] K. Ogura and I. Yoshida, *Electrochim. Acta* **32** (1987) 1191; K. Ogura and M. Fujita, *J. Mol. Catal.* **41** (1987) 303.

[220] Y. Hori, A. Murata, K. Kikuchi, and S. Suzuki, *J. Chem. Soc., Chem. Commun.* (1987) 728.

[221] H. Ishida, H. Tanaka, K. Tanaka, and T. Tanaka, *J. Chem. Soc., Chem. Commun.* (1987) 131.

[222] S. Daniele, P. Ugo, G. Bontempelli, and M. Fiorani, *J. Electroanal. Chem.* **219** (1987) 259.

6

Electrochemistry of Aluminum in Aqueous Solutions and Physics of Its Anodic Oxide

Aleksandar Despić

Faculty of Technology and Metallurgy, University of Belgrade, Belgrade, Yugoslavia

Vitaly P. Parkhutik

Department of Microelectronics, Minsk Radioengineering Institute, Minsk, USSR

I. INTRODUCTION

Anodic oxidation of valve metals, particularly, aluminum, has attracted considerable attention because of its wide application in various fields of technology. Traditionally, aluminum is "anodized" in order to protect the metal against corrosion, to improve its abrasion and adsorption properties, etc.[1] The more recent and rapidly growing applications of anodic aluminas in electronics are due to their excellent dielectric properties, perfect planarity, and good reproducibility in production. Finally, ways have recently been found to use the energy potential of aluminum oxidation for chemical power sources of the metal-air type[2,3] and other electrochemical applications.

Thus, interest in the electrochemical behavior of aluminum in aqueous solutions and anodic oxides, which, until recently, was stimulated entirely by attempts to cope with corrosion, has been enhanced by the wide new areas of application.

A vast body of literature tackles the different aspects of anodic oxidation, including the growth, structure, morphology, and properties of anodic oxides A number of review articles are available. These are widely cited in a review by Thompson and Wood published some five years ago.[4] The latter can be considered the most comprehensive review to date on the mechanism of barrier and porous alumina growth, with special emphasis on chemical composition and morphology. However, some important aspects seem not to have been given adequate attention and much new knowledge has accumulated since its publication.

Moreover, novel techniques of thin-film analysis (EXAFS, RBS, XPS, etc.) and improved sensitivity of traditional techniques (e.g., IR spectroscopy) have afforded a better understanding of anodic oxide growth and have even led to a reconsideration of commonly accepted concepts.

Also, the increasing application of alumina films in the electronics industry requires that attention be paid to their electrophysical properties (dielectric strength, conduction, etc.). However, since the work of Goruk, Young, and Zobel,[5] published as long ago as 1966 in this same series, no articles reviewing these problems have appeared. An attempt is made here to emphasize the correlation between the electrophysical properties of oxides and the history of their growth.

Finally, a large number of phenomena connected with active electrochemical dissolution of aluminum in the electrolyte, promoted by the presence of aggressive anions, are considered to deserve special attention, because understanding of these phenomena is far from complete, and it is hoped that a review of them will stimulate further research.

It is clear that the problem of anodic oxide films is increasingly of multidisciplinary interest, shared by specialists in different areas of physics and chemistry. It is felt, from reviewing the literature, that these researchers often tend to overlook some aspects which are somewhat removed from their immediate field. Thus, an otherwise excellent experiment or theory may lose significance because of some neglected or ill-defined detail. Hence, it is considered useful at the outset of this article to try and give an overview of the system as a whole, summarizing all the factors which contribute to its extreme complexity. Though some of the points in this overview may be considered only too well known by some scientists, it is hoped that they will arouse a new awareness in others.

II. OVERVIEW OF THE SYSTEM

Formation of aluminum oxide (alumina) upon contact of aluminum metal with pure water occurs because the reaction

$$2Al + 3H_2O = Al_2O_3 + 3H_2 \tag{1}$$

has a free energy of −864.6 kJ/mol under standard conditions. However, the reaction occurs spontaneously ("chemically") only until a compact layer of the virtually insoluble alumina has formed and separated the points at which hydrogen could evolve, from further supply of water. At that point the reactants in Eq. (1) become spatially separated. Two interfaces are formed—the metal/oxide (M/O) and the oxide/water (O/S)—and electrical potential differences, $\Delta\phi_{M/O} = (\phi_{Al}) - (\phi_0')_O$ and $\Delta\phi_{O/S} = (\phi_0'')_O - (\phi_S)$, are built up at both, as shown schematically in Fig. 1a, due to the

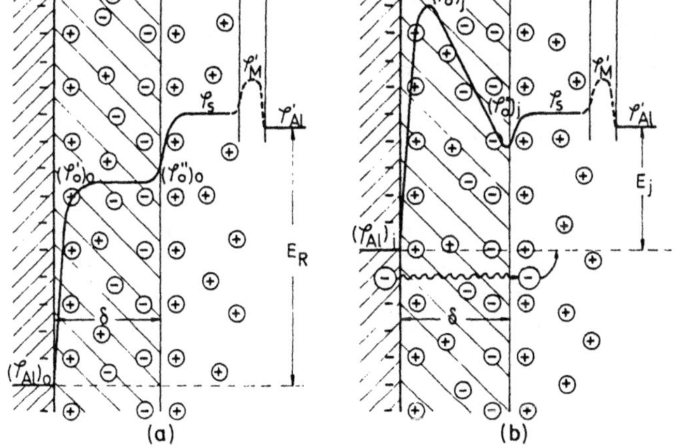

Figure 1. Schematic representation of potential profile and charge distribution across an anodic oxide film of thickness δ on aluminum: (a) hypothetical situation in the absence of any current; (b) in the presence of an anodic current caused by corrosion or by an external source. RE, reference electrode to which the potential of aluminum is referred.

transfer of an excess of aluminum ions and oxygen ions, respectively, across the two interfaces, which compensates for the further tendency for the formation of the oxide.†

These potential differences cannot be directly measured. Nevertheless, the underlying processes at the two interfaces

$$2Al + (3O^{2-})_{oxide} = Al_2O_3 + 6e^- \tag{2}$$

and

$$H_2O = 2H^+ + (O^{2-})_{aq} \tag{3}$$

$$(O^{2-})_{aq} = (O^{2-})_{oxide} \tag{4}$$

must make significant contributions to the overall free energy change of reaction (1). Hence, the corresponding potential differences must be relatively large.

If there were no metallic impurities at the aluminum surface penetrating through the oxide and if the oxide possessed no electronic conductivity and no permeability to water, the measured potential difference with respect to a hydrogen electrode in pure water (pH = 7) would correspond to the thermodynamic one of −1.90 V. However, some electronic connection between the outer and the inner interface is always established, allowing reaction (1) to proceed at some rate (corrosion). This changes the situation of the potential distribution inasmuch as some electric field must be created, large enough to provide for the corresponding rate of transport of the two reacting ions toward each other. Hence, in reality the situation is similar to that shown in Fig. 1b, and the metal immersed in water tends to acquire a significantly more positive rest potential [open-circuit potential (OCP)] than the thermodynamic one.

† The only determinable quantities are the potential difference $E = \phi_{Al} - \phi'_{Al}$, i.e., the reversible potential with respect to a reference electrode (RE), $E_R = (\phi_{Al})_O - \phi'_{Al}$ (from thermodynamic data), and the cell voltage at a given current density j, $E(j) = (\phi_{Al})_j - \phi'_{Al}$ (which can be measured experimentally). The potential differences at the two interfaces, M/O and O/S, cannot be known exactly because of the unknown potential difference $\phi_S - \phi_M$ and the volta potential difference between the reference electrode metal and aluminum, $\phi'_{Al} - \phi'_M$.

The same applies if reaction (1) is forced by anodic current from an external source. The potential must shift further in the positive direction with respect to the OCP.†

It should be noted that the three parts of the electric field—across the M/O interface, across the bulk of the oxide, and across the O/S interface—are determined by entirely different factors [cf. Section III(1)]. Hence, any imposed change of the potential level of the metal redistributes itself among the three parts in a manner which cannot be predicted without detailed knowledge of the kinetics of the processes taking place in each of them.

In considering the wide spectrum of phenomena observed in this system, one must keep in mind the following points, which make for an extremely complex situation:

1. Although the anhydrous oxide is the stable reaction product at room temperature, the free energies of dehydration are relatively small. Hence, species of different degrees of hydration (or protonation) could form during anodic oxidation up to $Al_2O_3 \cdot 3H_2O$ [or $Al(OH)_3$]. This is due to the possibility of stepwise splitting of water,

$$H_2O \rightarrow OH^- + H^+ \tag{5}$$

$$OH^- \rightarrow O^{2-} + H^+ \tag{6}$$

which may be kinetically more favorable than the direct splitting of two protons. Hence, a hydrated alumina should be expected at least at the O/S interface. [cf. Section IV (4(i))].

2. Anhydrous alumina can have a variety of structures, from an entirely disordered one (truly amorphous), through short-range ordered amorphous, to highly ordered in a tetrahedral or octahedral arrangement (γ,γ'-or α-alumina) [cf. Section IV(3)].

† When an external anodic current is applied, it is of interest to know the "faradaic efficiency," η_F, of anodic oxidation, defined as the ratio of the amount of metal actually oxidized to the amount which should be oxidized by the external current if Faraday's law is strictly applicable with three electrons obtained per atom. The fact that a part of the metal is oxidized by a corrosion process, without using the external current, makes the faradaic efficiency larger than 1. On the other hand, the "material efficiency," η_M, takes into account the quantity of electricity obtained per quantity of the metal, related again through Faraday's law (of importance for chemical power sources). Obviously, η_M is the reciprocal of η_F. Hence, while the corrosion process makes $\eta_F > 1$, it gives $\eta_M < 1$. Other reasons for deviations of η_F and η_M from unity will be discussed below.

3. The type of conductance exhibited by the oxide and its value are structure sensitive. The oxide is essentially an ionic conductor. One could maintain that it has a relatively high concentration of low-mobility ionic charge carriers. As far as electronic conductance is concerned although pure alumina is an insulator with a band gap of 8 to 9 eV, one has to bear in mind that when it is produced anodically as a thin film adhering firmly to the metal, an entirely different electronic situation may arise [cf. Section V(2)].

4. The metal virtually always possesses [even in the purest forms known today (99.9999%)] a microheterogeneity of the surface with respect to the ease of oxidation or the adsorption affinities for various species.

5. Even trace impurities have a profound effect on the open-circuit potential and the rate of corrosion. The electrochemical behavior is even more sensitive to alloying with small amounts of other elements [cf. Section III(5(v))].

6. The three electrons in the outer shell of the metal atoms are not identical (the electronic configuration is $2s^2 2p^1$). Hence, the anodic oxidation at the M/O interface is very likely to proceed stepwise, i.e.,

$$Al \rightarrow Al^{+} + e^{-} \tag{7}$$

$$Al^{+} \rightarrow Al^{3+} + 2e^{-} \tag{8}$$

forming the low-valency intermediate [cf. Section III(1(i))]. The latter may penetrate through the oxide by some valency transfer mechanism and react with water at the O/S interface to form hydrogen [cf. Section III(5(iv))].

7. Aluminum is known to undergo another reaction with water, making the hydride:

$$Al + \tfrac{3}{2}H_2O \rightarrow AlH_3 + \tfrac{3}{4}O_2 \tag{9}$$

This reaction has a large positive standard free energy change. However, under certain conditions, it could proceed electrochemically in the cathodic reaction

$$Al^{3+} + 3H^{+} + 6e^{-} \rightarrow AlH_3 \tag{10}$$

with formation of an intermediate hydride ion:

$$Al^{3+} + H^{+} + 2e^{-} \rightarrow AlH^{2+} \tag{11}$$

and

$$AlH^{2+} + H^{+} + 2e^{-} \rightarrow AlH_2^{+} \tag{12}$$

These reaction products have their domains of stability in certain ranges of potential and pH as shown in the Pourbaix diagram in Fig. 2,[6] which may have relevance in cases when open-circuit potentials are established at highly negative values [cf. Sections III(2) and III(5(v))].

8. At some negative potentials, one could expect implantation of metal atoms obtained by the reduction of, for example, alkali metal cations.[7]

9. The oxide is virtually insoluble only in pure water of neutral pH. Its solubility increases sharply in both acid and alkaline solution, because it undergoes chemical reactions of protonation

$$\tfrac{1}{2}Al_2O_3 + \tfrac{1}{2}H_2O + H^+ \rightarrow Al(OH)_2^+ \qquad (13)$$

$$Al(OH)_2^+ + H^+ \rightarrow Al(OH)^{2+} + H_2O \qquad (14)$$

$$Al(OH)^{2+} + H^+ \rightarrow Al^{3+} + H_2O \qquad (15)$$

and aluminate formation

$$\tfrac{1}{2}Al_2O_3 + \tfrac{3}{2}H_2O \rightarrow Al(OH)_3 \qquad (16)$$

$$Al(OH)_3 + OH^- \rightarrow AlO_2^- + 2H_2O \qquad (17)$$

All of the ionic species formed are highly soluble.

This phenomenon enables some aluminum ions to cross the O/S interface and go into the solution. If the efficiency of oxide formation, η_{ox}, is defined as the ratio of the amount of solid oxide actually formed to the amount which would be formed if no aluminum went into the solution, such a "solubilization" reduces this efficiency below 1.

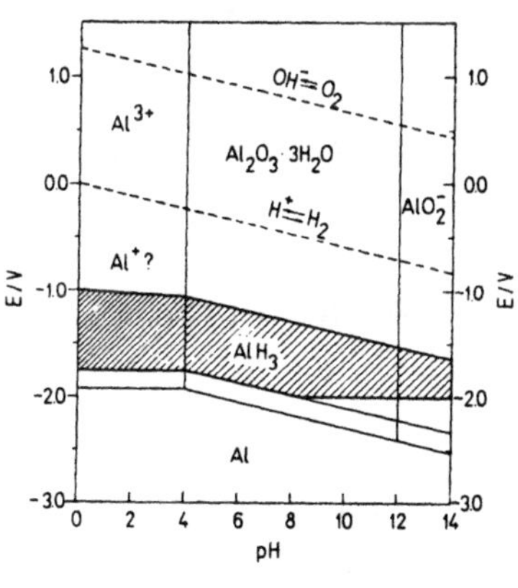

Figure 2. The potential-pH (Pourbaix) diagram for aluminum in aqueous medium, defining regions of thermodynamic stability of the different species.

10. At the O/S interface, for each molecule of alumina formed inside the oxide layer, i.e., three O^{2-} ions transferred across the O/S interface, six hydrogen ions are formed. Thus, the acidity at the interface tends to rise to an extent which depends on the rate removal of these ions by some mechanism. In view of Eqs. (13) to (15), this should lead to oxide dissolution and a further decrease of η_{ox}.

11. When the metal is immersed in a solution of a salt of a weak acid (e.g., boric or tartaric), the latter exhibits a buffering capacity and thus provides one mechanism for the removal of hydrogen ions from the interface [cf. Section III(3(iv))].

12. Some anions exhibit a complexing affinity toward aluminum ions. Thus, in such a situation, one should visualize the presence not only of a series of pure complexes but also of mixed ones with oxygen-containing ions, e.g., $Al(OH)_2Cl$, $Al(OH)Cl_2$, $AlCl_3$, $Al(OH)Cl^+$, and $AlCl^{2+}$. The ionic species are soluble, and thus the interaction with complexing ions may provide an additional mechanism for solubilization of the oxide.

13. The complexing affinity may cause penetration of anions into the oxide by some ion exchange mechanism. The presence of such species inside the oxide may have a profound effect on its conductivity.

14. Anions may exhibit a tendency toward specific adsorption at the O/S interface. This may be related in some way to the complexing affinity. This effect, occurring at the inner Helmholtz plane of the electrochemical double layer, may significantly change the charge transfer situation [cf. Section III(5(iii))].

15. Finally, one should note that a significant difference in behavior could exist between a very thin (<1 nm) and a thick oxide layer because of possible interface effects on the bulk oxide in the former case; such effects should be negligible in the latter case.

III. KINETICS OF ALUMINUM ANODIZATION

1. General Considerations

The variety of electrolytes and the wide range of their concentrations, temperatures, and anodization regimes provide for a variety of anodization kinetics, structures, compositions, and properties of anodic oxides.

Nonporous "barrier" oxides can be grown in neutral (pH 7-8) solutions of borates, citrates, tartrates, phosphates, etc.[4,8] They are limited to a thickness of several hundred nanometers by dielectric breakdown initiation during growth. Porous oxides are formed in electrolytes promoting oxide dissolution, i.e., aqueous sulfuric, oxalic, or phosphoric acid solutions, with a thickness of up to hundreds of microns.[9] Generally, it is assumed that the structure of porous oxides is a close-packed array of columnar hexagonal cells, each containing a central pore normal to the substrate surface and separated from it by a layer of hemispherically shaped barrier-type film as shown in Fig. 3.[10] Nonordered, fibrous-like porous oxides are also reported for anodization in chromic acid[11] or alkaline baths[12] or under pulse anodization.[13] Some electrolytes containing "aggressive" anions (halides) cause localized dissolution (pitting). In others, depending on the anodization regime, together with barrier or porous oxide formation, pitting,[14] uniform oxide dissolution,[15] or "burning"[16] can be produced as seen in Fig. 4, reproduced from the work of Fukushima *et al.*[9] Various types of corrosion can also arise.[17]

It was customary to study these different situations separately. The present state of the art, however, makes it reasonable to attempt considering all of these cases in a unified way. In fact, in quite a number of publications it has been shown that there is no sharp boundary between barrier and porous films. Prolonged anodization

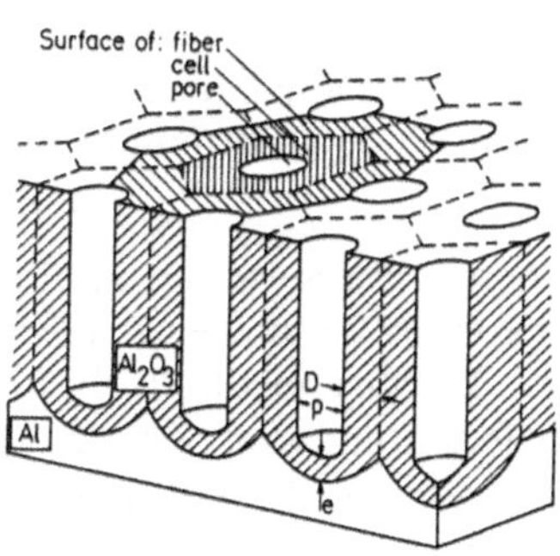

Figure 3. A model of a porous oxide film formed at 120 V in a phosphoric acid solution, according to Heber.[10]

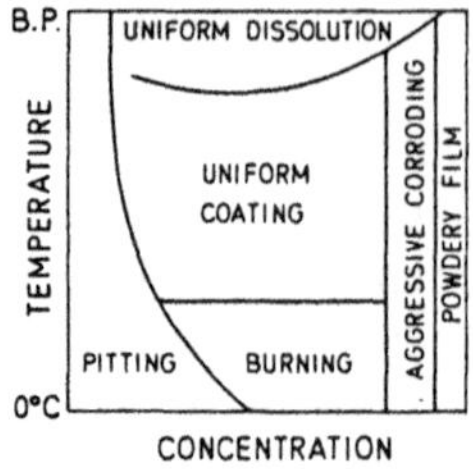

Figure 4. The temperature–electrolyte concentration diagram, defining regions of different anodic dissolution phenomena, according to Fukushima *et al.*[9]

of aluminum in barrier-forming electrolytes for 1 to 50 h leads to the classical porous structure of the oxide.[18,19] On the other hand, as mentioned above, porous oxide growth comprises, at the initial stage, formation of a barrier oxide. This barrier film is maintained during further oxide growth as a semispherical oxide layer at pore bottoms.[20]

It is obvious that some common processes have to take place during oxide formation, irrespective of how thick the oxide is or which type of electrolyte the metal is immersed in. The two interfaces and the bulk of the oxide will be considered separately.

(*i*) *M/O Interface*

The transfer of an aluminum atom from the metal phase into the oxide to form an ion should occur by a simultaneous charge transfer, in much the same way as in all electrochemical metal dissolutions. An electrochemical double layer should be established with aluminum oxide as a solid electrolyte. Since there are no other solvent molecules present, the likelihood of the presence of a layer analogous to the Helmholtz layer is small. The electrolyte part of the double layer is more likely to be of a Gouy–Chapman type, causing the appearance of a space charge and potential distribution in the oxide according to the Gouy–Chapman equation[21]

$$\phi_x = \phi_0 \exp(-\kappa x) \tag{18}$$

where κ is a constant containing the ionic charge and concentration as well as the dielectric permittivity of the oxide.

Because of the likely high ionic concentration and the small dielectric constant of the oxide, the diffuse layer thickness is expected to be small, and hence this space charge is limited to a few nanometers.

The two-step charge transfer [cf. Eqs. (7) and (8)] with formation of a significant amount of monovalent aluminum ion is indicated by experimental evidence. As early as 1857, Wholer and Buff discovered that aluminum dissolves with a current efficiency larger than 100% if calculated on the basis of three electrons per atom.[22] The anomalous overall valency (between 1 and 3) is likely to result from some monovalent ions going away from the M/O interface, before they are further oxidized electrochemically, and reacting chemically with water further away in the oxide or at the O/S interface.[23,24] If such a mechanism was operative with activation-controlled kinetics,[25] the current-potential relationship should be given by the Butler–Volmer equation

$$j = j_0\left[\exp\frac{\alpha_a F}{RT}\Delta(\Delta\phi_{M/O}) - \exp\frac{\alpha_c F}{RT}\Delta(\Delta\phi_{M/O})\right] \qquad (19)$$

where j_0 is the exchange current density for the equilibrium potential difference across the interface (Fig. 1a) and the "overpotential"

$$\Delta(\Delta\phi_{M/O}) = (\phi_{Al})_j - (\phi'_O)_j - [(\phi_{Al})_O - (\phi'_O)_O]$$

is the change of that potential difference needed to pass the current density j. The transfer coefficients could have two sets of values[25]:

(a) for the case of the first step rate determining

$$\alpha_a = 0.5 \quad \text{and} \quad \alpha_c = 2.5$$

and

(b) for the case of the second step [Eq. (8)] rate determining

$$\alpha_a = 2.0 \quad \text{and} \quad \alpha_c = 1.0$$

Considerations of the "negative difference effect" [see Section III(5(iv))] indicate the second case to be the likely one, but no direct experimental evidence has been obtained so far.

In any case, experiments reviewed below [see Section III(3)] indicate that j_0 is very large, i.e., significant currents can pass without much polarization. Hence, in such a case, the linearized form of Eq. (19) should be valid, i.e.,

$$\Delta(\Delta\phi_{M/O}) = \frac{3F}{RTj_0}j \qquad (20)$$

(*ii*) O/S Interface

(*a*) *Ion transfer across the interface*

The problem of ion transfer across the interface has been treated in detail by Sato,[26,27] Scully,[28] and also Valand and Heusler,[29] following the general theory of Vetter.[30] Valand and Heusler assumed the same type of activation-controlled charge transfer kinetics, except that the dominant charge here is that on the O^{2-} ions (or OH^- ions) obtained by splitting water at the interface. The electrochemical double layer here is of the usual type for aqueous systems and the equilibrium p.d. is determined by the main charge transfer reaction

$$_{sol}H_2O = {}_{ox}O^{2-} + 2\,{}_{sol}H^+ \tag{21}$$

or, alternatively,

$$_{sol}H_2O = {}_{ox}OH^- + {}_{sol}H^+ \tag{22}$$

and

$$_{sol}OH^- = {}_{ox}OH^- \tag{23}$$

The species entering the oxide do not, of course, stay as free entities but are likely to combine into hydrated oxide species of different charges and degrees of hydration. The equation describing the kinetics is

$$j_{O^{2-}} = k^+_{O^{2-}} a(OH^-)^y \exp\left[\frac{\alpha'_a F}{RT} \Delta(\Delta\phi_{O/S})\right] - k^-_{O^{2-}} a(OH^-)^{(2-y)} \exp\left[\frac{(2-\alpha'_a)F}{RT} \Delta(\Delta\phi_{O/S})\right] \tag{24}$$

where $\Delta(\Delta\phi_{O/S}) = (\phi''_O)_j - (\phi''_O)_0$ is the change of the p.d. from the equilibrium one to the one needed to drive j.

One could also expect protons to cross the interface causing protonation (hydration) of the oxide:

$$_{sol}H_3O^+ = {}_{ox}H^+ + {}_{sol}H_2O \tag{25}$$

Also, any protons formed in the oxide from OH^- ion splitting should cross the interface in the opposite direction (the Hoar-Yahalom mechanism of field-assisted proton transfer[31]).

Inasmuch as the protonation of the oxide can be favored by the direction of the field at the O/S interface (cf. Fig. 1) at equilibrium, the proton current in this direction should decrease exponentially with increasing anodic polarization, as the field strength is decreasing and can even change sign. Conversly, the deprotonation should be favored, becoming the main mechanism of formation of the anhydrous oxide.

However, the Hoar–Yahalom mechanism[31] has been questioned by a number of scientists.[4,19,32]

Transfer of aluminum ions from the oxide into the solution was considered as a statistically independent process, whose kinetics are governed by a rate equation similar to Eq. (24), i.e. (neglecting the return of the ions into the oxide),

$$j_{Al^{3+}} = k^{+}_{Al^{3+}} a(OH^{-})^{r} \exp\left[\frac{F\gamma}{RT}\Delta(\Delta\phi_{O/S})\right] \tag{26}$$

One should note that $j_{O^{2-}}$ is the oxide-forming current while $j_{Al^{3+}}$ is the dissolution current. Hence, their ratio determines η_{ox}.

Valand and Heusler[29] determined $j_{Al^{3+}}$ experimentally, by chemical analysis of the solution. Hence, the oxygen ion current could also be estimated from the total current and $j_{Al^{3+}}$.

The overpotential $\Delta(\Delta\phi_{O/S})$ could not be experimentally determined. However, taking only the first term in Eq. (24) (which is a reasonable assumption at any real anodic dissolution current density), one could derive the ratio of the Tafel slopes of the two currents as

$$(\partial \ln j_{O^{2-}}/\partial \ln j_{Al^{3+}})_{pH} = \alpha'_a/\gamma \tag{27}$$

Indeed, in electrolytes containing no "aggressive" anions (as are the halides), over a wide pH range between 0 and 12 (from sulfuric acid through acetate to phthalate and borate), a double logarithmic plot of $j_{O^{2-}}$ versus $j_{Al^{3+}}$, shown in Fig. 5, yielded straight lines, with slopes of 1.38 ± 0.14.

Similarly, pH dependences of $j_{O^{2-}}$ and $j_{Al^{3+}}$ (shown in Fig. 6) give

$$(\partial \log j_{O^{2-}}/\partial pH) = y - \alpha'_a \tag{28}$$

and

$$(\partial \log j_{Al^{3+}}/\partial pH) = r - \gamma \tag{29}$$

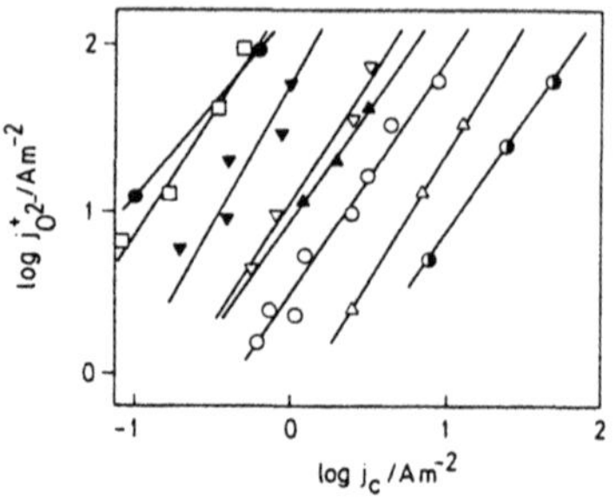

Figure 5. Double logarithmic plot of current density of oxygen ion incorporation into the oxide, $j^{+}_{O^{2-}}$, versus aluminum dissolution current density, j_c, at different pH values: ◑, pH 0.0; △, pH 1.55; ○, pH 4.63; ▽, pH 5.53; ▼, pH 6.9; ●, pH 8.9; □, pH 9.85; ▲, pH 11.0.

Figure 6 indicates a change in the charge transfer mechanism at a pH between 9 and 9.5, corresponding to the pH of zero charge of aluminum oxide.[33,34] Experimental results on the slopes enabled speculation on the values of transfer coefficients and reaction orders. From that, Valand and Heusler concluded that the most probable mechanism of oxygen ion transfer [reaction (21)] is

$$H_2O = {}_{ad}OH^- + {}_{sol}H^+ \tag{30}$$

$${}_{ad}OH^- = {}_{ad}O^{2-} + {}_{sol}H^+ \tag{31}$$

$${}_{ad}O_2 \rightarrow {}_{ox}O^{2-} \tag{32}$$

with the last step rate determining.

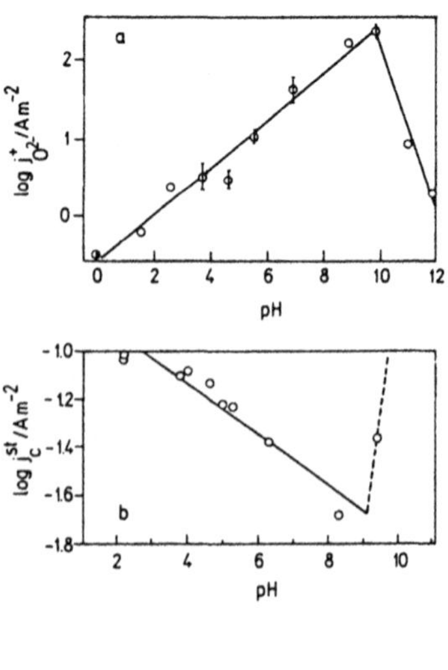

Figure 6. pH dependence of (a) current density (on log scale) of oxygen ion incorporation into the oxide, at a constant total current density of 0.1 mA/cm^2, and (b) the steady-state dissolution (aluminum ion) current density of oxide-covered aluminum at 4 V versus SCE.[29]

As far as the aluminum ion transfer is concerned, the indicated rate-determining step is

$$_{ox}AlOH^{2+} \rightarrow {}_{sol}AlOH^{2+} \qquad \text{at pH} < 9 \qquad (33)$$

and

$$_{ox}AlOH^{2+} + {}_{sol}OH^{-} \rightarrow {}_{sol}Al(OH)_2^{+} \qquad \text{at pH} > 9.5 \qquad (34)$$

It is obvious that such an ion transfer must be preceded by some association of the aluminum ion from the oxide lattice with OH^- ion (directly from the solution or adsorbed at the interface) [Eq. (22) or (23)] or by protonation with H^+ ions from the solution (Eq. (25)]. Valand and Heusler maintain the first case to be operative. This conclusion must, however, be taken as tentative, and further arguments of an experimental nature are warranted.

Additional effects are produced by the presence of "aggressive" anions in the electrolyte. They are treated in Section III(5).

(*b*) *Chemical dissolution of the oxide*

In the above considerations, the O/S interface was taken to be a clear-cut boundary between the oxide and the electrolyte. In reality, however, the outer part of the oxide is likely to be hydrated and penetrated by the electrolyte. Hence, the true O/S interface is likely to be withdrawn from the surface to a sufficient depth such that some oxide is left without any electric field imposed across it. This is especially true of thick porous oxide layers, but it can occur with compact layers as well. For example, Hurlen and Haug[35] found a duplex film in acetate solution (pH 7-10), composed of a dry barrier-type part and a thicker hydrated part consisting of $Al_2O_3 \cdot \frac{1}{2}H_2O$. Although the hydrated part becomes thinner with decreasing pH and seems to practically vanish at low pH, even a thickness of less than a nanometer is sufficient for the surface oxide to stay outside the electrochemical double layer.

In such a case, chemical interaction with the solution can take place. This is likely to be primarily the protonation (hydration) reaction leading to formation of some soluble complex ions which diffuse toward the bulk of the solution. Hence, this produces a mechanism for dissolving and thinning the oxide layer, and the rate of this process should be some function of the hydrogen ion concentration in solution. In such a case, the attainment of a "steady state" in which the oxide stops growing does not necessarily imply an equilibrium situation with respect to oxygen ion transfer across the O/S interface ($j_{O^{2-}} = 0$), as assumed by Valand and Heusler, but rather the one governed by material balance. Hence, $j_{O^{2-}}$ should, in such a case, be larger, the higher the rate of chemical dissolution.

(*c*) *Proton buildup effects*

The above considerations are based on the assumption that the pH at the O/S interface is constant and equal to that in the bulk of the solution. However, in view of the fact that the formation of each molecule of oxide is accompanied by liberation of six protons (see Section II, point 10), this need not be so, and this also appears to affect the extent to which the oxide layer will grow during anodization (the efficiency of oxide formation, η_{ox}) and the type of oxide that will be formed (compact "barrier type" or porous).

The problem of the changes in pH close to and at the O/S interface attracted attention primarily in relation to pit growth in localized (pitting) corrosion and in attempts to predict the hydrogen ion concentration inside the pit. Thus, in 1937, Hoar suggested an "autocatalytic" mechanism of pit propagation,[36] the basic reason behind which was a drop in pH inside the pit. Indeed, in a number of investigations,[37-42] the pH was found to decrease significantly inside pits. Galvele[43] has attempted to calculate the point, depending on a product of pit depth and current density, at which the pH falls below that of stability of the insoluble oxide (hydroxide). His model was unrealistic inasmuch as he assumed that the metal is dissolved directly in the electrolyte in the form of a free metal ion (zero current efficiency of oxide formation), the only cause of decreased pH being the hydrolysis of this ion. Nevertheless, he showed that for aluminum ions in solution, the hydrogen ion concentration could change, for this reason only, by four orders of magnitude for a product of current density and diffusion layer thickness ("depth of a pit") of 10^{-3} A/cm.

Inasmuch as this was denied as the possible cause of pit initiation (cf. Vetter and Strehblow[44]), there should be no doubt that, not only in pits, but wherever anodic dissolution of aluminum and oxide formation takes place, a significant effect of a pH change can be expected. Possible events are represented schematically in Fig. 7,[45] in which consumption of H^+ ions by buffering with anions is also envisaged.

With such a model, the rate of increase in oxide thickness is determined by the difference between alumina formation, strictly following Faraday's law, and its dissolution, the rate of which should be some function of hydrogen ion concentration at the interface, i.e.,

$$\frac{d\delta}{dt} = \frac{M}{\rho}\left[\frac{j}{3F} - kC(\mathrm{H}^+)^n\right] \tag{35}$$

where M and ρ are the molecular weight and density of the alumina, C represents concentration, and k may be an electrochemical or a chemical rate constant, but in view of the above considerations, it is more likely to be the latter.

In any case, the time dependence of the hydrogen ion concentration at the interface should be obtainable from a mass-balance equation as

$$\frac{dC(\mathrm{H}^+)}{dt} = \frac{j}{F} - kC(\mathrm{H}^+)^n - \vec{k}C(\mathrm{A}^-)C(\mathrm{H}^+) + \overleftarrow{k}C(\mathrm{HA}) - \Phi_D(t) \tag{36}$$

the source being proportional to the current density and the three sinks being (a) the reaction with the oxide, (b) the net association

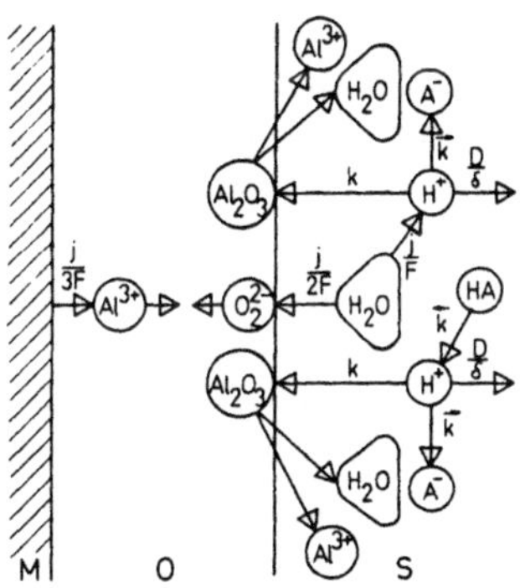

Figure 7. Schematic representation of events at the O/S interface in the presence of a buffering electrolyte (HA/A^-).

with the anion of the electrolyte (with rates of association and dissociation being determined by the rate constants $\vec{k}$ and $\overleftarrow{k}$, respectively), and (c) molecular (or convective) diffusion into the electrolyte represented by the diffusion flux Φ_D. The latter is time dependent in a complex way. Hence, Eq. (36) cannot be integrated in a straightforward manner. Nevertheless, different situations can be discussed:

1. If the buffering capacity of the electrolyte is significant and the association/dissociation rates are very fast so that the corresponding two terms dominate all the others, then

$$\frac{dC(\mathrm{H^+})}{dt} = -\vec{k}C(\mathrm{A^-})C(\mathrm{H^+}) + \overleftarrow{k}C(\mathrm{HA}) = 0 \tag{37}$$

i.e., the buffering equilibrium is virtually not disturbed by hydrogen ion formation, and $C(\mathrm{H^+})$ is equal to that in the bulk of solution:

$$C(\mathrm{H^+}) = \frac{\overleftarrow{k}}{\vec{k}}\frac{C(\mathrm{HA})}{C(\mathrm{A^-})} = K_a\frac{C(\mathrm{HA})}{C(\mathrm{A^-})} \tag{38}$$

Substituting this in Eq. (35), the rate of growth of the oxide is seen to be constant, i.e., the oxide layer thickness increases linearly with time.

2. If there is no buffering capacity [$\overleftarrow{k} \to \mathrm{O}$ and $C(\mathrm{HA}) \to \mathrm{O}$] and if the diffusion away is very slow, the concentration of $\mathrm{H^+}$ ions at the interface will grow until the second term in Eq. (36) becomes equal to the first one. Substituting such a condition in Eq. (35), one can see that the rate of growth of the oxide becomes zero, i.e., the oxide attains a constant thickness. (In fact, some hydrogen ions will always escape by diffusion and, hence, complete equality of the two terms in Eq. (35) can never be attained so that some growth will have to continue.)

A more detailed discussion of the problem, including some approximate solutions for the time dependence of oxide growth, is available in Ref. 46.

In reality, the increase of the rate of dissolution due to increasing hydrogen ion concentration should increase the overall rate over that of the dissolution via direct transfer of aluminum cation through the O/S interface discussed above.

(iii) *Transfer of Species through the Oxide*

A number of reviews are available on the transfer of species through the oxide.[5,47-49]

Ionic conduction studies in solids date back to the work of Gunterschultze and Betz,[50] who derived the empirical relationship between the electric field strength, $E = [(\phi_0')_j - (\phi_0'')_j]/\delta$ (cf. Fig. 1b), in an oxide and the nonohmic ionic current density, j,

$$j = A \exp(BE) \tag{39}$$

where A and B are constants depending on the temperature. Cabrera and Mott[51] have deduced this equation from a model assuming the influence of an electric field, E, on the barrier height for migrating ions at oxide interfaces:

$$j_i = \frac{DC_0}{2a} \exp\left(\frac{zeaE}{kT}\right) \tag{40}$$

where D is a diffusion coefficient, C_0 is a surface density of moving ions, and a is an activation distance of migration. (This equation cannot be directly applied in the case of nonplanar oxide geometry.)

Equations of this type are most often used by experimentalists to fit their data to theory,[52-54] as indications of an exponential $j(E)$ dependence are numerous. Vervey[55] has used a similar approach to consider the volume-limited processes of ionic migration and obtained the same equation. In order to explain deviations of experimental behavior from that predicted by Eq. (40), Vermilyea[56] has also taken into consideration the effect of electrostriction modifying the activation distance for migrating ions and obtained the following equation for the ionic current flow:

$$j_i = \frac{DC_0}{2a} \exp\left[\frac{ze(\alpha + \beta E)E}{kT}\right] \tag{41}$$

where α and β are constants.

An analogous expression assuming space charge effects and the double layer structure of the anodic oxide has been obtained by Goruk *et al.*[5] and Bray.[57]

Dewald[58] has introduced a mechanism in which space charge generation in the anodic oxide results from the difference between activation energies at the oxide surface and in the bulk. Winkel *et al.*[59] considered the influence of the amorphous structure of the oxide on ionic migration by assuming the oxide to consist of a series of potential barriers with Gaussian distributions of activation distances and heights. The following equation for the ionic flux resulted:

$$j_i = \exp\left(\frac{zea'L - W'}{kT}\right) \tag{42}$$

where $W' = \langle W \rangle + \delta_W/2kT$ and $a' = \langle a \rangle - ze\delta_a E/2kT$ are mean values, with δ_w and δ_a variations of activation heights and distances, respectively. Young and Zobel[60] have assumed ions to migrate over ramified trajectories and be captured by the oxide region locally charged by the charge of opposite sign, leading to the following expression for the ionic current:

$$j_i = j_{i0} \exp\left(\frac{W - \beta E^{1/2}}{kT}\right) \tag{43}$$

where j_{i0} is a constant characteristic of the material. Fromhold[61] has introduced space charge effects of moving ions disturbing the exponential $j(E)$ law. This approach has been developed further in his more recent work.[48,49]

All the above derivations are based on the assumption of a single ionic species moving through the oxide. The implications of such an approach have been considered most thoroughly by Dignam.[47] The present state of the art in the field of ionic conduction modeling needs improvement. The theory should include the following:

(a) Multi-ionic migration. The majority of metals and semiconductors exhibit both cationic and anionic migration, each of them in its turn able to be a multicomponent one.

(b) Amorphous structure of anodic oxides. Since the work of Winkel *et al.*,[59] there have been attempts to consider ionic conduction in disordered anodic oxides. One such attempt, based on hopping conduction of charged carriers, was made by Parkhutik and Shershulskii[62] [cf. Section V(2)].

(c) The heterogeneous chemical structure of anodic oxide films across the surface and perpendicular to it. The effects of a composition varying with depth, causing complications in modeling the ionic conduction, are currently attracting the attention of researchers. Fromhold and Fromhold[49] have modeled uni- and multipolar ionic migration in anodic oxides possessing varying stoichiometry and have shown that moving ions participate in oxidation processes in the bulk of the oxide, thus changing its stoichiometry. Parkhutik and Shershulskii[62] have modeled the heterogeneous chemical composition of the oxide by considering the inhomogeneous space charge of heterogeneously incorporated ionic species.

2. Open-Circuit Phenomena

A metal which has been handled in the presence of oxygen from the air is always covered by a thin protective layer of oxide.[63-65] As soon as it is immersed into an aqueous electrolyte, even if free of oxygen, a corrosion process starts, imposing after some time a steady-state potential difference between the metal and the solution [open-circuit potential (OCP)]. The corrosion process could be represented by a Wagner–Traud (Evans) model.[66] However, the partial current of the anodic dissolution of the metal has somewhat unusual characteristics. It remains constant on changing the potential in a positive direction and increases only with an increase in pH. It appears that the dissolution process is controlled by the transport of OH^- ions in solution.[67] (It should be noted, however, that suppression of the dissociation of water [reverse of reaction (30)] would lead to a similar pH effect.) This remains so until a certain potential is reached, which depends on the anion of the electrolyte. At that potential, a sudden rise in current density occurs [cf. Section III(5)]. Driving the potential negative from the OCP results again at a certain point in a sudden increase in the anodic dissolution rate (cathodic pitting corrosion). This resembles the passive–active transition at passivating interfaces. It is interpreted[68] as arising from the hydration of the oxide film due to hydroxyl ion formation accompanying increasing cathodic evolution of hydrogen. Hence, the Wagner–Traud diagram (in the absence of oxygen in solution) should resemble the one presented in Fig. 8.†

† It is noteworthy that cathodic polarization of aluminum in highly diluted acids can also cause oxide formation (so-called "cathodization").[69] This effect is due to the fact that proton discharge at the cathodic potentials results in a pH increase and, hence, coagulation of aluminum hydroxide near the cathode. This causes colloidal $Al(OH)_3$ deposition at the cathode, especially if it is made of aluminum.

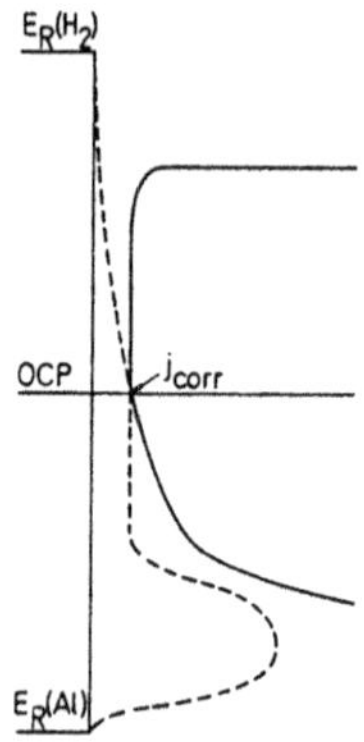

Figure 8. Wagner-Traud (Evans) diagram for aluminum in aqueous solution, in the absence of dissolved oxygen.

Thus, it is interesting to note that high-purity aluminum rests at a potential at which corrosion is at its minimum and is, indeed, relatively very small. It is also largely independent of the anions present in the electrolyte.[69] This may be attributed to the coulombic repulsion of anions away from the surface by the negative charge on the metal. The latter seems not to be completely compensated in a thin oxide film, as shown schematically in Fig. 9, so that the solution side of the double layer formed at the O/S interface contains excess cations, anions being repelled. The anions could approach the O/S interface either at thicker films or at potentials more positive than the OCP.

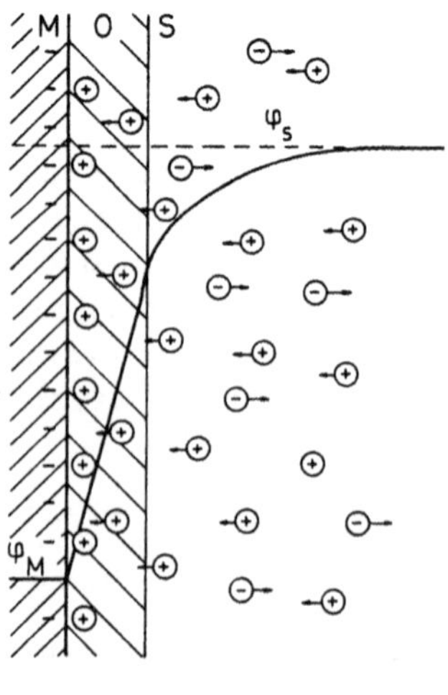

Figure 9. Schematic representation of potential profile and charge distribution in a thin anodic oxide film on aluminum.

When the oxide is formed by anodizing in acid solutions and the sample is then left to rest at the OCP, some dissolution can occur. This process has been studied by a numbers of authors,[70-75] especially in relation to porous oxides [cf. Section III(4)]. It was found that pore walls are attacked, so that they are widened and tapered to a trumpet-like shape.[70,71] Finally, the pore skeleton collapses and dissolves, at the outer oxide region. The outer regions of the oxide body dissolve at higher rates than the inner ones.[9,19] The same is true for dissolution of other anodic oxides of valve metals.[76] This thickness dependence is interpreted in terms of a depth-dependent vacancy concentration in the oxide[75] or by acid permeation through cell walls by intercrystalline diffusion, disaggregating the microcrystallites of γ-alumina.[4]

As for the thinning of the barrier film in such a case, it can be understood in terms of the effects discussed earlier [cf. Section III(1(ii))], as the relaxation of anodic polarization increases the rate of proton transfer. Thus, the hydration of the outer regions of the film takes place, resulting in double-layer withdrawal and chemical dissolution at the surface.

The open-circuit dissolution is very sensitive to temperature (unlike the field-stimulated dissolution). It was found that a temperature rise from 20 to 25°C doubles the rate of dissolution.[74] The rate of dissolution is, of course, pH dependent and is virtually nil at higher pH.

In the presence of oxygen in solution, the cathodic reaction becomes that of oxygen reduction. This can shift the OCP in the positive direction and bring it to, or close to, the potential of the sharp rise of anodic current ["pitting potential"; cf. Section III(5(i))].

3. Kinetics of Barrier-Film Formation

The compact, nonporous anodic alumina film is the most suitable for fundamental investigations. It is grown by anodization, mostly under constant-current (galvanostatic) conditions, in neutral solutions of borates, tartrates, citrates, and phosphates, all of which possess significant buffering capacity and hence do not allow significant dissolution of the oxide.

(*i*) *Dependence of Potential and Current on Time*

The growth of an anodic alumina film, at a constant current, is characterized by a virtually linear increase of the electrode potential with time, exemplified by Fig. 10, with a more or less notable curvature (or an intercept of the extrapolated straight line) at the beginning of anodization.[73] This reflects the constant rate of increase of the film thickness. Indeed, a linear relationship was found experimentally between the potential and the inverse capacitance[78] (the latter reflecting the thickness in a model of a parallel-plate capacitor under the assumption of a constant dielectric permittivity). This is foreseen by applying Eq. (38) to Eq. (35). It is a consequence of the need for a constant electric field on the film in order to transport constant ionic current, as required by Eqs. (39)-(43).

The intercept should reflect the unchanging activation polarization at the two interfaces, as well as some other effects (presence of a film before anodization, time lag in attainment of the steady state, etc.). Nevertheless, the fact that it is small or negligible indicates that charge transfer processes at the interfaces are fast and that the kinetics of the growth are entirely transport controlled.

The linearity leads to another important conclusion: a constant field for a constant current implies a constant overall conductivity throughout the film. Since the conductivity is very structure sensitive, this implies also that either (i) the film grows homogeneously

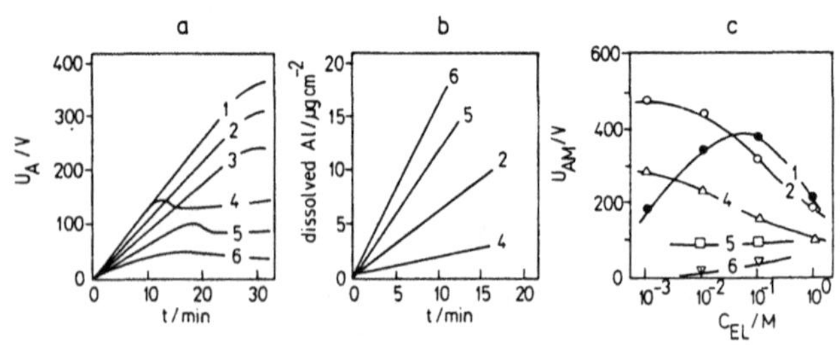

Figure 10. Experimental records of galvanostatic aluminum anodization ($j_a = 5\ A/m^2$) in various electrolytes: (1) adipate, (2) citrate, (3) tartrate, (4) phosphate, (5) oxalate, (6) borate. (a) Anode potential versus time; (b) dissolved aluminum versus time; (c) maximum forming voltage versus electrolyte concentration.[20]

throughout or (ii) any inhomogeneity expands in proportion to the thickness. (This applies also to the space charge since it affects the field in the interior of the oxide). It has been shown that, in at least some instances,[79] the second case is operative [cf. Section IV(2)].

(ii) *Current–Field Relationship*

The fact is that, on the one hand, a significant field strength, E, is needed to provide significant current. On the other hand, once in the practical current density range between 0.1 and 10 mA/cm^2, a relative insensitivity of the field to the current density is found. In fact, an inverse field of 1.3 to 1.8 nm/V is accepted in the literature as characteristic of the oxide growth, without mention of the current density used.

Equations (39)-(43) indicate that a Tafel-type relationship between the current and the field should be expected, i.e., from Eq. (40),

$$\log j = \frac{zea}{2.3kT} E + \log \frac{DC_0}{2a} \tag{44}$$

or, from Eq. (43),

$$\log j = \frac{\beta}{2.3kT} E^{1/2} - \left(\frac{W}{2.3kT} + \log j_0 \right) \tag{45}$$

Figure 11 exemplifies experimentally observed dependences of the current density on the imposed voltage at a constant oxide thickness. It is seen that they fit both equations fairly well. Tafel slopes [Eq. (44)] are in the range of 0.4 V cm^{-1} dec^{-1}. All other

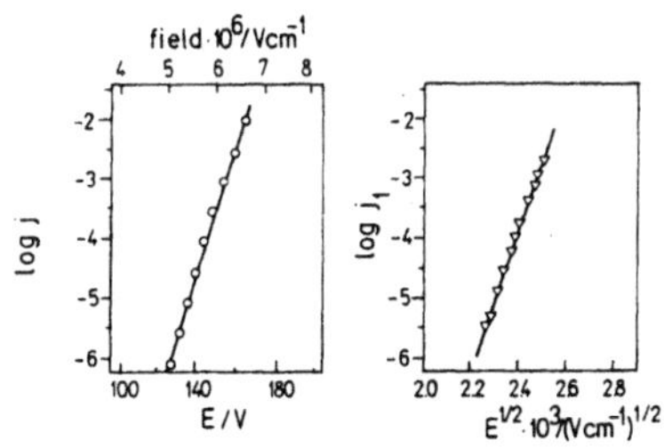

Figure 11. Steady-state current density as a function of imposed voltage at a constant thickness of a barrier-type oxide, in $\log j$ versus E and $\log j$ versus $E^{1/2}$ coordinates. (Based on data from Ref. 5.)

constants being known, the values of the slopes in Fig. 11 imply $za = 0.15$ or $\beta = 1.1 \times 10^{-19}$.

In a number of works, a potentiostatic regime has been used for the experimental and theoretical study of the anodization of aluminum and other valve metals.[80] Upon the application of a constant potential step, V_a, barrier-forming electrolytes are characterized by a sharp increase in the anodic current to a certain maximum. Both the slope and the maximum are determined by the impedance of the cell circuit. Subsequently, there is a continuous decrease in the anodic current, which is due to oxide growth. The decay of the anodic current can be described by the expression[81]

$$\frac{1}{j_a \ln(j_a/j_{a0})} - \frac{1}{j_0 \ln(j_0/j_{a0})} = \frac{kE^*}{V_a} t \tag{46}$$

where $j_0 = j_a$ at $t = 0$, $j_{a0} = DC_0/2a$, k is a constant, representing oxide volume per migrating ion, and E^* is a constant. Within a narrow time interval, Eq. (46) can be approximated by a hyperbolic $j_a t^{-1}$ dependence. It then leads to a logarithmic time dependence of oxide growth[52]

$$\delta' = \frac{1}{BE_0} \ln(BE_0 t' + 1) \tag{47}$$

where δ' and t' are dimensionless thickness and time, respectively, E_0 is an initial electric field, and B is a constant.

Satisfactory agreement of experiments with kinetic laws, described by Eqs. (44) and (45), are observed only for tantalum and niobium, when the current efficiency approaches 100%. Even for these metals, certain deviations occur which could be attributed to space charge effects,[82] electronic leakage currents,[83] or other factors. In the case of aluminum, these deviations are relatively large, as, even in barrier-forming electrolytes, some oxide dissolution takes place from the very beginning of voltage supply to an anodized sample.[32]

(iii) *Oxide Layer Thickness*

The oxide layer thickness can be determined in a number of ways. Direct microscopic observation has been demonstrated by Takahashi *et al.*[79] Results of other methods can be made to agree

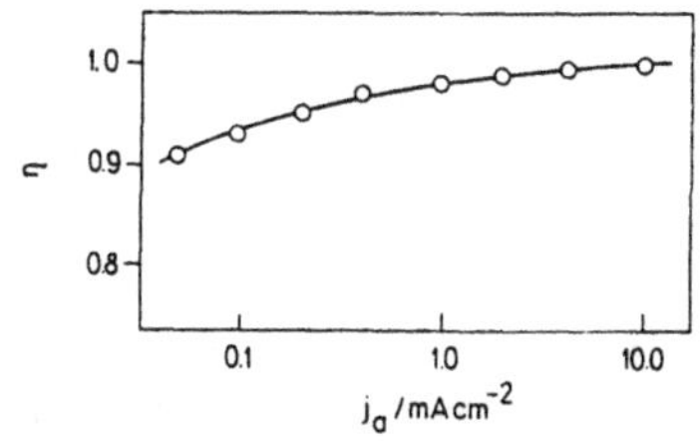

Figure 12. Efficiency of oxide formation as a function of anodic current density.[79]

with this direct evidence, by assuming certain values of some constants. Thus, capacitance measurements could render the same result if it is assumed that the dielectric constant is 9.8 (a value higher than the generally assumed one of 8.4, but explainable when a surface roughness of 1.2 is taken into account). The thickness determined by dissolving the oxide and determining the amount of aluminum ions analytically would agree with the value obtained by direct observation, if the density of the oxide is taken to be 2.95 and the weight fraction of Al^{3+} in the oxide, 0.51. Finally, from the voltage-to-field ratio, one could calculate the same thickness by taking the inverse field (thickness-to-field ratio) as 1.5 nm/V, which is in the usually observed range.

(*iv*) *Efficiency of Oxide Formation*

The efficiency of oxide formation can be found by comparing the amount of aluminum determined analytically upon dissolving the film to that indicated by Faraday's law. As seen in Fig. 12, the efficiency is found to be less than 1 and increases with increasing anodic current density. On the one hand, this is in line with the finding of Valand and Heusler[29] that the oxygen ion current increases more steeply with $\Delta(\Delta\phi_{O/S})$, and hence with the total current j, than the aluminum dissolution current (α_a' being larger than γ). On the other hand, this is also predicted by Eq. (35), reflecting chemical dissolution. In this instance, it is not possible to decide which of the two causes is operative.

(*v*) *Ionic Migration in the Oxide Film*

The anodic oxide is formed, generally speaking, by migration of the positive ion (of the oxidized metal) and/or of the negative ion(s) (oxygen ions, hydroxyl ions, or even anions of the electrolyte) inside the oxide toward the opposite interface under the influence of the very high electric field created by the externally applied potential. The question arises as to what contribution each of the ionic species makes to oxide formation, which reflects itself in the position of the original oxide layer, existing prior to the anodization. The problem amounts to transport number determination and can be solved by various methods.

Radiotracer techniques involving ^{18}O in the anodization process are used with subsequent neutron activation analysis[84] or SIMS.[85] Another method involves implantation of inert ion markers into the surface layer of the sample prior to anodization and examination of the position of the markers after the oxide film has grown to a certain thickness.[86] Assuming immobility of the inert species, the ratio of the cation to the anion transport number, t_+/t_-, should be equal to the ratio of the outer to the inner layer thickness. Numerous experimental determinations[72,87] suggest t_+ and t_- to be 0.4 and 0.6, respectively.

However, there are indications that these values depend on the conditions of ionization. Vermilyea[88] has interpreted the change from compressive to tensile stress, recorded in the oxide, to be due to the dependence of the transport number of aluminum on the electric field strength. Brown[89] has found this transport number to depend on the electrolyte used in anodization.

A systematic study by Khalil and Leach,[90] using α-spectrometry, has provided values for transport numbers in the valve metal oxides, which are interesting to compare. These are listed in Table 1.

Takahashi *et al.*,[79] in their work on the structure of the barrier layer [cf. Section IV(2)], have considered phosphate ions, which are found in the outer layer of the oxide, as immobile markers and, from the position of the boundary between the outer and the inner layer, deduced the transport number of the cation to vary between 0.73 and 0.81 in the current density range between 0.05 and 10 mA/cm^{-2}.

A similar method with similar results was used earlier by Randall and Bernard,[91] but the objection that the result may be too high because of the motion of the phosphate ions was raised by Davies *et al.*[92] and Pringle.[93] Smaller values independent of the

Table 1
Transport Numbers for Various Metals Anodized in Borate Solutions at 25°C[a]

Metal	Current density (mA/cm^2)	Oxide thickness (nm)	Electric field strength (V/nm)	t_+
Ti	6	42.1	0.427	0.35
	50	63.3	0.742	0.39
Zr	6	200	0.5	0.1
	50	179.2	0.588	0.12
Ta	6	160.6	0.623	0.28
	50	133.0	0.752	0.34
Al	6	110	0.91	0.4
	50	110	0.91	0.49

[a] Ref. 90.

marker were obtained when inert gas markers were used.[92,93] However, even in this case, objections were found in that the method was shown to depend on the film existing prior to the anodization[94] as well as on the acceleration voltage used to implant the marker.[95]

The above evidence, however, shows very clearly that in all barrier film making, both positive and negative ions contribute in comparable proportions to the building of the oxide and hence that the oxide grows in both directions leaving the original oxide buried somewhere inside, close to the center.

At a certain anodic potential, the compact film breaks down and lets electrons pass through without much resistance, causing oxygen evolution at a high rate. This "dielectric breakdown" is discussed in more detail in Section V.

4. Formation of Porous Oxides

(*i*) *Dependence of Potential and Current on Time*

Dibasic and tribasic acids, such as sulfuric, oxalic, malonic, and phosphoric acids,[96] cause the appearance and development of a very regular porous structure of the oxide (cf. Fig. 3). Here, the kinetics of galvanostatic anodization are characterized by an initial linear potential rise, followed, however, at relatively low anodic potentials, by curving to a maximum value and a decrease to a steady-state value, as exemplified by Fig. 13a.[74,97]

It should be noted here that the barrier-film-promoting electrolytes are also characterized by $V_A(t)$ curves similar to those of the pore-forming ones, if comparatively small current densities are used (less than $0.5 mA/cm^2$).[20]

The maximum and steady-state anode potentials depend on the pH of the electrolyte in a manner shown in Fig. 13b, which was obtained for anodization in sulfuric acid solutions with the pH adjusted to constant ionic strengths with Na_2SO_4. This dependence can be expressed as

$$V_{A,0} = A \ln(B - C[H^+]) \tag{48}$$

where A, B, and C are empirical constants.[93] The same behavior was observed by Fukushima *et al.* for a variety of electrolytes.[9]

The parameters of the $V_A(t)$ function are very sensitive to temperature, as seen in Fig. 13a. As the temperature rises, the position of the maximum shifts to lower anodization times, with the value of the potential maximum also decreasing.

In the case of a potentiostatic oxidation with different applied potentials, as shown in Fig. 14,[98] there is always a dip in the current density corresponding to the potential maximum in galvanostatic

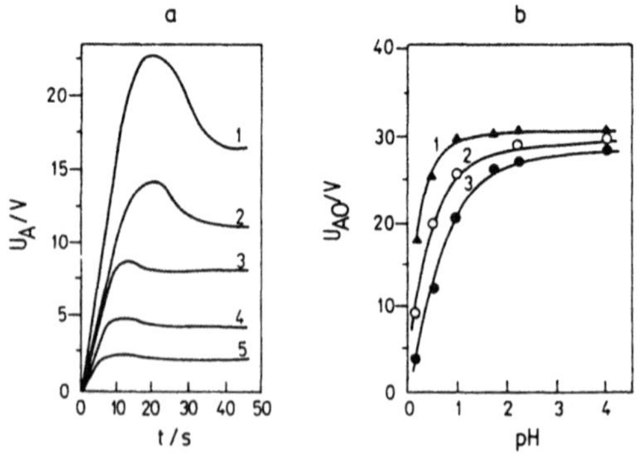

Figure 13. (a) Porous oxide growth kinetics for anodization in 15 wt % H_2SO_4 at $j_a = 5 mA/m^2$ and $T_e = 5°C$ (1), 15°C (2), 25°C (3), 35°C (4), and 45°C (5).[93] (b) Stationary anode potential versus electrolyte pH for anodization in H_2SO_4 solution at $j_a = 4 mA/cm^2$ (1), $2 mA/cm^2$ (2), and $1 mA/cm^2$ (3).[97]

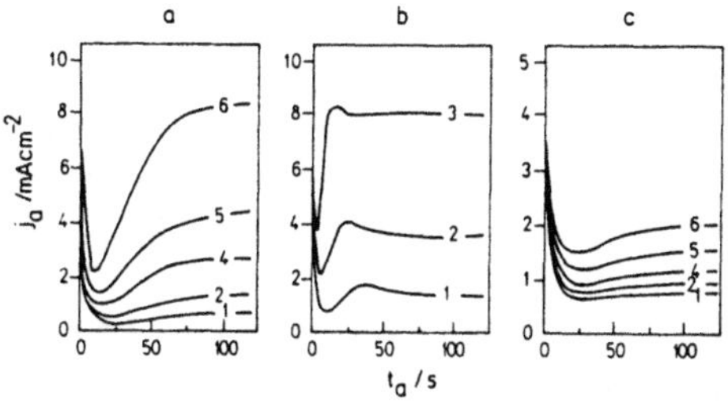

Figure 14. Anodic current versus time curves for potentiostatic Al anodization in oxalic (0.2 *M*) (a), sulfuric (0.5 *M*) (b), and phosphoric (0.4 *M*) (c) acid solutions. U_a = 5 V (1), 10 V (2), 15 V (3), 20 V (4), 30 V (5), and 40 V (6).

experiments, then a rise to a steady state, with the current density in some cases significantly larger than the initial current density.

The details of the $j_a(t)$ dependence are thoroughly examined in a number of works,[99,100] including, most recently, one by Hurlen *et al.*[101] His data indicate that aluminum anodization in slightly acidic electrolytes at low potentials (only slightly higher than that corresponding to the active-passive transition at about −1.0 V vs. SCE) reveals a similar current maximum, a sharp decrease, and further growth to a steady-state value, but at much higher potentials (Fig. 15).

Such a behavior of the $U_A(t)$ or $j_a(t)$ functions is consistent with a fairly well established pore growth mechanism.[4] According to this mechanism, the linear potential growth (and current density

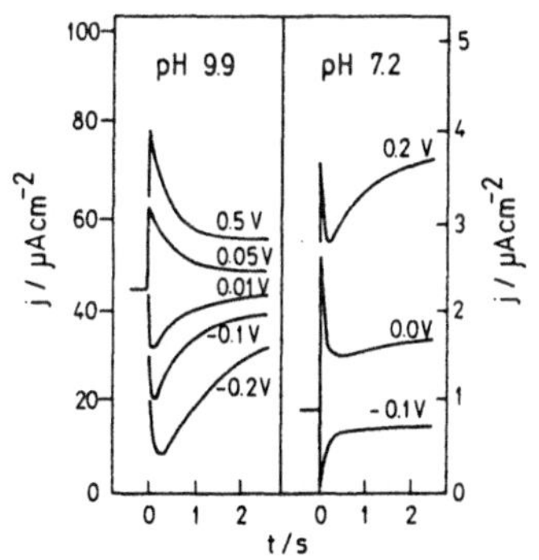

Figure 15. Kinetics of potentiostatic Al anodization in ammonium acetate at low potentials.[35]

fall) corresponds to the formation of a planar barrier film. After this film reaches a certain thickness, micropores are nucleated over the surface, which causes potential saturation and decrease. This actually means that, for some reason, microheterogeneity of the oxide comes into play, resulting in easier dissolution of the oxide at some points than at others [implying a varying dissolution rate constant in Eq. (35) over the surface]. This heterogeneity may be the same as that causing pitting phenomena under different conditions [cf. Section III(5)].

(ii) *Steady-State Potential–Current Density Relationship*

The steady-state potential (or current density) is related to a steady growth of the porous oxide into the solution, maintaining a constant number of pores and a constant pore radius. This scheme is supported by electron microscopic observations reported by Xu *et al.*[102]

Typical steady-state voltage–current characteristics in pore-forming electrolytes are shown in Fig. 16. A number of authors have attempted to interpret these dependences.[103,104] Ebihara *et al.*[105] used an equation based on a pore model and taking into account a rate-determining transport through the barrier part of

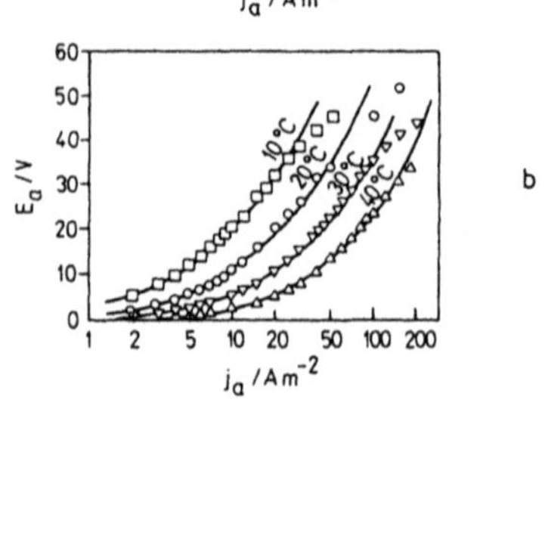

Figure 16. Relationship between anodic current density and steady-state voltage for aluminum in (a) sulfuric acid (2 *M*) (a) and oxalic acid (0.46 *M*), (b) solutions at different temperatures.[105]

the oxide:

$$j = A \exp\left(\frac{W - BE}{kT}\right) \tag{49}$$

They extracted values of the pre-exponential factor and of W in Eq. (49) for two electrolytes (sulfuric and oxalic acids) at different concentrations, A being in a reasonable range and W exhibiting remarkable constancy (0.847 and 0.751 for sulfuric acid and oxalic acid, respectively). It should be noted, however, that the model contained many adjustable parameters.

(*iii*) *Pore Nucleation Phenomena*

Two questions that arise are, why does pore formation occur only after the compact oxide has reached a certain value and, once the dissolution starts, why is it not even?

An answer to the first question may be found in noting that the electric field in a thin oxide film is different from that in a thick one and that weakening of electrostatic repulsion which prevents hydration and withdrawal of the O/S interface from the surface is a prerequisite for chemical dissolution.

One should note that dibasic and tribasic acids still have buffering capacity, since in the second (or third) dissociation step they behave as weak acids. Hence, it takes some time before the hydrogen ion concentration at the surface can increase sufficiently to start dissolving the oxide. Once this is achieved and a local attack on weaker oxide sites commences, additional buffering by the dissolved aluminum (oxo) ions prevents the further increase in hydrogen ion concentration needed for an equally fast dissolution of the "harder" oxide. The existence of different-quality oxides in a hexagonal arrangement, with an amorphous one in the middle and a crystalline one at the hexagon edges, was found by Franklin.[106]

5. Active Dissolution of Aluminum

Active anodic dissolution occurs when all the electrochemically oxidized aluminum passes into the aqueous phase and the oxide layer does not grow, i.e., the current efficiency of oxide formation

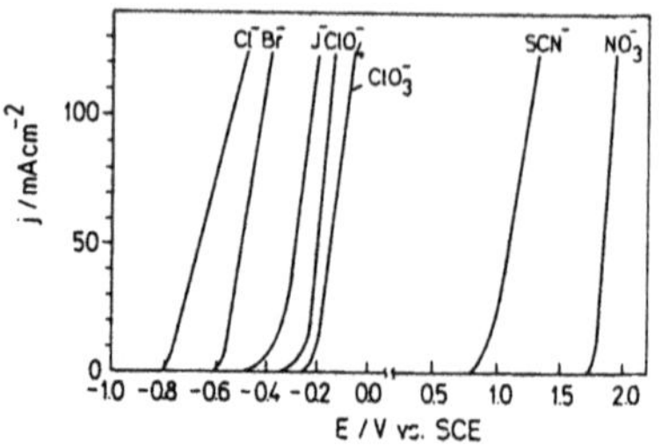

Figure 17. Increase in current obtained on aluminum upon sweeping the potential in the anodic direction in aqueous electrolytes containing different anions.[107]

falls virtually to zero. This phenomenon appears in the presence of a number of "aggressive" anions, in particular, halide ions and hydroxyl ion, although the mechanism of action of these two types of anions appears to be different.

In solutions containing different anions, as seen in Fig. 17, the sudden rise in the anodic current density mentioned earlier [see Section III(2)] and characteristic of initiation of active dissolution occurs at different potentials. It was shown[108] that, at least with halides, this potential is a linear function of the crystalline radius of the ion.

(i) Dissolution in the Presence of Halide Ions

If a well-defined compact oxide layer is grown to a certain thickness in a barrier-forming electrolyte (so that the electrode potential increases to very high values in order to maintain a constant current), when chloride ions are added, a dramatic decay of the potential results within milliseconds, as shown in Fig. 18.[77]

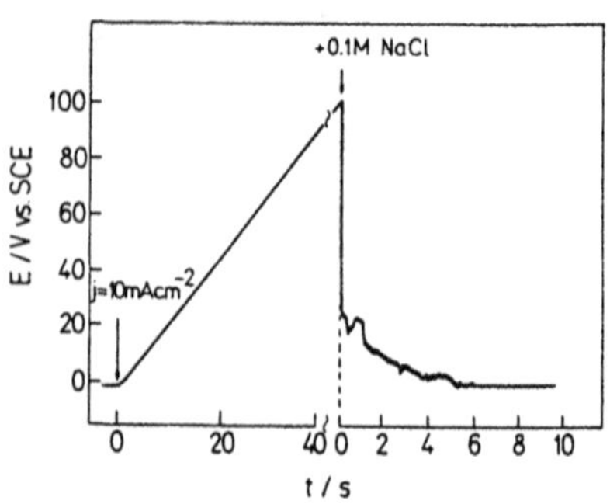

Figure 18. Decay of oxide formation potential at a constant current density of 10 mA/cm^2 upon addition of chloride to the electrolyte.

Further oxidation at the same current density takes place at a relatively low constant potential, indicating that the oxide has stopped growing, i.e., that the efficiency of oxide formation, η_{ox}, has changed from virtually 100% to zero.

As Kaesche pointed out for the example of chloride solutions, on which most of the studies were done,[67] at such a particular potential the current density rises with an infinite slope, i.e., at any current density applied galvanostatically, one and the same dissolution potential is obtained. Actually, careful determination and subtraction of the pseudo-ohmic potential drop between the Luggin capillary of the reference electrode and the electrode surface resulted in recording a negative slope for the true polarization curve, i.e., an increase in the current density led to a somewhat reduced anodic polarization.[109] This phenomenon was associated with the appearance of localized attack, or "pitting," and hence the potential plateau at which this occurs is termed the "pitting potential."[110] The pitting potential was found to be independent of pH (-485 ± 10 mV versus SHE for pH 2 to 11) and stirring, down to very low pH values, but was strongly dependent on chloride concentration down to 10^{-3} *M*. Of the two possible types of pitting found in corrosion of metals, shown in Fig. 19, aluminum in halide solutions develops the crystallographic, or "etch," pits.

The constancy of the potential with increasing current density could be explained in terms of an automatic adjustment of the number of pits while maintaining a constant current per pit. At potentials more positive than the pitting potential, Kaesche[67] has found the total current to increase with time. This complied very well with a model in which the true current density at the pit (found to be of the order of 300 mA/cm^2) and the number of pits,

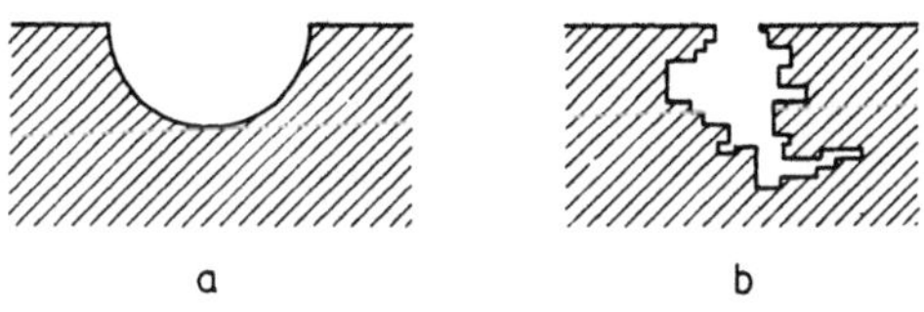

Figure 19. Schematic drawing of cross sections of two types of pits developing in pitting corrosion of passive metals: (a) geometric pit; (b) crystallographic or "etch" pit.

counted by optical microscopy, remain constant while the surface area of each pit increases as the third power of time. However, the fact that at the constant-potential plateau of active dissolution, the imposed constant current remains constant with time implies a decrease of the true current density at the increasing active surface inside the pit or a steady deactivation of some pits (reduction of the number of pits), or both. A constant current per pit could be caused either by ohmic control of the process at the pit mouth [which often remains unchanged in diameter while dissolution occurs underneath the surface ("undercutting")] or, more likely, by automatic adjustment of the surface film properties (e.g., thickness, by the hydrogen ion concentration effect). An interesting fact is that the dissolution potential also remains constant when the current density is decreased from any high value (up to 1 A/cm^2), even down to the lower limit at which the potential starts deviating toward the OCP (below 1 mA/cm^2). This implies an automatic deactivation of dissolution, i.e., either a reduction in the number of active pits or a decrease in the true current density at their bottom.

Recent studies of the processes of activation and deactivation[111] have shown, as seen in Fig. 20, that the time dependences of the potential, upon the application of current steps, resemble those characteristic of porous film formation and that the differences are of a quantitative nature. The initial part, representing a typical galvanostatic charging curve (with the initial jump due to the

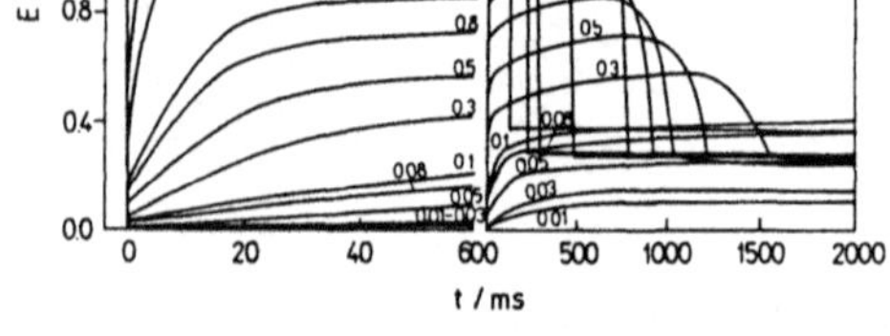

Figure 20. Initial change of potential of aluminum in 2 *M* NaCl solution upon application of different constant anodic current density steps.

pseudo-ohmic resistance and a subsequent charging of the interfacial capacitance), is not observed in porous film formation simply because the voltages involved are an order of magnitude higher than in the case of films obtained in halide-containing electrolytes.

The capacitance determined from the initial slopes of the charging curve is about $10\,\mu F/cm^2$. Taking the dielectric permittivity as 9.0, one could calculate that initially (at the OCP) an oxide layer of the barrier type existed, which was about 0.6 nm thick. A Tafelian dependence of the extrapolated initial potential on current density, with slopes of the order of 700–1000 mV/decade, indicates transport control in the oxide film. The subsequent rise of potential resembles that of barrier-layer formation. Indeed, the inverse field, calculated as the ratio between the change of oxide film thickness (calculated from Faraday's law) and the change of potential, was found to be about 1.3 nm/V, which is in the usual range. The maximum and the subsequent decay to a steady state resemble the behavior associated with pore nucleation and growth. Hence, one could conclude that the same inhomogeneity which leads to pore formation results in the localized attack in halide solutions.

The deactivation seems to be as fast as the activation. In the same recent work,[111] a period of time was measured between two

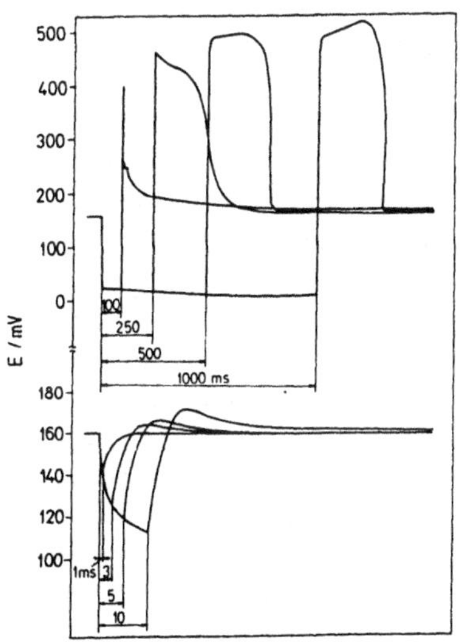

Figure 21. Change of potential of aluminum in 2 M NaCl solution upon application of two constant-current pulses ($10\ mA/cm^2$) with shortening of the time interval between them.

pulses of equal current such that the second pulse started at the same potential at which the previous pulse ended, implying that no noticeable deactivation occurred. The record of the shortening of the time between the pulses is seen in Fig. 21. The time until the restoration of the film to the OCP conditions begins is seen to be very short indeed (1 ms). The kinetics of restoration can be followed from the increase of the potential maximum with the increase of duration of the off period, toward the potential maximum obtained after 500 ms, which is similar to that of a nonactivated electrode resting at the OCP.

That the number of sites, or film structure or thickness, depends on the current density is also seen if the current is increased or decreased stepwise.[112] An additional activation peak of potential is needed at any increase in the current density. Conversely, upon decreasing the current density, an inverted potential peak is obtained, indicating that for some moments the current passes under previous film conditions.

(ii) pH Dependence and Dissolution in Alkaline Solutions

The potential plateau (pitting potential) is said to be insensitive to pH changes in the medium-pH region.[67] This is in line with the model suggesting the accumulation of hydrogen ions at the O/S interface to very high concentrations, when the very small initial concentration (at pH above 2) becomes irrelevant.

At higher pH (above 8), however, an effect is even noted on the OCP, shifting to increasingly negative values. According to Hurlen and Haug,[35] the anodizing of aluminum in acetate buffered by ammonia exhibits a change in the potentiostatic activation with an increase in pH from 7.2 to 9.9, as shown in Fig. 15. Current transients become relatively slow and the current values high, which indicates an increasing role of alkaline dissolution processes. Similar results have been obtained by others.[113]

Sysoeva *et al.*[114] made a systematic potentiostatic investigation of anodization in KOH solutions in the concentration range between 0.1 and 12.5 *M* and in the potential range between −1.5 and 0.5 V vs. SCE. They found a maximum in the aluminum dissolution rate at a KOH concentration of 3–5 *M*. This is interpreted in terms of a change in the mechanism of passivation: At low KOH concentra-

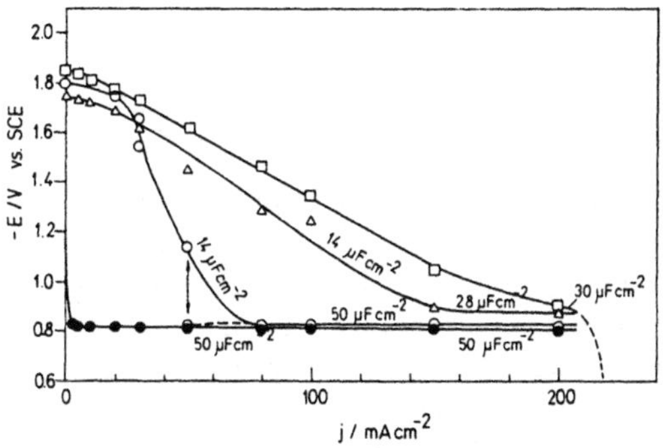

Figure 22. Steady-state polarization curves of aluminum in pure and mixed NaOH + NaCl solutions: □, 4 *M* NaOH; △, 4 *M* NaOH + 2 *M* NaCl; ○, 1 *M* NaOH + 2 *M* NaCl; ●, 2 *M* NaCl (pH 1 to 13). Labels on the lines denote measured capacitances of the interface.

tions, aluminum passivation is due to hydrargillite formation, whereas high KOH concentrations cause precipitation of aluminate in the form of $K_2O{\cdot}Al_2O_3{\cdot}3H_2O$ from the supersaturated solution. It should be noted that the effect of chloride is the same in alkaline solutions as in neutral ones.[115] As seen in Fig. 22, increasing passivation is caused by increasing current density in pure KOH solution. However, when chloride is present, a characteristic free increase in current density is observed at a virtually constant potential plateau corresponding to that in neutral chloride solutions.

(*iii*) *Mechanism of Active Dissolution of Aluminum*

Inasmuch as dissolution at a constant-potential plateau at any current density can be ascribed to continuous activation and deactivation of pitting, the latter could be considered a consequence, rather than the cause, of setting conditions of active dissolution. It only reflects the very subtle inhomogeneity of the surface, covering a continuous spectrum of energies of surface sites. Energetic microheterogeneity with respect to adsorption centers evenly distributed over the surface is a concept known and used in heterogeneous catalysis.[116] An improved Langmuir adsorption isotherm could be derived as

$$\theta = \frac{[A]}{K_m - K_0} \ln\left(\frac{[A] + K_m}{[A] + K_0}\right) \tag{50}$$

where the surface coverage, θ, depends not only on adsorbate concentration in solution, [A], but also on adsorption equilibrium constants ranging continuously between K_0 and K_m. It is the mechanism of active dissolution (at any active site, or, for that matter, at the bottom of a pit) that presents a problem.

Three basic mechanisms of pit initiation have been advocated in the literature (see, e.g., Strehblow[117]) as applying to pitting processes at any passive metal. They are shown schematically in Fig. 23.

(a) The penetration mechanism introduced by Hoar *et al.*[118] assumes fast migration of aggressive anions through the oxide, stimulated by a high electric field, as a rate-determining step in pit initiation.

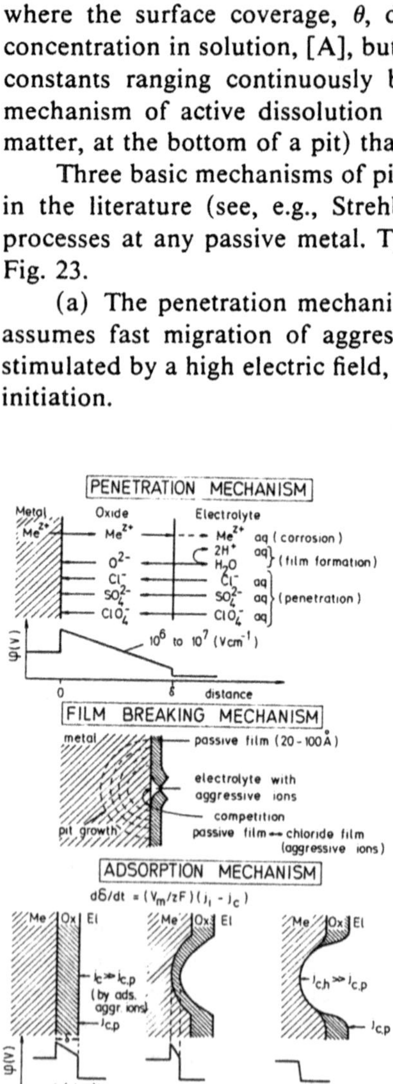

Figure 23. Models of mechanisms of pitting initiation at the surface of passive metals.

(b) Film cracking as a consequence of mechanical stress produced by adsorbing anions was suggested by Vetter and Strehblow[119] as well as by Sato.[120]

(c) The adsorption mechanism, suggested first by Kolotyrkin,[121] involves not only adsorption of anions at particular surface sites, but also their complexing with metal ions from the oxide, producing soluble species. Once such species leave the oxide, it starts thinning locally, with a resulting increase in the electric field strength, which accelerates the process until the oxide is more or less completely dissolved.

The fact to be noted here is that halide ions (or any other "aggressive anions") do not possess any buffering capacity and, hence, when the oxide starts being formed, a high hydrogen ion concentration can be achieved at the interface, leading to a virtual cessation of further oxide growth even in neutral halide solutions. The existence of a hydrogen ion concentration gradient well outside any pits can be observed experimentally, by the fact that flocculation of $Al(OH)_3$ in a cell is found to occur at a significant distance from the electrode, where the pH is sufficiently high for the hydrolysis of aluminum ions to take place. (Quantitative differences between these observations and those in porous film formation may be ascribed to different film thicknesses due to different steady-state hydrogen ion concentrations, as a consequence of some buffering capacity of dibasic acids.)

While the above effect must play a significant role in the active dissolution under the influence of halide ions, there are reasons to believe that some additional effects must be involved. They are:

(a) Active dissolution does not occur until the pitting potential is reached;

(b) The decay of potential after the short induction period (cf. Fig. 20) is too sudden to be a consequence of establishing a high hydrogen ion concentration; and

(c) The decay of potential upon sudden addition of halides to barrier-forming buffering electrolytes is equally fast even at relatively thick films (sustaining over 100 V) and is thus independent of film thickness[77]; similarly, a sudden increase in the current, associated with the appearance of active dissolution, was observed at a barrier film at the OCP in a barrier-forming electrolyte upon addition of chloride.[122]

One can see no way in which the halides could affect the buffering capacity of these electrolytes. There, indeed, seems to be little doubt that adsorption of the halides at the oxide surface takes place and plays a significant role.

The linear dependence of the pitting potential on ionic radius is likely a reflection of the similarly linear relationship between the latter and the free energy of formation of aluminum halides.[108] It is reasonable to assume that the energy of adsorption of a halide on the oxide is also related to the latter. Hence, one could postulate that *the potential at which active dissolution takes place is the potential at which the energy of adsorption overcomes the energy of coulombic repulsion* so that the anions get adsorbed.

It is much less clear how the adsorption leads to such a dramatic change as a potential decay of several hundred volts, occurring within milliseconds. This short time is difficult to associate with film thinning, as assumed in the "adsorption mechanism" of pit initiation. It is not only that the mechanism of dissolution changes so much that the current efficiency falls from virtually 100% to virtually zero, but also that the resistance of the oxide decreases by orders of magnitude. The control of the process is, to a great extent, taken over by the events at the O/S interface, judging from the capacitance values measured,[115] which approach those typical of the electrochemical double layer (cf. Fig. 22).

The depth profiling technique used on samples with a barrier film before and after the addition of chloride to the buffering borate electrolyte showed no indication of either chloride penetration or significant reduction of the average oxide layer thickness.[123] This, of course, does not rule out the possibility of the formation, by any of the mechanisms suggested above, of pinholes with radii much smaller than that of the ion-gun beam, through which the entire active dissolution could take place, or the possibility that the beam missed pits formed sporadically across the surface. If pinholes which are not visible were formed, the dissolution should proceed in them with extremely high true current densities.

Three possible effects of halide adsorption are envisaged:

(a) The specifically adsorbed halide ions must affect the electric field in the double layer at the O/S interface as well as in the oxide as shown in Fig. 24. The center of the negative charges, resting at the inner Helmholtz plane, should accelerate both the transport

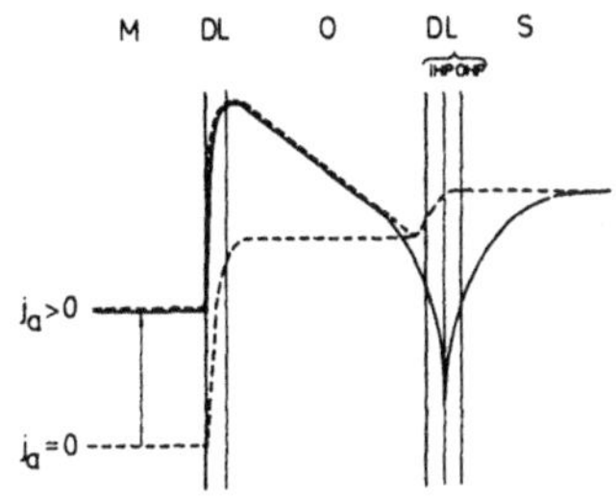

Figure 24. Schematic representation of the effect of anion adsorption on the potential profile at the O/S interface, showing the potential profile before (dashed line) and after adsorption (solid line).

of ions through the oxide and the charge transfers across the interface. While the first effect may be significant in thin oxide films, it must lose importance with increasing film thickness. It is difficult to envisage this effect having a major influence on the oxide which already sustains potential differences of the order of 100 V. Hence, this effect cannot explain the major change in the resistance of the oxide layer. On the other hand, if the surface coverage by halide ions approaches unity, the water supply to the O/S interface could be virtually cut off, so that oxygen ion transfer becomes strongly inhibited whereas aluminum ion transfer is accelerated, following an exponential dependence on the field. Thus, the cessation of oxide growth can be explained in this manner.

One should note that the field in the outer part of the double layer should repel OH^- ions and attract hydrogen ions. This, however, is not expected to have a major effect on the kinetics (the "Frumkin effect") since the main reactant is likely to be water.

(b) Halides are known to form soluble complexes with aluminum ions. These include neutral ones, such as $Al(OH)_2Cl$, $Al(OH)Cl_2$, and $AlCl_3$. Hence, these could pass through the O/S interface into the solution without any effect of the electric field in the double layer. This adds up to the partial current of aluminum transfer.

(c) The halide ions, once at the oxide surface, can be sucked into it by the high positive electric field, disrupting the oxide structure and suddenly increasing the concentration of charge carriers.

A complete dissolution of the oxide at the rate corresponding to extreme current densities in the pits seems very unlikely since it would have to involve too much mass transport inside the pinholes.

Deactivation can be understood in terms of the mechanism based on adsorption of the anions. Although a lower current density would need a less positive potential *if, for example, chloride ions stayed at the surface*, as soon as the potential shifts negative, desorption of chloride should take place, with a corresponding loss of activity.

Although a qualitative picture can thus be drawn, the model must still be considered as tentative until some quantitative relationships are developed and proven experimentally.

(iv) The "Negative Difference Effect"

When aluminum is anodically dissolved in halide solutions, the rate of hydrogen evolution linearly increases with increasing current density as shown in Fig. 25. This phenomenon is historically, and somewhat misleadingly, termed the "negative difference effect"[124] (NDE). It is contrary to what one would normally expect, for hydrogen evolution should subside with the potential going positive (as indeed is observed in alkaline solutions) or at least stay constant at a constant-potential plateau.

This phenomenon, however, is not difficult to understand in view of the mechanism of dissolution under such conditions. Since the number of active sites increases linearly with current density and these sites are characterized by a film structure (or thickness or both) different from that at the OCP, one could expect corresponding increases in the corrosion rate. However, as was mentioned earlier, the active surface area in the pits increases with time, and hence one should expect the corrosion rate to increase correspondingly. Therefore, since the effect is not time dependent, one

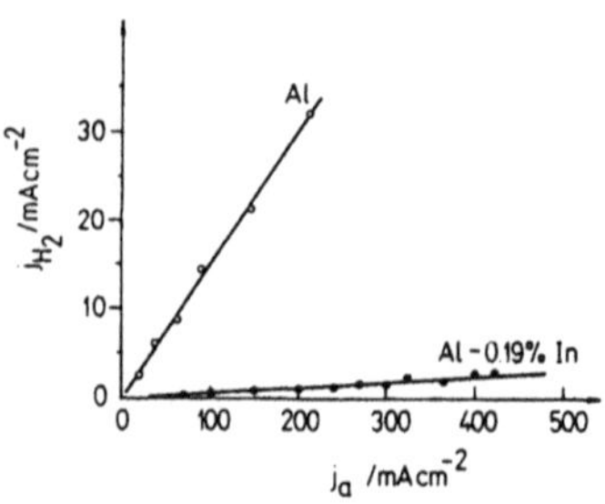

Figure 25. Rate of hydrogen evolution (expressed in terms of equivalent current density) as a function of anodic current density for aluminum and an aluminum-0.19% indium alloy immersed in 2 *M* NaCl solution.

must look for another explanation. This was indeed found in relating the NDE with the formation of subvalent Al^+ ions at the M/O interface. It is likely that the Al^+ ions are formed at the M/O interface in a concentration which is in some proportion to the current density [if the assumption of the second exchange of two electrons as the rate-determining step is accepted; cf. Section III(1(i))]. If so, their flux through the oxide layer to the O/S interface should retain this proportionality, and so should the rate of their reaction with water, forming hydrogen. As the true current density inside the pit is reduced with time, maintaining a constant current per pit, so should be the rate of hydrogen evolution.

In fact, the NDE can be interpreted along similar lines without invoking Al^+ ions: any changes that occur in the oxide layer, automatically reducing the true current density inside the pit, could cause a corresponding decrease in the corrosion rate, *whatever their origin.* Nevertheless, it is difficult to envisage any other mechanism of hydrogen evolution at the surface of the oxide apart from that involving subvalent ions.

(*v*) *Effect of Alloying Elements and Impurities on Electrochemical Activity of Aluminum*

A consideration of the electrochemical behavior of the large variety of aluminum alloys used in practice surpasses by far the scope of this chapter. Nevertheless, we consider it useful to review here the effect of some elements that have a profound effect on this behavior.

Reding and Newport[125] have pointed out that small amounts of a number of elements (Mg, Ba, Zn, Cd, Hg, Sn, Ga, In) added to aluminum cause significant shifts of the OCP in chloride solutions in the negative direction. Despić *et al.*[126] have shown, for the example of low-content (about 0.1%) gallium, indium, and thallium alloys, that this increased activity is maintained up to a very high anodic current density (up to 1 A/cm^2) and that the effect amounts to shifting the entire potential plateau in the negative direction.

The NDE in some alloys was found to be larger than in pure aluminum, but in others (with In) to be reduced to very small values, leading to a corrosive loss of the metal of only 0.5% (cf. Fig. 25).

Tuck *et al.*[127] have ascribed this effect to the accumulation of the alloying element inside the pit, forming a separate metal phase in intimate contact with the base metal. At the surface of such a phase, aluminum coming through it by diffusion could dissolve without difficulties imposed by the oxide film. Support for such a model was found[128] (a) in electron microscopic observation and electron probe detection of a separate metal phase consisting of the virtually pure alloying element, and (b) in the fact that some time is needed, after anodic dissolution starts, to attain increased activity. After the usual process of activation, the same as in pure aluminum, some time lag is recorded,[128] as shown in Fig. 26, before "superactivation" to the negative potential plateau takes place. Similar phenomena are recorded in alkaline solutions, in which some elements (Sn) have the additional effect of suppressing otherwise very high hydrogen corrosion rates. A process of superactivation appears to depend on some so far unknown property of the alloying element.

It is interesting to note that, as far as superactivation is concerned, a "hierarchy effect," rather than a simple additive or synergic effect, is found,[127] i.e., the "activators" act in the order Sn, Ga, In. Hence, when Sn is present, the superactivation occurs as though the other elements were not present at all.

Many other elements affect the electrochemical activity of aluminum even at a trace (ppm) level. Thus, copper, zinc, and iron are found to counteract the effect of the activating elements. So

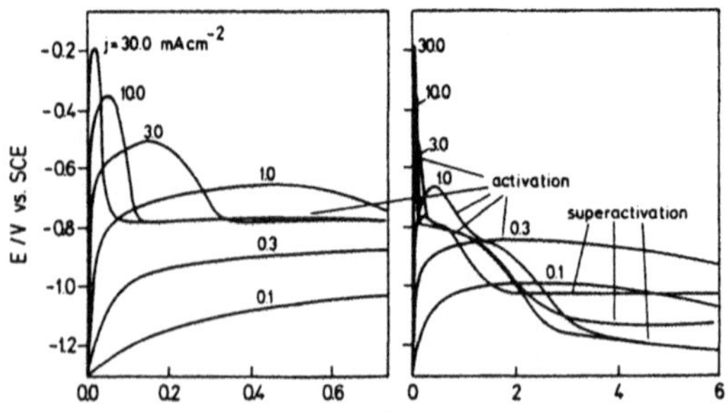

Figure 26. Activation and superactivation of an aluminum-gallium (0.2%) alloy at different current densities.[128]

far, there is no basis on which to even speculate about the reasons for such effects.

In nonalloyed metal, impurities affect the OCP and the corrosion behavior, while they have little effect on the potential plateau of active dissolution.

As shown by Bond *et al.*,[129] the microsegregation of impurities has been proven to be a more important factor than their content. Thus, on relatively fast cooling, producing cellular substructure, impurities segregate at the nodes even at a total impurity level in the ppm range in 99.9993% pure samples. The hydrogen overvoltage, being smaller at the impurity phase than at aluminum, shifts the OCP to a sufficiently positive potential to induce pitting.

A sample of 99.993% Al containing 10-fold higher Fe and Cu concentrations, but cooled in such a way as to produce noncellular structure and prevent the segregation of impurities into a separate phase, has maintained the OCP in the same 0.5 *M* NaCl solution well below the pitting potential and no pitting has been recorded.

Similarly, zone-refined aluminum, which has too low an impurity content (in the few hundred ppb range) for development of cellular structure, has not been detectably affected by exposure to the pitting solution for periods of up to 260 h.

The fact that impurities do not affect the active dissolution in chloride solutions at current densities larger than 0.01 mA/cm^2 shows that the inhomogeneity resulting in a pitting mechanism of dissolution is unrelated to impurities and is an inherent property of the metal.

IV. STRUCTURE AND MORPHOLOGY OF ANODIC ALUMINUM OXIDES

1. Methods of Determining Composition and Structure

All methods of surface analysis are based on primary particle irradiation of analyzed samples, causing primary flux disturbance or emission of secondary particles from the surface. Table 2 presents a classification of the most popular methods of analysis based on

Table 2
Excitation–Emission Matrix of Thin-Film Analysis[a]

	Excitation		
Emission	Photons	Electrons	Ions
Photons	IR EXAFS NMR ESR	EDX	
Electrons	XPS SR	AES ELS SEM, TEM LEED	IAS
Ions	LAMMA PID	EID	SIMS ISS RBS NMA

[a] Abbreviations: AES, Auger electron spectroscopy (Refs. 141–143); EDX, energy-dispersive analysis of X rays (Refs. 135 and 136); ELS, energy loss spectroscopy (Ref. 145); EID, electron-induced ion desorption; ESR, electron spin resonance (Ref. 138); EXAFS, extended X-ray absorption fine structure (Ref. 15); IAS, ion Auger spectroscopy; IR, infrared spectroscopy (Refs. 133 and 134); ISS, ion surface scattering (Ref. 150); LAMMA, laser microprobe mass analysis; LEED, low-energy electron diffraction (Ref. 147); NMR, nuclear magnetic resonance (Ref. 137); PID, photoinduced ion desorption; NMA, nuclear microanalysis (Ref. 152); RBS, Rutherford backscattering spectroscopy (Refs. 86 and 151); SEM, scanning electron microscopy (Ref. 146); SIMS, secondary ion mass spectroscopy (Refs. 148 and 149); SR, synchrotron radiation spectroscopy (Ref. 144); TEM, transmission electron spectroscopy (Ref. 146); XPS, X-ray photoelectron spectroscopy (Refs. 139 and 140).

the types of excited and emitted particles. According to Yeager[130] and other authors, these methods could be very helpful in studying solid–electrolyte interfaces, although they are mostly *ex situ* techniques, and hence the possibility of changes during transfer from electrochemical cells to the analyzing apparatus has to be borne in mind.

A comparative analysis of the existing analytical techniques is presented in a number of works (see, e.g., Refs. 131 and 132). As can be seen in Table 3, all problems of anodic alumina analysis

Table 3
Comparison of Characteristic Features of Analytic Methods

	XPS	AES	SIMS	RBS	IR	ISS	EDX
Sensitivity	*	**	***	*	*	**	*
Speed of analysis	—	***	**	***	**	**	***
Lateral resolution	—	***	**	*	—	*	**
Depth resolution	*	***	***	**	—	**	*
Chemical bonding information	***	*	—	—	**	—	—
Information on structure	—	—	—	*	*	**	—
Damage ability	***	*	*	**	***	*	—
Detection of hydration	*	—	***	—	*	—	—
Computer control	**	***	*	**	*	***	**
Availability of interpretation	**	**	*	***	*	*	***

cannot be solved by a single method. If lateral resolution is important, electron beam methods, such as AES, should be used, as electron beams can be focused down to several nanometers, to provide spot sizes lower than characteristic grain sizes at oxide surfaces. SIMS is preferable for the analysis of oxide hydration, as it offers a unique sensitivity to light elements, particularly hydrogen. In quantitative analysis of the oxide, where the stoichiometry and chemical state of the elements composing the oxide are of interest, the XPS method is to be preferred. To determine the depth profile (composition) of the oxide, the surface analysis method should be combined with a sample sectioning technique. Usually, inert ion sputter profiling methods are employed, and most analyzers are equipped with sputtering facilities. However, ion bombardment causes a large number of artifacts, such as preferential sputtering of some components in multicomponent oxides,[153] bombardment-stimulated chemical reactions,[154] and redistribution of elements due to knock-on effects. Hence, the results obtained should be checked by nondestructive analytical methods, such as RBS, and other methods of depth analysis, such as the chemical sectioning technique.[155]

An important problem in analyzing anodic aluminas is their porous structure. Ion sputter methods are useless in examining anodic oxides with well-developed porosity, since integral sputtering of the entire surface would obviously result in an average, quasi-homogeneous depth composition of the oxide, without resolving the microstructural features (pore center and pore walls).[156] Microbeam[135] or chemical sectioning[19] techniques should be used in this case.

2. Chemical Composition of Anodic Aluminum Oxides

The chemical composition of anodic aluminas, with special emphasis on the depth-dependent incorporation of electrolyte species (protons, anions, etc.), has been extensively studied.

The major elements composing the oxides are oxygen and aluminum. It has been shown in a number of works that they are distributed quasi-homogeneously.[141,143] The stoichiometry of anodic aluminas, as determined by XPS[139,157] and RBS,[86,151] corresponds to nearly perfect Al_2O_3, although there are indications of both oxygen[158] and aluminum[159] deficiency. Deviations from perfect stoichiometry are most significant at the O/S and M/O interfaces.[160] The transition layers observed at interfaces increase in thickness as the oxide grows,[143,160] due to the developing roughness of the interfaces.

Nonstoichiometry of the oxides can be due to a number of reasons, such as hydration,[159] incomplete oxidation,[158] and the generation of defects at interfaces.[157] An important factor affecting the chemical composition of the oxides is the incorporation of electrolyte species into the growing alumina. There have even been suggestions to use this for impurity doping of oxides and modifying their properties.[161] Various kinds of anion distributions and mechanisms of anion incorporation and their influence on oxide properties have been reported. The problems attracting attention are:

- What are the kinetics of anion incorporation into the growing oxide and how is this process influenced by anodization conditions?
- Does anion incorporation influence the oxide formation, and what is the mechanism of anion pickup?

Table 4
Data on Anion incorporation into Anodic Alumina

Anodizing bath	Method of study	Chemical state of anion and impurities	Mean anion content	Type of anion distribution[a]	$k = L_A/L$	Ref.
$NH_4H_2PO_4$(pH 7)	XPS	PO_4^{3-}		1	0.7	79
1 *M* Phosphate, 1 *M* chromate	SIMS	PO_2^{4-}, Cr^+		1	0.7, 0.16	162
0.001–0.1 *M* NaH_2PO_4	NMA	Phosphorus	1.6 wt %	1	0.66	163
0.4 *M* phosphate	STEM/EDX	Phosphorus	2.6 wt %	1	0.6	102
10–20 wt % H_3PO_4	AES	Phosphorus	1–2%	1	0.6	156
4% H_3PO_4, phosphate (pH 5.2–9.3)	Chem. section.	PO_4^{3-}		2	(phosphate)	19
					0.7	
				3	(H_3PO_4)	
0.4 *M* H_3PO_4	STEM/EDX	Phosphorus	1.6 at. %	3		135
0.01–5.0 *M* H_3PO_4	AES	PO_4^{3-}		2		163
4% H_3PO_4	AES	PO_4^{3-}		2		160
NH_4 pentaborate	RBS	B_2O_3	0.6 wt %	1	0.33–0.4	86
0.5 *M* H_3BO_3 + 0.05 *M* $Na_2B_4O_7$ (pH 7.4)	XPS	B_2O_3	7%	2	0.33	159
0.1–1.0 maleic acid	ESR + IR	$C_2O_4^{2-}$				138
0.2 *M* $H_2C_2O_4$	IR	$C_2O_4^{2-}$		3	0.7	165
0.2 *M* $H_2C_2O_4$	AES	$C_2O_4^{2-}$		3	0.9	160
0.5 *M* H_2SO_4	AES	SO_4^{2-}		3	1.0	160
0.6 *M* H_2SO_4 + 0.4 *M*	IR	SO_4^{2-}		3	1.0	166
$MgSO_4$ H_2SO_4	XPS	SO_4^{2-}, SO_3^{2-}		2	1.0	157
56% HNO_3	XPS	NO_3^{2-}, NO_2^-, NO_2		3	0.9	167
HCl + NH_4 tartrate	XPS	Cl^-		3		168
Na_2WO_3	EDX/STEM	Tungsten	0.72 at. %	1	0.3	136
0.1 *M* molybdate	EDX/EELS/RBS	Molybdenum	0.61 at. %	1	0.2	102

- In what manner are the oxide structure, morphology, and properties modified by incorporated anions and other electrolyte species?

Table 4 reviews electrolyte anion distributions in anodic oxides. In the following sections, the items of special interest are summarized.

(*i*) *Chemical State of Incorporated Electrolyte Constituents*

IR and XPS measurements have shown that electrolyte species incorporated into the growing oxide are mostly anions of the acid or salt used. There are indications that the molecular form of incorporated anions may depend on the depth of their location in the oxide. Fukuda and Fukushima[165,166] have established, using IR and ESR techniques, that the outermost layer of anodic alumina formed in oxalic acid contains oxalate anions, $C_2O_4^{2-}$, whereas species incorporated deeper than 10 nm are carboxylates, COO^-. Yaniv *et al.*[169] have reported that the molecular form of species incorporated in oxides formed in H_2SO_4 changes from sulfate (SO_4^{2-}) at the surface to elemental sulfur in the body of the oxide. Analogous behavior was observed for aluminum anodized in $H_2C_2O_4$.[170] This effect is presumably due to field-enhanced reactions in the oxide during its growth.

(*ii*) *Distribution of Electrolyte Species in Oxides*

The anion distribution in anodic oxides is usually determined by ion bombardment or chemical sectioning of alumina samples with subsequent analysis by AES or XPS methods, or by the use of the depth-resolving techniques, such as RBS.[150] Different types of concentration profiles are shown in Fig. 27.

Generally speaking, the distribution of anions in alumina films formed in neutral electrolytes (presumably borates and phosphates) can be regarded as homogeneous within an "outer" oxide layer comprising about two-fifths to four-fifths of the total thickness of the film[4] (Fig. 27, type 1 or 2). Relevant data have been obtained in a number of studies by the Manchester group,[86,143] as well as by Nagayama and Takahashi and co-workers.[19,79,171]

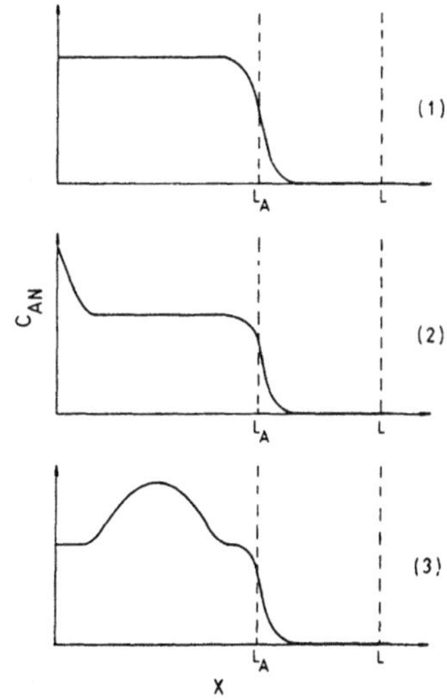

Figure 27. Types of anion concentration profiles in anodic oxide layers on aluminum, from the O/S interface (O) inwards to the M/O interface (L). L_A: front of anion penetration.

Another type of anion distribution is observed at anodic oxides formed in acidic electrolytes. It has been examined by Fukuda[165] and can be generally characterized as a single-lump pattern (Fig. 27, type 3). This distribution exhibits a maximum shifted inwards, as has been shown in a number of works by various authors[160,166,172] for a variety of acidic electrolytes. Figure 28 illustrates the distribution of sulfur and carbon in thickening oxides formed in H_2SO_4 and $H_2C_2O_4$ solutions, as obtained by AES measurements. Sulfate

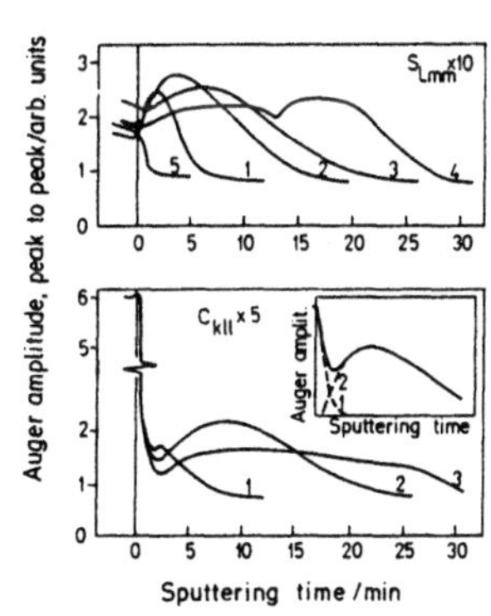

Figure 28. Distribution of sulfur and carbon in anodic aluminas corresponding to the different stages of porous structure growth, as determined by Auger spectroscopy.[160]

and oxalate species exhibit a maximum concentration that shifts deeper into the oxide during its growth. The carbon distribution is believed to consist of two components—one a sharply decreasing hydrocarbon contamination and the other due to anion incorporation. The anion distributions of ultrathin (5–7 nm) oxides formed in HNO_3 solutions and in HCl have the same single-lump shape.[167,168]

A special case of anodizing in chromic acid is characterized by the absence of incorporated anions.[162,173] It has been shown in a number of studies[4] that chromate anions are accumulated at the outer oxide surface and do not penetrate into the oxide body.

Recent data provided by Cocke *et al.*[174,175] in an RBS study of the distribution of heavy anions (tungstates, molybdates, manganates) yield unusual oscillatory anion profiles.

Sokol and co-workers[176–178] have studied doping of anodic alumina by rare-earth complex anions. The latter are formed by dissolving rare-earth oxides in citric acid, which exhibits pronounced chelating properties.[179] Citrate and polyphosphate complexes of rare earths possess either anionic (at electrolyte $pH < 2.5$) or cationic ($pH > 3$) properties,[180] and so, by adjusting the electrolyte pH to an appropriate value, one may provoke adsorption of rare-earth complexes at the surface of anodized samples. It has been shown by means of luminescence analysis and SIMS that a wide range of rare earths (Eu, Sc, Er, Y, Nd, La, and Ho) are picked up by growing alumina films with the concentration of complex anions exponentially decreasing inwards (Fig. 29).[176] It has also been found that when added to a phosphoric acid electrolyte, minor complex anion impurities diminish the pore growth, presumably

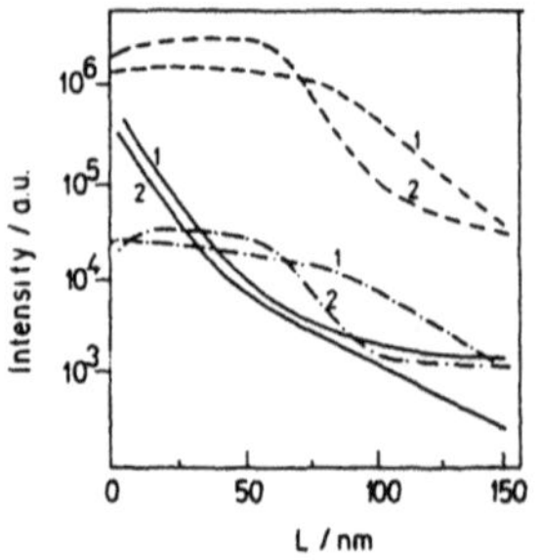

Figure 29. SIMS distributions of AlO_2^+ (— — —), PO_2^+ (— · — ·), and Y^+ or Nd^+ (———) ions in anodic alumina films formed in H_3PO_4 containing yttrium (1) and neodymium (2) complex anions.[176]

by the replacement of phosphate anions at the oxide surface and preferential incorporation. The mechanism of large complex anion incorporation into growing alumina and its influence on the oxide structure needs further elucidation.

Very interesting behavior of incorporating anions can be observed when a multicomponent electrolyte is used for oxide formation. Here, anion antagonism or synergism can be observed, depending on the types of anions used. The antagonism of hydroxyl ions and acid anions has been observed in a number of cases. Konno *et al.*[181] have observed, in experiments on anodic alumina deterioration and hydration, that small amounts of phosphates and chromates inhibit oxide hydration by forming monolayer or two-layer films of adsorbed anions at the oxide surface. Abd-Rabbo *et al.*[162] have observed preferential incorporation of phosphate anions from a mixture of phosphates and chromates.

In conclusion, one can say that most anodic oxide films are of a duplex, or even triplex, character, with only the inner portion being composed of a pure anhydrous oxide. In the duplex films, the outer layer contains anions and often a degree of hydration. There could exist a third thin oxide layer at the surface, again with somewhat different properties, which may have a role in the kinetics of oxide growth.

(*iii*) *Kinetics of Anion Incorporation into Growing Alumina Films*

The rate of anion pickup during the constant-current growth of barrier alumina films is constant but smaller than the rate of film growth, according to the reports of many authors, beginning with that of Randall and Bernard[163] and ending with the recent work of Skeldon *et al.*[86] and Takahashi *et al.*[87] An example of such linear kinetics is presented in Fig. 30. This fact leads to a constant ratio between the inner and outer layer thicknesses. An increase in the current density causes some growth of anion content and relatively deeper penetration.[87]

Data on anion incorporation into a growing porous oxide were obtained Fukuda and Fukushima.[165,166] Their study was the first to demonstrate a correlation between the kinetics of accumulation of oxalate[165] or sulfate[166] anions and the change of porous oxide growth stages. The results of galvanostatic and potentiostatic

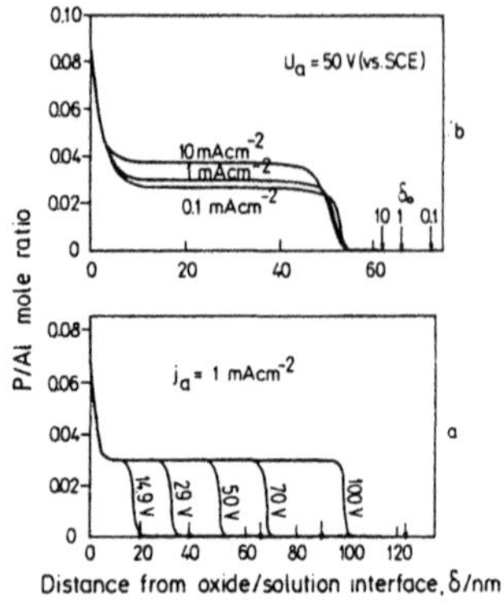

Figure 30. Kinetics of phosphorus incorporation into growing alumina at constant current density (a) and dependence of anion pickup on j_a value (b).[79]

anodization regimes are given in Fig. 31. The integral anion content attains a maximum value somewhat later than the transition to stationary pore growth. Further anodization does not increase the amplitude of anion concentration. Similar results have been obtained by Parkhutik and co-workers[160,182] for the oxides formed in sulfuric and oxalic acids (see Fig. 28), whereas anodizing in an H_3PO_4 bath did not yield such a correlation. Again, the results obtained for nitric acid anodization generally resemble those of the former two cases.[172] It seems justified to conclude that, in most cases, the anion concentration in the growing oxide reaches a maximum value at the moment when intensive pore growth starts.

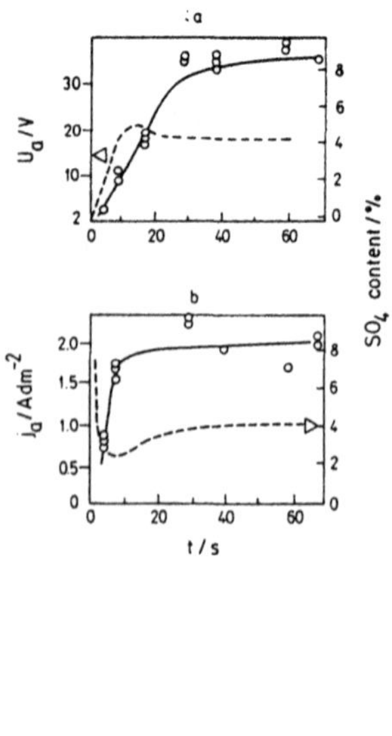

Figure 31. Correlation of the kinetics of oxide growth with kinetics of sulfate incorporation into the oxide during galvanostatic (a) and potentiostatic (b) anodization of Al in H_2SO_4 solutions.[166]

(iv) *Role of Anions in Anodic Oxide Growth and Their Effect on Oxide Properties*

The true role of incorporation of anions in the formation of anodic alumina is being intensively discussed. Baker and Pearson[183] have considered the anion effect in modifying the structure of anodic oxides to be due to the coordinative ability of anions to replace alumina tetrahedra in the body of the oxides. Dorsey[184,185] has postulated that in porous oxides, anions stabilize the network of alumina tetrahedra and octahedra.

Thompson, Wood, and co-workers, in a series of papers (cf., e.g., Ref. 186) have established a correlation between the nature of anions incorporated into an oxide and the features of the porous oxide formed. They assumed that the electric field applied to the oxide during its growth is inhomogeneous: higher in the inner layer of pure alumina, and lower in the outer part contaminated by anion species, as has indeed been observed.[79] The outer layer allows the easy passage of charged particles due to its imperfections. The imposed anodic voltage is divided between the contaminated and pure alumina regions in proportion to their thicknesses and conductivities. For a given voltage, the larger the thickness of the anion-containing outer layer, the higher is the electric field strength at the inner layer and the larger is the oxide growth rate. As the ratio of the thickness of the anion-containing layer to that of the pure oxide layer increases in the order chromic acid < phosphoric acid < oxalic acid < sulfuric acid,[186] Thompson and co-workers have concluded that the rate of oxide growth and dissolution increases in the same order.

When the results for oxide growth and anion incorporation[172,160] are compared with the kinetics of space charge accumulation in barrier and porous alumina films [see Section IV(1)], it can be concluded that anion incorporation modifies the electrostatics of the external oxide interface, thus influencing oxide dissolution and pore formation.[172]

3. Crystal Structure of Anodic Aluminas

Anodic aluminas are reported in the literature to have both an amorphous and a crystalline structure. The majority of anodic aluminas exhibit an amorphous structure with a short-range crystalline order corresponding to octahedral or tetrahedral coordination of aluminum ions.[187-189] According to the existing view,[187] amorphous aluminum oxides should be presented as close-packed arrays of Al_4O_6 molecular units (Fig. 32). Stacked sheets of these units (Fig. 32c) give the appropriate admixture of oxtahedral and tetrahedral sites occupied by Al^{3+}.[187] The oxide crystallinity is observed for thick oxides and those possessing composite structure.[189] The extent of crystallinity and the coordination number of aluminum effect the stability and mechanical properties of oxides and have been found to depend strongly on electrolyte and anodization regimes.

Thus, El-Mashri *et al.*[190] have recently studied the Al—O bond length in thin (50–100 nm) alumina films formed in sodium tartrate and phosphoric acid electrolytes. The average bond length was established to be 0.19 nm (tartrate) and 0.18 nm (phosphate). An analysis of these data have yielded the ratio of octahedral (AlO_6) to tetrahedral (AlO_4) aluminum ion coordination to be 80% : 20% and 30% : 70%, respectively. Popova[191] has shown, by transmission electron diffraction, a 100% tetrahedral coordination for alumina films formed in borates. Oka *et al.*[192] have reported AlO_6 : AlO_4 ratios of 30% : 70% and 40% : 60% for films formed in H_2SO_4 at ac and dc regimes, respectively. Parkhutik *et al.*,[193] by measuring the relative intensities of FT IR peaks of vibrating Al—OAl bonds, have determined the same ratio in oxides grown in oxalic acid solution.

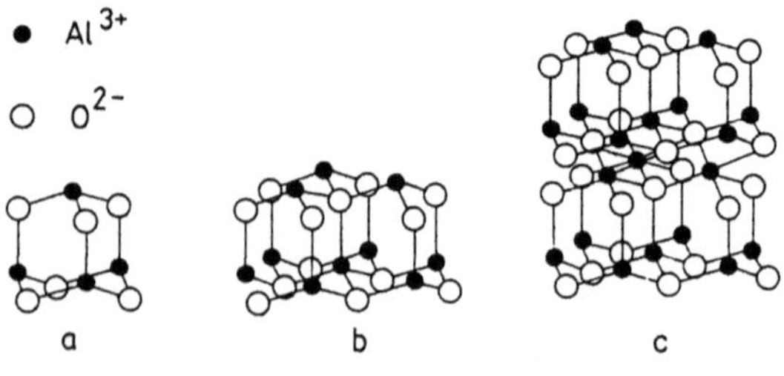

Figure 32. Structure of amorphous alumina showing a single Al_4O_6 group (a), a sheet of these (b), and a stack of sheets (c).[189]

As yet, no explanation has been advanced for such specific anion or pH effects.

According to El-Mashri *et al.*,[190] the $AlO_6:AlO_4$ ratio determines the hydration capacity of anodic oxides. Tetrahedral sites are hydrated easily to form a boehmite-like structure, which is known to be composed of double layers of Al-centered octahedra, weakly linked by water molecules to other layers.[184] As the oxide formed in H_3PO_4 contains about 70% tetrahedral aluminum bonds, its hydration ability should be higher than that of the oxide formed in tartrate solution. However, this has not been found in practice, which is interpreted by El-Mashri *et al.* as being due to some reduction of AlO_4 by incorporated phosphate species.

Besides the amorphous alumina films formed in the majority of acidic electrolytes (except those formed in chromic acid and exhibiting traces of γ-Al_2O_3[194]), there are possibilities of forming oxides with a more or less pronounced crystallinity. These oxides are formed in alkaline solutions[195] and especially in sodium carbonate baths.[196] According to the data provided by Hiroshi and Yoshimura,[197] these oxides contain a γ-Al_2O_3 phase easily hydrated and converted into a bayerite-like substance.

Specific structural features are observed in the formation of composite oxides. Kobayashi, Shimizu, and their co-workers have, in a series of papers, reported studies of the structure of barrier alumina films, anodically formed on aluminum covered by a thin (5 nm) layer of thermal oxide.[198,199] Their experiments have shown that the thermally oxidized thin layer generally contains γ-alumina crystals of about 0.2 nm size. This layer does not have a pronounced effect on ionic transport in the oxide during anodization. Also, islands of γ'-alumina are formed around the middle of anodic barrier oxides. They are nucleated and developed from tiny crystals of γ'-Al_2O_3 and grow rapidly in the lateral direction under prolonged anodization.[198,199]

The rate of γ'-alumina island formation essentially depends on the nature of the electrolyte used. If "outwards migrating" (in the terms of Xu *et al.*[102]) anions, such as tungstates and molybdates, are used in the anodization process, γ-alumina seed crystals are surrounded by pure alumina and crystallization occurs easily. In the case of "inwards migrating" anions (e.g., citrates, phosphates, tartrates), the oxide material surrounding the γ-nuclei is enriched in the incorporated anions, hindering structural rearrangement at the amorphous–crystalline boundary. Instead of the thermal treatment, samples can be immersed in an anodizing bath at 85°C for several minutes to provoke the formation of γ-alumina islands, which are inherited by alumina films under prolonged anodization.[200]

Recent data reported by Bernard and Florio[201] generally confirm such a behavior and the appearance of a bilayer oxide structure, with the outer layer amorphous and the inner one beneath it composed of γ-Al_2O_3.

Composite crystalline–amorphous films are also obtained by combining the anodization of aluminum with its hydration, as has been shown by Kudo and Alwitt[202] and recently confirmed by Takahashi and co-workers[192,203,204] as well as by Kobayashi *et al.*[205] The anodizing of samples initially covered by hydroxide with a pseudoboehmite structure proceeds at higher rates than without the hydrous oxide layer. During anodization, barrier oxide grows underneath the hydrous oxide layer, consuming its inner part. Incorporation of pseudoboehmite into the anodic oxide leads to the formation of γ-Al_2O_3 microcrystallites near the middle of the barrier layer. Under prolonged anodization, these crystal nuclei impinge upon one another and aggregate to form a band of γ-crystalline region, growing rapidly toward the barrier/hydrous oxide interface.

The γ-modification of alumina is the only one reported for anodic aluminum oxide. However, thermodynamic data[206] and the results of gravimetric analysis[207,208] indicate that α-Al_2O_3 is also possible if the oxide is annealed at temperatures of about 1200°C.

4. Hydration of Growing and Aging Anodic Aluminum Oxides

Two aspects of oxide hydration are generally considered. One is hydration during the growth of the oxide. The other is the interaction with water of aging oxides immersed therein. This is important for improving aging stability of oxides and their corrosion resistance.[209]

(*i*) *Hydration of Growing Oxides*

A number of researchers have assumed that oxide growth involves inward migration of OH^- groups from the electrolyte and their reaction with Al^{3+} ions at the M/O interface.[210,211] This mechanism is in contradiction with that based on transport number determination [cf. Section III(3(v))]. Nevertheless, if there is a penetration of water all the way to the M/O interface, this should yield a distribution of OH^- and H^+ species in the oxide, and so, in principle, the validity of this mechanism can be verified by direct measurements of hydroxyl and proton profiles in oxides.

Hydration of growing alumina films was studied by SIMS and XPS methods in the case of barrier oxides[159,212,213] and by IR spectroscopy and derivatography[207,208] for porous ones. Takahashi and co-workers[79,159] have interpreted the results of XPS analysis of barrier alumina formed in neutral phosphate solution in terms of oxide hydration. It has been established that the O/Al mole ratio is about 1.7 : 1 for the outer oxide layer contaminated by acid anions. The excess of oxygen with respect to the 1.5 : 1 ratio for Al_2O_3 is attributed to the hydration of this layer. It should be noted, however, that XPS cannot be considered a direct method for measuring oxide hydration. The O_{1s} lines corresponding to oxide, Al_2O_3, and hydroxide, AlOOH, practically coincide with one another[140] and cannot be resolved separately. Hence, if hydrated, the oxide should exhibit an O_{1s} peak with a half-width only slightly larger than that for pure Al_2O_3.[213] Besides, Takahashi *et al.*[79] have used a chemical sectioning technique for sample profiling and this procedure itself may cause some oxide hydration.

SIMS measurements by Abd Rabbo and co-workers[212] have seemingly presented more direct evidence of hydration of oxides formed in tartrate solution. Hydrogen was detected throughout an oxide film with a concentration depending on electrolyte pH. The outer regions of the oxide were found to be more hydrated than the inner ones. The results were consistent with the Hoar–Mott theory for barrier oxide growth[210] involving OH^- movement into the growing film and movement in the opposite direction of the protons released at the oxide–aluminum interface. This point of view is shared by many others.[211,213] However, SIMS, combined with ion sputter profiling, should not either be considered a direct method for observing oxide hydration since residual moisture in the chamber of the analyzer can alter the results obtained.[212]

To overcome the shortcomings of interpreting the SIMS data on hydrogen distribution in anodic aluminas, Lanford *et al.*[214] have used a nuclear reaction technique, based on the measurement of the gamma-ray emission output of the reaction of accelerated ^{15}N nuclei with ^{1}H according to the scheme $^{15}N + ^{1}H \rightarrow ^{12}C + ^{4}He + 4.43$ MeV. The important feature of this method is that the region being analyzed for the presence of hydrogen is not exposed to the atmosphere of the analyzer chamber, and background moisture can be excluded from consideration. According to the data obtained, the proton-enriched portion of 180-nm oxide films formed in phosphate, tartrate, and glycol-borate solutions does not exceed 50 nm and the hydrogen content amounts to 0.02–0.17 at. %. Thus, both the penetration and concentration of water are considerably less than found by Takahashi *et al.*[159] Besides, the protonation of the growing oxides is more than four times lower than the anion contamination.[159,214] All this suggests that protons and anions are introduced into oxides by independent processes. Hence, the small extent of oxide hydration opposes the models of barrier oxide growth by OH^- inward migration as well as by the dissolution–precipitation mechanism[210,211] and rather supports the earlier discussed mechanism, based on aluminum and oxygen ion movements.

However, this does not preclude the possibility that in a portion of the oxide at least (the outer layer), the OH^- transport mechanism is operative, with the release of protons at the interface between the two oxide layers. Hence, in such a case, some field-assisted proton transfer is likely to take place through the outer layer while chemical dissolution should be operative at the outer O/S interface.

The hydration of oxides formed in acidic electrolytes, and thus possessing a porous structure, occurs in the same manner as in the case of barrier oxides. It is generally recognized that porous oxides grown in acidic electrolytes have no bonded water in their bulk. Only chemisorbed OH groups and H_2O molecules are detected at the oxide surfaces. This is established by carefully conducted XPS and IR measurements[172,215] as well as by derivatography.[208] IR spectroscopy appears not to be very sensitive to oxide protonation, and there has been a good deal of controversy over the interpretation of IR data, concerning the assignment of observed spectral lines.[215]

As for the porous oxides formed in alkaline solutions, there is evidence that they are heavily hydrated. Hurlen and Haug[35,216] have recently shown that the chemical composition of the nonbarrier part of an oxide formed in an ammonium acetate buffer at pH 9.9 corresponds to $Al_2O_3 + \frac{1}{2}H_2O$. These results are supported by the results of derivatography.[113] Belov and Lebedeva[217] have established different degrees of hydration of oxides formed in different alkaline electrolytes.

(ii) *Hydration of Aging Oxides*

Anodic oxides placed in aqueous media increase their weight by picking up water molecules and hydroxyl ions. The ability of an oxide to be hydrated during aging at various temperatures depends on the conditions of oxide formation. Figure 33[218] illustrates the hydration capacity of porous oxides formed in various electrolytes, as well as the capacity of the same oxides to absorb various anions during aging at room temperature and at 95°C. The oxide formed in phosphate solution appears to be the most stable one. This is in good agreement with the data of Konno *et al.*[181] on oxide hydration in phosphate and chromate solutions. The data of Belov[218] show that the different aging behaviors of oxides formed in different electrolytes are determined by the coordinative ability

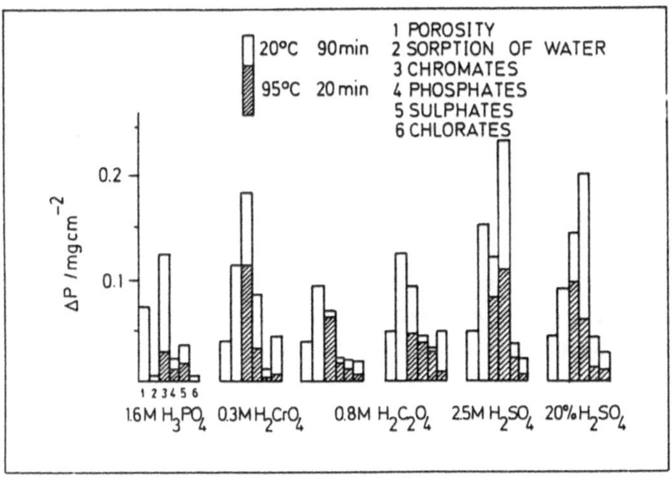

Figure 33. Hydration capacity of porous oxides formed in various electrolytes. The capacity to absorb various anions is also shown.

Table 5
Activation Energy of Hydration of Oxides Formed in Various Electrolytes

Electrolyte	Activation energy of hydration (eV)
Sulfamic acid	177.9
CrO_3	100.5
$H_2C_2O_4$	45.1
Na_3PO_4	46.1
H_2SO_4	32.1
Na_2BO_7	22.1
Na_2CO_3	7.7
NaOH	2.8-6

of the corresponding anions to form bonds with the oxide surface. According to Belov,[218] this ability decreases in the order

$$H_3PO_4^-(HPO_4^{2-}) > HCrO_4^- > HSO_4^-(SO_4^{2-}) > CrO_4^{2-} > HCO_3^-(CO_3^{2-}) > MnO_4^- > F^- > OH^-(H_2O) > Cl^- > NO_3^- > Br^- > I^- > ClO_4^-.$$

Very useful information concerning the tendency of oxides to undergo hydration in the presence of various anions is presented by Alwitt and Dyer.[209]

Various acid anions possess different abilities to compete with OH groups in adsorbing at oxide surfaces, as illustrated by Table 5.[219]

The apparent large differences in the activation energy of hydration for oxides formed in acidic and alkaline solutions reflect the basic differences in the mechanism of oxide growth in these two cases.

5. Morphology of Porous Anodic Aluminum Oxides

Numerous publications have been devoted to the investigation of the morphology of porous oxides of aluminum. Pores of virtually tubular shape with semispherical bottoms have been known to form in a more or less regular way inside alumina cells in hexagonal arrangements such as that shown schematically in Fig. 3. Such a formation is a logical consequence of expanding circles, evenly distributed over the surface in a (111) type of arrangement (starting from active sites), merging after their perimeters hit each other. The question arises as to why the oxide structure changes and the oxide becomes less soluble along the lines of the merger.

Early stages in porous structure development were well documented in the work of Csokan.[220] Especially revealing was the work of the Manchester group, conducted with the help of ultramicrotomy and ion-beam thinning techniques. It enabled the visualization of pore structure development in anodic aluminas.[4,146] In recent work by Nagayama *et al.*[70] and Ebihara *et al.*,[221] general trends in porous oxide growth have been summarized and their dependence on the conditions of anodization elucidated. This work as well as earlier investigations reviewed elsewhere[222,223] has yielded a large body of information concerning the geometry, the size, and other morphological features of oxides formed in a variety of electrolytes and in various regimes. Hence, only a brief summary will be given here. Table 6 reviews available information on various aspects of anodic alumina morphology.

It should be noted that anodization regimes have a major effect on oxide geometry. Palibroda[233,234] has summarized the empirical dependences of the pore diameter, d, the density of pores, n, and the lifetime of initial barrier growth, τ, on the ratio of the anodic voltage to the limiting voltage, $U_{a\max}$, as follows:

$$d = 4.986 + 0.709\,U_a = 3.64 + 18.89\,U_a/U_{a\max} \tag{51}$$

$$n = 1.6 \times 10^{12} \exp(-4.764\,U_a/U_{a\max}) \tag{52}$$

$$\tau = \tau_0(1 - U_a/U_{a\max}), \qquad \text{where } \tau_0 = \tau \text{ for } U_a/U_{a\max} \to 0 \tag{53}$$

Ebihara *et al.*[224] have put forward a similar dependence of cell size on the anodic voltage:

$$2R = \begin{cases} 14.5 + 2.0\,U_a & U_a < 20\text{ V} \quad (54) \\ -1.7 + 2.81\,U_a & U_a > 20\text{ V} \quad (55) \end{cases}$$

The universal character of this rule is illustrated by Fig. 34.

Table 6
Review of Different Aspects of Aluminum Oxide Morphology

Electrolyte used	Method of analysis	Morphological parameters studied	Reference
$H_2C_2O_4$	TEM	$N(U_a, C_e, T_e)$, (U_a, C_e, T_e), $r(t_d)$, r, $R(U_a, C_e, T_e)$, (j_a, t_a, T_e), $l(U_a)$	224
0.25–1.0 M $H_2C_2O_4 \cdot 2H_2O$	TEM	r, R, $N(t_a, j_a, U_a, T_e)$, $GM(j_a, T_e)^a$	225
0.16 M $H_2C_2O_4$	Pore filling	$N(U_a)$	
180 g/liter $H_2C_2O_4 \cdot 2H_2O$	TEM		
	Pore filling	N, $r(j_a, T_e)$	226
2–4% $H_2C_2O_4$	TEM	$r(t_d)$, $GM(t_d)$	70
3% $H_2C_2O_4$	TEM	D, $l(U_a)$	227
0.5–4.0 M H_2SO_4	TEM	N, r, $R(U_a, T_e, C_e)$, (U_a), $N(j_a)$	221
0.6 M H_2SO_4	TEM	N, $r(t_a)$	167
$H_2SO_4/H_2C_2O_4$	TEM	l, r, $R(U_a)$	105
3% Ammonium tartrate	TEM	r, $R(t_a, U_a)$	18
0.4 M H_3PO_4	TEM	R, N, $GM(t_a)$	228
44 g/liter CrO_3	TEM	$GM(T_e)$	229
CrO_3	TEM	$GM(U_a)$	11
0.1 M NaOH, Na_2CO_3 Na_3PO_4, NH_4OH–NH_4F	SEM	$H(t_a, T_e)$	230
0.3–1.0 M Na_2CO_3 + NaF	SEM	$l(t_a)$	231
HCOOH, CH_3COOH + $Na_4P_2O_7$	SEM	$H(t_a)$	232

[a] GM: "general morphology" of the oxides registered by TEM–SEM.

Analysis of experimental data shows that the dependence of the geometrical parameters of oxides on the temperature and concentration of electrolyte is different for galvanostatic and potentiostatic conditions (Fig. 35).[221] It appears that potentiostatic anodization is limited mainly by processes in the bulk of the oxide and thus is not influenced by temperature (Fig. 35b), whereas the galvanostatic anodization regime involves oxide dissolution processes at the O/S interface depending both on T_{el} and C_{el}.

Characteristic of both dependences is a decrease in the number of pores with an increase in either the current density or the steady-state voltage. To date, no clear explanation for this phenomenon is available.

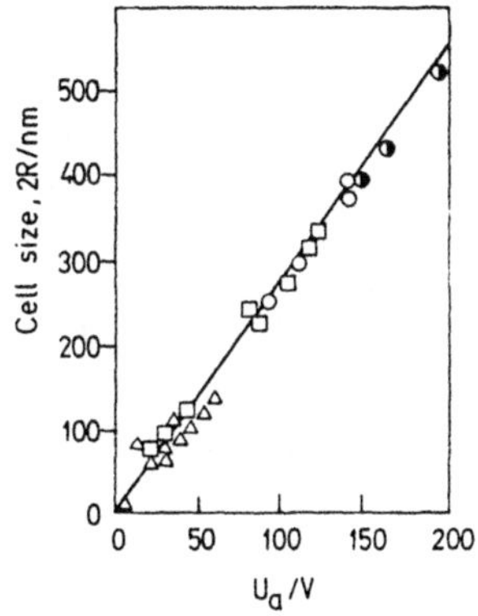

Figure 34. Cell size in porous oxides versus anode potential for different electrolytes: △, oxalic acid; □, phosphoric acid; ○, glycolic acid; ◑, tartaric acid.[224]

V. ELECTROPHYSICAL PROPERTIES OF ANODIC ALUMINUM OXIDE FILMS

1. Space Charge Effects

Every dielectric film, irrespective of the technology used for its formation, possesses a more or less pronounced space charge. A significant space charge is generated in oxide films produced by the thermal oxidation of materials,[235] plasma deposition,[236] and

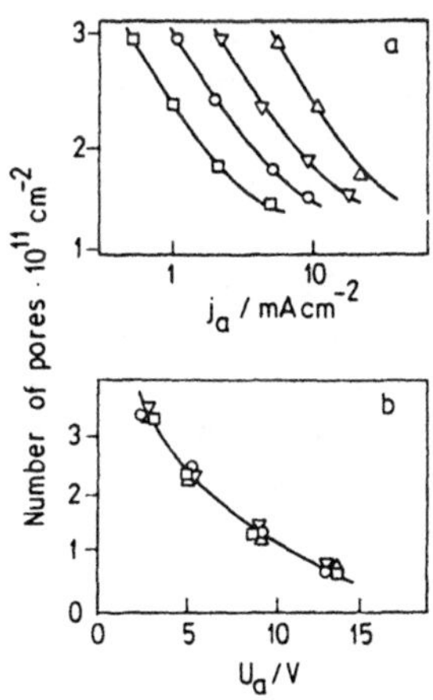

Figure 35. Number of pores versus j_a (a) and U_a (b) at different electrolyte temperatures: □, 10°C; ○, 20°C; ▽, 30°C; △, 40°C.[221]

plasma anodization.[237] Its value is especially high in the case of anodic oxidation of metals and semiconductors in electrolytes.[238] The incorporated space charge has a negative influence on the parameters of MOS structures with anodic oxide dielectrics[239] and causes a long-term drift of oxide properties.[238] It could, however, be useful in preparing electret films, i.e., dielectrics exhibiting an external electric field.[240] The most important problems in this field are:

- the nature of the incorporated space charge.
- the distribution of the space charge in an oxide.
- the impact of the space charge on the oxide properties.

(*i*) *The Nature of Space Charge in Anodic Aluminum Oxides*

Space charge accumulation in anodic alumina is closely related to the electrochemical processes taking place at the metal–solution contact, as discussed at the beginning of this review (cf. Section II). This is largely overlooked by physicists considering these phenomena.

Hence, there is a good deal of controversy about how to understand the nature of this space charge. Lobushkin and co-workers,[241,242] Zudov and co-workers,[243,244] and some others[245,246] assume that the space charge is generated in anodic oxides by electrons injected therein and captured at deep traps. Dyakonov and co-workers[247,248] have developed an approach assuming that the space charge is associated with lower-valency cationic defects generated in anodic oxides with depth-dependent stoichiometry. It is assumed by some that the space charge is caused by the decomposition of water molecules at the O/S interface, which is closer to electrochemical reality.

Another approach to the investigation of the nature of the space charge has been developed in the works of Dewald[58] and Fromhold and Fromhold[49,61] and has been further pursued by Parkhutik and co-workers.[62,249] The space charge is assumed to be generated by ionic defects, incorporated and moving in the oxide during its growth. Such an approach is consistent with the electrochemical nature and kinetics of oxide growth, the structural features of oxides, and the specific electric properties of anodic aluminas.[62]

Parkhutik *et al.*[172] have recently shown that there is a direct correlation between the kinetics of acid anion pickup in growing oxides and the kinetics of space charge accumulation in porous and barrier alumina films. This effect was interpreted in terms of the space charge being associated with the electrolyte anions incorporated into the oxide. This assumption appears to be useful in explaining the electret properties of anodic alumina films [see Section V(3)], the asymmetry of the dielectric strength [Section V(4)] and dc electronic conductivity [Section V(2)], and the transient behavior of anodic oxides [Section V(5)] and also in the theoretical modeling of space charge distribution in oxides, as is presented in the following section.

(*ii*) *Distribution of Space Charge*

There are a number of papers dedicated to the topic of space charge distribution in anodic oxides. Zudova and co-workers,[250,251] interpreting thermodepolarization measurements, have claimed that the negative space charge centroid is located at the interface separating the outer hydrated layer of the anodic oxide and the inner layer of pure Al_2O_3. At the same time, Morgan *et al.*[252] have assumed that the body of the oxide has a uniform positive charge, whereas the oxide boundaries are negatively charged. Berlicki *et al.*[253] have postulated that the space charge is distributed exponentially with the maximum value located at the oxide boundary.

Parkhutik and Shershulskii[249] have modeled the distribution of the space charge of ionic defects inside oxides (sufficiently far from the interfaces that the charge distribution near them can be neglected) based on the following assumptions:

(a) The oxide boundaries are permeable to space charge species, because of their high solubility in the electrode material.

(b) There is only one type of ionic defect creating the space charge.

(c) Migration of ionic defects in the oxide is determined by the classical high-field mechanism of ionic jumps over a series of potential barriers.

The details of modeling the space charge distribution in oxides are presented elsewhere.[249] Figure 36 presents the resulting space charge distribution in as-formed (curve 1) and aging oxides (curves

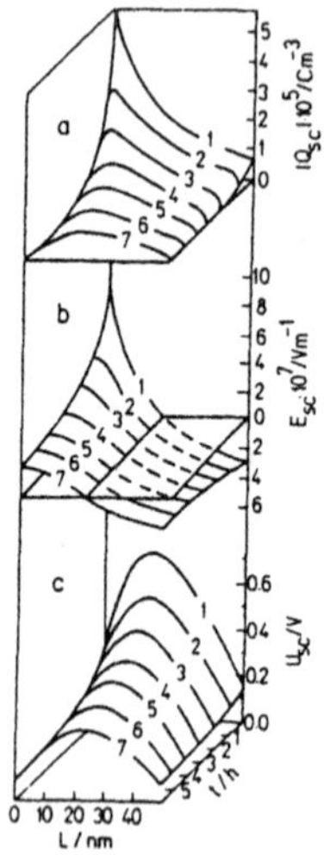

Figure 36. Stored negative space charge (a), coupled electric field (b), and internal voltage (c) in anodic oxide during anodization (curve 1) and during isothermal aging at various times (curves 2–7).[61]

2–7 correspond to different times of aging). The internal electric field coupled with this space charge is also given in Fig. 36. This field is essentially inhomogeneous and amounts to very high values, sufficient to cause anion redistribution after cutting off the external electric field.

One must keep in mind two important points:

1. The continuity of the current inside the oxide requires that the concentration of mobile charge carriers varies with the variation of the field with distance from the interface, so that their product remains constant.

2. The linearity of the change in anodization voltage with change in oxide thickness [cf. Section III(3(i))] requires that the space charge distribution inside the oxide remains constant during oxide growth, i.e., the space charge distribution profile widens in proportion to the thickening of the oxide.

2. Electronic Conduction

Electronic conduction plays a limited role, if any, in anodic oxide formation, since under the anodization conditions and with a high positive field, any electrons should be pulled into the metal. However, the growing prospects for anodic alumina films being applied outside the electrolyte, in electronics, presuppose knowledge of their electrophysical properties and, above all, of the electronic conduction mechanism. Analyzing the literature published on this subject, one can conclude that the state of the art at present is not very different from that reviewed by Goruk *et al.*[5] The traditional approach to studying the electronic conduction of thin film dielectrics, and anodic aluminum oxides in particular, is based on fitting experimental current-voltage characteristics to one or two classical mechanisms of electronic conduction, such as electron tunneling, Schottky or Poole–Frenkel emission, or the space charge current.[254,255] However, the amorphous structure of anodic oxides, their depth-dependent chemical composition, and the influence of surface states and other features make the validity of the traditional theoretical assumptions questionable in modeling oxide electronic conductivity.

It appears that the majority of specific features of anodic oxides can be covered by an approach based on the application of the mechanism of hopping electron conductivity through localized sites in a disordered dielectric. Mott and Twose[256] were the first to consider electron jumps through a random array of impurity centers and to take into account the amorphous structure of anodic oxides. Recent work by Bryksin *et al.*[257] has shown the validity of percolation hopping transport in anodic tantalum pentoxide, with the localized sites attributed to lower-valency cations, such as Ta^{3+} and Ta^{4+}. Important work in the field of the hopping transport of carriers through localized sites (impurity levels, structure defects, small polarons, etc.) has been done by Mott and Davis,[258] Jonscher and Hill,[259] Bonch-Bruevitch,[260] Shklovskii and Efros,[261] Bottger and Bryksin,[262] Firsov,[263] Austin and Mott,[264] and Emin.[265]

However, most of this work has avoided consideration of some complicating factors, arising especially from the limited thickness and real structure of anodic dielectric films. The latter causes the following effects:

(a) Ionic conduction in a dielectric can be of the same order as or even higher than electronic conduction.[266] Hence, the ionic current should be modeled as a modification of hopping conduction.[49,249]

(b) Very high electric fields are generated by relatively low voltages.

(c) Boundary conditions play an important role as they determine the carrier injection and ejection, image forces, etc.

(d) Carriers injected into the dielectric, localized at deep defect levels, generate an inhomogeneous and nonstationary space charge exhibited as an electret effect in the dielectric.[267]

Both classical[268] and more recent[269,270] papers have paid little, if any, attention to these complications.

The modeling of hopping conductivity of real amorphous dielectrics of limited thickness, with or without the incorporated space charge, has recently been done by Parkhutik and Shershulskii.[62]

The modeling of conductivity of thin disordered dielectrics was based on the following assumptions:

(i) All charged species participating in the conduction process are localized in the dielectric. Localized carriers are adiabatically separated into two subsystems with significantly differing mobilities. One comprises weakly localized carriers, and these carriers participate in hopping conduction. The other group of carriers is considered to be localized at deep centers in the dielectric and possess such low mobility that one can neglect their role in the conduction process.

(ii) The space charge effects introduced by the subsystem of strongly localized carriers are taken into consideration. As the mobility of these carriers is very low, the space charge distribution in the dielectric is assumed to be quasi-stationary, and only slow variations of the space charge are possible under the influence of various aging and relaxation processes.[238,271] The space charge distribution in the dielectric is determined by the nature of the process causing its accumulation (oxide formation, UV radiation,[272] corona discharge treatments,[273] etc.).

(iii) The electric field in the dielectric is a result of the superposition of an external (applied) field and the internal field of the space charge, and, hence, it is essentially inhomogeneous. There can be regions in the dielectric where the electric current is directed against the external electric field.[62] So, both the drift and diffusion modes of hopping transport have to be taken into consideration.

(iv) Dielectric interfaces are included by assuming carrier exchange between surface states randomly distributed in energy and the localized states in the bulk of the dielectric by the hopping mechanism or any other process (e.g., Fowler-Nordheim tunneling).

The resulting scheme of localized site distribution in a thin-film structure with disordered dielectrics is schematically illustrated in Fig. 37 with notations following those introduced by Jonscher and Hill.[259] The shape of the space charge region, Q (and adiabatically coupled with it, the hopping conduction region, W), is chosen to correspond to the case of an anodic oxide with an inhomogeneous space charge [see Section V(1(ii))].

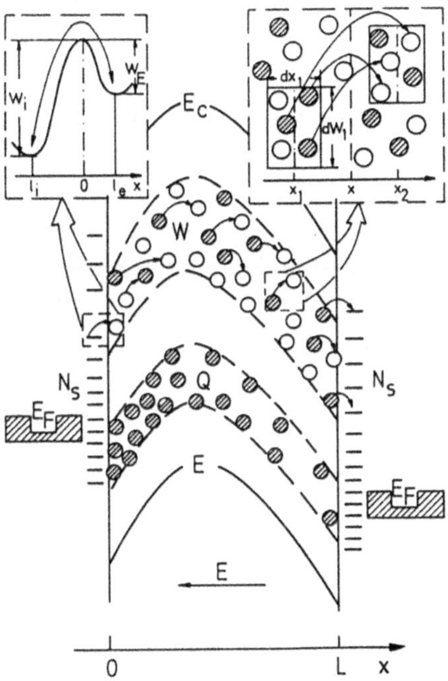

Figure 37. Schematic energy diagram of biased disordered dielectric. W, energy zone for hopping electrons; Q, energy zone for strongly localized species forming space charge; N_s, surface states.[61]

Using this approach, the hopping transport was modeled as a quasi-Marcovian process. The details of the analytical formulas forming the basis of the modeling and the numerical simulation procedure are given elsewhere.[62] The values of parameters included in the hopping transport model are listed in Table 7.

The current-voltage characteristics of the anodic oxide, as derived from the model, are given in Fig. 38. The different curves correspond to various stages of oxide aging, with the numbering following that in Fig. 36. The inserts in Fig. 38 illustrate linearization of the j-U curves in Schottky and Poole-Frenkel coordinates. Both linearizations, if conducted within the limited range of voltages, are rather satisfactory. This is a good illustration of the difficulties involved in correct identification of the mechanism of anodic oxide conduction. All j-U curves, linearized in logarithmic coordinates, exhibit two changes of slope at increasing external voltage. Such behavior, while being well established experimentally,[274,275] causes a good deal of controversy in interpretation. A polar assumption has been made to the effect that the slope decrease is caused by

Table 7
Parameters of the DC Electronic Conduction Model Based on Hopping Transport[a]

Symbol	Definition	Standard value	Variation interval	Units
T	Temperature	300	—	K
ε	Dielectric constant	10	—	
L	Thickness of dielectric	50	20–120	nm
W	Hopping activation energy	0.3	0.2–0.8	eV
l_h	Hopping distance	1	0.8–6.0	nm
W_i	Activation energy for carrier injection from the contact into the dielectric	0.7	0.3–1.0	eV
l_i	Activation length for carrier injection	0.8	0.4–1.2	nm
W_e	Activation energy for carrier ejection from the dielectric into the contact	0.1	0.01–0.6	eV
l_e	Activation length for carrier ejection	1.6	0.4–2.0	nm
N_s	Interface state density	5×10^{16}	—	m^{-2}

[a] Ref. 62.

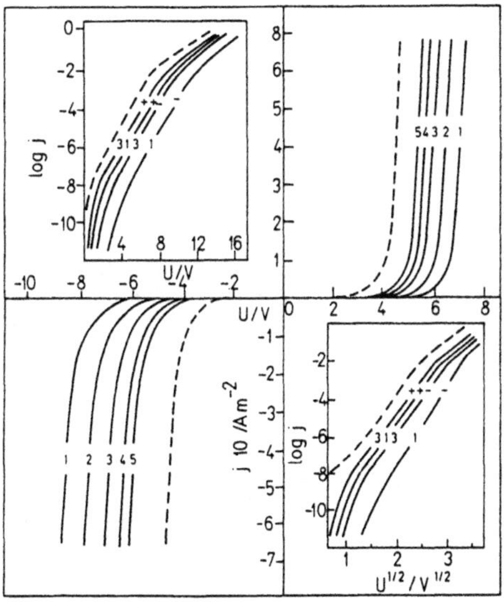

Figure 38. Calculated j-U curves for dielectric with negative space charge (solid curves) and reference uncharged one (dashed curve). The insets illustrate curve linearization in $\log j$-U and $\log j$-$U^{1/2}$ coordinates.[62]

transition from the surface to the volume-limited conduction mode[276,277] and by a reverse mode.[274,278] Parkhutik and Shershulskii[62] have shown that the j-U slope change is caused by the transition from bulk-limited hopping conduction to surface-limited conduction.

The fitting of theoretical curves to experimental data on aluminum[278] and tantalum[279] anodic oxides is illustrated in Fig. 39. The degree of agreement between theory and experiment is reasonable. The calculated j-U curves for a dielectric with negative space charge are essentially asymmetric (Fig. 38). Experimental data show that such asymmetry is most pronounced in the case of anodic oxides. This polarity has been interpreted in terms of the heterogeneous structure of anodic oxides ($p - i - n$ structure according to, for example, Gubanski and Hughes[271]) with the real nature of such a structure remaining unspecified. Polarization measurements[252] show that negative space charge is incorporated

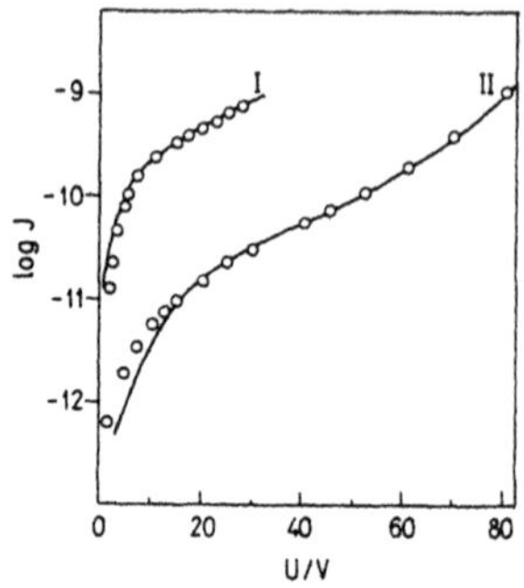

Figure 39. Comparison of theoretical j–U curves with experimental data on electrical conduction of aluminum (curve I) and tantalum (curve II) oxides.[62]

into anodic oxides with a maximum positioned in the vicinity of an external oxide boundary, which is in line with the electrochemical concept of potential profile (cf. Fig. 1). This charge causes asymmetric electric field generation in the manner presented in Fig. 36. The response of such an oxide to an external electric field must depend on the polarity of the field. In fact, the modulus of E_{sc} is higher to the left of the turning point (where $E = 0$). Therefore, the strength of the external electric field sufficient to neutralize E_{sc} in the region adjacent to the $x = L$ boundary is lower than that neutralizing E_{sc} at $x = 0$, as is schematically illustrated in Fig. 40. The value of the external voltage used for the figure ensures a higher conductivity of the oxide for positive bias of the $x = 0$

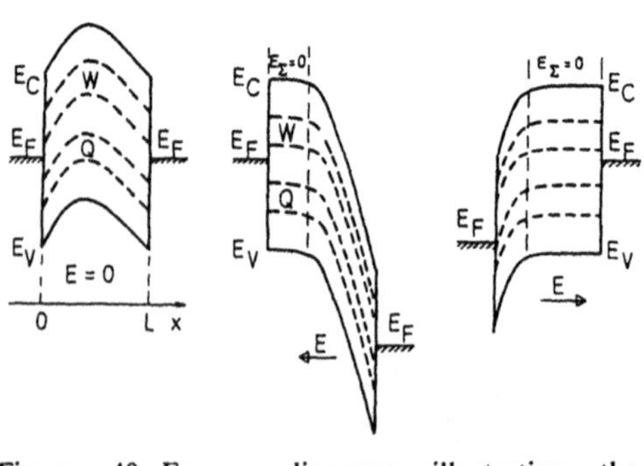

Figure 40. Energy diagrams illustrating the difference in values of external electric field of both polarities able to neutralize the internal electric field of asymmetric space charge.[62]

electrode (the electrode contacting the external oxide boundary) and a lower conductivity for the opposite bias. This kind of polar behavior is, indeed, exhibited in experiments.[271,280] The possibility arises of influencing the polar behavior and transient phenomena in anodic oxides by manipulating the space charge distribution through the judicious variation of experimental factors (electrolyte composition, anodization regime, etc.).

3. Electret Effects

When an anodic oxide grown on a metal is taken out of the solution in which anodization was carried out at a few hundred volts, it acts as a charged capacitor, but with the charge distributed inside the oxide (space charge). If one takes an average capacitance of the order of 1 F/cm^2 and a voltage of the order of 100 V across it, one could expect an integral space charge, Q_e, of about 10^{-4} C/cm^2 to be buried inside the oxide. Thus, the oxide is capable of exhibiting an electret effect.[267]

Since the work of Gunterschultze and Betz,[50] who pioneered the investigation of space charge effects in anodic oxides, the attention by researchers to this property has constantly been growing. In the original work, freshly obtained anodic oxides were short-circuited by a galvanometer, and an exponentially decreasing discharge current was registered. Later investigations by other authors[271,272] have confirmed these observations. It has also been established that thermal treatment of oxide samples causes thermally stimulated currents (TSC) directed like the isothermal discharge current. The similarity to electrets[267] has given rise to the term "anodoelectrets." Another electret property exhibited by anodoelectrets is an external electric field, registered by measuring the so-called electret (or surface) potential U_s of the oxide by a number of methods (e.g., screened probe, dynamic capacitor). The essential feature of anodoelectrets is their low thickness (at least an order of magnitude lower than that of polymer electrets), which makes them very promising for application in acoustic wave transducers[240] and other important fields.

There have been a number of investigations of the electret behavior of anodic oxides formed on various valve metals in different electrolytes and anodization regimes. The purpose of these

Table 8
Electret Parameters of Various Anodoelectrets

Material	Q_s (C/m^2)	T_{max} (°C)
Aluminum oxide	2×10^{-4}	90
Tantalum oxide	9.4×10^{-4}	110
Niobium oxide	1.2×10^{-5}	120
Titanium oxide (solution)	6×10^{-5}	105
Silicon oxide	2×10^{-7}	95

studies was to obtain electrets with the highest possible external electric field and the best long-term stability.[240,251] The results obtained by various authors can be briefly summarized as follows:

(i) All the anodic oxides formed on various materials exhibit more or less pronounced electret properties. Table 8 presents the integral charges and temperatures corresponding to the TSC maxima of various anodic oxides.[241]

(ii) Porous oxides exhibit lower electret parameters than barrier ones.

(iii) Electret parameters depend strongly on the electrolyte concentration, temperature, and other parameters, as is illustrated in Fig. 41. Higher electrolyte concentrations, temperatures (but not exceeding a certain limiting temperature), and current densities ensure higher values of electret parameters.[250,251]

(iv) The kinetics of accumulation of electret properties in growing oxides with a barrier structure are superlinear in the galvanostatic regime and exhibit saturation in the potentiostatic regime.[172,242] At the same time, during porous oxide formation, U_s

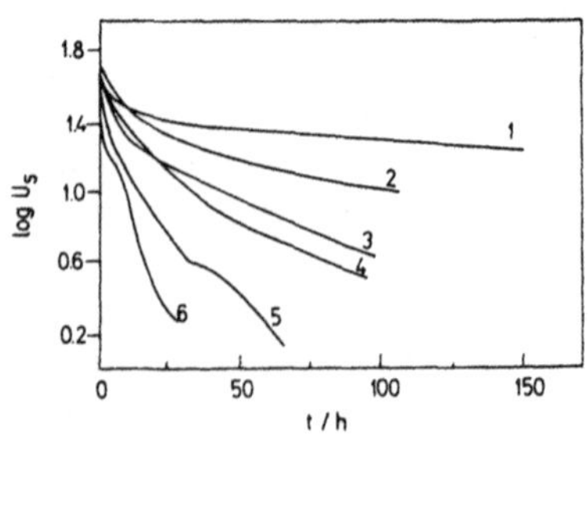

Figure 41. Kinetics of electret potential dissipation for anodic aluminas formed in (1) 1 wt % H_3PO_4, (2) 2 wt % H_3PO_4, (3) 4 wt % H_3PO_4, (4) 6 wt % H_3PO_4, (5) 0.1 wt % H_3PO_4 + 1.0 wt % APB, and (6) 0.1 wt % APB.[240]

exhibits a sharp maximum corresponding to the moment of commencement of pore development. This clearly indicates that the electret parameters of anodic oxides are coupled with their chemical, compositional, and structural features.

(v) The electret behavior of anodic oxides is affected by finishing treatments of freshly formed oxides: it is higher when samples are ejected from an anodizing bath, rinsed in water, and dried while kept at the potential of anodization.

(vi) The electret effect in naturally aging or thermally treated samples decreases monotonously, with the rate of decrease depending on the humidity of the atmosphere, temperature, and other factors.[241,277] Depolarized anodic oxides can be recharged by reanodization, UV irradiation, and corona discharge treatments.[272,273]

Various mechanisms for electret effect formation in anodic oxides have been proposed. Lobushkin and co-workers[241,242] assumed that it is caused by electrons captured at deep trap levels in oxides. This point of view was supported by Zudov and Zudova.[244,250] Mikho and Koleboshin[272] postulated that the surface charge of anodic oxides is caused by dissociation of water molecules at the oxide-electrolyte interface and absorption of OH^- groups. This mechanism was put forward to explain the restoration of the electret effect by UV irradiation of depolarized samples. Parkhutik and Shershulskii[62] assumed that the electret effect is caused by the accumulation of incorporated anions into the growing oxide. They based their conclusions on measurements of the kinetics of U_s accumulation in anodic oxides and comparative analyses of the kinetics of chemical composition variation of growing oxides.

From the electrochemical point of view, the anodoelectret effect can be described simply as a residual charge resulting from charging the capacitance of the oxide layer in order to create the field needed for ionic motion in the process of oxide growth. Such a view is supported by the order of magnitude of the charges recorded in anodoelectrets (Table 8). In addition, the anodoelectrets could also be considered as poorly defined galvanic cells with solid electrolytes. The M/O interface represents one well-defined electrode, but the interface between the oxide and the other contact represents an electrode where the electrochemical process depends largely on the actual environmental conditions, the presence of oxygen from the air enabling oxygen reduction to become the potential-determining process, and adsorbed water, depending on the humidity of the surrounding atmosphere, supplying oxygen ions and releasing hydrogen, thus acting as a kind of hydrogen electrode, etc. This could lead to some additional charge.

4. Electric Breakdown of Anodic Alumina Films

Electric breakdown of growing alumina films limits their maximum attainable thickness. It also causes degradation of thin film electronic devices with alumina films. Hence, it is a subject of intensive research and a large number of papers have been published and further reviewed. A good deal of controversy exists on various aspects of alumina oxide breakdown, and a variety of models have been proposed in attempts to fit experimental findings.

Generally, the phenomenon of breakdown exhibits the following features: during galvanostatic anodization of valve metals in barrier-oxide-forming electrolytes, when the anodic voltage attains a certain value (the so-called "first-spark voltage" according to Ikonopisov[284]), a single spark appears on the anode surface and further growth of U_a is arrested (Fig. 42).

Each breakdown is accompanied by some sound effect and is followed by a steady degradation of properties.[284] It can also lead to a complete destruction of the oxide with visible fissures and cracks.[286] The particular behavior observed depends on a large number of factors (electrolyte concentration,[287] defect concentration in the oxide,[288] etc.). The breakdown of thin-film systems (M-O-M and M-O-S structures) as a rule leads to irreversible damage of oxide dielectric properties.[289]

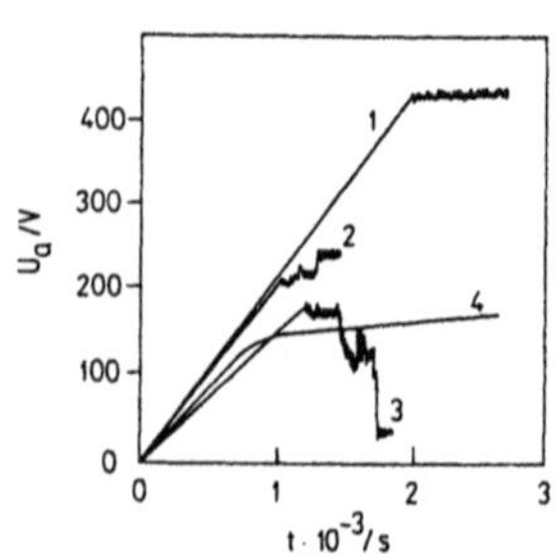

Figure 42. Breakdown voltages for Al anodized in (1) ammonium borate, (2) ammonium adipate, (3) potassium hydrogen phthalate, and (4) ammonium dihydrogen phosphate.[285]

There are a number of papers offering explanations of the breakdown phenomenon. Suggestions have been made that it is due to the presence of macro- and microdefects in oxides (electrolyte-filled fissures, micropores, flaws, etc.).[286] Joule heating effects were also considered[289] as well as volume increase and the resulting increase of internal stresses during anodization,[290] electrostriction effects,[291] or field-assisted ionic migration.[292]

Ikonopisov[284] has conducted a systematic study of breakdown mechanisms in growing anodic oxides. He has enumerated factors significantly affecting the breakdown (nature of the anodized metal, electrolyte composition and resistivity) as well as those of less importance (current density, surface topography, temperature, etc.). By assuming a mechanism of avalanche multiplication of electrons injected into the oxide by the Schottky mechanism, Ikonopisov has correctly predicted the dependence of U_b on electrolyte resistivity and other breakdown features.

Klein and co-workers have included, in the consideration of electronic avalanche breakdown, the effects of charge trapping in the oxide[293,294] and a stochastic nature of the avalanches.[295] According to these authors, trapped electron charge varies strongly with the field strength, temperature, oxide thickness, trap density, and depth distribution. All of this accounts for the encountered irreproducibility of the experimental results. They have also shown that breakdown characteristics of anodic alumina films strongly depend on the polarity of the applied voltage (in thin-film metal oxide/metal systems) and are not influenced by the material used for the conducting cover.[293] When the polarity of the substrate on which the oxide is grown is negative, the breakdown strength is twice as low as it is for positive polarity. This is in line with the observed asymmetry of dc conductivity [see Section V(2)] and generally supports the hypothesis of the influence of negative space charge on the properties of anodic oxides.

Further support for this hypothesis was presented by Albella and co-workers.[296-298] They reported evidence that electrolyte species incorporated into oxides act as a source of avalanching electrons. This assumption has yielded the well-known logarithmic dependence of breakdown voltage on electrolyte concentration:

$$V_b = A - B \ln[\text{Anion}] \tag{56}$$

Albella and co-workers have also explained the prebreakdown deviation of the $U_a(t)$ dependence in galvanostatic anodization from linearity. This is ascribed to increasing participation of electrons in the anodic current, resulting in the loss of current efficiency.[296]

There are also models assuming the electrostrictive input of incorporated anions into the breakdown initiation,[285,299] ionic drift models,[300] and many others reviewed elsewhere.[283,293] However, the majority of specialists agree that further work is necessary in order to properly understand the physics of the electric breakdown in growing oxide films and that caused by electric stress in thin-film structures.

5. Transient and Aging Phenomena in Anodic Alumina Films

Both naturally aging[253] and electrically[301] or thermally[302] stressed anodic oxides exhibit characteristic nonstationary behavior. One can distinguish the following transient effects:

1. Thermally stimulated currents in unbiased short-circuited oxides.
2. Transient currents in biased samples at room temperature.
3. Electroforming effects and negative resistivity of oxides.

Thermal treatment of short-circuited oxides causes so-called "thermally stimulated currents" (TSC) due to redistribution of excess charge incorporated into the oxide. TSC measurements are important for examining the oxide properties (namely, space charge effects). Thermal treatment also presents a method for symmetrization of electrophysical properties.[274]

Figure 43 illustrates the possible current transients during thermal treatment of Al-anodic Al_2O_3-Au structures at linearly increasing temperature (a) and during isothermal annealing (b). The first case is characterized by a TSC maximum at ~400 K followed by a change in current direction and a second maximum (Fig. 43a). In the case of isothermal treatment, j_{TSC} follows a t^{-n} dependence, where n is close to unity. These findings are usually interpreted in terms of a release from deep traps of those electrons that were initially captured there in the process of anodization. There are no clear ideas as to the physical nature of these traps. Parkhutik and Shershulskii[249] have postulated that traps are associ-

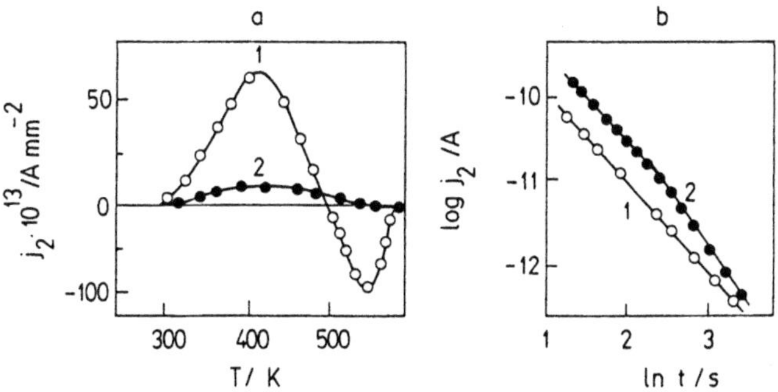

Figure 43. Thermodepolarization currents in Al–Al_2O_3–Au structures[244]: (a) isochromical annealing of oxide formed up to U_a = 180 V (1) and 100 V (2); (b) isothermal (20°C) resorption current in atmosphere (1) and vacuum (2).

ated with incorporated ionic defects. On the basis of this assumption, they carried out a theoretical modeling of TSC. The results of the modeling are illustrated in Fig. 44. They are seen to reproduce the experimental behavior (cf. Fig. 43) reasonably well.

Transient effects in naturally aging samples occur with some delay and are very slow. Nazar and Ahmad[274] have observed a slow decrease of Al–Al_2O_3–Al capacitance that was attributed to neutralization of Al^{3+} cations in the vicinity of the internal boundary and a corresponding increase of the effective oxide thickness. However, the same effect may be due to neutralization of negatively

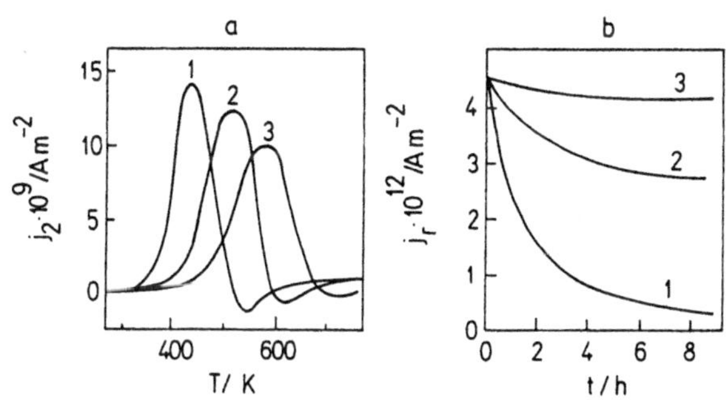

Figure 44. Resorption currents theoretically modeled for activation energy[249]: Vt = 0.9 eV (1); 1.0 eV (2); 1.1 eV (3). (a) and (b) correspond to Fig. 43.

charged ionic species at the outer oxide interface according to the results of Section V(1).

A group of scientists have studied current transients in biased M-O-M structures.[271,300] The general behavior of such a system may be described by classic theoretical work.[268,302] However, the specific behavior of current transients in anodic oxides made it necessary to develop a special model for nonsteady current flow applicable to this case. Aris and Lewis have put forward an assumption that current transients in anodic oxides are due to carrier trapping and release in the two systems of localized states (shallow and deep traps) associated with oxygen vacancies and/or incorporated impurities.[301] This approach was further supported by others,[271,279] and it generally resembles the oxide band structure theoretically modeled by Parkhutik and Shershulskii[62] (see. Fig. 37).

The negative resistance effect is observed when anodic oxides are subjected to so-called electroforming (i.e., annealing in vacuum).[93] Such a treatment removes the special features of the anodic oxides (asymmetry of conduction and electric strength, electret effect, etc.), and the negative resistance effect may be explained using the general approach developed for amorphous dielectrics.[5]

6. Electro- and Photoluminescence†

A hundred years ago, Sluginov[302] discovered a weak light emission during the anodic oxidation of aluminum. This phenomenon (also found in other valve metals) has been extensively studied by Gunterschulze,[303] Guminski,[304] Anderson,[305] Ruzievich,[306] van Geel,[307] Vermiliyea,[308] Smith,[309] Ganley and Mooney,[310] and many other investigators. Shimizu and Tajima[311] explained the effect as an "impact type" *electroluminescence* (EL) of the oxide films. Electrons injected into the conduction band of the oxide at the electrolyte-oxide interface are accelerated in high electric fields (10^6–10^7 V/cm) and produce nondestructive avalanches, which are quenched at some distance from the interface as a result of the opposing electric field created by slowly moving positive charges. Some of the

† This section was written by Vladeta Urosevic, Institute of Physics, University of Belgrade, Yugoslavia.

avalanche electrons have sufficient kinetic energy to excite luminescence centers. EL is always produced in barrier layers (even in the case of porous alumina), i.e., in the high-field region.

Several types of EL centers have been found:

1. For pure Al in inorganic electrolytes which form barrier oxide films, such as boric acid-borax solution, *surface defects* ("flows") have a dominant role.[308] It has been shown[312] that in this case only, EL vanishes for electropolished samples.

2. For Al *doped with 3d or 4f elements* (Mn, Cr, Cu, Fe, Zn, Mg, Nd, etc.), sharp and intense emission lines of the dopant are dominant.[310]

3. For Al in organic electrolytes (for instance, aliphatic acids and their salts) and in inorganic electrolytes which form porous oxide films (e.g., sulfuric acid), a broad and rather intense emission maximum is found,[313,314] whose origin is probably connected with *incorporated anions.*

A representative example of an EL spectrum is shown in Fig. 45. The energy levels from which the emission starts are always inside the forbidden band of Al_2O_3.

Photoluminescence (PL) of anodic aluminum oxides was first investigated in films formed in organic acids, the most intense PL being in those formed in oxalic acid. Tajima, in his comprehensive review[315] on electro- and photoluminescence in anodic oxide films, concluded that PL centers are carboxylate anions incorporated into the oxide. On the other hand, Eidel'berg and Tseitina[316] proposed

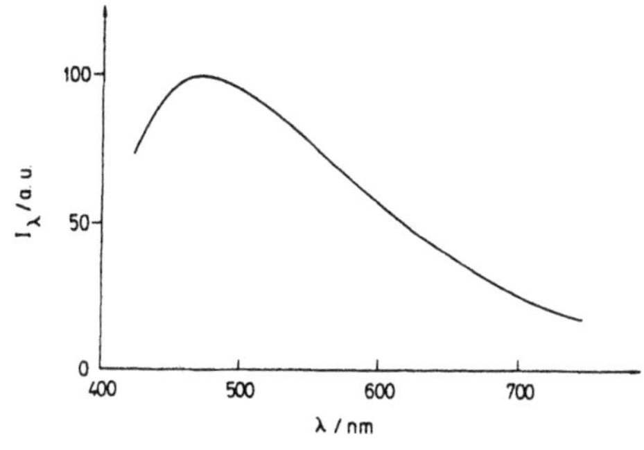

Figure 45. EL spectrum of anodic oxide film formed in an ammonium tartrate solution.

Table 9
Positions of Excitation and Emission Maxima of PL in Anodic Oxide Films Formed in Various Electrolytes by DC Anodization and in Boiling Water

Electrolyte	Excitation maximum (nm)	Emission maximum (nm)
Oxalic acid	250	420
	310	420
Formic acid	370	465
Phosphoric acid	270	340, 465
	340	460
Sulfuric acid	265, 340	330, 460
	340	460
Chromic acid	320	400
Sodium carbonate	355	460
Boiling water (bidistilled)	250	430

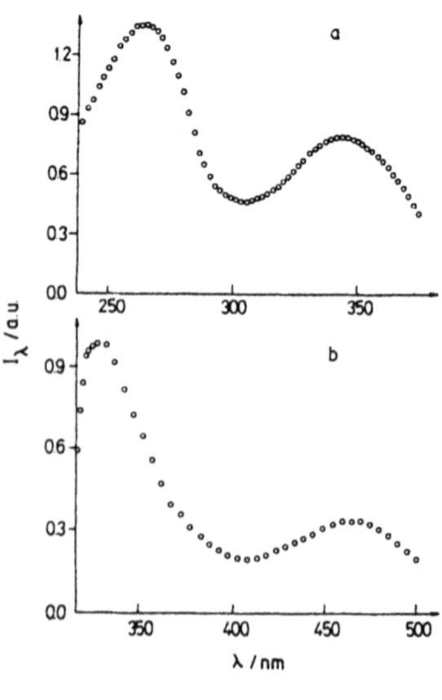

Figure 46. Pl excitation (a) and emission spectrum (b) of anodic oxide film formed in a sulfuric acid solution.

that these centers are formed by adsorption of water molecules, or OH radicals, at specific points (defects) in oxide films. The results of some recent investigations of PL in anodic oxide films are summarized in Table 9, and examples of excitation and emission spectra are given in Fig. 46. Contrary to Tajima's conclusion, it has been found (see Table 9) that inorganic anions and even water molecules (or OH radicals) can also produce PL centers. Calculations based on ligand field theory and infrared adsorption measurements seem to confirm this conclusion.[314]

If measurements are made in thin oxide films (of thickness less than 5 nm), at highly polished Al, within a small acceptance angle ($\alpha < 5°$), well-defined additional maxima and minima in excitation (PL) and emission (PL and EL) spectra appear.[322] This structure has been explained as a result of *interference* between monochromatic electromagnetic waves passing directly through the oxide film and EM waves reflected from the Al surface. In a series of papers,[318-320] this effect has been explored as a means for precise determination of anodic oxide film thickness (or growth rate), refractive index, porosity, mean range of electron avalanches, transport numbers, etc.

It is to be expected that a more profound investigation of EL and PL can give important information concerning electronic structure of anodic oxide films.

VI. TRENDS IN APPLICATION OF ANODIC ALUMINA FILMS IN TECHNOLOGY

Anodic alumina oxides find steadily growing application in various spheres of technology. Traditionally, they are most popular in civil industrial engineering for producing protective and decorative surface finish in panels and different objects. These applications are well reviewed in the literature.[321] Anodic alumina is also widely used in the aircraft and aerospace industry for adhesive bonding of aluminum structures,[322-324] composite materials, etc.

Alumina membranes are also produced and used in separation processes.[325,326]

The past decade has been very fruitful for applications of anodic alumina films in the electronics industry. These applications have not been presented adequately in the review literature, although they might be of interest to specialists engaged in aluminum anodization either as a science or as a technology.[327] Hence, a brief review here of applications of anodic aluminum oxides in electronics is considered worthwhile.

1. Electrolytic Capacitors

Aluminum foil capacitors occupy an important position in circuit applications due to their unsurpassed volumetric efficiency of capacitance and low cost per unit of capacitance.[328] Together with tantalum electrolytic capacitors, they are leaders in the electronic discrete parts market. Large capacitance is provided by the presence of extremely thin oxide layers on anodes and cathodes, and high surface areas of electrodes could be achieved by chemical or electrochemical tunnel etching of aluminum foils. The capacitance of etched eluminum can exceed that of unetched metal by as much as a factor of 50.[328]

Barrier anodic oxides covering the surface of aluminum etched foil are usually formed in borate or phosphate solutions. To improve capacitor characteristics, high-purity aluminum is desirable with as low a concentration of impurities as is acceptable in terms of cost.

A significant development in the sphere of aluminum foil capacitors involves forming the anodic oxide layer after a hot-water treatment of bare aluminum.[329,330] The hydrous oxide layer formed is converted during anodization, in its inner part, into a dense barrier oxide film of high dielectric strength and capacitance.[307] Further improvements of aluminum capacitors can be expected in the direction of increasing breakdown characteristics, long-term stability, and working temperatures.

One way to achieve such improvements is by doping of aluminum oxide with properly selected impurities. These could be introduced by implantation into aluminum and subsequent transfer into the oxide during anodization.[331] Alternatively, complex anions containing impurity atoms could be introduced into the anodizing bath [see Section IV(2)]. The incorporated anions influence the dielectric permittivity, E, of the oxide.[176] Hence, one can manipulate the E value by changing the electrolyte concentration and anodization regime.[91] According to the published data, rare-earth-doped alumina capacitors exhibit very good dielectric characteristics, approaching those of the bulk rare-earth metal oxides.[332]

2. Substrates for Hybrid Integrated Circuits

Alumina ceramic substrates were traditionally used for hybrid microelectronics due to their good dielectric properties.[333] At present, these ceramic substrates are being substituted by metallic plates covered by a dielectric insulating layer so as to improve the mechanical strength, the heat transfer ability, and other parameters.[334] This is especially important for high-power hybrid integrated circuits. The problem here is to choose an appropriate metal-dielectric couple providing for good adhesion, absence of various surface defects in the dielectric (flaws, pores, cracks, etc.), and a good quality of thin films formed at the substrate.

Table 10 presents the parameters of various metallic substrates. It can be seen that the most appropriate are those made of aluminum covered by anodic oxides.

Usually, an aluminum-magnesium alloy is used for plate formation, as it provides good mechanical strength (not less than 20 GPa) and uniformity of the porous anodic oxide layer. Optimal alloy composition is 3.2-3.8% Mg, 0.3-0.6% Mn, and 0.5-0.8% Si.[335] Oxalic and phosphoric acid solutions are used as electrolytes because they render thick (40-60 micron) porous oxide layers. Good results are also obtained with a chromic acid electrolyte.[336] The existence of pores in the oxide is desirable to avoid cracking at high temperatures. The pores act as a buffer opposing substrate damage. There is a relationship between the allowed damageless temperature rise and the porosity of the oxide[334]:

$$T_{cr} = 2\left[\frac{\eta_1 E_1(1-\mu_1)}{\pi l_{cr}(1+\mu_1)}\right]^{1/2}[E_1(\alpha_2-\alpha_1)]^{-1} \qquad (57)$$

where E is the Young's modulus, η is the Poisson coefficient, μ is the surface tension coefficient, α is the linear expansion coefficient, and the indices 1 and 2 correspond to aluminum oxide and aluminum metal, respectively. E_1 and μ_1 depend on oxide porosity N and pore radius r as follows:

$$E_1 = (E_1)_0(1-r^2N); \qquad \mu_1 = (1-r^2N) \qquad (58)$$

Table 10
Parameters of Various Metallic Substrates[a]

Substrate material	Price per unit area (arb. units)	Dielectric permittivity of insulator	Maximum working temperature (K)	Resistivity of dielectric layer ($\Omega\cdot$cm)	Density (g/cm^3)	Linear expansion coefficient $\times 10^{-6}$ (K^{-1})	Thermal conductivity [$W/(m\cdot K)$]
Al with epoxy resin	0.006	4	473–523	10^6–10^9	2.7	24	49
Al with anodic oxide	0.003	7	673–723	10^{12}–10^{13}	2.8	16–18	200
Enameled steel	0.001	10–12	1073–1273	10^{11}–10^{12}	6.4	10–16	40
Covar covered by dielectric	0.01	4.6	673–773	10^8–10^{10}	8.5	16	180
Ti with Al anodic oxide	0.12	8–9	723–773	10^{12}–10^{13}	4.8	20	29
Steel with epoxy resin	0.008	10–12	473–523	10^6–10^9	7.8	12	1.1

[a] Ref. 335.

Hence, T_{cr} is seen to increase with pore density and pore radius. However, a problem appears at a porous substrate when thin films are to be deposited during metallization to form interconnections, thin-film capacitors, etc.[335] Sputtered material falls deep into the pores, which affects the planarity of the deposited layer and the electrical resistivity of the oxide layer underneath.[335] To cope with this effect, the porous oxide should be padded by inorganic (Al_2O_3 and SiO_2) or organic (polyimide, negative photoresist) layers.

Aluminum plates covered by anodic oxides are also used in manufacturing magnetic recording disks.[336,337]

3. Interconnection Metallization for Multilevel LSI

Large-scale integration (LSI) of semiconductor and hybrid integrated circuits requires the use of multilevel interconnection metallization. The possible ways of producing multilevel metallization are reviewed in a number of publications.[338,339] Here again, aluminum and its anodic oxide provide a very good planarity of metallization patterns, fair electric parameters, low cost, and good reproducibility.[338] There are numerous patents dealing with multilevel aluminum metallization for LSI, differing from one another in anodizing procedure, geometry, and other features.

Aluminum metallization in combination with tantalum thin films is used for manufacturing thin-film capacitors built into the metallization pattern.[340]

4. Gate Insulators for MOSFETs

Native oxides, grown thermally or anodically at semiconductor surfaces, are not suitable for producing gate insulators for metal-oxide-semiconductor field effect transistors (MOSFETs). Oxide-semiconductor interfaces exhibit high densities of surface states and fixed charge, especially in the case of compound semiconductors.[341] Hence, aluminum deposition onto semiconductors and its further anodization to complete oxidation of the metal film is considered as a way to improve the situation. There are papers indicating that MOSFETs with anodic aluminum oxide can be formed on silicon substrates[342] and A_3P_5 semiconductors.[343] Best results are obtained in producing *n*-channel enhancement-mode

InP transistors for high-speed memory and optoelectronic applications.[344] Further work is necessary to investigate the use of anodic alumina in producing high-frequency MOSFETs at other compound semiconductors.

5. Magnetic Recording Applications

Polished aluminum-based alloys are used as substrates for producing hard magnetic disks.[336] The ferromagnetic layer is usually formed by sputter deposition onto an aluminum plate covered by a thin anodic oxide.[337] However, very promising results have been obtained with electrodeposition of the ferromagnetic metal and alloys in the pores of thick aluminum oxides. The obtained magnetic media exhibit high coercive force (3200 Oe according to Ref. 345) and good squareness ratios and wear resistance.[346] Films with both vertical and horizontal anisotropy may be formed, and reading density is supposed to reach about 15,000 BPI.[345]

6. Photolithography Masks

The use of aluminum-based masks in photolithography has been proposed.[347] According to the scheme employed, aluminum is deposited onto a polished glass sheet. The regions of the mask that should be light transparent are then converted into porous oxide. As the operation of aluminum anodization exhibits a much better vertical anisotropy than chemical etching, the masks obtained reproduce the parameters of standard masks more precisely than the chromium masks usually used.

There are also several proposals to use anodic aluminum oxides in producing optoelectronic devices. Porous oxides may find use as antireflecting coatings for optical pathways. Anodic alumina films doped by Eu and Tb are promising for application in electroluminescent cells for TEELs.[28]

7. Plasma Anodization of Aluminum

The process of plasma anodization takes place when liquid electrolyte is replaced by low-temperature oxygen plasma. It was first reported by Dankov[348] and Nazarova,[349] but Miles and Smith are considered to have pioneered the investigation of the process in their work of 1963.[350] Plasma anodization belongs among the "dry" technological processes and is suitable for other vacuum processes in electronics technology. It provides a high-purity oxidation atmosphere so that the oxides formed are not contaminated by impurities usually occurring in wet anodization. Hence, it is quite promising for applications in various spheres of technology and is being widely investigated. Still, only a few reviews of this area are available.[237,351,352] Plasma anodic oxides of aluminum are suitable for application as dielectrics for MOSFETs,[353,354] magnetic memory disks,[355] capacitors,[351] etc. However, plasma anodization has several shortcomings which have to be overcome, including low efficiency of oxide growth and the small thickness of the oxides formed[356] and high fixed-charge density at the oxide–substrate interface.[357] Various kinds of discharge (dc glow discharge, high-frequency and microwave discharges, arc discharge, corona discharge) have been used for oxygen plasma excitation and various constructions of plasma generators have been developed. Hence, the results obtained by various authors differ very much.[237] Besides, the plasma anodization process is very sensitive to the parameters of the plasma (particle composition of the plasma, energy and concentration of electrons and oxygen ions, discharge current, and so on). As a result, the reproducibility of the results is often poor. To realize the potential advantages of the process, it is necessary to establish the proper values of these parameters.

ACKNOWLEDGMENTS

The authors are indebted to Dr. Vladimir Jovic and Mrs. Ljiljana Gajic-Krstajic for their help in the preparation of this manuscript. One of the authors (P.V.) wishes to acknowledge the encouraging support of Professor Vladimir Labunov and Dr. Eugenija Matweeva.

REFERENCES

[1] S. Wernick and R. Pinner, *Finishing of Al*, Ed. by R. Draper, London, 1976.
[2] E. L. Littauer and J. F. Cooper, in *Handbook of Batteries and Fuel Cells*, Ed. by David Linden, McGraw-Hill, New York, 1984, pp. 10–30.
[3] N. P. Fitzpatrick and G. M. Scamans, *New Scientist* **111** (1517) (1986) 34.

[4] G. E. Thompson and G. C. Wood, in *Treatise on Materials Science and Technology*, Vol. 23, Ed. by J. C. Scully, Academic, New York, 1983, p. 205.
[5] W. S. Goruk, L. Young, and F. G. R. Zobel, in *Modern Aspects of Electrochemistry*, No. 4, Ed by J. O'M. Bockris and B. E. Conway, Plenum Press, New York, 1966, p. 176.
[6] L. N. Antropov, G. G. Vrzhosek, and Yu. Fateev. *Zashch. Metal.* **11** (1975) 300.
[7] B. N. Kabanov, I. I. Astakhov, and I. G. Kiseleva, *Electrochim. Acta* **24** (1979) 167.
[8] I. A. Ammar, S. Darwish, and P. Khalil, *Z. Werkstofftechn.* **12** (1981) 421.
[9] T. Fukushima, Y. Fukuda, G. Ito, and A. Shimizu, *J. Met. Finish. Soc. Jpn.* **25** (1974) 542.
[10] K. V. Heber, *Electrochim. Acta* **23** (1978) 127.
[11] D. J. Arrowsmith and D. A. Moth, *Trans. Inst. Met. Finish.* **64** (1986) 91.
[12] M. Hirochi and C. Yoshimura, *J. Met. Finish. Soc. Jpn.* **28** (1977) 634.
[13] K. Yokoyama, H. Konno, H. Takahashi, and M. Nagayama, *Plat. Surf. Finish.* **69** (1982) 62.
[14] A. Ben Rais and J. Sohm, *Corros. Sci.* **25** (1985) 1035.
[15] W. A. Badawy, M. M. Ibrahim, M. M. Abon-Romia, and M. S. El-Basionny, *Corrosion* **42** (1986) 324.
[16] S. Hoshino, T. Imamura, S. Matsumoto, and K. Kojima, *J. Met. Finish. Soc. Jpn.* **28** (1977) 167.
[17] P. A. Malachesky, in *Encyclopedia of Electrochemistry of the Elements*, Vol. 6, Ed. by A. J. Bard, Marcel Dekker, New York, 1976, p. 64.
[18] Y. H. Choo and O. F. Devereux, *J. Electrochem. Soc.* **122** (1975) 1645.
[19] M. Nagayama, H. Takahashi, and M. Koda, *J. Met. Finish. Soc. Jpn.* **30** (1979) 438.
[20] H. Takahashi, H. Saito, and M. Nagayama, *J. Met. Finish. Soc. Jpn.* **33** (1982) 225.
[21] J. O'M. Bockris and A. K. N. Reddy, *Modern Electrochemistry*, Vol. 2, Plenum Press, New York, 1973, p. 729.
[22] Wholer, Buff, *Liebig's Ann.* **103** (1857) 218.
[23] H. Aida, J. Epelboin, and M. Garreau, *J. Electrochem. Soc.* **118** (1971) 244.
[24] M. Garreau and P. L. Bonora, *J. Appl. Electrochem.* **7** (1977) 197.
[25] J. O'M. Bockris and A. R. Despic, in *Physical Chemistry, Vol. 9b, Electrochemistry*, Ed. by H. Eyring, W. Jost, and J. N. Linett, Academic, New York, 1970, p. 628.
[26] N. Sato, in Proceedings of the 4th International Symposium on Passivity of Metals, Princeton, 1978, p. 29.
[27] N. Sato, *J. Electrochem. Soc.* **129** (1982) 255.
[28] J. C. Scully, in *Atomic Fraction Proc. NATO*, Adv. Res. Inst., Calcatogglio, 1983, p. 637.
[29] T. Valand and K. E. Heusler, *J. Electroanal. Chem.* **149** (1983) 71.
[30] K. J. Vetter, *J. Electrochem. Soc.* **110** (1962) 597; *Electrochim. Acta* **16** (1971) 1923.
[31] T. P. Hoar and J. Yahalom, *J. Electrochem. Soc.* **110** (1963) 614.
[32] M. Nagayama, K. Tamura, and H. Takahashi, in Proceedings of the 5th International Congress on Metallic Corrosion, Houston, 1974, p. 175.
[33] S. M. Ahmed, in *Oxides and Oxide Films*, Vol. 1, Ed. by J. Diggle, Marcel Dekker, New York, 1972, p. 319.
[34] D. E. Yates and T. W. Healy. *J. Chem. Soc., Faraday Trans. 1*, **76** (1980) 9.
[35] T. Hurlen and A. T. Haug, *Electrochim. Acta* **29** (1984) 1133.
[36] T. P. Hoar, *Electrode Processes*, Butterworths, London, 1961, p. 299.
[37] G. Sandoz, C. T. Fujii, and B. F. Brown, *Corros. Sci.* **10** (1970) 839.
[38] J. A. Smith, M. H. Peterson, and B. F. Brown, *Corrosion* **26** (1970) 539.
[39] J. A. Davis, in *Localized Corrosion*, Ed. by R. W. Stahle, B. F. Brown, J. Krufger, and A. Agrawal, Natl. Assoc. Corr. Eng., Houston, 1974, p. 168.
[40] B. E. Wilde and E. Williams, *Electrochim. Acta* **16** (1971) 1971.

[41] R. Piccini, M. Marek, A. J. E. Pourbaix, and R. F. Hochman, in *Localized Corrosion*, Ed. by R. W. Staehle, B. F. Brown, J. Kruger, and A. Agrawal, Natl. Assoc. Corr. Eng., Houston, 1974, p. 179.
[42] T. Suzuki, M. Yamabe, and Y. Kitamura, *Corrosion* **29** (1973) 18.
[43] J. R. Galvele, *J. Electrochem. Soc.* **123** (1976) 464.
[44] K. J. Vetter and H. H. Strehblow, in *Localized Corrosion*, Ed. by R. W. Staehle, B. F. Brown, J. Kruger, and A. Agrawal, Natl. Assoc. Corr. Eng., Houston, 1974, p. 240.
[45] A. R. Despic, *J. Electroanal. Chem.* **191** (1985) 417.
[46] A. R. Despic, *J. Serb. Chem. Soc.* **50** (1985) 375.
[47] M. J. Dignam, in *Oxides and Oxide Films*, Vol. 1, Ed. by J. W. Diggle, Marcel Dekker, New York, 1972, p. 319.
[48] A. T. Fromhold, *Theory of Metal Oxidation*, Vol. 1, 2, North-Holland, Amsterdam, 1977.
[49] A. T. Fromhold and R. G. Fromhold, in *Comprehensive Chemical Kinetics*, Vol. 21, Ed. by C. H. Bamford, Elsevier, Amsterdam, 1984, p. 1.
[50] A. Gunterschultze and H. Betz, *Z. Phys.* **92** (1934) 367.
[51] N. Cabrera and N. F. Mott, *Rept. Prog. Phys.* **12** (1948–49) 163.
[52] J. M. Albella, I. Montero, O. Sanchez, and J. M. Martinez-Duart, *J. Eléctrochem. Soc.* **133** (1986) 876.
[53] V. Macagno and J. W. Schultze, *J. Electroanal. Chem.* **180** (1984) 157.
[54] L. L. Odynets, *Elektrokhimiya* **20** (1984) 463.
[55] E. L. W. Verv́ey, *Physica* **2** (1935) 1059.
[56] D. A. Vermilyea, *J. Electrochem. Soc.* **102** (1955) 655.
[57] A. R. Bray, *Proc. Phys. Soc.* **71** (1958) 405.
[58] J. F. Dewald, *J. Electrochem. Soc.* **102** (1955) 1.
[59] J. F. Winkel, C. A. Pistorius, and W. C. Van Geel, *Philips Res. Rept.* **13** (1958) 277.
[60] L. Young and F. G. R. Zobel, *J. Electrochem. Soc.* **113** (1966) 277.
[61] A. T. Fromhold, *J. Electrochem. Soc.* **124** (1977) 538.
[62] V. P. Parkhutik and V. I. Shershulskii, *J. Phys. D: Appl. Phys.* **19** (1986) 623.
[63] R. K. Hart, *Trans. Faraday Soc.* **53** (1957) 1020.
[64] M. J. Prior. *Z. Electrochem.* **62** (1958) 788.
[65] M. S. Hunter and P. Fowle, *J. Electrochem. Soc.* **103** (1956) 482.
[66] W. H. Smyrl, in *Comprehensive Treatise of Electrochemistry*, Vol. 4, Ed. by J. O'M. Bockris, B. E. Conway, E. Yeager, and R. E. White, Plenum Press, New York, 1981, p. 97.
[67] H. Kaesche, *Z. Phys. Chem.* (*N.F*) **26** (1960) 138; **34** (1962) 87.
[68] K. Nisancioglu and H. Holtan, *Electrochim. Acta* **24** (1979) 1229.
[69] J. Radosevic, M. Kliskic, and A. R. Despić, Xth Meeting of Chemists of Croatia, Zagreb, 16–18 Feb. 1987, Abstracts.
[70] M. Nagayama, K. Tamura, and H. Takahashi, *Corros. Sci.* **10** (1970) 617.
[71] J. H. Manhart and F. A. Mozelewski, *Plat. Surf. Finish.* **66** (1979) 54.
[72] T. N. Nguen and R. T. Foley, *J. Electrochem. Soc.* **129** (1982) 27.
[73] V. A. Sokol and A. I. Vorobjeva, *Izv. Akad. Nauk Beloruss. SSR, Ser. Khim., no.* 2 (1983) 39.
[74] H. Torii, H. Majima, and T. Tanaka, *Denki Kagaku* **45** (1977) 218.
[75] M. M. Hefny, A. A. Mazhlar, and M. S. El-Basiounhy, *Brit. Corros. J.* **17** (1982)
[76] W. A. Badawy, M. M. Ibrahim, M. Abou-Romania, and M. S. El-Basiouny, *Corrosion* **42** (1986) 324.
[77] A. R. Despic, D. M. Drazic, and L. J. Gajic-Krstajic, *J. Electroanal. Chem.* **242** (1988) 303.
[78] J. Bessone, C. Mayer, K. Juttner, and W. J. Lorenz, *Electrochim. Acta* **28** (1983) 171.

[79] H. Takahashi, K. Fujimoto, H. Konno, and M. Nagayama, *J. Electrochem. Soc.* **131** (1984) 1856.
[80] C. J. Dell'Oka, D. L. Pulfrey, and L. Young, in *Physics of Thin Films*, Vol. 6, Ed. by M. H. Francombe, Academic, New York, 1971.
[81] V. P. Parkhutik, *Izv. Akad. Nauk Beloruss. SSR, Ser. Khim.* **1** (1984) 113.
[82] P. A. Bakhtin, A. B. Emelyanov, and R. A. Suris, *Elektrokhimiya* **12** (1976) 1834.
[83] L. L. Odynetz and E. Ya. Khanina, *Elektrokhimiya* **9** (1973) 1378.
[84] C. Cherki and J. Siejka, *J. Electrochem. Soc.* **120** (1973) 786.
[85] J. Perriere, J. Siejka, and S. Rigo, *Corros. Sci.* **20** (1980) 91.
[86] P. Skeldon, K. Shimizu, G. E. Thompson, and G. C. Wood, *Surf. Interface Anal.* **5** (1983) 247.
[87] H. Takahashi, H. Kumagai, and M. Nagayama, *J. Met. Finish. Soc. Jpn.* **36**, (1985) 478.
[88] D. A. Vermilyea, *J. Electrochem. Soc.* **110** (1963) 345.
[89] F. Brown, *J. Electrochem. Soc.* **109** (1962) 999.
[90] N. Khalil and J. S. L. Leach, *Electrochim. Acta* **31** (1986) 1279.
[91] J. J. Randall, Jr. and W. J. Bernard, *Electrochim. Acta* **20** (1975) 653.
[92] J. A. Davies, B. Domeij, J. P. S. Pringle, and F. Brown, *J. Electrochem. Soc.* **112** (1965) 675.
[93] J. P. S. Pringle, *J. Electrochem. Soc.* **120** (1973) 398.
[94] F. Brown and W. D. Mackintosh, *J. Electrochem. Soc.* **120** (1973) 1096.
[95] J. P. Thomas, M. Fallavier, P. Spender, and E. Francois, *J. Electrochem. Soc.* **127** (1980) 685.
[96] C. J. Dell'Oka and P. J. Fleming, *J. Electrochem. Soc.* **123**, (1976) 1487.
[97] V. P. Parkhutik, in Extended Abstracts, 5th International Symposium on the Passivity of Metals, ISE, Bombannes, 1983, p. 253.
[98] G. E. Thompson, *Phil. Mag.*, in press.
[99] L. V. Andreeva, S. M. Ikonopisov, and D. D. Tokarev, *Elektrokhimiya* **7** (1971) 1698.
[100] P. L. Cabot, F. A. Centellas, E. Perez, and J. Virgili, *Electrochim. Acta* **30** (1985) 1035.
[101] T. Hurlen, H. Lian, O. S. Odegard, and T. Valand, *Electrochim. Acta* **29** (1984) 579.
[102] Y. Xu, G. E. Thompson, and G. C. Wood, *Trans. Inst. Met. Finish.* **63** (1985) 98.
[103] E. Palibroda, *Electrochim. Acta* **23** (1978) 835.
[104] P. L. Cabot and E. Perez, *Electrochim. Acta* **31** (1986) 319.
[105] K. Ebihara, H. Takahashi, and M. Nagayama, *J. Met. Finish. Soc. Jpn.* **35** (1984) 205.
[106] R. W. Franklin, *Nature* **180** (1957) 1470.
[107] D. M. Drazic, S. K. Zecevic, R. T. Atanasoski, and A. R. Despic, *Electrochim. Acta* **28** (1983) 751.
[108] A. R. Despic, *J. Electroanal. Chim.* **184** (1985) 401.
[109] A. R. Despic, D. M. Drazic, S. K. Zecevic, and R. T. Atanasoski, *Electrochim. Acta* **26** (1981) 173.
[110] H. Kaesche, *Korroziya Metallov* (Russian translation), Metallurgiya, Moscow, 1984.
[111] R. M. Stevanovic, A. R. Despic, and D. M. Drazic, *Electrochim. Acta* **33** (1988) 397.
[112] S. K. Zecevic, Master's thesis, University of Belgrade, 1979.
[113] A. Kheribech and J. Pagetti, *Metaux* **59** (1983) 361.
[114] V. V. Sysoeva, E. D. Artyugina, V. G. Gorodilova, and T. Berkman, *Zh. Prikl. Khim.* **58** (1985) 921.

[115] D. M. Drazic, A. R. Despic. L. Gajic, and L. Atanasoska, 34th ISE Meeting, Erlangen, Germany, September 1983, Abstract 1105.
[116] S. W. Benson, *Foundations of Chemical Kinetics*, McGraw-Hill, New York, 1960, p. 625.
[117] H. H. Strehblow, *Werkst. Korros.* **27** (1976) 792; **35** (1984) 437.
[118] T. P. Hoar, D. C. Mears, and G. P. Rothwell, *Corros. Sci.* **5** (1965) 279.
[119] K. J. Vetter and H. H. Strehblow, *Ber. Bunsenges. Phys. Chem.* **74** (1970) 1024.
[120] N. Sato, *Electrochim. Acta* **16** (1971) 1683.
[121] Ya. M. Kolotyrkin, *Corrosion* **19** (1964) 261.
[122] Z. A. Foroulis and M. J. Thubrikar, *J. Electrochem. Soc.* **122** (1975) 129.
[123] L. D. Atanasoska, D. M. Drazic, A. R. Despic, and A. Zalar, *J. Electroanal. Chem.* **182** (1985) 179.
[124] A. Thiel, *Z. Elektrochem.* **33** (1927) 370.
[125] I. T. Reding and J. J. Newport, *Materials Protection* **5** (1966) 15.
[126] A. R. Despic, D. M. Drazic, M. M. Purenovic, and N. Cikovic, *J. Appl. Electrochem.* **6** (1976) 527.
[127] C. D. S. Tuck, J. A. Hunter, and G. M. Scamans, Electrochemical Society Meeting, Boston, May 1986, Extended Abstract 638.
[128] R. Stevanovic and A. R. Despic, unpublished results.
[129] A. P. Bond, G. F. Bolling, H. A. Domian, and H. Biloni, *J. Electrochem. Soc.* **113** (1966) 773.
[130] E. Yeager, *Surf. Sci.* **101** (1980) 1.
[131] C. C. Chang, *J. Vac. Sci. Technol.* **18** (1981) 276.
[132] M. P. Seah, *Surf. Interface Anal.* **2** (1980) 222.
[133] V. A. Labunov, V. P. Parkhutik, and V. A. Sokolov, *Inorg. Mater.* **19** (1983) 2015.
[134] V. T. Belov and I. A. Kopylova, *Zh. Neorg. Khim.* **24** (1979) 291.
[135] G. E. Thompson, G. C. Wood, and K. Shimizu, *Electrochim. Acta* **26** (1981) 951.
[136] K. Shimizu, G. E. Thompson, and G. C. Wood, *Thin Solid Films* **81** (1981) 39.
[137] R. M. Pearson, *J. Catal.* **23** (1971) 388.
[138] M. Shimura, *J. Electrochem. Soc.* **125** (1980) 190.
[139] T. L. Barr, *J. Phys. Chem.* **82** (1978) 1801.
[140] V. I. Nefedov, *X-Ray Photoelectron Spectroscopy of Chemical Compounds*, Khimiya, Moscow, 1984.
[141] T. Smith, *Surf. Sci.* **55** (1976) 601.
[142] D. Briggs and M. P. Seah (Eds.), *Practical Surface Analysis by Auger and Photoelectron Spectroscopy*, Wiley, Chichester, 1983.
[143] Y. Xu, G. E. Thompson, G. C. Wood, and B. Bethune, *Corros. Sci.* **27** (1987) 83.
[144] D. W. Lynch, *J. Electroanal. Chem.* **150** (1983) 229.
[145] Y. Xu, G. E. Thompson, and G. C. Wood, *J. Electrochem. Soc.* **130** (1983) 2395.
[146] G. E. Thompson, R. C. Furneaux, and G. C. Wood, *Corros. Sci.* **18** (1978) 481.
[147] A. T. Hubbard, J. L. Stickney, S. D. Rosasco, M. P. Soriaga, and D. Song, *J. Electroanal. Chem.* **150** (1983) 165.
[148] A. Benninghoven, *Thin Solid Films* **39** (1976) 3.
[149] F. G. Rudenauer, *Int. J. Mass Spectrom. Ion Phys.* **45** (1982) 335.
[150] H. Puderbach and H. J. Gohansen, *Spectrochim. Acta* **39B** (1984) 1547.
[151] R. C. McCune, R. L. Shilts, and S. M. Fergusson, *Corros. Sci.* **22** (1982) 1049.
[152] H. M. Hindam and W. N. Smeltzer, *Oxid. Met.* **17** (1980) 337.
[153] K. S. Robinson and P. M. A. Sherwood, *Surf. Interface Anal.* **6** (1984) 261.
[154] J. D. Venables, D. K. McNamara, J. M. Chen, T. S. Sun, and R. L. Hopping, *Appl. Surf. Sci.* **3** (1979) 88.
[155] I. P. Shelpakova, I. G. Yudelevitch, and B. M. Ayupov, *Depth Analysis of Electronic Materials*, Nauka, Novosibirsk, 1984, p. 173.

[156] T. S. Sun, D. K. McNamara, J. S. Ahearn, J. M. Chen, and J. D. Venables, *Appl. Surf. Sci.* **5** (1980) 406.
[157] J. A. Treverton and N. C. Davis, *Electrochim. Acta* **25** (1980) 1571.
[158] E. Szontag, A. B. Kiss, and E. Kocsardy, *Aluminium* **59** (1983) 696.
[159] H. Konno, S. Kobayashi, H. Takahashi, and M. Nagayama, *Electrochim. Acta* **25** (1980) 1667.
[160] V. P. Parkhutik, *Corros. Sci.* **26** (1986) 296.
[161] J. Antula, *J. Appl. Phys.* **42** (1971) 2081.
[162] M. F. Abd-Rabbo, J. A. Richardson, and G. C. Wood, *Corros. Sci.* **88** (1976) 689.
[163] J. Randal and W. J. Bernard, *Electrochim. Acta* **20** (1975) 653.
[164] R. Bador, G. Bonyssoux, M. Ramand, and J. C. Sololon, *Mater. Res. Bull.* **12** (1977) 197.
[165] Y. Fukuda, *Trans. Natl. Res. Inst. Met. Finish.* **17** (1975) 239.
[166] Y. Fukuda and T. Fukushima, *Bull. Chem. Soc. Jpn.* **63** (1980) 3125.
[167] V. P. Parkhutik, Yu. E. Makushok, V. I. Kudrjavstev, and V. A. Sokol, *Elektrokhimiya* **23** (1987) 1538.
[168] G. S. Nadkarni, S. Radhakrishnan and S. V. Maduskar, *J. Phys. D: Appl. Sci.* **17** (1984) 209.
[169] A. E. Yaniv, N. Fin, F. N. Doduk, and I. E. Klein, *Appl. Surf. Sci.* **20** (1985) 638.
[170] Y. Yamamoto and N. Baba, *Thin Solid Films* **101** (1983) 329.
[171] M. Nagayama, *Denki Kagaku* **53** (1985) 862.
[172] V. P. Parkhutik, Yu. E. Makushok, and V. I. Shershulskii, in *Proceedings of the Symposium on Aluminum Finishing Technology*, Proceedings Vol. 86-11, The Electrochemical Society, Pennington, New Jersey, 1986, p. 406.
[173] G. E. Thompson, G. C. Wood, and R. Hutchings, *Trans. Inst. Met. Finish.* **58** (1980) 21.
[174] D. L. Cocke, C. A. Polansky, and D. E. Halverston, *J. Electrochem. Soc.* **132** (1985) 3065.
[175] D. L. Cocke, O. J. Murphy, S. M. Kormali, and D. Halverston, in *Proceedings of the International Symposium on Aluminum Surface Treatment Technology*, Proceedings Vol. 86-11, The Electrochemical Society, Pennington, New Jersey, 1986, p. 418.
[176] V. A. Sokol, M. M. Pinaeva, A. I. Vorobjeva, and I. I. Chekalova, *Elektronnaya Tekhnika, Ser. Materialy* **53**(7) (1985) 206.
[177] V. A. Labunov, V. A. Sokol, M. M. Pinaeva, and A. I. Vorobjeva, *Dokl. Akad. Nauk BSSR* **26**(3) (1982) 215.
[178] V. A. Labunov, V. A. Sokol, and M. M. Pinaeva, *Dokl. Akad. Nauk BSSR* **28**(3) (1984) 215.
[179] G. Kortum, W. Vogel, and K. Andrussow, *Dissociation Constants of Organic Acids in Aqueous Solution*, International Union of Pure and Applied Chemistry, Butterworths, London, 1961.
[180] H. B. Herman and J. R. Rairden, in *Encyclopedia of Electrochemistry of the Elements*, Ed. by A. Bard, Marcel Dekker, New York, Vol. 6, 1976, p. 33.
[181] H. Konno, S. Kobayashi, H. Takahashi, and M. Nagayama, *Corros. Sci.* **22** (1982) 913.
[182] V. P. Parkhutik, V. P. Bondarenko, V. A. Labunov, and V. A. Sokol, *Elektrokhimiya* **20** (1984) 530.
[183] B. R. Baker and R. M. Pearson, *J. Electrochem. Soc.* **119** (1972) 160.
[184] G. A. Dorsey, *Plat. Surf. Finish.* **11** (1970) 1117.
[185] G. A. Dorsey, *J. Electrochem. Soc.* **116** (1969) 466.

[186] G. E. Thompson, R. C. Furneaux, G. C. Wood, J. A. Richardson, and J. S. Goode, *Nature* **274** (1978) 433 and references cited therein.
[187] S. M. El-Mashri, A. J. Forty, L. A. Freeman, and D. J. Smith, *Inst. Phys. Conf. Ser.* no. 61, (1981) 395.
[188] V. E. Cosslett, *Proc. Roy. Soc.* **A370** (1980) 1.
[189] S. M. El-Mashri, A. J. Forty, and R. G. Jones, *Scanning Electron Microscopy* **11** (1983) 569.
[190] S. M. El-Mashri, R. G. Jones, and A. J. Forty, *Phil. Mag. A.* **48** (1983) 665.
[191] I. A. Popova, *Inorg. Mater.* **14** (1978) 1503.
[192] Y. Oka, T. Takahashi, K. Okada, and S. Iwai, *J. Non-Cryst. Growth* **30** (1979) 349.
[193] V. P. Parkhutik, Yu. E. Makushok, V. A. Labunov, and V. A. Sokol, *Zh. Prikl. Spektrosk.* **44** (1986) 494.
[194] A. B. Viharev, *Zashch. Metal.* **4** (1985) 601.
[195] M. A. Chernykh, *Zh. Prikl. Khim.* **58** (1985) 2442.
[196] I. Mita and M. Yamada, *J. Met. Finish. Soc. Jpn.* **33** (1980) 421.
[197] M. Hirochi and C. Yoshimura, *J. Met. Finish. Soc. Jpn.* **31** (1980) 596.
[198] K. Kobayashi and K. Shimizu, *Proceedings of the International Symposium on Aluminum Surface Treatment Technology*, Proceedings Vol. 86-11, The Electrochemical Society, Pennington, New Jersey, 1986, p. 380.
[199] K. Kobayashi, K. Shimizu, and D. Teranishi, *Jpn. J. Inst. Light Met.* **36** (1986) 81.
[200] G. A. Hutchings and C. T. Chen, *J. Electrochem. Soc.* **133** (1986) 1332.
[201] W. J. Bernard and S. M. Florio, *J. Electrochem. Soc.* **132** (1985) 2319.
[202] T. Kudo and R. S. Alwit, *Electrochim. Acta* **23** (1978) 341.
[203] H. Takahashi, M. Mukai, and M. Nagayama, *Proceedings of the 9th International Congress on Metal Corrosion*, Natl. Rec. Counc. Can., Ottawa, Toronto, 1984, p. 155.
[204] H. Takahashi, Y. Umerhara, and M. Nagayama, *Proceedings of the International Symposium on Aluminum Surface Treatment Technology*, Proceedings Vol. 86-11, The Electrochemical Society, Pennington, New Jersey, 1986, p. 367.
[205] K. Kobayashi, K. Shimizu, and A. Fujiwara, *Jpn. J. Inst. Light Met.* **35** (1985) 611.
[206] K. Okuba, *J. Met. Finish. Soc. Jpn.* **28** (1977) 212.
[207] V. T. Belov, *Zashch. Metal.* **4** (1986) 597.
[208] A. Vikharev, N. Botshkareva, and A. Dozortseva, *Zashch. Metal.*, no. 1 (1982) 125.
[209] R. S. Alwit and C. K. Dyer, *Electrochim. Acta* **23** (1978) 335.
[210] T. P. Hoar and N. F. Mott, *J. Phys. Chem. Solids* **9** (1969) 97.
[211] K. V. Heber, *Electrochim. Acta* **23** (1978) 135.
[212] M. Abd Rabbo, J. Richardson, and G. C. Wood, *Electrochim. Acta* **22** (1977) 1375.
[213] H. Takahashi and M. Nagayama, *J. Met. Finish. Soc. Jpn.* **36** (1985) 96.
[214] W. A. Lanford, R. S. Alwit, and C. K. Dyer, *J. Electrochem. Soc.* **127** (1980) 405.
[215] A. J. Maeland, R. Rittenhouse, and K. Bird, *Plating* **63** (1976) 56.
[216] T. Hurlen, *Electrochim. Acta* **29** (1984) 1161.
[217] V. T. Belov and M. P. Lebedeva, *Zh. Prikl. Khim.* **56** (1983) 673.
[218] V. T. Belov, *Zh. Prikl. Khim.* **50** (1977) 1725.
[219] V. T. Belov, *Zh. Prikl. Khim.* **48** (1975) 2836.
[220] P. Czokan, in *Advances in Corr. Sci. Technology*, Vol. 7, Ed. by M. G. Fontana and R. W. Staehle, Plenum Press, New York, 1980, p. 239.
[221] K. Ebihara, H. Takahashi, and M. Nagayama, *J. Met. Finish. Soc. Jpn.* **33** (1982) 4.
[222] G. C. Wood, in *Oxides and Oxide Films*, Vol. 2, Ed. by J. W. Diggle, Marcel Dekker, New York, 1973, p. 283.
[223] J. W. Diggle, T. Downie, and C. Goulding, *Chem. Rev.* **69** (1969) 365.
[224] K. Ebihara, H. Takahashi, and M. Nagayama, *J. Met. Finish. Soc. Jpn.* **34** (1983) 548.

[225] G. Bailey and G. C. Wood, *Trans. Inst. Met. Finish.* **52** (1974) 187.
[226] H. Takahashi and M. Nagayama, *Corros. Sci.* **18** (1970) 617.
[227] S. Ono and T. Sato, *J. Met. Finish. Soc. Jpn.* **32** (1981) 184.
[228] G. E. Thompson, R. C. Furneaux, J. S. Goode, and G. C. Wood, *Trans. Inst. Met. Finish.* **56** (1978) 159.
[229] A. Kwakernaak, R. Exalto, and H. A. Hoff, in *Adhesive Joints: Formation, Characteristics, Testing,* Kanzas, 1982, p. 103.
[230] C. Yoshimura and H. Noguchi, *J. Met. Finish. Soc. Jpn.* **25** (1978) 414.
[231] C. Yoshimura, H. Noguchi, and S. Yamada, *J. Met. Finish. Soc. Jpn.* **35** (1984) 428.
[232] H. Noguchi and C. Yoshimura, *J. Met. Finish. Soc. Jpn.* **34** (1983) 137.
[233] E. Palibroda, *Electrochim. Acta* **28** (1983) 1185.
[234] E. Palibroda, *Surf. Technol.* **23** (1984) 353.
[235] W. A. Pliskin and P. A. Gdula, in *Handbook on Semiconductors,* Vol. 3, Ed. by S. P. Keller, North-Holland, Amsterdam, 1980, p. 642.
[236] J. A. Amick, G. L. Schable, and J. L. Wossen, *J. Vac. Sci. Technol.* **14** (1977) 1053.
[237] V. A. Labunov and V. P. Parkhutik, in *Reviews in Electronics,* Centr. Res. Inst. Electronica, no. 1 (557), 1978, pp. 1–78.
[238] V. P. Parkhutik, in *Proceedings of the International Conference "Telavi-84," Electrodynamics and Quantum Phenomena at Interfaces,* Tbilisi, Mecniereba, 1984, p. 385.
[239] R. K. Smeltzer and C. C. Chen, *Thin Solid Films* **56** (1979) 75.
[240] J. J. Bernstein and R. White, *J. Electrochem. Soc.* **132** (1985) 1140.
[241] V. N. Lobushkin, I. M. Sokolova, and B. N. Tairov, *Elektrokhimiya* **12** (1976) 392; **12** (1976) 778.
[242] V. N. Lobushkin, *Elektrokhimiya* **13** (1977) 117.
[243] A. I. Zudov, L. A. Zudova, G. P. Sadakova, and S. I. Najmushina, *Elektrokhimiya* **19** (1983) 187.
[244] A. I. Zudov and L. A. Zudova, *Izv. Vuzov SSSR (Fizika)* **8** (1967) 15.
[245] J. Siejka, A. Moravski, J. Lagowski, and H. C. Gatos, *Appl. Phys. Lett.* **38** (1981) 552.
[246] I. Tatsuyo and K. Futoshi, *J. Jpn. Inst. Light Met.* **25** (1975) 237.
[247] M. N. Dyakonov, V. M. Muzhdaba, and M. D. Khanin, *Reviews in Electronics,* Centr. Res. Inst. Electronica, Ser. 5, no. 3 (886), 1982, p. 2.
[248] M. N. Dyakonov and V. M. Muzhdaba, *Electronic Industry* (Russ.), **4** (64), (1978) 31.
[249] V. P. Parkhutik and V. I. Shershulskii, in *Proceedings of the 1st International Conference on Conduction and Breakdown in Solid Dielectrics,* IEEE Publ., Toulouse, 1983, p. 40.
[250] A. I. Zudov and L. A. Zudova, *Elektrokhimiya* **9** (1973) 331; **11** (1975) 1239; **14**, (1978) 1044.
[251] L. A. Zudova, A. I. Zudov, and G. P. Sadakova, *Elektrokhimiya* **22** (1986) 1034.
[252] D. Morgan, A. Guile, and Y. Bektore, *J. Phys. D: Appl. Phys.* **13** (1980) 307.
[253] T. M. Berlicki, E. Murawski, and S. J. Osadnik, *Thin Solid Films* **120** (1984) 1.
[254] V. B. Lazarev, V. G. Krasov, and I. S. Shaplygin, *Conductivity of Oxides and Thin Film Systems,* Nauka, Moscow, 1979.
[255] J. G. Simmons, in *Handbook of Thin Film Technology,* Ed. by L. I. Maissel and R. Glang, McGraw-Hill, New York, 1970.
[256] N. F. Mott and G. Twose, *Adv. Phys.* **10** (1961) 107.
[257] V. V. Bryksin, M. Dyakonov, and M. Hanin, *Fiz. Tverd. Tela* **22** (1980) 1403.
[258] N. F. Mott and E. A. Davis, *Electron Processes in Non-Crystalline Materials,* Clarendon, Oxford, 1979.

[259] A. K. Jonscher and R. M. Hill, in *Physics of Thin Films*, Vol. 8, Ed. by G. Hass and M. H. Francombe, Academic, New York, 1975.
[260] V. L. Bonch-Bruevitch, *Electronic Theory of Disordered Semiconductors*, Nauka, Moscow, 1981.
[261] B. I. Shklovskii and A. I. Efros, *Electronic Properties of Disordered Semiconductors*, Nauka, Moscow, 1979.
[262] F. Bottger and V. V. Bryksin, *Phys. Status Solidi B* **18** (1976) 416.
[263] Yu. A. Firsov, *Polarony*, Nauka, Moscow, 1975, p. 207.
[264] T. G. Austin and N. F. Mott, *Adv. Phys.* **18**, (1969) 41.
[265] D. J. Emin, *Polycrystalline and Amorphous Thin Films and Devices*, Ed. by L. L. Kazmerski, Academic, New York, 1980.
[266] J. Hladik, *Physics of Electrolytes*, Vol. 1, Academic, London, 1972.
[267] G. M. Sessler, Ed., *Electrets*, Springer, Berlin, 1980.
[268] M. A. Lampert and P. Mark, *Current Injection in Solids*, Academic, New York, 1970.
[269] S. J. Radautsan, *Transient Injection Current in Solids*, Stinitza, Kishinev, 1983.
[270] K. C. Kao and W. Hwang, *Electrical Transport in Solids with Partial Reference to Organic Semiconductors*, Pergamon, Oxford, 1981.
[271] S. M. Gubanski and D. M. Hughes, *Thin Solid Films* **52** (1978) 119.
[272] V. V. Mikho and V. Ya. Koleboshin, *Elektrokhimiya* **12** (1980) 1841.
[273] A. Goldman and R. S. Sigmond, *J. Electrochem. Soc.* **132** (1985) 282.
[274] K. M. Nazar and N. Ahmad, *Int. J. Electron.* **47** (1979) 81.
[275] J. C. Schug, A. C. Lilly, and D. A. Lowitz, *Phys. Rev.* **131** (1973) 4811.
[276] G. S. Nadkarni and J. G. Simmons, *J. Appl. Phys.* **43** (1972) 374.
[277] M. Stuart, *Phys. Status. Solidi* **23** (1967) 595.
[278] G. Dittmer, *Thin Solid Films* **9** (1972) 141.
[279] D. M. Hughes and M. W. Jones, *J. Phys. D: Appl. Sci.* **7** (1974) 2081.
[280] G. Eftekhari, D. de Cogan, and B. Tuck, *Phys. Status Solidi A* **76** (1983) 331.
[281] N. Klein, *Thin Solid Films* **50** (1978) 223.
[282] J. J. O'Dwyer, *The Theory of Electrical Conduction and Breakdown in Solid Dielectrics*, Oxford University Press, London, 1973.
[283] H. R. Zeller, *IEEE Trans. Electric. Insulat.* **EI-22** (1987) 115.
[284] S. Ikonopisov, *Electrochim. Acta* **22** (1977) 1077.
[285] M. Kato, N. Fukushima, K. Yokoi, and T. Kudo, *J. Met. Finish. Soc. Jpn.* **34** (1983) 384.
[286] K. Shimizu, G. E. Thompson, and G. C. Wood, *Thin Solid Films* **92** (1982) 231.
[287] F. di Quarto, S. Piazza, and C. Sunseri, *J. Electrochem. Soc.* **131** (1984) 2901.
[288] D. V. Morgan, M. J. Howes, R. D. Pollard, and D. G. P. Waters, *Thin Solid Films* **15** (1974) 123.
[289] J. Yahalom and T. P. Hoar, *Electrochim. Acta* **15** (1970) 877.
[290] J. S. L. Leach and B. R. Pearson, *Proceedings of the* 7th *International Congress on Metal Corrosion*, Rio de Janeiro, ABRACO, 1978, p. 151.
[291] N. Sato, *Electrochim. Acta* **16** (1971) 1683.
[292] A. H. M. Shousha, *J. Non-Cryst. Solids* **17** (1975) 100.
[293] N. Klein and M. Albert, *J. Appl. Phys.* **53** (1982) 5840.
[294] N. Klein, *Thin Solid Films* **100** (1983) 335.
[295] N. Klein, V. Moskovici, and V. Kadary, *J. Electrochem. Soc.* **127** (1980) 152.
[296] J. M. Albella, I. Montero, and J. M. Martinez-Duart, *Electrochim. Acta* **32** (1987) 255.
[297] J. M. Albella, I. Montero, and J. M. Martinez-Duart, *J. Electrochem. Soc.* **131** (1984) 1101.

[298] I. Montero, J. M. Albella, and J. M. Martinez-Duart, *J. Electrochem. Soc.* **132** (1985) 814.
[299] M. Kato, E. Uchida, and T. Kudo, *J. Met. Finish. Soc. Jpn.* **35** (1984) 475.
[300] P. Solomon, *J. Vac. Sci. Technol.* **14** (1977) 1122.
[301] F. S. Aris and T. J. Lewis, *J. Phys. D: Appl. Phys.* **6** (1973) 1067.
[302] N. P. Sluginov, *Zh. Ross. Fiz-Khim. Obsh.* **15** (1983) 232.
[303] A. Gunterschulze, *Ann. D. Physik* **14** (1906) 601.
[304] K. Guminski, *Bull. Int. Acad. Pol. Sci.* (*Ser. A*), No. 3–4 (1936) 145.
[305] S. Anderson, *J. Appl. Phys.* **14** (1943) 601.
[306] Z. Ruzievich, *Bull. Acad. Pol. Sci.* (Cl.III) **4** (1956) 537.
[307] W. C. van Geel, *J. Phys. Radium* **17** (1956) 714.
[308] D. A. Vermilyea, *Acta Met.* **1** (1953) 282.
[309] A. W. Smith, *Can. J. Phys.* **37** (1959) 591.
[310] W. P. Ganley and P. M. Mooney, *Bull. Am. Phys. Soc.* **13** (1968) 128.
[311] K. Shimizu and S. Tajima, *Electrochim. Acta* **24** (1979) 309.
[312] L. D. Zekovic, Ph. D. thesis, University of Belgrade, 1982.
[313] G. E. Thompson, K. Shimizu, and G. G. Wood, *Nature* **286** (1980) 471.
[314] B. R. Jovanic, Ph. D. Thesis, University of Belgrade, 1987.
[315] S. Tajima, *Electrochim. Acta* **22** (1977) 905.
[316] M. I. Eidel'berg and T. Z. Tseitina, *Izv. Vyssh. Ucheb. Zaved., Fiz.* **13**(2) (1970) 133.
[317] L. D. Zekovic and V. V. Urosevic, *Thin Solid Films* **78** (1981) 278; **86** (1981) 347.
[318] L. D. Zekovic, V. V. Urosevic, and B. R. Jovanic, *Appl. Surf. Sci.* **11/12** (1982) 90.
[319] L. D. Zekovic, V. V. Urosevic, and B. R. Jovanic, *Thin Solid Films* **105** (1983) 169; **109** (1983) 217; **139** (1986) 109.
[320] L. D. Zekovic, B. R. Jovanic, L. Ristovski, G. Davidovic-Ristovski, and V. V. Urosevic, *Thin Solid Films* **157** (1988) 59.
[321] J. M. Kape, E. Survila, and S. Wernick, *Trans. Inst. Met. Finish.* **62** (1984) 41.
[322] P. J. Thompson and H. B. Heaton, *Trans. IMF* **58** (1980) 81.
[323] F. J. Arrowsmith and A. W. Clifford, *Int. J. Adhesion* **3** (1980) 193.
[324] K. E. Herfert, Northrop Corp., Report AD-A038068, Aug. 1976, p. 111.
[325] K. Itaya, S. Sugawara, K. Arai, and S. Saito, *J. Chem. Eng. Jpn.* **17** (1984) 514.
[326] N. Baba and T. Yoshino, *Kiuki Arunin. Hyiom. Shori* **114** (1985) 1.
[327] I. Mizuki, in *Advanced Metal Finishing Technology*, Fuji MRC Ltd., Tokyo, 1980, p. 141.
[328] W. J. Bernard, *J. Electrochem. Soc.* **124** (1977) 403C.
[329] D. Altenphol, U.S. Patent 2, 859, 148.
[330] R. S. Alwitt and W. J. Bernard, *J. Electrochem. Soc.* **121** (1974) 1019.
[331] A. M. Hunts, G. Ben Abderrazik, and G. Moulin, *Appl. Surf. Sci.* **28** (1987) 345.
[332] A. T. Fromhold and W. D. Forster, *Electrocomponent Sci. Technol.* **3** (1976) 51.
[333] P. Y. Dalvi, D. D. Upadhyaya, and S. Kulkarni, *Trans. Indian Ceram. Soc.* **45** (1986) 492.
[334] I. N. Vozjenin, G. A. Blinov, and L. A. Koledov, *Microelectronic Devices with Unpacked Integral Circuits*, Radio i Svjaz, Moscow, 1985.
[335] G. A. Blinov, *Hybrid Integral Functional Devices*, Visshaja Shkola, Moscow, 1987, p. 60.
[336] Y. Hirayama, Y. Oka, T. Hajiyama, and K. Yoshida, in *Proceedings of the International Symposium on Aluminum Surface Treatment Technology*, Proceedings Vol. 86–11, The Electrochemical Society, Pennington, New Jersey, 1986, p. 147.
[337] A. H. Eltokkhy, *J. Vac. Sci. Technol.* **A4**(3), Part 1 (1986) 539.
[338] G. C. Schwartz and V. Platter, *J. Electrochem. Soc.* **123** (1976) 34.

[339] W. Leimbrock and S. Gunter, *Wiss. Z. Tech. Hochsch. Karl-Marx St.* **21** (1979) 835.
[340] V. A. Sokol and A. I. Vorobjeva, *Elektronnaya Teknika, Ser. Radiodetali*, **3** (48), (1982) 3.
[341] W. E. Spicer, P. W. Chyle, and P. R. Sheath, *J. Vac. Sci. Technol.* **16** (1979) 1422.
[342] F. B. Micheletti, P. E. Norris, and K. Zaininger, *PCA Rev.* **31** (1970) 330.
[343] T. Hwang, P. R. Chang, K. M. Geib, and C. W. Wilmsen, *J. Vac. Sci. Technol.* **A4**(3), part 1, (1986) 1018.
[344] T. Sawada, K. Ishii, and H. Hasegawa, *Jpn. J. Appl. Phys., Suppl.* **21** (1981) 397.
[345] S. Kawai and I. Ishiguro, *J. Electrochem. Soc.* **123** (1976) 1047.
[346] N. Tsuya, T. Tokushima, M. Shiraki, and Y. Wakui, *IEEE Trans. Magn.* **MAG-22** (1986) 1140.
[347] U.S. Patent 3, 720, 143, 1973.
[348] P. Dankov, *Electron-Microscopic Study of Oxide Films on Metals*, USSR Academy of Science Publ., Moscow, 1958, p. 131.
[349] R. I. Nazarova, *Zh. Fiz. Khim.* **32** (1958) 79; **36** (1962) 1001.
[350] J. L. Miles and P. H. Smith, *J. Electrochem. Soc.* **110** (1963) 1240.
[351] J. O'Hanlon, in *Oxides and Oxide Films*, Vol. 6, Marcel Dekker, New York, 1976, p. 105.
[352] P. Friedel and S. Gourrier, *J. Phys. Chem. Solids* **44** (1983) 353.
[353] Y. Hirayama, F. Koshiga, and T. Sugano, *Thin Solid Films* **103** (1983) 71.
[354] T. Matsuda, H. Niu, and M. Takai, *J. Vac. Soc. Jpn.* **27** (1984) 901.
[355] R. W. Adama-Acquah and J. G. Swanson, *J. Electrochem. Soc.* **134** (1987) 2585.
[356] K. Ando and K. Matsumura, *Mem. Fac. Eng. Osaka City Univ.*, **18** (1977) 63.
[357] J. B. Theetan, S. Gourrier, and P. Friedel, in *Materials Research Symposium Proceedings*, Vol. 38, Materials Research Society, Boston, 1985, p. 499.

Index

GPSR Compliance

The European Union's (EU) General Product Safety Regulation (GPSR) is a set of rules that requires consumer products to be safe and our obligations to ensure this.

If you have any concerns about our products, you can contact us on ProductSafety@springernature.com

In case Publisher is established outside the EU, the EU authorized representative is:

Springer Nature Customer Service Center GmbH
Europaplatz 3
69115 Heidelberg, Germany

Batch number: 10682427

Printed by Printforce, the Netherlands